Bacteriology: Principles and Practices

Bacteriology: Principles and Practices

Edited by Ricky Parks

New York

Published by Syrawood Publishing House,
750 Third Avenue, 9th Floor,
New York, NY 10017, USA
www.syrawoodpublishinghouse.com

Bacteriology: Principles and Practices
Edited by Ricky Parks

International Standard Book Number: 978-1-68286-593-4 (Hardback)

Cataloging-in-Publication Data

Bacteriology : principles and practices / edited by Ricky Parks.
p. cm.
Includes bibliographical references and index.
ISBN 978-1-68286-593-4
1. Bacteriology. 2. Microbiology. I. Parks, Ricky.
QR46 .B33 2018
616.01--dc23

TABLE OF CONTENTS

PREFACE

Over the recent decade, advancements and applications have progressed exponentially. This has led to the increased interest in this field and projects are being conducted to enhance knowledge. The main objective of this book is to present some of the critical challenges and provide insights into possible solutions. This book will answer the varied questions that arise in the field and also provide an increased scope for furthering studies.

Bacteriology studies bacteria, their taxonomy, genetics, biochemistry, etc. Bacteria are classified according to the substances they produce, the optimum temperature required for their growth and the nutrients required to sustain them. Bacteriology is a sub-field of microbiology. Applications of bacteriology include antibiotics, biomining, bioremediation, nitrification, wastewater treatment, etc. This book on bacteriology strives to provide a fair idea about this discipline and to help develop a better understanding of the latest advances within this field. Those in search of information to further their knowledge will be greatly assisted by this book.

I hope that this book, with its visionary approach, will be a valuable addition and will promote interest among readers. Each of the authors has provided their extraordinary competence in their specific fields by providing different perspectives as they come from diverse nations and regions. I thank them for their contributions.

Editor

Cationic Synthetic Peptides: Assessment of Their Antimicrobial Potency in Liquid Preserved Boar Semen

Stephanie Speck[1*¤], Alexandre Courtiol[1], Christof Junkes[2], Margitta Dathe[2], Karin Müller[1], Martin Schulze[3]

1 Leibniz Institute for Zoo and Wildlife Research, Berlin, Germany, **2** Leibniz Institute of Molecular Pharmacology, Berlin, Germany, **3** Institute for Reproduction of Farm Animals Schoenow e. V., Bernau, Germany

Abstract

Various semen extender formulas are in use to maintain sperm longevity and quality whilst acting against bacterial contamination in liquid sperm preservation. Aminoglycosides are commonly supplemented to aid in the control of bacteria. As bacterial resistance is increasing worldwide, antimicrobial peptides (AMPs) received lively interest as alternatives to overcome multi-drug resistant bacteria. We investigated, whether synthetic cationic AMPs might be a suitable alternative for conventional antibiotics in liquid boar sperm preservation. The antibacterial activity of two cyclic AMPs (c-WWW, c-WFW) and a helical magainin II amide analog (MK5E) was studied *in vitro* against two Gram-positive and eleven Gram-negative bacteria. Isolates included ATCC reference strains, multi-resistant *E. coli* and bacteria cultured from boar semen. Using broth microdilution, minimum inhibitory concentrations were determined for all AMPs. All AMPs revealed activity towards the majority of bacteria but not against *Proteus* spp. (all AMPs) and *Staphylococcus aureus* ATCC 29213 (MK5E). We could also demonstrate that c-WWW and c-WFW were effective against bacterial growth in liquid preserved boar semen *in situ*, especially when combined with a small amount of gentamicin. Our results suggest that albeit not offering a complete alternative to traditional antibiotics, the use of AMPs offers a promising solution to decrease the use of conventional antibiotics and thereby limit the selection of multi-resistant strains.

Editor: Axel Cloeckaert, Institut National de la Recherche Agronomique, France

Funding: The study was funded by AiF e.V. (www.aif.de) PRO INNO II Grant KF0376101MD6 and the Leibniz Institute of Zoo and Wildlife Research (www.izw-berlin.de). The funders had no role in study design, data collection and analysis, decision to publish, or preparation of the manuscript.

Competing Interests: The authors have declared that no competing interests exist.

* Email: stephanie.speck@vetmed.uni-leipzig.de

¤ Current address: Institute of Animal Hygiene and Veterinary Public Health, University of Leipzig, Leipzig, Germany

Introduction

Artificial insemination (AI) is the most commonly used assisted reproductive technology in swine industry [1]. For AI, short- or long-term semen extenders are used to process and store semen while maintaining sperm viability over days at 15 to 17°C. Bacteria are frequently found in freshly retrieved boar ejaculates but are detrimental to sperm quality and longevity particularly in liquid-preserved semen [2–4]. Up to 10^9 colony forming units/mL ejaculate have been reported [5–7]. The most prevalent bacteria were Gram-negative with the majority belonging to the family *Enterobacteriaceae* [7,8]. Bacterial contamination seems to have little effect on fecundity under natural mating conditions. However, processing and storage of extended semen for AI might facilitate bacterial growth and concentration-dependent spermicidal effects [9]. Besides a proper sanitation and hygiene management, antimicrobial substances, such as Aminoglycosides, are commonly supplemented to aid in the control of bacteria [7,9,10].

Bacteria are highly effective in adapting to changing environments [11] and due to an increasing spread of resistance to classic antibiotics there is a need for new antimicrobial alternatives [3,12]. In recent studies, antimicrobial peptides (AMPs) have received considerable attention as candidates to overcome bacterial resistance [13]. AMPs are naturally occurring molecules with a broad spectrum of antimicrobial activity that rapidly kill their target cells [14]. Well-known AMPs are mammalian defensins, amphibian magainins, and insect cecropins but even bacteria and fungi produce cationic AMPs (lantibiotics, bacteriocins) [14]. Roughly 5,500 AMPs have been discovered, predicted or synthesized so far [15]. Fortunately, most cationic peptides do not induce resistant mutant strains *in vivo* [14]. Among the large variety of AMPs, short arginine (R)- and tryptophan (W)-rich cyclic peptides demonstrated high antimicrobial activity and low toxic effects against eukaryotic cells [16]. Furthermore, the interaction of these R- and W-rich cyclic hexapeptides with *E. coli* rapidly permeabilised the outer membrane of *E. coli* [16,17].

The aim of our study was to evaluate whether selected synthetic AMPs are useful as substitutes for conventional antibiotics used in liquid boar sperm preservation. We describe the antimicrobial activity of two cationic cyclic peptides (c-WWW, c-WFW) [16] and a cationic helical magainin II amide analog (MK5E) [18] *in vitro* and in liquid preserved boar semen.

Materials and Methods

Synthetic cationic antimicrobial peptides

A helical magainin II amide derivative (MK5E) and two cyclic hexapeptides (c-WWW, c-WFW) were used in this study. The antimicrobial activity of these peptides against *E. coli* DH5α and *Bacillus subtilis* subsp. *spizizenii* DSM 347 (further referred to as *B. subtilis*) and their interaction with eukaryotic cells have been described in detail previously [16–18]. Peptides (Table 1) were obtained lyophilized from Biosyntan, Berlin, Germany. Stock solutions (400 μM) prepared in sterile distilled water were stored at –80°C until further use. The peptide synthesis was previously described in detail [16].

Antimicrobial susceptibility testing

***In vitro* antimicrobial activity of c-WFW, c-WWW, and MK5E.** For the determination of *in vitro* Minimum Inhibitory Concentrations (MICs), broth microdilution was performed according to the Clinical and Laboratory Standards Institute (CLSI) standard M31-A3 [19] using cation-adjusted Mueller-Hinton-II-Bouillon (MHIIB; Merck, Darmstadt, Germany). All antimicrobial substances were tested in 96-well plates in triplicate. These experiments were independently repeated twice. Selected Gram-negative bacteria isolated from native boar semen in preceding studies (unpublished data) were used: *Enterobacter cloacae*, hemolytic *E. coli* (further referred to as *E. coli* HE), *Klebsiella* (*K.*) *pneumoniae*, *Proteus* (*P.*) *myxofaciens*, *P. vulgaris*. In addition, AMPs were tested on *E. coli* DH5α, *B. subtilis* DSM 347, and four gentamicin-resistant *E. coli* (kindly provided by Stefan Schwarz, FLI, Mariensee, Germany). All strains were grown on Columbia sheep-blood (5%) agar (CSBA; Oxoid, Wesel, Germany). Briefly, MHIIB containing 5×10^5 CFU/mL was prepared for subsequent inoculation into 96-well plates containing the different peptide dilutions. The final peptide concentrations ranged from 100 μM−0.05 μM (1:2 serial dilutions) as previously described [16]. Plates were sealed and incubated at 37°C for 18 to 24 h. The MIC of each tested AMP was defined as the lowest concentration exhibiting no visible growth compared to drug-free control wells. Turbidity was monitored with unaided eyes and a microplate reader at 600 nm. Gentamicin MICs were also determined (final concentration 0.113 μg/mL–116 μg/mL). As a quality control (QC) for broth microdilution, *E. coli* ATCC 25922 and *Staphylococcus* (*S.*) *aureus* ATCC 29213 were used as reference strains as recommended by CLSI [19]. Results were compared to the MIC QC ranges for broth microdilution (μg/mL) given by CLSI [19]. The test results were considered valid only when MICs for reference strains were within the QC ranges accepted by CLSI [19].

Evaluation of potency-enhancing effects: application of c-WWW and MK5E combined to gentamicin. The combination of AMPs and classical antibiotics has the potential to enhance the potency and target selectivity of AMPs [20]. We therefore combined c-WWW (2 μM) and MK5E (1 μM) but not c-WFW (as the latter was most promising for a stand-alone application) to gentamicin. AMP-concentrations were chosen according to sperm toxicity data as c-WWW and MK5E even at their lowest MIC (see results) would be harmful to boar spermatozoa (unpublished data). Gentamicin concentrations (i.e. 0.025 μg/mL–1 μg/mL) were selected according to MIC values defined in the first experiments and combined with c-WWW and MK5E. Determination of bacterial *in vitro* susceptibility was performed according to CLSI [19] and as outlined before. In addition, MICs were determined for gentamicin as a QC. The four multi-resistant *E. coli* were not included in these experiments.

Detection of bacteria in preserved semen

Ejaculates were collected from mature Pietrain boars housed at an EU-approved commercial insemination center during routine semen production and not as an animal experiment. The approval number according to Directive 90/429/EEC is KBS 085-EWG. Samples originated from a total of 39 boars and were retrieved by the gloved-hand technique. The gelatinous ejaculate fraction was removed using gauze. Boar ejaculates were diluted in Beltsville Thawing Solution (BTS) without additives (Minitüb, Tiefenbach, Germany), split, adjusted to 2×10^9 spermatozoa/portion (90 mL), and slowly cooled to 16°C over a 5 h-period.

The standard extender BTS containing 250 μg/mL gentamicin (BTS+G) was used as the control for all experiments. Ejaculates of ten individuals were comparatively investigated using BTS + c-WWW (2 μM) and BTS + c-WFW (4 μM). Samples of nine other individuals were prepared using BTS + MK5E (1 μM). In addition, a preparation using BTS without antimicrobial additive (BTS only) was available from three of these nine individuals. For the combined application of gentamicin (G) and AMPs, ejaculates from another 20 boars were prepared. BTS+G (16 μg/ml) was combined with c-WWW (2 μM), c-WFW (4 μM), and MK5E (1 μM), respectively. BTS+G (16 μg/mL) served as additional control. The latter concentration corresponded to the two-fold MIC breakpoint for gentamicin-resistant *Enterobacteriaceae* [19].

Each preparation was stored for 96 h at 16°C. Counting of bacteria and determination of bacterial species was performed after 12 h, 48 h, and 96 h of storage, respectively. To identify the different bacteria, a 50 μL-aliquot of the respective sample was each plated onto CSBA, Gassner medium (Oxoid), and McConkey agar (Oxoid). Plates were incubated for 48 h at 37°C. Bacterial species identification was carried out based on growth characteristics, Gram-staining, catalase- and oxidase-reaction, and conventional as well as commercially available (API® test system, bioMérieux, Nürtingen, Germany) biochemical tests. In addition, a serial dilution (10^{-1} to 10^{-5}) was prepared from each preparation after the respective storage time. 100 μL of each dilution were plated onto two nutrient agar plates (Oxoid), respectively. Plates were inspected after 24 h and 48 h of

Table 1. Cationic synthetic peptides used in this study.

Abbreviation	Peptide sequence	MW (g/mol)
c-WFW	Cyclic (RRWFWR)	989.5
c-WWW	Cyclic (RRWWWR)	1027.2
MK5E	Ac-GIGKF IHAVK KWGKT FIGEI AKS-NH2	2515.1

alanine (A), arginine (R), glutamic acid (E), glycine (G), histidine (H), isoleucine (I), lysine (K), phenylalanine (F), serine (S), threonine (T), tryptophan (W), valine (V), MW – molecular weight. The linear peptide, MK5E is N-terminally acetylated (Ac) and C-terminally amidated (NH_2).

Table 2. Minimum inhibitory concentrations (MICs) determined for synthetic cationic peptides.

Bacteria	MICs (μM) determined for c-WFW	c-WWW	MK5E
Escherichia coli ATCC 25922	6.3–12.5	50	25–50
Escherichia coli DH5α	6.3	12.5–25	25–50
Escherichia coli (hemolytic)	6.3–12.5	50	25–50
Escherichia coli 26	12.5	25–50	50
Escherichia coli 629	6.3	25	25
Escherichia coli 2078	12.5	25	50
Escherichia coli 2715	12.5	25	25
Enterobacter cloacae	25	25	25
Klebsiella pneumoniae	12.5–25	25–50	50
Proteus myxofaciens	>100	>100	>100
Proteus vulgaris	>100	>100	>100
Bacillus subtilis DSM 347	6.3	6.3	6.3–12.5
Staphylococcus aureus ATCC 29213	25	50	>100

incubation at 37°C. Colony forming units (CFU)/mL were calculated after 48 h of incubation.

Statistical analysis

To study the influence of AMPs on bacterial growth in preserved semen, we used the non-parametric test for longitudinal data in factorial experiments by Brunner *et al.* (2002) [21]. This test has specifically been designed to analyze time-dependent outcomes of an experiment performed on a small number of subjects. Analyses were implemented using the package nparLD version 2.1 [22] for the free statistical software R version 3.0.2 [23]. Following authors' terminology, our experiment setting corresponded originally to a F0-LD-F2 design. This means that for each semen sample, that we consider as subjects, we have no between-subject covariate and two within-subject covariates: time and treatment. The response variable was the number of CFU/mL.

In order to compare the effect of BTS only, gentamicin and the three AMPs on bacterial growth, we pooled the ten ejaculates treated with BTS+G (250 μg/mL), BTS + c-WWW (2 μM) and BTS + c-WFW (4 μM) and the nine ejaculates treated with BTS+G (250 μg/mL), BTS + MK5E (1 μM) and BTS only (for three of those nine ejaculates) in a first analysis. As preserved semen from each animal was not treated by all five treatments, we could not run the analysis as a F0-LD-F2 design. Instead, we randomly selected one treatment for each animal, making sure that the random sampling always included at least one sample for each treatment, and considered treatment as a between-subject covariate (F1-LD-F1 design). When testing of the effect of treatments on bacterial growth, the outcome is subject to variation due to the random sampling procedure. Therefore, we replicated the analysis 1000 times and report the median of all 1000 p-values obtained (hereafter reported *simulated p-value*). Importantly, making a separate analysis for each experiment and respecting the initial F0-LD-F2 study design led to same qualitative conclusions but precludes one to compare all treatments together (analysis performed without the treatment BTS only as this latter was not applied on all ejaculates, data not shown).

We also reran this analysis excluding the preparation BTS+G (250 μg/mL) to study differences between AMPs. Then, we performed a second analysis for the 20 ejaculates treated with BTS+G (250 μg/mL), BTS+G (16 μg/mL), BTS+G (16 μg/mL) + c-WWW (2 μM), BTS+G (16 μg/mL) + c-WFW (4 μM) and BTS+G (16 μg/mL) + MK5E (1 μM) to study the effect of a combined application of gentamicin and AMPs. For this latter analysis, directly fitting a F0-LD-F2 model was possible because each subject received all treatments.

Results

Antimicrobial susceptibility testing using c-WWW, c-WFW, and MK5E

MICs (μg/mL) defined for gentamicin using *S. aureus* ATCC 29213 (i.e. 0.225–0.7 μg/mL) and *E. coli* ATCC 25922 (i.e. 0.45–0.9 μg/mL) were within QC range recommended by CLSI (*S. aureus* ATCC 29213 0.12-1 μg/mL, *E. coli* ATCC 25922 0.25-1 μg/mL) [19]. Test results were reproducible in all experiments. Hence, systematic errors could be excluded. MICs determined for AMPs are given in Table 2. For most bacteria, the lowest MICs were defined for c-WFW followed by c-WWW and the linear magainin derivative MK5E. Using *Proteus*, MIC values for all peptides exceeded 100 μM and were not further specified. *Enterobacter cloacae* revealed identical values for all three AMPs. MIC values determined for a certain bacteria/peptide combination did not differ within one experiment but small variation was observed between experiments. This has been expected as approved QC MIC values for standard antibiotics also span over a range of concentrations in broth microdilution [19].

Combination of c-WWW, MK5E and gentamicin

Addition of 2 μM c-WWW or 1 μM MK5E to varying concentrations of gentamicin resulted in MIC values that did not considerably differ from those obtained solely for gentamicin (Table 3). Compared to the latter, a slight increase of MICs was noticed with the exception of *B. subtilis* DSM 347 as test organism.

Table 3. Minimum inhibitory concentrations (MICs) determined for gentamicin when combined with c-WWW or MK5E.

Bacteria	MIC (μg/mL) determined for gentamicin	MIC (μg/mL) determined for gentamicin when combined with	
		c-WWW (2 μM)	MK5E (1 μM)
Escherichia coli ATCC 25922	0.45–0.9*	0.6–0.7	0.6–0.8
Escherichia coli DH5α	0.113	0.3–0.5	0.2–0.5
Escherichia coli (hemolytic)	0.45	0.8–0.9	0.9
Enterobacter cloacae	0.113–0.225	0.2–0.4	0.3–0.4
Klebsiella pneumoniae	0.225–0.45	0.4–0.7	0.5–0.6
Proteus myxofaciens	0.45–0.9	0.7–0.9	0.7–0.9
Proteus vulgaris	0.45	0.6–0.8	0.5–0.8
Bacillus subtilis DSM 347	0.113	0.05–0.1	0.1
Staphylococcus aureus ATCC 29213	0.225–0.7*	0.6–0.7	0.5–0.6

*QC ranges as recommended by CLSI [19]: *S. aureus* (ATCC 29213) 0.12–1 μg/mL and *E. coli* (ATCC 25922) 0.25–1 μg/mL.

Effect of synthetic antimicrobial peptides on bacterial contamination in liquid preserved boar semen

Ejaculates of ten boars prepared with BTS+G (250 μg/mL), BTS + c-WWW (2 μM) and BTS + c-WFW (4 μM) and of nine boars prepared with BTS+G (250 μg/mL) and BTS+MK5E (1 μM) were investigated. In addition, BTS only-preserved samples from three boars were studied. As shown in Figure 1, treatments with AMPs or gentamicin presented fewer bacteria than the BTS only control. The number of CFU/mL did not significantly change with time for any preparations but BTS only and MK5E (simulated p-value for Anova Type Statistic [ATS] of the effect of time: for BTS only $p = 0.021$; for MK5E $p < 0.001$; for all other treatments: $p > 0.38$). Meanwhile, there was significantly less CFU/mL observed when using BTS+G (i.e. the standard semen extender) compared to when using any of the three AMP preparations (Figure 1). The comparison of AMPs showed that all three preparations did not differ significantly when the entire length of the experiment is considered ($p = 0.11$), but as bacteria grew with time for the MK5E treatment, once the bacteria count at 12 h is discarded the difference between treatments becomes significant (simulated p-value for modified ATS of the effect of preparation: $p = 0.015$). At 48 h and 96 h, MK5E was a less effective treatment against bacteria than c-WWW and c-WFW ($p < 0.001$) and lost the initial improvement it had over the BTS only control observed at 12 h. During the entire experiment, c-WWW and c-WFW did not differ between each other in CFU/mL observed ($p = 0.8$).

Ejaculates of 20 boars were prepared to evaluate the effects of BTS+G (16 μg/mL) + c-WWW (2 μM), BTS + G + c-WFW (4 μM), and BTS + G + MK5E (1 μM) compared to the standard BTS+G (250 μg/mL) and BTS+G (16 μg/mL). Figure 2 shows that the amount of CFU/mL did not seem to change with time for any of the combined AMP/gentamicin-preparations (ATS for main effect of time: 0.57, $df = 1.58$, $p = 0.52$; ATS for time interacting with treatment: 0.65, $df = 4.22$, $p = 0.63$). In contrast, the number of CFU/mL was influenced by the preparation (ATS: 9.51, $df = 3.33$, $p < 0.0001$) with BTS+G (16 μg/mL) being the less effective treatment, followed by BTS+G (16 μg/mL)+MK5E. Best results were obtained from preparations containing BTS+G (16 μg/mL)+c-WFW, BTS+G (16 μg/mL)+c-WWW, and BTS+ G (250 μg/mL). There was no significant difference in CFU/ml when using the latter three preparations (ATS: 1.63, $df = 1.86$, $p = 0.20$).

The amount of bacteria determined in different sperm preparations over time clearly varied between individuals. The CFU/mL counted for all preparations can be found in Table S1.

In total, 151 samples were investigated for bacterial growth. In the majority of samples ($n = 125$) more than one bacteria species was found. Scant growth of non-specific bacteria including mainly Gram-positive skin flora and Gram-negative bacteria commonly known as contaminants of distilled and stored water was found in 34% of all samples. Besides the non-specific bacteria, *Stenotrophomonas (S.) maltophilia* was predominant in samples treated solely by BTS+G (250 μg/mL, 16 μg/mL). Between three and five different Gram-negative and Gram-positive bacteria were isolated from the three BTS only preparations and identified as *S. maltophilia, Acinetobacter* sp., *Proteus (P.) vulgaris, Proteus* sp., *Serratia marcescens, Providencia rettgeri*, and *Staphylococcus* species. Preparations made of AMPs revealed ten different Gram-negative bacteria including *P. mirabilis, P. penneri, P. vulgaris, P. myxofaciens, Providencia alcalifaciens, Providencia rettgeri* or the non-fermentative bacteria *S. maltophilia, Ralstonia pickettii, Burkholderia cepacia*, and *Delftia acidovorans*. Of 29 samples treated solely with the single use of c-WWW, c-WFW, and MK5E, 21 (72%) revealed *Proteus* spp. and six (21%) were positive for *S. maltophilia*. In contrast, among the 60 samples obtained after the combined AMP/gentamicin treatment, we obtained eight (13%) *Proteus* spp.-positive specimens and 21 (35%) *S. maltophilia*-positive samples. Therefore, combining gentamicin (16 μg/mL) with an AMP significantly decreased the prevalence of *Proteus* spp. (proportion test: $X^2 = 28.4$, $df = 1$, $p < 0.0001$), but did not significantly influence *S. maltophilia* counts ($X^2 = 1.28$, $df = 1$, $p = 0.26$).

Discussion

Alternatives to conventional antibiotics are in urgent need to combat multidrug-resistant bacteria. Because of their effectiveness, antimicrobial peptides have been suggested for antimicrobial therapy [24]. The aim of our study was to investigate whether cationic AMPs are effective against bacteria often found in boar semen and therefore might be a suitable alternative to antibiotics currently used in liquid sperm preservation.

MICs could be determined *in vitro* for c-WFW, c-WWW, and MK5E using eleven bacterial strains with the exception of *Proteus* spp. (all AMPs) and *S. aureus* ATCC 29213 (MK5E). These latter bacteria are known to produce proteases that cleave naturally

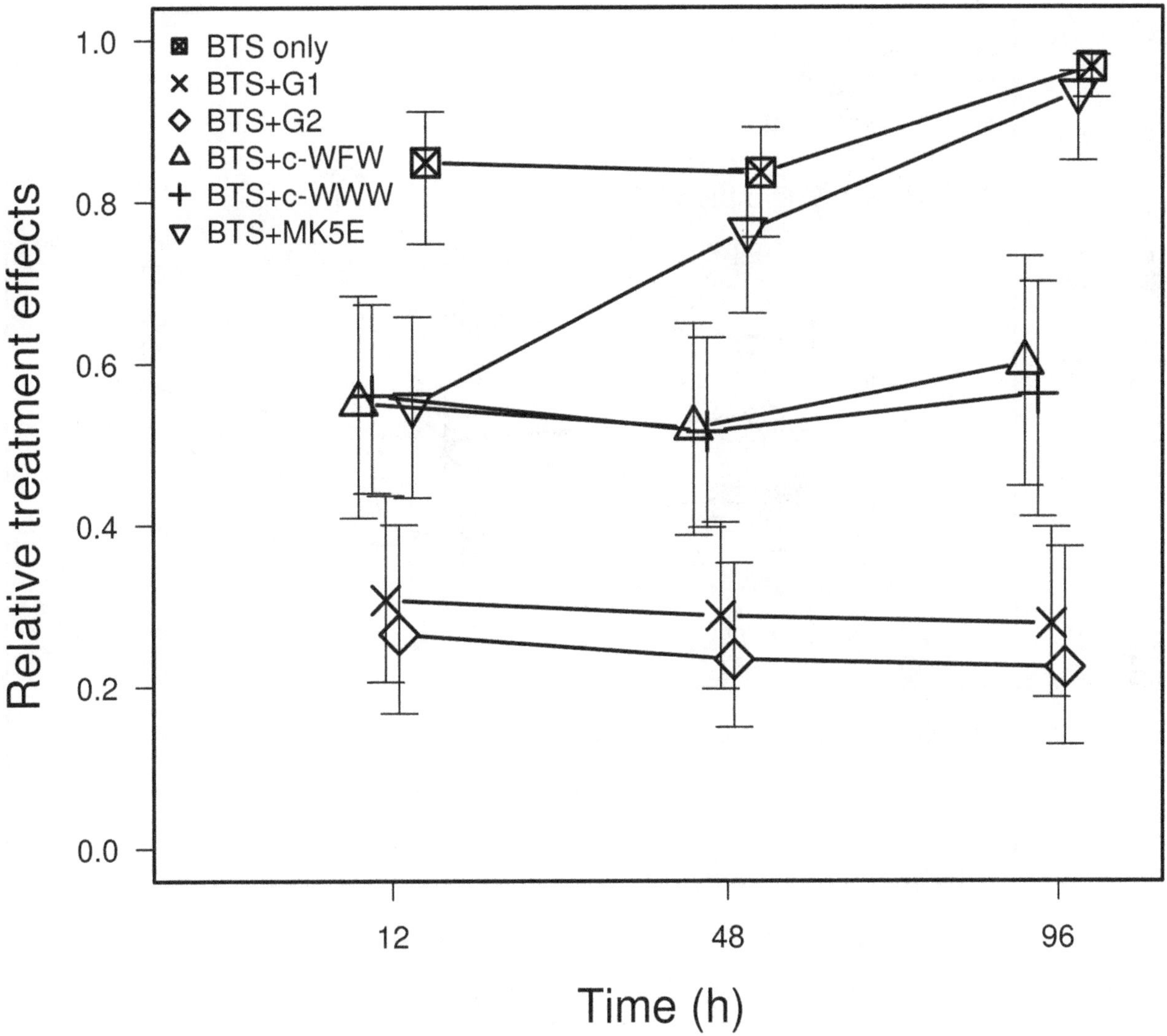

Figure 1. Relative effect of gentamicin or AMPs on the amount of bacteria in sperm preparation over time. Sperm preparations were made of BTS+G (250 µg/mL), BTS + c-WWW (2 µM) and BTS + c-WFW (4 µM) for ejaculates from ten individuals, and of BTS+G (250 µg/mL) and BTS + MK5E (1 µM) for ejaculates from nine other boars. Controls involving only BTS were also prepared from three of these nine individuals. The treatment BTS+G (250 µg/mL) is labeled BTS+G1 and BTS+G2 for the first and second experiment, accordingly. BTS+G1 and BTS + G2 were not distinguished in the analyses. The y-axis is the conventional graphical representation of the nonparametric method we used (see methods). It represents the relative marginal effect of the different treatments across time, i.e. the probability that the value being considered presents more CFU/mL than a random observation. The higher is the value on the y-axis, the higher is the corresponding value of CFU/mL, and the less effective is the treatment. Intervals represent 95% confidence intervals of the relative marginal effects and can here be used to compare treatments as the sample size is relatively similar for each point.

occurring linear cationic AMPs [24] and this mechanism might contribute to the results obtained in our experiments. Of the three peptides investigated, c-WFW resulted in lowest MIC values followed by c-WWW and MK5E. In former studies, hemolytic activity as well as toxicity against human cells at peptide concentrations up to 200 µM was negligible (c-WWW, c-WFW) [16] to non-existent (MK5E) [18]. However, our MIC-results revealed that only c-WFW might be applicable in liquid sperm preservation as negative effects on boar spermatozoa appeared at peptide concentrations higher than the MIC determined in this study (unpublished data). In contrast, even the lowest c-WWW and MK5E MIC determined for Gram-negative and -positive bacteria would be harmful to boar spermatozoa (unpublished data). We further investigated, whether a combined application of gentamicin and AMPs would result in enhanced antimicrobial effectiveness. For these experiments sperm-compatible concentrations of c-WWW and MK5E but not c-WFW (as the latter was most promising for a stand-alone application) were used. Results of the combined application revealed bactericidal activity when c-WWW (2 µM) and MK5E (1 µM) were combined with gentamicin at a concentration of <1 µg/mL. However, MIC values defined for gentamicin in the combined application were slightly higher than those obtained solely for gentamicin. Hence, we cannot deduce an enhancing effect from the results of these experiments *in vitro*. In fact, the increase of gentamicin MICs in the presence of AMPs would rather indicate an antagonistic effect. Cell membrane interaction is the first and most crucial step for the antimicrobial activity of AMPs [17]. Cationic charge and amphipathicity of AMPs constitute the structural prerequisite for an initial electrostatic interaction with negatively charged lipid

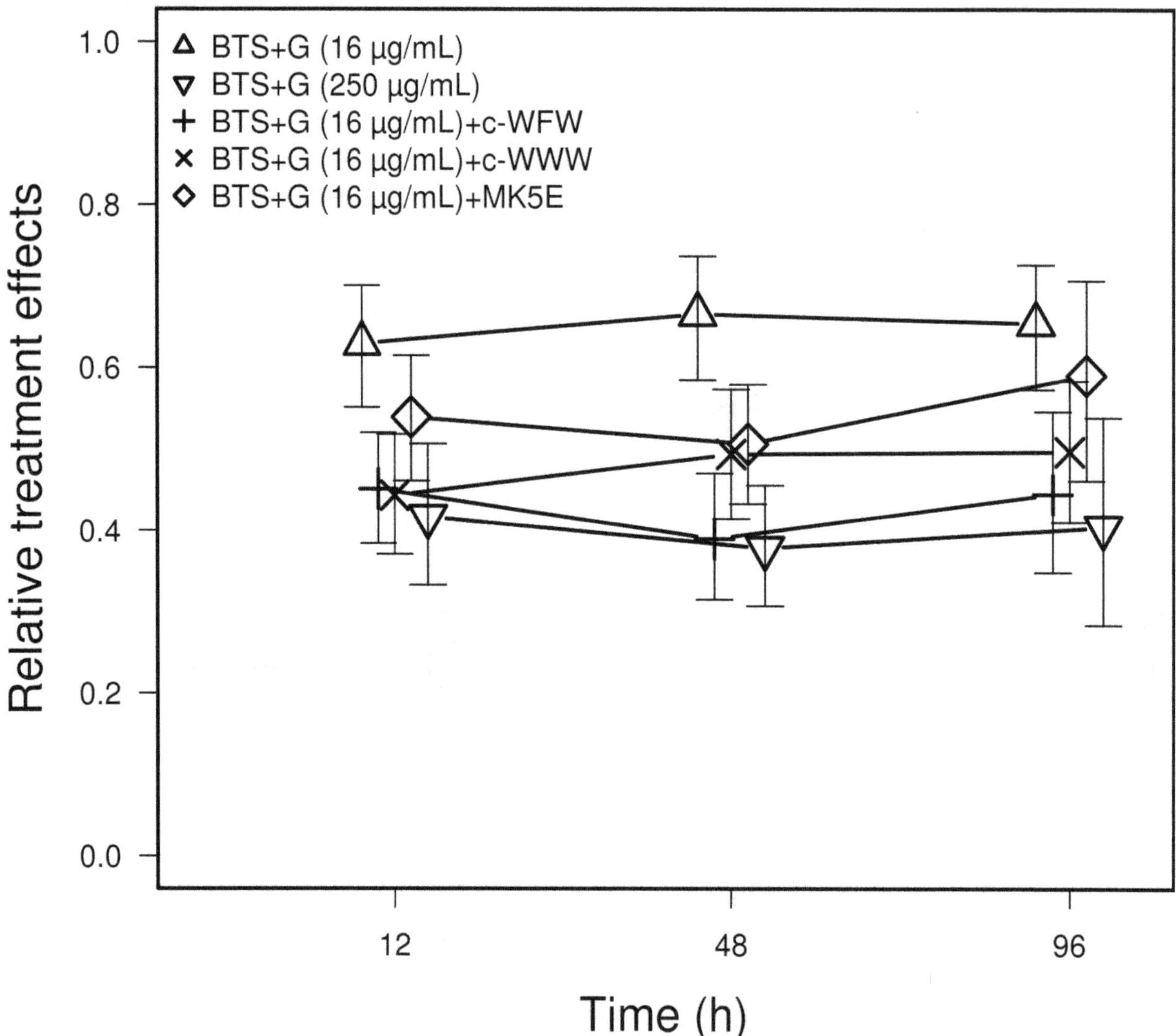

Figure 2. Relative effect of gentamicin or gentamicin combined with AMPs on the amount of bacteria in sperm preparation over time. Sperm preparations were made of BTS+G (250 µg/mL), BTS+G (16 µg/mL), BTS+G (16 µg/mL)+c-WWW (2 µM), BTS+G (16 µg/mL)+c-WFW (4 µM), and BTS+G (16 µg/mL)+MK5E (1 µM). See Figure 1 for legend details.

systems [25]. Electrostatic interactions are also the first step in aminoglycoside (e.g. gentamicin) action [26], hence a competing effect between both molecules might be assumed resulting in apparently higher MICs *in vitro*.

Based on the fact that, with the exception of c-WFW, MIC values determined for c-WWW and MK5E would be detrimental to spermatozoa, we decided to use sperm-compatible AMP concentrations to investigate whether their use in liquid sperm preservation would have any effect on bacterial contamination *in situ*. Treatments with AMPs or gentamicin presented fewer bacteria than the BTS only control. Interestingly, although used at concentrations below MIC determined *in vitro,* the different AMPs influenced the number of CFU/mL in liquid-preserved semen *in situ*. CFU/mL in preparations made of standard extender BTS containing 250 µg/mL gentamicin did not seem to change over time as was also the case for c-WFW and c-WWW that presented both the same antibacterial power (Figure 1). In contrast, preparations containing MK5E (1 µM) were less efficient and no longer prevented bacteria growth after 12 h. Enhancement of AMP-potency and target selectivity when combined to conventional antibiotics has been described [20] and might be affirmed by our data regarding AMP/gentamicin-preserved sperm *in situ* although this is not supported by our *in vitro* data. Figure 2 clearly demonstrates that the combination of gentamicin and c-WFW as well as c-WWW enhanced the antimicrobial effectiveness *in situ*. In fact, the standard BTS+G (250 µg/mL) was as effective as preparations made of gentamicin (16 µg/mL) + c-WFW as well as gentamicin (16 µg/mL) + c-WWW. This effect cannot be attributed to gentamicin alone because BTS containing 16 µg/mL gentamicin without AMPs was significantly less effective than all other preparations in this study. Therefore, our results suggest that albeit not offering a complete alternative to traditional antibiotics, the use of adequate AMPs may allow for a substantial reduction in concentration of antibiotics used for semen preservation.

The bacteria isolated from liquid extended boar semen confirmed findings reported by others [7,9,27]. In their studies, also *Enterobacteriaceae*, *Xanthomonadaceae*, *Alcaligenaceae*, and *Burkholderiaceae* accounted for most of the Gram-negative contaminants. Most of the bacteria we isolated originate from the boar or occur ubiquitously and are often associated with water.

Many of them have an inherent ability to form biofilms and possess intrinsic or acquired resistance mechanisms. Overall, approximately one third of all samples contained *Enterobacteriaceae or S. maltophilia*. Althouse *et al.* [9] stated that ejaculates contaminated by bacteria only have little effect on fecundity under natural mating conditions. However, the presence of *S. maltophilia* was directly correlated to sperm agglutination and decreased gross motility [9]. Other Gram-negative bacteria may also act spermicidal thus negatively affecting litter size, when sows are inseminated with contaminated semen [27].

The usage of AMPs in liquid semen preservation was hindered by their sperm-toxicity at higher concentrations (unpublished data). Unexpectedly, we found AMPs effective *in situ* at concentrations that deemed to be ineffective during screening *in vitro*. We chose performance standards for antimicrobial dilution susceptibility tests according to CLSI [19] for quality assurance. Cation-adjusted MHIIB is recommended when using gentamicin as a reference [19] but may affect AMP properties. Cation-adjusted MHIIB contains 20 to 25 mg/L Ca^{2+} and 10 to 12.5 mg/L Mg^{2+} who might influence AMP-target structure-interactions. With regard to the magainin II amide analog MK5E this is supported by results of Matsuzaki *et al.* (1999) [28] who reported that Mg^{2+} tightens the lipopolysaccharide (LPS) packing by crosslinking adjacent phosphate groups. Their studies showed that 10 mM Mg^{2+} blocked the bactericidal action of magainin 2 on membrane models *in vitro* [28]. The *in situ* effect seen in our study might be explained by the finding that the antimicrobial activity of AMPs depends on an ionic milieu comparable to that in mammalian body fluids [29]. This was demonstrated on a structurally diverse panel of AMPs [29]. The presence of $NaHCO_3$ (27 mM) significantly enhanced antimicrobial activity against Gram-positive and -negative bacteria [29]. It has also been suggested that carbonate enhances AMP activity due to alterations in bacterial susceptibility [29]. Besides other components to preserve sperm metabolic activity (e.g. 3.7 mM EDTA), the standard extender BTS we used contained 15 mM $NaHCO_3$ thus possibly enhancing microbial susceptibility to AMPs in liquid-preserved semen.

Conclusions

Our results demonstrate activity of synthetic cationic antimicrobial peptides against different Gram-negative and Gram-positive bacteria *in vitro*. Furthermore, c-WWW and c-WFW suppressed bacterial growth in semen preparations *in situ*, especially when combined with a small concentration of gentamicin. As we also examined that AMPs did not impede the quality of sperm (unpublished data), they offer a promising solution to decrease the use of conventional antibiotics and thereby limit the selection of multi-resistant strains. In order to achieve comparable data for *in vitro* susceptibility testing and *in situ* studies, the implementation of a valid standardized method is in need. With regard to the application of AMPs in liquid boar sperm preservation further investigations should include the reduction of sperm toxicity, detection of possible enhancing effects using other conventional antibiotics, and analyses of peptide-stability in different standard semen extenders.

Supporting Information

Table S1 Bacterial counts given in CFU/mL in different sperm preparations determined after 12 h, 48 h, and 96 h of storage at 16°C.

Acknowledgments

The authors are deeply grateful to Lars Konkel and Nadine Jahn for their excellent technical assistance and laboratory work.

Author Contributions

Conceived and designed the experiments: SS MS KM MD CJ. Performed the experiments: SS MS. Analyzed the data: SS AC. Contributed reagents/materials/analysis tools: SS AC MS KM. Contributed to the writing of the manuscript: SS AC MS KM CJ MD. Peptide design: MD CJ.

References

1. Althouse GC, Rossow K (2011) The potential risk of infectious disease dissemination via artificial insemination in swine. Reprod Domest Anim 46: Suppl 2: 64–67.
2. Sone M, Ohmura K, Bamba K (1982) Effects of various antibiotics on the control of bacteria in boar semen. Vet Rec 111: 11–14.
3. Althouse GC, Lu KG (2005) Bacteriospermia in extended boar semen. Theriogenology 63: 573–584.
4. Maes D, Nauwynck H, Rijsselaere T, Mateusen B, Vyt P, et al. (2008) Diseases in swine transmitted by artificial insemination. Theriogenology 70: 1337–1345.
5. Dagnall GJR (1986) An investigation of the bacterial flora of the preputial diverticulum and of the semen of boars. London: Royal Veterinary College. M. Ph. Thesis.
6. Danowski KM (1989) Qualitative und quantitative Untersuchung zum Keimgehalt von Ebersperma und zur Antibiotikaempfindlichkeit des überwiegenden Keimspektrums (unter dem Aspekt der Samenkonservierung). Hannover: Tierärztliche Hochschule, Diss. med. vet. 87 p.
7. Althouse GC, Pierdon MS, Lu KG (2008). Thermotemporal dynamics of contaminant bacteria and antimicrobials in extended porcine semen. Theriogenology 70: 1317–1323.
8. Okazaki T, Mihara T, Fujita Y, Yoshida S, Teshima H, et al. (2010) Polymyxin B neutralizes bacteria-released endotoxin and improves the quality of boar sperm during liquid storage and cryopreservation. Theriogenology 74: 1691–1700.
9. Althouse GC, Kuster CE, Clark SG, Weisiger RM (2000) Field investigations of bacterial contaminants and their effects on extended porcine semen. Theriogenology 53: 1167–1176.
10. Althouse GC (2008). Sanitary procedures for the production of extended semen. Reprod Domest Anim 43 (Suppl. 2): 374–378.
11. Theuretzbacher U (2011) Resistance drives antibacterial drug development. Curr Opin Pharmacol 11: 433–438.
12. Wolska KI, Grzes K, Kurek A (2012) Synergy between novel antimicrobials and conventional antibiotics or bacteriocins. Pol J Microbiol 2: 95–104.
13. Arouri A, Dathe M, Blume A (2009) Peptide induced demixing in PG/PE lipid mixtures: a mechanism for the specificity of antimicrobial peptides towards bacterial membranes? Biomed Biochim Acta 1788: 650–659.
14. Hancock REW (1997) Peptide antibiotics. Lancet 349: 418–422.
15. Zhao X, Wu H, Lu H, Li G, Huang Q (2013) A database linking antimicrobial peptides. PLoS One 8(6): e66557.
16. Junkes C, Harvey RD, Bruce KD, Dölling R, Bagheri M, et al. (2011) Cyclic antimicrobial R-, W-rich peptides: the role of peptide structure and *E. coli* outer and inner membranes in activity and the mode of action. Eur Biophys J 40: 515–528.
17. Junkes C, Wessolowski A, Farnaud S, Evans RW, Good L, et al. (2008) The interaction of arginine- and tryptophan-rich cyclic hexapeptides with *Escherichia coli* membranes. J Pept Sci 14: 535–543.
18. Dathe M, Nikolenko H, Meyr J, Beyermann M, Bienert M (2001) Optimization of the antimicrobial activity of magainin peptides by modification of charge. FEBS Lett 501: 146–150.
19. Clinical and Laboratory Standards Institute (CLSI) (2008) Performance standards for antimicrobial disk and dilution susceptibility tests for bacteria isolated from animals; approved standard. 3rd ed. M31-A3. Vol. 28 No. 8. Wayne Pennsylvania: CLSI. 99 p.
20. Anantharaman A, Rizvi MS, Sahal D (2010) Synergy with rifampin and kanamycin enhances potency, kill kinetics, and selectivity of de novo-designed antimicrobial peptides. Antimicrob Agents Chemother 54(5): 1693–1699.
21. Brunner E, Domhof S, Langer F (2002) Nonparametric Analysis of Longitudinal Data in Factorial Experiments. New York: Wiley. 288 p.
22. Noguchi K, Gel YR, Brunner E, Konietschke F (2012) nparLD: An R Software Package for the Nonparametric Analysis of Longitudinal Data in Factorial Experiments. J Stat Softw 50(12): 1–23. Available: http://www.jstatsoft.org/v50/i12/.

23. R Core Team (2013) R: A language and environment for statistical computing. Vienna, Austria: R Foundation for Statistical Computing. URL http://www.R-project.org/.
24. Kraus D, Peschel A (2006) Molecular mechanisms of bacterial resistance to antimicrobial peptides. Curr Top Microbiol Immunol 306: 231–250.
25. Dathe M, Wieprecht T, Nikolenko H, Handel L, Maloy WL, et al. (1997) Hydrophobicity, hydrophobic moment and angle subtended by charged residues modulate antibacterial and haemolytic activity of amphipathic helical peptides. FEBS Lett 403: 208–212.
26. Magnet S, Blanchard JS (2005) Molecular insights into aminoglycoside action and resistance. Chem Rev 105: 477–497.
27. Maroto Martin LO, Cruz Munoz E, De Cupere F, Van Driessche E, Echemendia-Blanco D, et al. (2010) Bacterial contamination of boar semen affects the litter size. Anim Reprod Sci 120: 95–104.
28. Matsuzaki K, Sugishita K, Miyajima K (1999) Interactions of an antimicrobial peptide, magainin 2, with lipopolysaccharide-containing liposomes as a model for outer membranes of Gram-negative bacteria. FEBS letters 449: 221–224.
29. Dorschner RA, Lopez-Garcia B, Peschel A, Kraus D, Morikawa K, et al. (2006) The mammalian ionic environment dictates microbial susceptibility to antimicrobial defense peptides. FASEB J 20: 35–42.

2

Analysis of the Relationship between Genomic GC Content and Patterns of Base Usage, Codon Usage and Amino Acid Usage in Prokaryotes: Similar GC Content Adopts Similar Compositional Frequencies Regardless of the Phylogenetic Lineages

Hui-Qi Zhou[☯], **Lu-Wen Ning**[☯], **Hui-Xiong Zhang, Feng-Biao Guo***

Center of Bioinformatics and Key Laboratory for NeuroInformation of the Ministry of Education, University of Electronic Science and Technology of China, Chengdu, China

Abstract

The GC contents of 2670 prokaryotic genomes that belong to diverse phylogenetic lineages were analyzed in this paper. These genomes had GC contents that ranged from 13.5% to 74.9%. We analyzed the distance of base frequencies at the three codon positions, codon frequencies, and amino acid compositions across genomes with respect to the differences in the GC content of these prokaryotic species. We found that although the phylogenetic lineages were remote among some species, a similar genomic GC content forced them to adopt similar base usage patterns at the three codon positions, codon usage patterns, and amino acid usage patterns. Our work demonstrates that in prokaryotic genomes: a) base usage, codon usage, and amino acid usage change with GC content with a linear correlation; b) the distance of each usage has a linear correlation with the GC content difference; and c) GC content is more essential than phylogenetic lineage in determining base usage, codon usage, and amino acid usage. This work is exceptional in that we adopted intuitively graphic methods for all analyses, and we used these analyses to examine as many as 2670 prokaryotes. We hope that this work is helpful for understanding common features in the organization of microbial genomes.

Editor: Ren Zhang, Wayne State University, United States of America

Funding: This study was supported by the National Natural Science Foundation of China (grant number 31470068); the Program for New Century Excellent Talents in University (grant number NCET-11-0059); Sichuan Youth Science and Technology Foundation of China (grant number 2014JQ0051); the Fundamental Research Funds for the Central Universities of China (grant numbers ZYGX2013J100 and ZYGX2013J101); and the China Postdoctoral Science Foundation (grant number 2013M540705). The project was also sponsored by Scientific Research Foundation for the Returned Overseas Chinese Scholars, State Education Ministry. The funders had no role in study design, data collection and analysis, decision to publish, or preparation of the manuscript.

Competing Interests: The authors have declared that no competing interests exist.

* Email: fbguo@uestc.edu.cn

☯ These authors contributed equally to this work.

Introduction

With the wide application of high-throughput sequencing technology, a large number of prokaryotic genomes have been published. This makes it very convenient to mine rules or new patterns from the sequences using comparative analysis methods. Among them, the effect of the genomic GC content on nucleotide or amino acid composition has received special attention [1–4]. Especially, some research showed that the GC content in bacterial genomes ranges from about 25% to 75% [5–7]. The range is proposed to extend to 0.211 and 0.789 based on theoretical induction [8]. It is a well-known fact that usage of synonymous codons for amino acids is not equal [9–11]. Deep analysis [12] showed that codon usage seems compatible with the idea that the genome, not the individual gene, is the unit of selection. That is to say, each gene in a genome tends to conform to its species' usage of the codon catalog. It was also shown that [13] the genomic GC content of bacteria is related to their phylogeny. The GC content of microorganism genomes is one of the recommended characteristics for the standard description of bacterial species [14], where a low GC difference within 10–12% probably indicates homogeneity and a high GC difference indicates heterogeneity. Indeed, GC content is linked to the codon usage pattern. Previous studies using varied numbers of prokaryotic genomes [15–18] showed that the genomic GC content is linearly correlated with the G+C content or single base frequencies of genes. Also, numerous studies illustrated that a similar relationship appeared between the frequency of amino acids and genomic GC content [19–22].

Though it was indicated that the GC content has a large impact on base usage at the three positions of a codon, codon usage, and amino acid usage, previous work only considered a limited number of species. Thus, we decided to investigate the influence of genomic GC content on the three usage patterns in a wider range

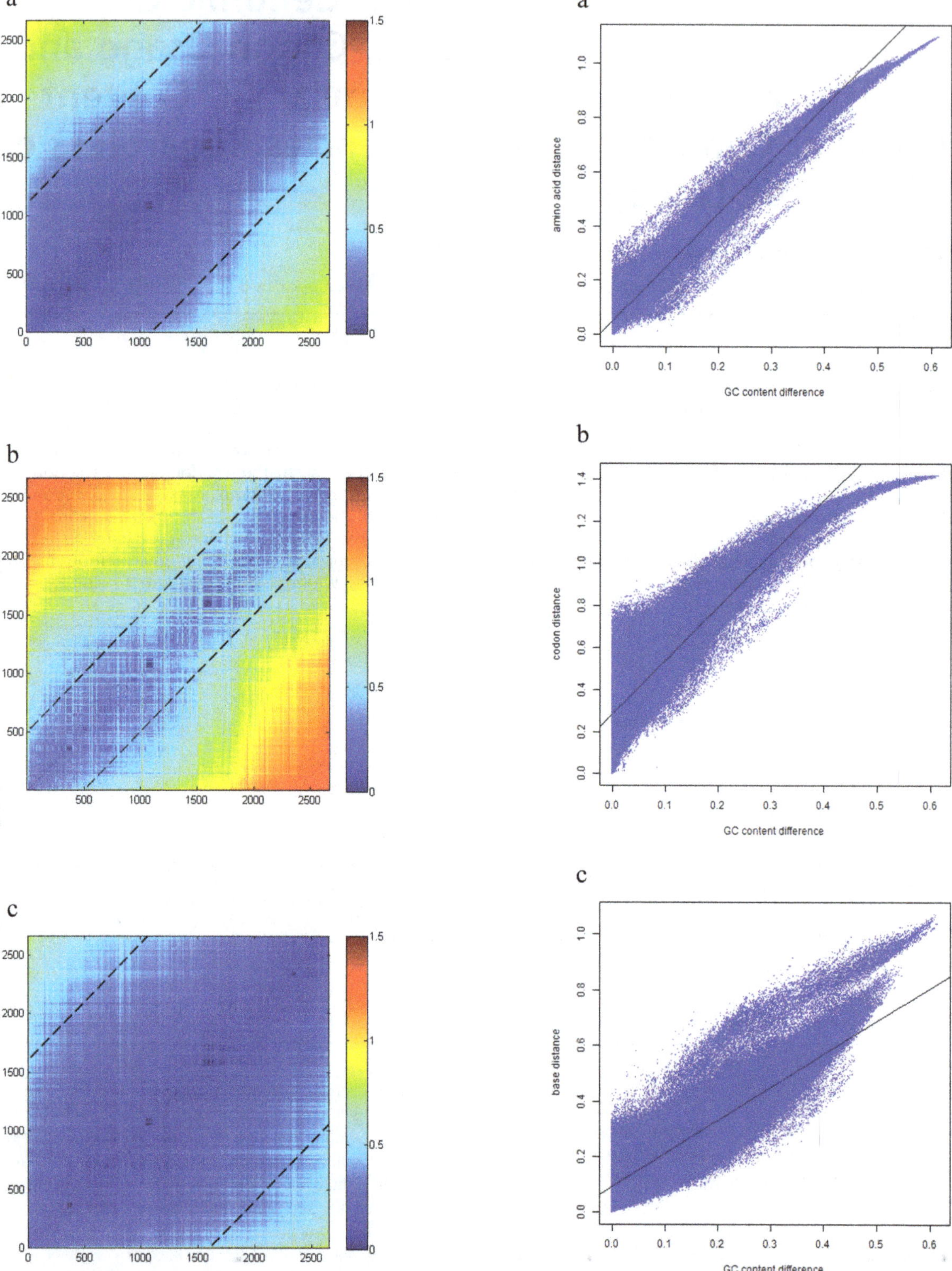

Figure 1. Base, codon, and amino acid frequency-based heat maps. The x-axis and y-axis represent 2670 prokaryotic genomes with GC content arranged from smallest to largest. a) Base frequency-based heat map; maximum base distance = 1.0986. b) Codon frequency-based heat map; maximum codon distance = 1.4199. c) Amino acid frequency-based heat map; maximum amino acid distance = 1.0725.

Figure 2. Vector distance of two genomes plotted against their GC content difference. a) Y: Base distance; X: D-value of GC content. b) Y: Codon distance; X: D-value of GC content. c) Y: Amino acid distance; X: D-value of GC content.

Table 1. Results of least squares fitting between vector distance of genome pairs and their GC content difference.

	Slope	Intercept	SSE	R-square
Base distance	1.968	0.05285	4332	0.9732
Codon distance	2.532	0.2871	2.556e+004	0.9108
Amino acid distance	1.185	0.09203	1.356e+004	0.8069

of species. In this paper, the influence is analyzed by regression analysis and intuitively graphic methods within all sequenced bacteria and archaea using the frequencies of bases A, T, C, and G at three codon positions, frequencies of 64 codons, and frequencies of 20 amino acids. We confirmed the existence of a linear relationship between the genomic GC content and amino acid usage [22] using the data of over 2600 sequenced prokaryotic genomes. Also, we confirmed that the genomic GC content has more influence on base usage, codon usage, and amino acid usage than phylogenetic lineage. To do this, we analyzed phylum-divided groups and GC content-divided groups. As expected, the distance variance of the phylum-divided groups is much larger than that of the GC content-divided groups.

Materials and Methods

Database

The data used in this paper are sequenced bacterial and archaeal genomes that were available as of September 2013. In total, 2670 prokaryotic genomes along with their annotation information were downloaded from GenBank (ftp://ftp.ncbi.nlm.nih.gov/genbank/genomes/Bacteria). We want to state that theoretically, using a much larger data set may have the possibility to introduce bias in some rare cases although this issue does not appear in this work The corresponding genomic length and GC content information of all these prokaryotic genomes are presented in Table S1.

Protein coding regions

Protein coding regions were generated using the genome-wide sequence from.fna files, and the information about base location, strand direction, and length were derived from.ptt files. In addition, the coding regions with a length that was not a multiple of 3 were eliminated.

Base distance, codon distance, and amino acid distance

For notation, we designate matrix $B_i(l)$ ($l = 1, 2\ldots12$) as base usage vectors with 12 dimensions in the ith genome, where B(1) to B(12) indicate A, T, C, and G usage frequencies at the first, second, and third codon positions, respectively. Matrix $C_i(l)$ ($l = 1, 2\ldots64$) is codon usage vectors with 64 dimensions in the ith genome, and each dimensional vector $C_i(l)$ is the frequency of the ith codon. Matrix $A_i(l)$ ($l = 1, 2\ldots20$) is amino acid usage vectors with 20 dimensions in the ith genome. They are frequencies of the 20 amino acids. $B_i(l)$, $C_i(l)$, and $A_i(l)$ in the ith genome is the average of $B(l)$, $C(l)$, and $A(l)$ values among all genes contained. The base, codon, and amino acid usage frequencies are presented in Table S2, S3, and S4, respectively.

These three matrices can be recognized as patterns of base usages, codon usages, and amino acid usages for each species. To compare pattern similarities between different pairs of genomes, the vector distance is required. We used the Euclidean distance as the vector distance in this paper with the following equations (where i corresponds to one genome, and j corresponds to another genome):

Base distance

$$D_B = \frac{2\times\sqrt{\sum_{l=1}^{12}(B_i(l)-B_j(l))^2}}{\sqrt{\sum_{l=1}^{12}B_i(l)^2}+\sqrt{\sum_{l=1}^{12}B_j(l)^2}}$$

Codon distance

$$D_C = \frac{2\times\sqrt{\sum_{l=1}^{64}(C_i(l)-C_j(l))^2}}{\sqrt{\sum_{l=1}^{64}C_i(l)^2}+\sqrt{\sum_{l=1}^{64}C_j(l)^2}}$$

Amino acid distance

$$D_A = \frac{2\times\sqrt{\sum_{l=1}^{20}(A_i(l)-A_j(l))^2}}{\sqrt{\sum_{l=1}^{20}A_i(l)^2}+\sqrt{\sum_{l=1}^{20}A_j(l)^2}}$$

Linear regression

Linear regression analysis of the relationship model between one or more independent variables and the dependent variable was performed using the least squares function. The function is a linear combination of the model parameters of several regression coefficients, which is named simple linear regression when there is only one independent variable, with the following regression equation:

$$Y = \alpha + \beta X$$

The goodness of fit using analysis of variance includes the following equations:

Sum of squares for error (SSE)

$$SSE = \sum_i (y_i - \hat{y}_l)^2$$

R-square

$$R^2 = \frac{\sum_i(\hat{y}_l-\bar{y})^2}{\sum_i(y_i-\bar{y})^2}$$

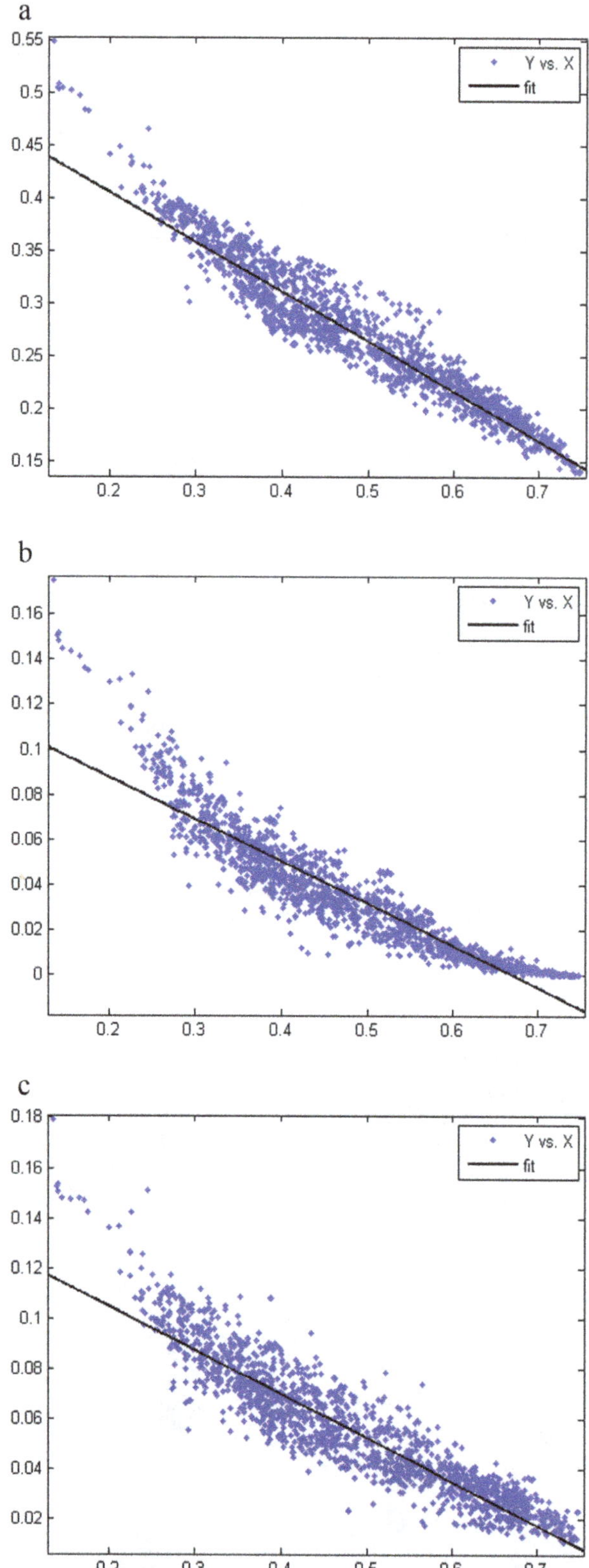

Figure 3. Base, codon, and amino acid frequencies of 2670 prokaryotic genomes plotted against the genomic GC content. a) Y: Base A frequency at the first codon position; X: genomic GC content. b) Y: Codon AAA frequency; X: genomic GC content. c) Y: Amino acid Lys frequency; X: genomic GC content.

R software

Related analyses and calculations in our work were accomplished through a tool called R, which is a static and computing language built with language S. R software can be downloaded (http://www.r-project.org/) and used freely.

Results and Discussion

Base, codon, and amino acid frequency-based heat maps

The base, codon, and amino acid frequency-based heat maps are presented in Fig. 1. The heat maps are based on pair-wise comparisons of the base, codon, and amino acid usage vectors that were created for each two prokaryotic genomes. These vectors, which have been sorted by increasing genomic GC content, were clustered using the vector distance described above. The color button, which is the vector distance, distinguishes the codon, base, and amino acid usage patterns of the two genomes with different GC content. The x-axis and y-axis are not the GC contents of each species; instead, they represent the detailed identity of the species by their GC contents. For example, the genome *Candidatus Zinderia insecticola* CARI, uid52459, whose GC content is 13.5%, is the lowest among the 2670 species and was placed at the leftmost side of the x-axis and the lowest position on the y-axis. The genome *Anaeromyxobacter dehalogenans* 2CP-C, uid58135, whose GC content is 74.9%, is the highest among the 2670 species and was put at the rightmost side of the x-axis and the highest position on the y-axis. On the other hand, each unit length on the x-axis and y-axis represent the same number of genomes.

As seen from the blue areas of the heat maps, within a limited range of GC content difference, the genomes have a small distance between each other. For instance, In Fig. 1a, the first bacterium (*Candidatus Zinderia insecticola* CARI, uid52459) contains a GC content of 0.135388, while genome 441 (*Thermoanaerobacter* X513, uid53065) has a GC content of 0.345191. The GC content difference and base distance between both are 0.209803 and 0.3995, respectively. In Fig. 1b, the 16th genome (*Candidatus Sulcia muelleri* SMDSEM, uid59393) contains a GC content of 0.225952, which has a difference of 0.090564 from the 1st genome. Correspondingly, their codon distance is 0.3683. In Fig. 1c, the 30th genome (*Candidatus Portiera aleyrodidarum* TV, uid195460) has a GC content of 0.246937, which has an amino acid distance of 0.3838 relative to the 1st genome; the GC content between the 1st and the 30th genomes differ by 0.111549. The GC content difference range of the dark blue area (distance smaller than 0.4) of Fig. 1a, b, and c are 0.0000–0.2986, 0.0000–0.1740, and 0.0000–0.4140, respectively.

However, the maximum codon distance is near 1.5 (Fig. 1b) when the genomic GC content gap increases, as shown by the deep red areas. Hence, Fig. 1b indicates that different prokaryotic species with a similar GC content have a similar codon usage pattern. The distance similarity presented in base frequency- and amino acid frequency-based heat maps (Fig. 1a and c). The maximal distances of Fig. 1a, b, and c are 1.0986, 1.4199, and 1.0725, respectively. We found that codon usage has the largest distance with the same GC content difference because the dark blue area (distance smaller than 0.4) is the least among the three figures, and base usage has the least distance as indicated by the largest blue area. The differences found among the three heat maps is possibly attributed to increasing vector dimensions among base usage, amino acid usage, and codon usage and the fact that all amino acids except Met and Trp are encoded by more than one codon. The undisputed indication is that base, codon, and amino acid usage change with genomic GC content.

Table 2. Results of least squares fitting between base, codon, and amino acid frequencies and the genomic GC content.

	Slope	Intercept	SSE	R-square
A_1*	−0.4728	0.5011	0.9880	0.9091
AAA	−0.1865	0.1251	0.2231	0.8733
Lys	−0.1748	0.1399	0.2287	0.8549

*A_1 denotes base A at the first codon position.

Base distance, codon distance, and amino acid distance vs. GC content difference

To analyze how the base usage, codon usage, and amino acid usage change with genomic GC content, the frequency vector distances were plotted against the genomic GC content difference in Fig. 2. The linear regression model for each scatter plot was established, and the corresponding results are listed in Table 1.

The linear regressions of all the three models are obvious and positive (the slopes are 1.968, 2.532, and 1.185), and the R-square of the base distance regression is the highest (0.9732). This means that the genomic GC content has a stronger impact on base usage than the other two usages, which have R-square values of 0.9108 and 0.8069. However, the points are more dispersive in the amino acid distance regression than others. This phenomenon indicates that the contribution of the genomic GC content to the amino acid composition is the weakest. Such a situation may be attributed to the fact that amino acid bias is associated with base bias [2,23] and is not a direct effect of the GC content.

In Fig. 2, we noticed that the ranges of the base usage, codon usage, and amino acid usage distance corresponding to a specific range of GC content differences decrease monotonically with an increasing GC content difference. For instance, in Fig. 2a, among the whole points, the GC content difference between 0.0000 and 0.0100 has a range of base usage distance from 0.0000 to 0.1802, which contains 172,973 points. However, a GC content difference between 0.6000 and 0.6100 has a range only from about 1.0815 to 1.0915, which contains only 35 points (Fig. 2a). In other words, pairs with a small GC difference may have a relatively large composition distance, but pairs with a large GC difference do not have a small composition distance. We believe that the large distance between some pairs with a small GC difference is caused by some composition differences among species that are not only determined by GC content. For example, previous results indicated that genes with a high GC content have a high gene expression level [24,25]. Furthermore, codon usage can maintain a force balance between mutational bias and translational selection [26]. Thus, gene expression level, gene function, and origination are additional fundamental factors that shape the pattern of biased codon usage.

Base, codon, and amino acid frequencies vs. genomic GC content

To directly analyze whether the base, codon, and amino acid usage are correlated with the genomic GC content, we checked the usages of all bases (4 types×3 positions of a codon = 12), codons (64 types), and amino acids (20 types). Here, we picked out the frequencies of base A at the first codon position, codon AAA, and amino acid Lys, which is translated from AAA, as an example. The frequencies against genomic GC content are shown in Fig. 3, and the results of fitting are shown in Table 2. As seen from Fig. 3 and Table 2, we found that 1) the usage of A at the first codon position, AAA, and Lys decreased almost linearly with increasing genomic GC content, and 2) genomes with both a high GC content and a low GC content adopt a similar pattern. Furthermore, our work shows that the almost linear relationship between the genomic GC content and the base usage, codon usage, and amino acid usage is consistent across all sequenced genomes of different species.

To obtain a quantitative measurement of the frequencies of bases, codons, and amino acids, we also calculated the slope of the best-fitting line for each scatter plot. The slope for base A is −0.4728, which means that if one bacterial genome has a 10% higher GC content than another, the percentage of base A at the first codon position would decrease approximately 4.728%. The results of codon AAA frequencies and amino acid Lys frequencies showed a similar, but smaller, effect with slopes of −0.1865 and −0.1748, respectively. Lightfield et al. reported that the usage percentage of amino acids encoded by three low-GC codon families including Lys and genomic GC content of the representative genomes showed a negative linear relationship, which was roughly consistent with our work [22].

Phylum and GC content

Although the results described above indicate that the GC content has a strong biased mutation pressure, we need to know whether the pressure is stronger than the phylogenetic distribution.

In the next analysis, all 2670 bacterial and archaeal genomic sequences were divided into 34 sections based on phyla (Group 1) and genomic GC content (Group 2). In Group 1, each section includes one phylum, and five unclassified bacteria and archaea (*Halophilic archaeon* DL31, uid72619; *Candidatus Cloacamonas acidaminovorans* Evry, uid62959; *Candidatus Saccharobacterium alaburgensis*, uid203361; *Candidatus Methylomirabilis oxyfera*, uid161981; and *Thermobaculum terrenum* ATCC BAA-798, uid42011) were excluded. In Group 2, the GC content range of each section was averaged, which has a range among x = (0.749053−0.135388)/34 (genomic GC content ranges from 0.135388 to 0.749053). Thus, the GC content of the *N*th section has a range from 0.135388 to 0.135388+*N*×x. Number of genomes in each section of the two groups are shown in Table S5 and S6, respectively. The base distance, codon distance, and amino acid distance were investigated for every section except the sections with data from only one and two genomes. Thus, if there are data for M (M>2) genomes in one section, then there are M×(M-1)/2 distance data. The variance of the M×(M-1)/2 distance data was calculated for each section to reflect the impact of the phylum or GC content on the base, codon, and amino acid usage. When M≤2, the variance value of this section was set to 0. The line charts of variances of the two groups were plotted, and nonzero values were plotted increasingly in Fig. 4.

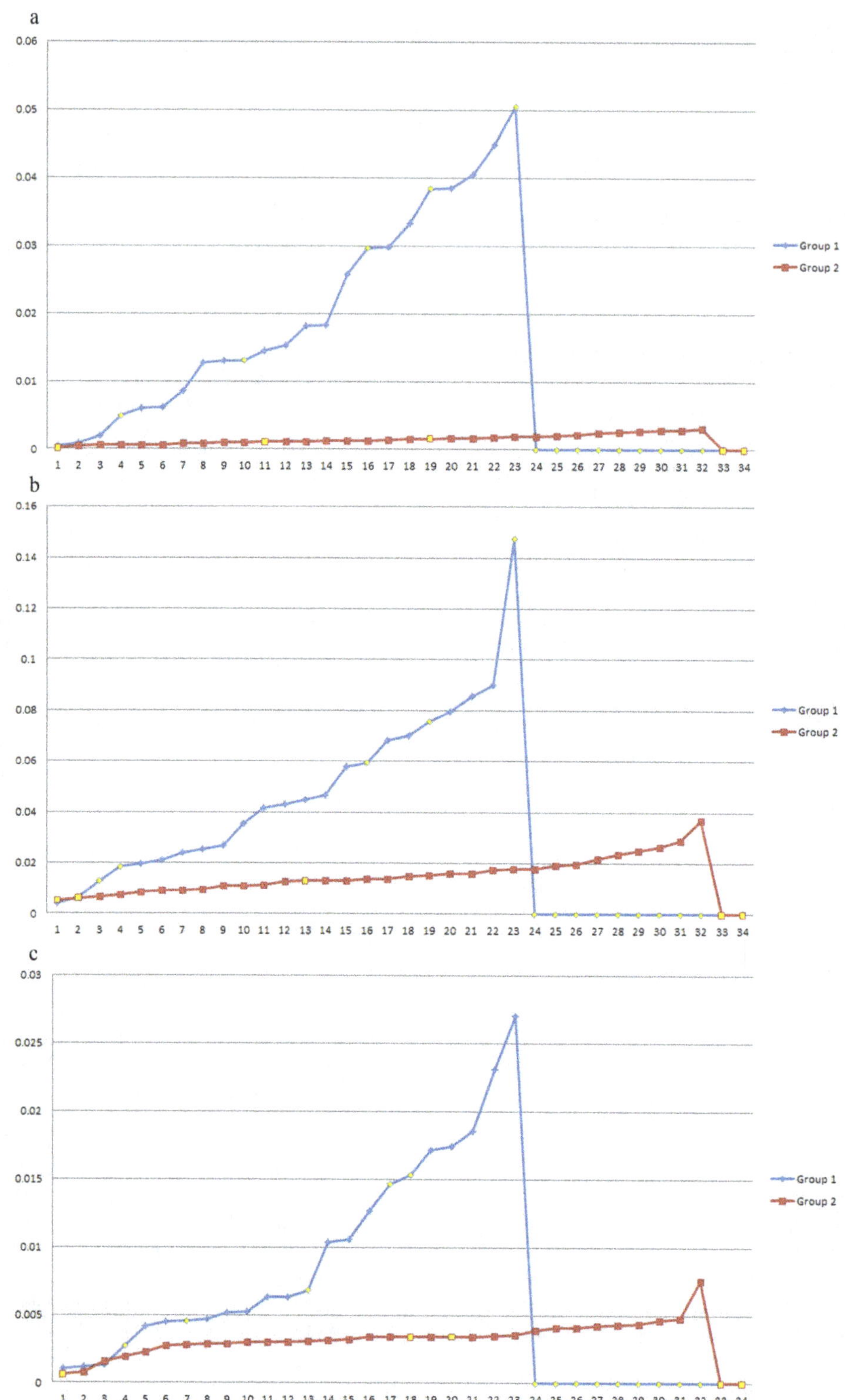

Figure 4. Line charts of variances of base distance, codon distance, and amino acid distance for phylum-divided groups and genomic GC content-divided groups. Group 1: the data were divided based on phylum; Group 2: the data were divided based on genomic GC content. a) Y: Base distance variances; X: sections. b) Y: Codon distance variances; X: sections. c) Y: Amino acid distance variances; X: sections.

Fig. 4 shows that the base distance variances of Group 1 vary widely from 0.000477 to 0.050552. Compared with Group 1, the base distance variances of Group 2 range within a smaller scale from 0.000169 to 0.003120. Additionally, the codon distance variances in Group 1 and Group 2 range from 0.004812 to 0.147581 and from 0.005235 to 0.036949, and the amino acid distance variances of the two groups are from 0.001018 to 0.027048 and from 0.000627 to 0.007536, respectively. Nevertheless, several particularly high values appear in both Group 1 and Group 2. This may be caused by the fact that the genome sequence data were too few in these sections. We marked data with fewer than 5 sequences as yellow points in Fig. 4 so that these data did not affect the analysis.

A previous report indicated the relationship between the bacterial genomic GC content and phylogeny through a phylogenetic tree [13]. Here, our results from graphic analysis illustrate that GC content is more important than phylogenetic lineages in general for their base, codon, and amino acid usage biases because the variances from the GC content are much smaller than those from phylogenetic lineages. This conclusion is consistent with that obtained for codon usage patterns in microbial genomes with a high GC content [17] and is also consistent with that obtained for amino acid frequencies [22]. However, the conclusion in this paper is based on a much larger data set and different analyzing methods.

Conclusions

In this paper, we analyzed the base usages, codon usages, and amino acid usages with respect to the genomic GC contents of a large number of prokaryotic genomes. Our work has demonstrated that in prokaryotic genomes: a) base usage, codon usage, and amino acid usage changes with GC content with a linear correlation; b) the distance of each usage has a linear correlation with the GC content difference; and c) the GC content has a larger impact on base usage, codon usage, and amino acid usage than the phylogenetic lineage. We believe that our work will be helpful to better understand the role that GC content plays in prokaryotic genomes.

Supporting Information

Table S1 Genomic length and GC content of all samples included in the analyses.

Table S2 Mean usage frequencies of base A, T, C, and G at the first, second, and third codon positions of all samples included in the analyses.

Table S3 Mean usage frequencies of 64 codons of all samples included in the analyses.

Table S4 Usage frequencies of 20 amino acids of all samples included in the analyses.

Table S5 Number of genomes, mean and variance of the distance for nucleotide frequencies, codon usages and amino acid compositions in each section of Group 1, that is to say, each section corresponds to one phyla.

Table S6 Number of genomes, mean and variance of the distance for nucleotide frequencies, codon usages and amino acid compositions in each section of Group 2, that is to say, each section corresponds to one specific G+C interval.

Author Contributions

Conceived and designed the experiments: FBG. Performed the experiments: LWN HQZ. Analyzed the data: FBG. Contributed reagents/materials/analysis tools: FBG. Contributed to the writing of the manuscript: HQZ FBG HXZ.

References

1. Banerjee T, Gupta S, Ghosh TC (2005) Role of mutational bias and natural selection on genome-wide nucleotide bias in prokaryotic organisms. Biosystems 81: 11–18.
2. Singer GA, Hickey DA (2000) Nucleotide bias causes a genomewide bias in the amino acid composition of proteins. Mol Biol Evol 17: 1581–1588.
3. Wilquet V, Van de Casteele M (1999) The role of the codon first letter in the relationship between genomic GC content and protein amino acid composition. Res Microbiol 150: 21–32.
4. Ohama T, Muto A, Osawa S (1990) Role of GC-biased mutation pressure on synonymous codon choice in Micrococcus luteus a bacterium with a high genomic GC-content. Nucleic Acids Res 18: 1565–1569.
5. Sueoka N (1962) On the genetic basis of variation and heterogeneity of DNA base composition. Proc Natl Acad Sci USA 48: 582.
6. Fleischmann RD, Adams MD, White O, Clayton RA, Kirkness EF, et al. (1995) Whole-genome random sequencing and assembly of Haemophilus influenzae Rd. Science 269: 496–512.
7. Ikemura T (1985) Codon usage and tRNA content in unicellular and multicellular organisms. Mol Biol Evol 2: 13–34.
8. Zhang CT, Zhang R (2004) A nucleotide composition constraint of genome sequences. Comput Biol Chem 28: 149–53.
9. Grantham R, Gautier C, Gouy M (1980) Codon frequencies in 119 individual genes confirm consistent choices of degenerate bases according to genome type. Nucleic Acids Res 8: 1893–1912.
10. Sharp PM, Cowe E, Higgins DG, Shields DC, Wolfe KH, et al. (1988) Codon usage patterns in Escherichia coli, Bacillus subtilis, Saccharomyces cerevisiae, Schizosaccharomyces pombe, Drosophila melanogaster and Homo sapiens; a review of the considerable within-species diversity. Nucleic Acids Res 16: 8207–8211.
11. Shields DC, Sharp PM, Higgins DG, Wright F (1988) "Silent" sites in Drosophila genes are not neutral: evidence of selection among synonymous codons. Mol Biol Evol 5: 704–716.
12. Grantham R, Gautier C, Gouy M, Mercier R, Pave A (1980) Codon catalog usage and the genome hypothesis. Nucleic acids research 8: 197–197.
13. Gupta RS (2000) The phylogeny of proteobacteria: relationships to other eubacterial phyla and eukaryotes. FEMS Microbiol Rev 24: 367–402.
14. Goodfellow M, O'Donnell AG (1993) Roots of bacterial systematic. London: Academic Press. 3–54 p.
15. Muto A, Osawa S (1987) The guanine and cytosine content of genomic DNA and bacterial evolution. Proc Natl Acad Sci USA 84: 166–169.
16. Knight RD, Freeland SJ, Landweber LF (2001) A simple model based on mutation and selection explains trends in codon and amino-acid usage and GC composition within and across genomes. Genome Biol 2: research0010.
17. Chen LL, Zhang CT (2003) Seven GC-rich microbial genomes adopt similar codon usage patterns regardless of their phylogenetic lineages. Biochem Biophys Res Commun 306: 310–317.
18. Wan XF, Xu D, Kleinhofs A, Zhou J (2004) Quantitative relationship between synonymous codon usage bias and GC composition across unicellular genomes. BMC Evol Biol 4: 19.
19. Sueoka N (1961) Correlation between base composition of deoxyribonucleic acid and amino acid composition of protein. Proc Natl Acad Sci USA 47: 1141.
20. Lobry J (1997) Influence of genomic G+C content on average amino-acid composition of proteins from 59 bacterial species. Gene 205: 309–316.
21. Gu X, Hewett-Emmett D, Li WH (1998) Directional mutational pressure affects the amino acid composition and hydrophobicity of proteins in bacteria. Genetica 102: 383–391.
22. Lightfield J, Fram NR, Ely B (2011) Across bacterial phyla, distantly-related genomes with similar genomic GC content have similar patterns of amino acid usage. PLOS ONE 6: e17677.
23. Tan H, Wang J, Yang F, Zhao ZK (2008) Genome-wide analysis of coding DNA and amino acid variation in Saccharomyces cerevisiae. Yeast 25: 29–39.
24. Pan A, Dutta C, Das J (1998) Codon usage in highly expressed genes of Haemophillus influenzae and Mycobacterium tuberculosis: translational selection versus mutational bias. Gene 215: 405.
25. Gupta S, Ghosh T (2001) Gene expressivity is the main factor in dictating the codon usage variation among the genes in Pseudomonas aeruginosa. Gene 273: 63–70.
26. Sharp PM, Devine KM (1989) Codon usage and gene expression level in Dictyosteiium discoidtum: highly expressed genes do 'prefer' optimal codons. Nucleic Acids Res 17: 5029–5040.

3

Plasma-Mediated Inactivation of *Pseudomonas aeruginosa* Biofilms Grown on Borosilicate Surfaces under Continuous Culture System

Kurt G. Vandervoort[1], Graciela Brelles-Mariño[2*¤]

1 Physics and Astronomy Department, California State Polytechnic University, Pomona, California, United States of America, 2 Biological Sciences Department, California State Polytechnic University, Pomona, California, United States of America

Abstract

Biofilms are microbial communities attached to a surface and embedded in a matrix composed of exopolysaccharides and excreted nucleic acids. Bacterial biofilms are responsible for undesirable effects such as disease, prostheses colonization, biofouling, equipment damage, and pipe plugging. Biofilms are also more resilient than free-living cells to regular sterilization methods and therefore it is indispensable to develop better ways to control and remove them. The use of gas discharge plasmas is a good alternative since plasmas contain a mixture of reactive agents well-known for their decontamination potential against free microorganisms. We have previously reported that *Pseudomonas aeruginosa* biofilms were inactivated after a 1-min plasma exposure. We determined that the adhesiveness and the thickness of *Pseudomonas* biofilms grown on borosilicate were reduced. We also reported sequential morphological changes and loss of viability upon plasma treatment. However, the studies were carried out in batch cultures. The use of a continuous culture results in a more homogenous environment ensuring reproducible biofilm growth. The aim of this work was to study plasma-mediated inactivation of *P. aeruginosa* biofilms grown on borosilicate in a continuous culture system. In this paper we show that biofilms grown on glass under continuous culture can be inactivated by using gas discharge plasma. Both biofilm architecture and cell culturabilty are impacted by the plasma treatment. The inactivation kinetics is similar to previously described ones and cells go through sequential changes ranging from minimal modification without loss of viability at short plasma exposure times, to major structure and viability loss at longer exposure times. We report that changes in biofilm structure leading to the loss of culturability and viability are related to a decrease of the biofilm matrix adhesiveness. To our knowledge, there has been no attempt to evaluate the inactivation/sterilization of biofilms grown in a continuous system.

Editor: Anthony George, University of Technology Sydney, Australia

Funding: This work was supported by the National Institutes of Health Grant SCORE SC3 # 1SC3GM088070 to GBM, National Science Foundation Nanotechnology Undergraduate Education Program, award # 0406533 to KGV and Agencia Nacional de Promoción Científica y Tecnológica Proyectos de Investigación Científica y Tecnológica Argentina PICT-2009-0047 to GBM. The funders had no role in study design, data collection and analysis, decision to publish, or preparation of the manuscript.

Competing Interests: The authors have declared that no competing interests exist.

* Email: gbrelles@biotec.quimica.unlp.edu.ar

¤ Current address: Center for Research and Development of Industrial Fermentations, Facultad de Ciencias Exactas, Universidad Nacional de La Plata, La Plata, Argentina

Introduction

In the past, the microbial world was thought as composed of isolated microorganisms growing apart. However, this model is more the exception than the rule. Most microbes are "social" and prefer to live and thrive as part of communities where interactions take place [1]. A biofilm is an example of this type of community where cooperative effects become important. Biofilms are microbial communities that grow attached to a surface and embedded in a viscous matrix composed of exopolysaccharides together with proteins and excreted nucleic acids. Biofilms are present almost everywhere and impact all aspects of our life. Not all bacterial biofilms are detrimental but in many cases their presence is responsible for expensive and undesirable effects such as disease, prostheses colonization, product contamination, biofouling and equipment damage, pipe plugging, tooth decay, and dental plaque. About 90 percent of infections in humans and 65 percent of nosocomial infections are due to biofilms according to the National Institutes of Health (NIH) and The Center for Disease Control (CDC) respectively. Biofilms also contaminate water sources and cause pipe plugging. *Helicobacter pylori*, a microorganism responsible for gastric ulcers, have been found in pipes in drinking water systems. Therefore not only is the industrial contamination a concern but also the possibility of spreading disease by contaminated water [2].

Studies on microbial growth and its control using free-living, planktonic microorganisms have provided a good deal of information. However, these results cannot always be easily

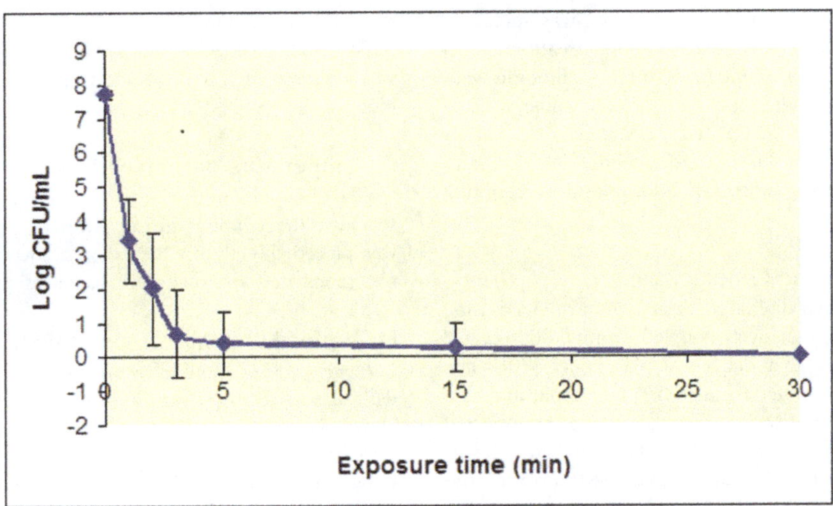

Figure 1. Survivor curve for *P. aeruginosa* plasma-treated biofilms. Log of the number of *P. aeruginosa* CFU/mL vs plasma exposure time (0 to 30 minutes). Results are the average of five independent experiments. Each experiment was performed in duplicate. Bars represent the standard error of the mean.

extrapolated to microbial communities such as biofilms, more resilient to standard killing methods. The use of moist heat in the autoclave is still an inexpensive method for many applications but it cannot be applied to all situations, such as prosthetic devices or thermosensitive materials. Low-temperature disinfection can be achieved using chemicals; among these, ethylene oxide is both mutagenic and carcinogenic and chlorine pose an environmental hazard and risks to human health. Radiation can be used in some but not all the cases. Therefore, it is indispensable to develop better ways to control and remove biofilms. The use of gas discharge plasmas is a good alternative since plasmas contain a mixture of reactive species, free radicals, and UV photons well-known for their decontamination potential against free microorganisms [3]. Investigations aimed at elucidating the effects of plasmas on bacterial cells were initiated, mostly in the United States since the mid nineties [4]. Most of those studies were carried out with microorganisms in the free-living state or with spores. By the middle of the last decade, our group and a few others reported the use of plasma for biofilm disinfection or inactivation [5–9].

More recently we reported the use of plasma to inactivate *Pseudomonas aeruginosa* biofilms [10]. *P. aeruginosa* is a Gram-negative opportunistic pathogen that preys on victims with compromised immune systems, patients on respirators, and causes infections of burned tissue and colonization of catheters and medical devices. It also co-colonizes, together with *Burkholderia cenocepacia*, lung tissue and is the main cause of mortality in cystic fibrosis patients [11]. *Pseudomonas* biofilm inactivation/sterilization have been intensively studied by different approaches such as the use of biocides, antibiotics or a combination of both [12,13]; the use of chelators [14]; and compounds such as furanone and *N*-acyl homoserine lactones [15,16]; and the modification of surfaces [17,18], among others.

In our previous contributions, *P. aeruginosa* biofilms were grown on borosilicate, polycarbonate, and staineless-steel surfaces in batch culture. We showed that almost 100% of the cells were inactivated after a 5-min plasma exposure. Through atomic-force-microscopy (AFM) we determined that the adhesiveness to borosilicate and the thickness of the *Pseudomonas* biofilms grown on borosilicate were reduced and we reported sequential morphological changes and loss of viability upon plasma treatment [10,19]. However, all the above studies were carried out with biofilms grown in a batch culture system. The way biofilms grow in nature differs from the way they are grown in batch in the laboratory. A batch culture is a closed system starting with an inocculum that grows until nutrients are depleted or toxic products accumulate in the reactor. Therefore, bacterial concentration varies with time. A continuous culture is a way of growing microorganisms in which there is a continuous flow of nutrients in the environment and growth does not depend on time as a variable. In nature, biofilms are surrounded by an aqueous environment in an open system and therefore the continuous culture better mimicks biofilm growth in the real world. The use of the continuous culture results in a more homogenous environment ensuring reproducible biofilm growth.

In this paper we present data on plasma-mediated inactivation of *P. aeruginosa* PAO1 biofilms grown on borosilicate in continuous culture. We show that the inactivation kinetics is similar to previously described ones [10,19] and that cells go through sequential changes ranging from minimal modification without loss of viability at short plasma exposure times, to major structure and viability loss at longer exposure times. We report that changes in biofilm structure leading to the loss of culturability and viability are related to a decrease of the biofilm matrix adhesiveness. To our knowledge, there has been no attempt to evaluate the inactivation/sterilization of biofilms grown in a continuous system.

Materials and Methods

A. Biofilm growth

Pseudomonas aeruginosa strain PAO1 biofilms were produced in continuous culture using the CDC biofilm reactor (BioSurface Tech., MT) until constant optical density for 24 hours. The biofilms were grown on borosilicate (glass) coupons in TSB (Tryptic Soy Broth) at 37°C with agitation. After the selected growth time, the coupons were aseptically removed from the reactor and unbound bacteria were removed by rinsing the coupons twice with sterile saline. Coupons were air-dried prior to being subjected to gas discharge plasma for various exposure times (1, 2, 3, 5, 15, and 30 minutes) under sterile conditions. A control without plasma treatment (0-min exposure time) was included. Coupons were placed in a wet chamber after treatment and incubated with 50 µL of sterile saline for 10 minutes. Biofilms were then scraped off and suspended in 1 mL of sterile saline, serially diluted, and suspensions were plated in duplicate on TSA (Trypctic Soy Agar) medium. Plates were incubated at 37°C and evaluated for colony-forming-units (CFU) formation by counting the colonies.

B. Plasma Generation and Conditions

Atmospheric-pressure gas discharge plasma was produced using a commercially available inductively-coupled Atomflo 300 reactor (Surfx Technologies, CA) that delivers a plasma jet [20]. The reactor consists of two perforated rectangular plates separated by a gap 1.6-mm across. The upper aluminum electrode is connected to a 100-W RF power supply (13.56 MHz), and the lower electrode is grounded. The size of the plasma showerhead is 0.63-cm wide by 2.54-cm across. For the experiments, an atmospheric-pressure plasma jet was generated using a He flow of 20.4 L/min, a secondary gas flow (N_2) of 0.135 L/min, and an input power of 35 W. Both gases were industrial grade. The plasma applicator was mounted such that the showerhead was 4 mm away from the biofilm.

C. Temperature determination

Glass coupons were exposed to plasma to test the temperature reaching the coupon during plasma treatment. A thermometer was placed directly on the coupon surface and the temperature was monitored and recorded once a minute for 10 minutes.

D. Atomic Force Microscopy (AFM)

P. aeruginosa biofilms were grown on glass coupons, treated with plasma for 0, 1, and 30 minutes, and processed as indicated above. The coupons were rinsed twice, air-dried, and AFM images were obtained in air in contact mode using a Quesant Instruments Universal Scanning Probe Microscope. Commercial silicon cantilevers from MikroMasch were employed with spring constants from 0.1 to 0.5 N/m. For each coupon, at least four

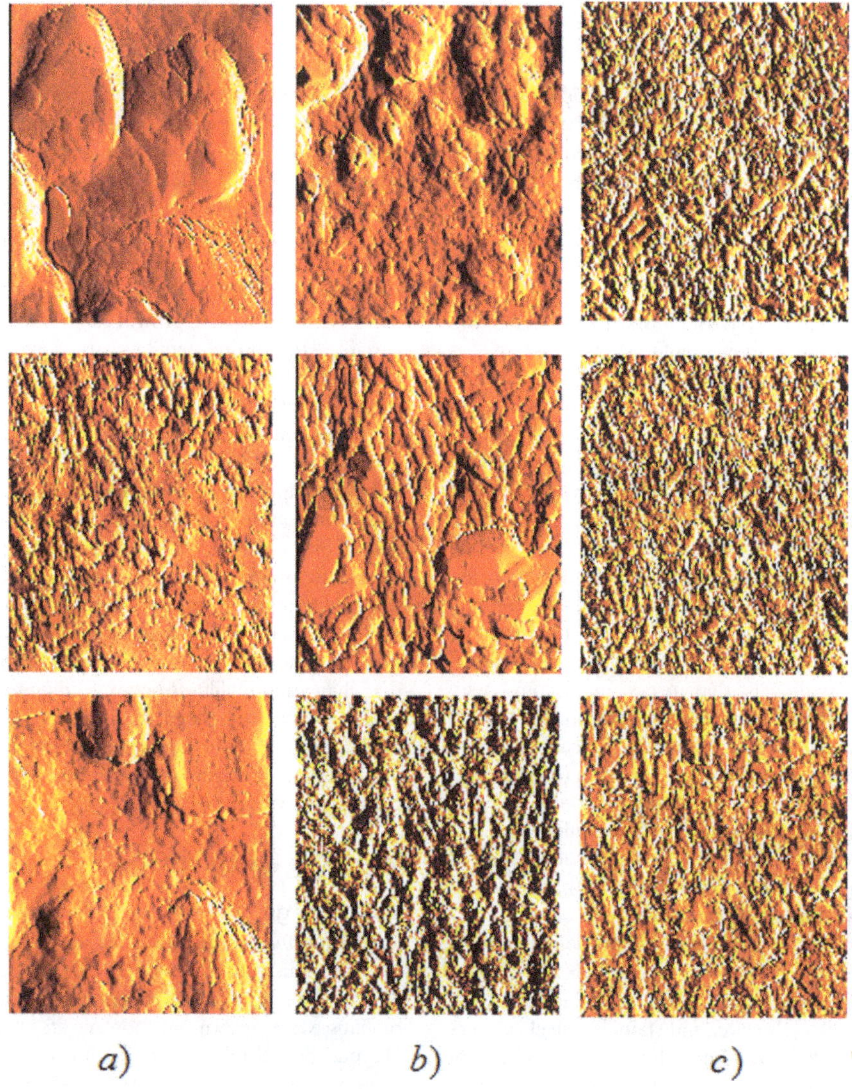

Figure 2. Atomic force microscope images of *P. aeruginosa* biofilms treated with plasma for 0 minute (column a), 1 minute (column b) and 30 minutes (column c). Data for samples 1, 2, and 3 are displayed in the top, middle, and bottom rows, respectively. Image areas are 10×10 μm^2.

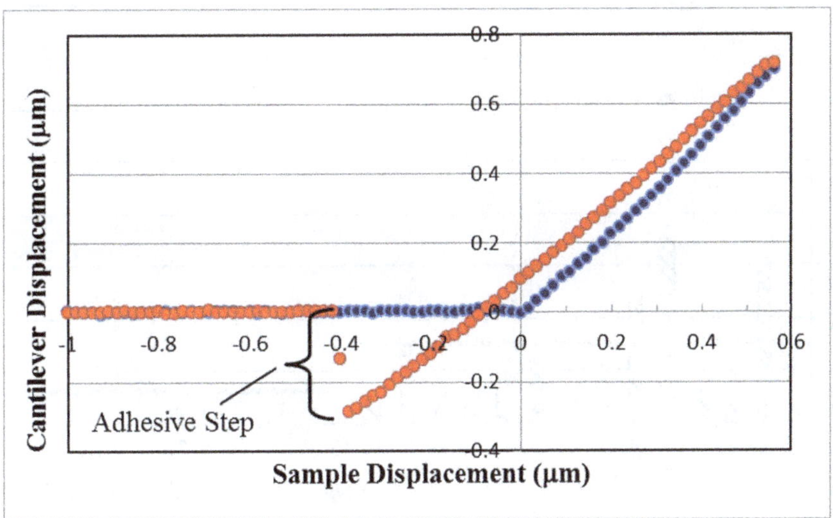

Figure 3. *P. aeruginosa* biofilm force-displacement curve for the control sample on a glass coupon. Data for tip-sample approach (blue circles) and tip-sample retraction (red circles) are shown. The negative displacement of the cantilever that occurred due to tip adhesion to the biofilm upon retraction is designated as the adhesive step, and is measured to be 0.288 µm for this curve.

widely separated regions were imaged to obtain a representative sample and ensure reproducibility. Images consisted of 500 lines of 500 points per line for a total of 250,000 pixels of data.

To assess micromechanical properties of the biofilm, force-displacement curves were obtained at widely separated locations on the coupon, to achieve representative results. The procedure consisted of bringing the AFM tip in contact with the sample and then moving the sample upward a set distance while monitoring the deflection of the cantilever. At each sampling location where force-displacement curves were obtained, the tip was brought in and out of contact at a rate of 0.5 Hz to the maximum set sample deflection (0.6 µm) and the displacement curve was recorded upon the fifth trial. This technique helped to reduce hysteresis that was often observed in the first few trials. The process was then repeated so that at least five force-displacement curves were recorded at each sampling location. Similar techniques for measuring the micromechanical properties of bacteria and bacterial biofilms have been employed successfully in previous studies by us [10,19] and by others [21,22]. For comparing samples with different plasma treatments, all of the force-displacement data were recorded on the same day using the same cantilever. This method ensured

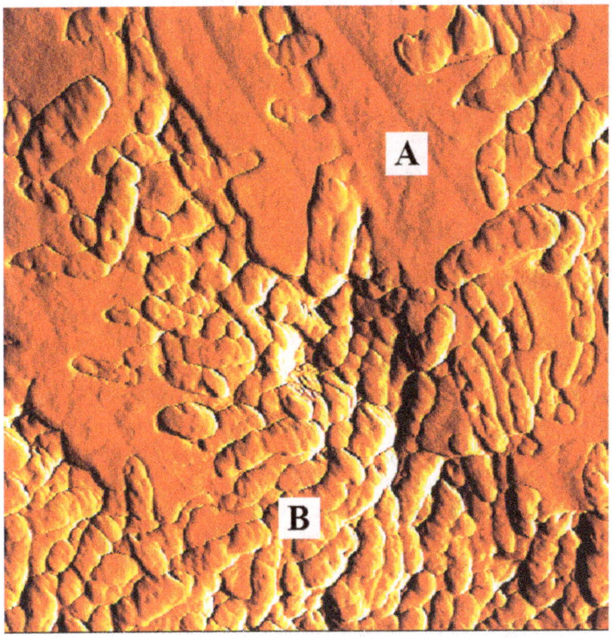

Figure 4. AFM images. 10×10 µm² area AFM image of *P. aeruginosa* biofilm for the control sample indicating locations for obtaining force-displacement curve data. Force-displacement curves obtained at locations similar to point A were designated as areas of predominately matrix material and locations similar to point B were designated as areas of predominately bacteria.

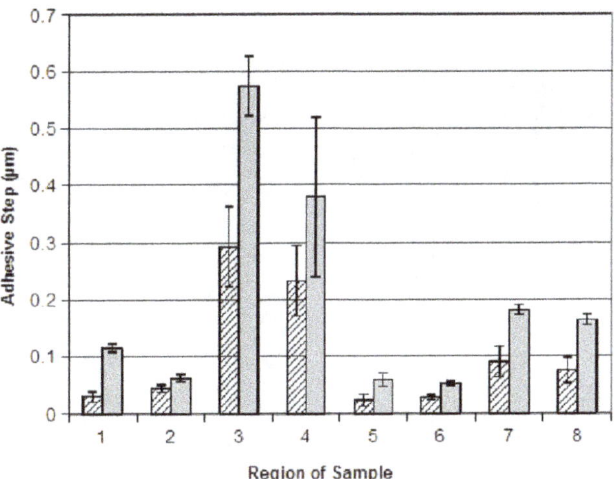

Figure 5. *P. aeruginosa* biofilm force-displacement curve adhesive step data for 0 minute (control) plasma-treated sample 1. The height of each bar on the graph corresponds to the mean adhesive step height of five curves obtained for each region. Error bars represent the standard error of the mean. Hatched bars and solid bars represent measurements on mostly bacteria and mostly matrix areas, respectively.

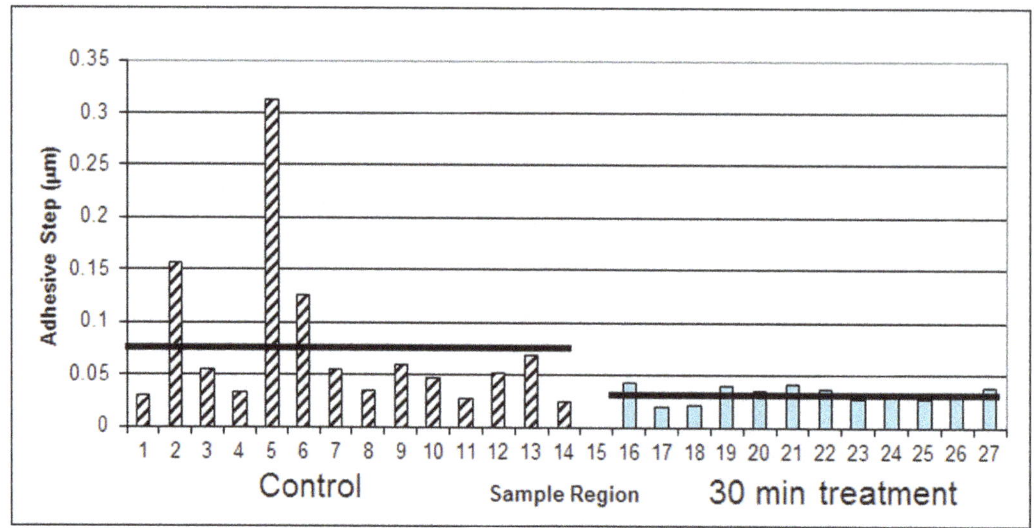

Figure 6. *P. aeruginosa* biofilm force-displacement curve adhesive step data (sample 3), 0 minute (control, hatched bars) and 30 minute plasma-treatment (solid bars). The height of each bar on the graph corresponds to the mean slope of five curves obtained for each region. The horizontal black bars represent the mean adhesive step values.

Figure 7. *P. aeruginosa* virulence tests on lettuce leaves. Lettuce leaves were injected with *P. aeruginosa* biofilm-forming cells treated with plasma for 0, 1 or 30 minutes (panels "a", "b" and "c" respectively) as described in Materials and Methods. A control injected with saline (panel "d") was included as a negative control.

control for humidity and cantilever dependent factors (such as spring constant) that can influence the shape of these curves [23].

E. Virulence Tests

A plant assay was developed to test the virulence of biofilm-derived bacteria treated with plasma for various exposure times [19]. The mid-vein of surface-sterilized leaves of Romaine lettuce was inoculated with a 1 mL-suspension of *P. aeruginosa* cells obtained after treating biofilms with plasma as indicated above and diluted with sterile saline to OD = 1. Unexposed *P. aeruginosa* was used as a positive control. A leaf injected with saline as well as a leaf that had not undergone any injections were also included as controls. Leaves were then aseptically placed into UV sterilized- plastic bags. The bags were sealed, laid flat at 37°C and incubated for three additional days. Tissue damage was assessed and photos of leaves were taken on the fourth day.

Results and Discussion

Survival Curves and Inactivation Kinetics

The concentration of remaining culturable cells versus plasma exposure time for borosilicate-grown *P. aeruginosa* biofilms is shown in Figure 1. At time 0 the concentration of culturable cells corresponds to the control without plasma treatment. The graph shows that *P. aeurginosa* biofilms grown in continuous culture were almost completely removed after a one-minute treatment with plasma. A good measure of the removal efficiency is the determination of the decimal reduction time (D value) [24,25], the time required to reduce an original concentration of microorganisms by 90%. This parameter has been originally defined for thermal killing of microorganisms by autoclaving. The results show a rapid decline in CFU/mL (first phase) followed by a much slower subsequent decline. As the kinetics of biofilm inactivation shows a double-slope curve, two D values were determined: one for short exposure times (D_1) and another one for the portion of the curve showing a slower decline (D_2). D_1 is 27 sec and D_2 is 6.13 min. We previously reported double-slope kinetics and D-values of 2.3 min (D_1) and 37.4 min (D_2) for *Chromobacterium violaceum* biofilms grown on polyesterene microtiter plates [5].

We also reported double-slope kinetics for *P. aeurginosa* biofilm grown in batch cultures although D values were not calculated [10]. In that previous work, we showed that regardless of the biofilm age, there was a clear decrease in the percentage of surviving cells versus time. Similar results were reported for *C. violaceum* biofilms grown for 4 or 7 days on polystyrene microtiter plates [5,26]. In the case of *P. aeruginosa* biofilms, the decrease in the percentage of viable cells was more dramatic since there are almost no culturable cells after a 5-minute treatment with plasma. As previously reported for *P. aeruginosa* biofilm of different maturity grown on borosilicate coupons, most of the inactivation occurs before a biofilm exposure to plasma of less than one minute.

To rule out the effect of temperature on the inactivation of the biofilm we measured the temperature of the gas reaching the coupon surface. Equilibrium temperatures of 35 °C were reached within a few minutes and remained constant over time confirming that temperature is not responsible for biofilm inactivation since *P. aeruginosa* is a human pathogen that prefers living a t 37°C.

To rule out effects of gas flow on biofilms, plasma-treated biofilms were compared to biofilms exposed to a flow of gas in the absence of plasma (plasma source turned off) as described elsewhere [26]. Although the flow of gas dried out cells and caused a decrease in the number of CFUs of about 5 to 10%, this decrease is not significant compared to the decrease produced by plasma. Therefore, cell inactivation is due to plasma treatment and not due to excessive cell drying from the gas flow.

Other authors have reported bi and triphasic behaviors for plasma-assisted killing of free-living microorganisms and spores. Two-slope behavior with a smaller D value for shorter exposure times and a higher D value for longer exposure times, were reported by Kelly-Winterberg et al. for *Staphylococcus aureus and Escherichia coli* on polypropylene samples, and by Laroussi *et al.* for *P. aeruginosa* in liquid suspension [27,28]. The biphasic curves were explained by the damaging of the cell membrane due to plasma reactive species in the first phase. Once the membrane is compromised, the reactive species can easily penetrate causing rapid cell death in the second phase. We first hypothesized that the initial rapid decline of CFU/mL might be due to the killing of the upper layers of microorganisms in the biofilm, that were more exposed to plasma. After this initial killing, plasma has to penetrate layers of cell debris and dead cells before reaching the inner portion of the biofilm [5]. However, no experimental evidence was obtained in our laboratories to support the hypothesis. Further results from our laboratory showed that plasma-mediated biofilm inactivation proceeds through a first step in which bacterial cells are not culturable but still alive, followed by a second step, characterized by a higher D-value, in which cells are killed. Results are consistent with a first step in which bacteria enter a viable-but-non-culturable (VBNC) state and/or they result in spheroplasts that are smaller in size due to damage and further removal of cell walls after plasma treatment. These spheroplasts are non-culturable but still alive since they retain an intact cell membrane [26]. The VBNC state is a dormant state that represents a survival mechanism of bacteria facing one or more environmental stresses which might otherwise be ultimately lethal to the cell and that has been reported for many gram-negative organisms [29]. Regardless of the mechanism, it was clear from our results that treating biofilm cells with plasma for short exposure times resulted in cells that were not culturable but still alive. These findings paved the way to a change in the paradigm that implied that cells were killed based only on their lack of culturability. Since then, we always carried out viability experiments before concluding that bacteria are killed just because we cannot count colonies [26,10,19].

AFM images and force-displacement curves

Figure 2 displays a series of AFM images of the biofilm, before and following plasma treatment. The control (0 minute-treatment) images in the left column of Figure 2 show the typical tridimensional structure of a *P. aeruginosa* biofilm. The 1 minute-treated samples in the middle column of Figure 2 depict some morphological changes in the biofilm but still show distinctly discernable bacteria. In the right column of Figure 2, the 30 minute-treated samples display a flatter, more disrupted structure that less convincingly resembles a tridimensional biofilm although there are still discernable bacteria. We have previously reported sequential structural changes of the biofilm after plasma treatment. Those changes ranged from minimal ones at short exposure times to very noticeable modifications at longer exposure times for both *C. violaceum* and *P. aeruginosa* biofilms [10,19,26,30]. We also reported a decrease in biofilm thickness after plasma treatment for *P. aeruginosa* biofilms grown on glass coupons in batch culture [10]. In the case of *Chromobacterium* biofilms, no recognizably intact cells but only debris or cell remnants were obtained after a 60-minute plasma treatment [26,30]. However, in those experiments cells were dettached from the biofilm by sonication before processing and also, exposure times were longer than the ones used for the experiments reported here. If bacterial envelopes were damaged by plasma treatment, cells might have become weaker and more prone to become disrupted/broken by the sonication treatment. In our present case, cells remain within the biofilm structure and attached to the surface. Also, although both *Chromobacterium* and *Pseudomonas* are gram-negative bacteria, they might exhibit slightly different responses to plasma treatment. Results from this work confirm our previous observations with other types of biofilms and demonstrate that the effect of plasma on the biofilm does not depend on the whether microorganisms are grown in a closed system (batch) or an open one (continuous culture).

Column (a) of Figure 2 clearly displays distinguishable bacteria together with large smooth areas of biofilm matrix presumably composed of exopolysaccharide. However, as displayed in columns (b) and (c) in the same figure, matrix areas become reduced/smaller upon exposure to plasma. For 30 minutes of plasma treatment (column (c)) no matrix areas can be seen. In fact, of the seven samples measured, matrix areas were never observed in any of the 30-minute plasma treatments. This qualitative observation suggests that plasma treatment may reduce the areas of biofilm matrix probably through oxidation/peroxidation of the exopolysaccharide by the reactive agents, mostly free radicals, present in the plasma. We hypothesized that the decrease and eventual loss of the biofilm matrix would reduce the adhesiveness of the biofilm to the surface to which it is anchored and would lead to disorganization of the tridimensional structure of the biofilm. To test this hypothesis we assessed micromechanical properties of the biofilm through force-displacement curves as described in Materials and Methods. Figure 3 shows a typical force-displacement curve for *P. aeruginosa* biofilms grown on glass coupons in continuous culture and not treated with plasma (control). The curve displays the same general features that were exhibited in all of our force-displacement curves and is similar to the ones we previously reported [10]. The tip encounters the sample surface upon approach (blue data, at the origin of the graph) and deflects upward with an increasing slope. Upon retraction (red data), the tip only roughly retraces the approaching curve with moderate hysteresis. Upon further retraction, the tip adheres to the surface until it breaks free, and the points retrace the approaching data along the negative x axis of the graph. For this study, to ascertain surface adhesion, only one section of the curve was analyzed, the height of the adhesive step upon retraction, as defined in Figure 3.

Force-displacement curves were obtained from two distinct sample locations, areas of mostly matrix material or areas of mostly bacteria, as indicated in Figure 4. Figure 5 displays adhesive step results for sample 1 obtained over a number of regions on the sample. It is apparent from this graph that there is a wide variability in adhesive step values over the various regions of the sample. Even considering this variability, for each region, there is more adhesion for the mostly matrix areas than for areas of mostly bacteria. This general result was also confirmed for the other three samples measured. For each sample, the average value of the heights of the adhesive steps was always greater when measured at mostly matrix areas than for areas of mostly bacteria.

In addition to examining adhesion differences due to the matrix versus the bacteria, force-displacement curves were obtained on two of the samples comparing the 0 minute to the 30-minute plasma treatment. For these comparisons, curves were obtained only on areas of predominately bacteria since none of the 30-minute treatments yielded any areas of mostly matrix, as described previously. Figure 6 displays these results for one of the samples. Although there is again variability over different regions of the sample, on average, the adhesion to the plasma treated areas is significantly less than for the control. The mean adhesive step for this sample was 0.077 μm for the control and 0.032 μm for the 30-minute plasma treatment. For another sample (data not shown), the mean adhesive step was 0.030 and 0.022 μm for the control and 30 minute- plasma treatment, respectively.

Taken together with the results from about 300 images and approximately 700 force displacement curves analyzed, we can conclude that plasma treatment of *P. aeruginosa* biofilms results in little change in cell morphology for short exposure times while longer exposures results in significant loss of the biofilm structure resulting in cell death. The adhesiveness of the biofilm varies across its structure and is higher in areas with larger amounts of matrix. Plasma treatment removes or at least reduces the matrix, presumably by oxidation/peroxidation due to the presence of free radicals, and the areas of predominantly bacteria are less adhesive after the treatment.

Virulence tests

P. aeruginosa is an opportunistic pathogen for humans but also a plant pathogen that produces tissue damage. Therefore we developed a plant assay that allowed us to test for viability. Figure 7 shows the results of the virulence tests on lettuce leaves. It is clear that the mid-vein of leaves in panels "c" and "d" show no damage whereas both the leaf treated with bacteria not subjected to plasma treatment (panel "a" positive control) or treated with plasma for one minute (panel "b") show tissue damage. Although bacterial cells treated with plasma for one minute do not yield appreciable numbers of colonies on a petri dish, it is obvious that those cells are still not only viable but also virulent.

Conclusions

Our results demonstrate that *P. aeruginosa* bacterial biofilms grown on glass under continuous culture can be inactivated by using gas discharge plasma. Both biofilm architecture and cell culturabilty are impacted by the plasma treatment. These results are evidence of the potential of plasma as an alternative sterilization method against biofilms.

One of the issues of more concern regarding plasma-assisted cell inactivation is that many authors still assess the lethality of plasma solely based on the number of colonies that can be counted after the treatment. However, bacterial cells can respond to one or more environmental stresses by entering a viable-but-non-culturable (VBNC) state [29] and although they are unable to produce colonies on an agarized medium they are still alive and may retain pathogenicity. The virulence assay using lettuce plants for *P. aeruginosa* biofilms indicates that cells retain viability after short exposures to plasma. Therefore, and in agreement with previously reported results from our group [10,19,26], viability experiments should always be carried out before drawing the conclusion that plasma is useful to kill cells based solely on measurement of culturable cells.

AFM results show that plasma treatment of *P. aeruginosa* biofilms results in little change in cell morphology for short exposure times while longer exposures results in significant loss of the biofilm structure. The adhesiveness of the biofilm varies across its structure and is higher in areas with larger amounts of matrix. Plasma treatment removes or at least reduces the matrix, presumably by oxidation/peroxidation due to the presence of free radicals, and the areas of predominantly bacteria are less adhesive after the treatment. An interesting observation from the AFM experiments is that biofilms treated with plasma for 30 minutes resulted in images that consistently had a "noisier" background than the images corresponding to the control or to 1-minute exposure to plasma. These results suggest changes in the biofilm surface chemistry that may affect the AFM tip interaction with the surface. In a previous work we studied the chemistry of the generated plasma by spectroscopy and we reported the presence of OH and NO radicals in the plasma [5]. These reactive species have direct impact on microorganisms, especially on the cell wall and cell membrane compromising their function and viability and altering the surface chemistry. It is well known that free radicals produce oxidation and peroxidation of lipids of the bacterial outer envelopes and cytoplasmic membrane. It is also possible that the exopolisaccharide that composes the biofilm matrix and anchors bacteria to the surface may become oxidized by plasma as well. Therefore, it can be speculated that the background "noise" observed in images obtained from plasma treated biofilms results from changes in the biofilm surface chemistry after plasma exposure. Further HPLC experiments will be carried out to study putative changes in the biofilm matrix exopolysaccharide.

Acknowledgments

The authors thank Anna J. Zelaya for technical help.

Author Contributions

Conceived and designed the experiments: GBM KGV. Performed the experiments: GBM KGV. Analyzed the data: GBM KGV. Contributed reagents/materials/analysis tools: GBM KGV. Contributed to the writing of the manuscript: GBM KGV.

References

1. Brelles-Mariño G (2010) Bacterial Biofilm Inactivation by Gas-Discharge Plasmas. In: Biological and Environmental Applications of Gas Discharge Plasmas. Brelles-Mariño G, editor. Nova Science Publisher. ISBN: 978-1-60741-945-7, 75–98.
2. Park SR, Mackay WG, Reid DF (2001) *Helicobacter* sp recovered from drinking water biofilm samples from a water distribution system. Water Res 35: 1624–1626.
3. Laroussi M (1996) Sterilization of contaminated matter with an atmospheric pressure plasma. IEEE Trans Plasma Sci 24: 1188–1191.

4. Laroussi M (2010) Foreword. In: Biological and Environmental Applications of Gas Discharge Plasmas. Brelles-Mariño G, editor. Nova Science Publisher. ISBN: 978-1-60741-945-7. 3–4.
5. Abramzon N, Joaquin JC, Bray JD, Brelles-Mariño G (2006) Biofilm Destruction by RF High-Pressure Cold Plasma Jet. IEEE Trans Plasma Sci 34: 1304–1309.
6. Akishev YS, Grushin ME, Karalnik VB, Monich AE, Pan'kin MV, et al. (2005) Sterilization/decontamination of physiological solution and dry surface by non-thermal plasma created in bubbles and jet. In: Proc. of the 2nd International Workshop on Cold Atmospheric Pressure Plasmas ISBN:908086692X. 69–72.
7. Akishev Y, Grushin M, Karalkin V, Trushkin N, Kholodenko V, et al. (2008) Atmospheric-pressure, nonthermal plasma sterilization of microorganisms in liquids and on surfaces. Pure Appl Chem 80: 1953.
8. Becker K, Koutsospyros A, Yin SM, Christodoulatos C, Abramzon N, et al. (2005) Environmental and Biological Applications of Microplasmas. Plasma Phys Control Fusion. 47: B513–B523.
9. Brelles-Mariño G, Joaquin JC, Bray JD, Abramzon N (2005) Gas discharge plasma as a novel tool for biofilm destruction. In: Proc. 2nd Int. Workshop Cold Atmospheric Pressure Plasmas. ISBN: 908086692X, 69–72.
10. Zelaya A, Stough G, Rad N, Vandervoort K, Brelles-Mariño G (2010) *Pseudomonas aeruginosa* Biofilm Inactivation: Decreased Cell Culturability, Adhesiveness to Surfaces, and Biofilm Thickness upon High-Pressure Non-Thermal Plasma Treatment. IEEE Trans Plasma Sci 38: 3398–3403.
11. Tümmler B, Kiewitz C (1999) Cystic fibrosis: an inherited susceptibility to bacterial respiratory infections. Mol Med Today 5: 351–358.
12. Gillis RJ, Iglewski BH (2004) Azithromycin Retards *Pseudomonas aeruginosa* Biofilm Formation. J Clin Microbiol 42: 5842–5845.
13. Tanaka G, Shigeta M, Komatsuzawa H, Sugai M, Suginaka H, Usui T (2000) Effect of Clarithromycin on *Pseudomonas aeruginosa* Biofilms. Chemotherapy 46: 36–42.
14. Banin E, Vasil ML, Greenberg EP (2005) Iron and *Pseudomonas aeruginosa* biofilm formation. Proc Nat Acad Sci 102: 11076–11081.
15. Hentzer M, Riedel K, Rasmussen TB, Heydorn A, Andersen JB, et al. (2002) Inhibition of quorum sensing in *Pseudomonas aeruginosa* biofilm bacteria by a halogenated furanone compound. Microbiology 148: 87–102.
16. Davies D, Parsek M, Pearson J, Iglewski B, Costerton JW, Greenberg EP (1999) The use of signal molecules to manipulate the behavior of biofilm bacteria. Clin Microbial Infect 5: 5S7–5S8.
17. Balazs DJ, Triandafillu K., Wood P, Chevolot Y, van Delden C, et al. (2003) Inhibition of bacterial adhesion on PVC endotracheal tubes by RF-oxygen glow discharge, sodium hydroxide and silver nitrate treatments. Biomaterials 11: 2139–2151.
18. Bryers JD, Ratner BD (2004) Bioinspired Implant Materials Befuddle Bacteria. ASM News. 70: 232–237.
19. Zelaya A, Vandervoort K, Brelles-Mariño G (2012) Battling Bacterial Biofilms with Gas Discharge Plasma. In: NATO Science for Peace and Security Series. Plasma for bio-decontamination, medicine and food security Machala Z Hensel K, Akishev Y, editors. Springer. ISBN 978–94–007–2909–4 135–148.
20. Schutze JY, Jeong SE, Park J, Selwyn GS, Hicks RF (1998) The atmospheric plasma jet: A review and comparison to other plasma sources. IEEE Trans Plasma Sci 26: 1685–1694.
21. Oh YJ, Lee N R, Jo W, Jung WKL, Lim JS (2009) Effects of substrates on biofilm formation observed by atomic force microscopy. Ultramicro 109: 874–880.
22. Zhao L, Schaefer D, Marten MR (2005) Assessment of Elasticity and Topography of *Sperfillus nidulans* Spores via Atomic Force Microscopy. Appl Environ Mircrobiol 71: 955–960.
23. Jones R, Pollock HM, Cleaver JAS, Hodges CS (2002) Adhesion forces between glass and silicon surfaces in air studied by AFM: effects of relative humidity, particle size, roughness, and surface treatment. Langmuir 18 : 8045–8055.
24. International Standard, ISO 14937 (2000) Sterilization of Health Care Products. General Requirements for Characterization of a Sterilizing Agent and the Development, Validation and Routine Control of a Sterilization Process for Medical Devices.
25. International Standard, ISO 11134 (1994) Sterilization of Health Care Products. Requirements for Validation and Routine Control – Industrial Moist Heat Sterilization.
26. Joaquin J, Kwan C, Abramzon N, Vandervoort K, Brelles-Mariño G (2009) Is Gas-Discharge Plasma a New Solution to the Old Problem of Biofilm Inactivation? Microbiology. 175: 724–732.
27. Kelly-Wintenberg K, Montie TC, Brickman C, Roth JR, Tsai APPY (1998) Room temperature sterilization of surfaces and fabrics with one atmosphere uniform glow discharge plasma. J Ind Microbiol Biotechnol 20: 69–74.
28. Laroussi M, Alexeff I, Kang W (2000) Biological Decontamination By Non-Thermal Plasmas. IEEE Trans Plasma Sci 28: 184–188.
29. Rozak DB, Colwell RR (1987) Survival strategies of bacteria in the natural environment. Microbiol Rev 51: 365–379.
30. Vandervoort K, Abramzon N, Brelles-Mariño G (2008) Plasma Interactions with Bacterial Biofilms as Visualized through Atomic Force Microscopy. IEEE Trans Plasma Sci 36: 1296–1297.

4

The Predatory Bacterium *Bdellovibrio bacteriovorus* Aspartyl-tRNA Synthetase Recognizes $tRNA^{Asn}$ as a Substrate

Ariel Alperstein, Brittany Ulrich, Denise M. Garofalo, Ruth Dreisbach, Hannah Raff, Kelly Sheppard*

Chemistry Department, Skidmore College, Saratoga Springs, New York, United States of America

Abstract

The predatory bacterium *Bdellovibrio bacteriovorus* preys on other Gram-negative bacteria and was predicted to be an asparagine auxotroph. However, despite encoding asparaginyl-tRNA synthetase and glutaminyl-tRNA synthetase, *B. bacteriovorus* also contains the amidotransferase GatCAB. *Deinococcus radiodurans*, and *Thermus thermophilus* also encode both of these aminoacyl-tRNA synthetases with GatCAB. Both also code for a second aspartyl-tRNA synthetase and use the additional aspartyl-tRNA synthetase with GatCAB to synthesize asparagine on $tRNA^{Asn}$. Unlike those two bacteria, *B. bacteriovorus* encodes only one aspartyl-tRNA synthetase. Here we demonstrate the lone *B. bacteriovorus* aspartyl-tRNA synthetase catalyzes aspartyl-$tRNA^{Asn}$ formation that GatCAB can then amidate to asparaginyl-$tRNA^{Asn}$. This non-discriminating aspartyl-tRNA synthetase with GatCAB thus provides *B. bacteriovorus* a second route for Asn-$tRNA^{Asn}$ formation with the asparagine synthesized in a tRNA-dependent manner. Thus, in contrast to a previous prediction, *B. bacteriovorus* codes for a biosynthetic route for asparagine. Analysis of bacterial genomes suggests a significant number of other bacteria may also code for both routes for Asn-$tRNA^{Asn}$ synthesis with only a limited number encoding a second aspartyl-tRNA synthetase.

Editor: John R. Kirby, University of Iowa, United States of America

Funding: This work was supported by the National Science Foundation (MCB-1244326 to KS, www.nsf.gov) and the Skidmore Faculty/Student Summer Research Program (to AA, DG, BU, and KS, http://www.skidmore.edu/dof-vpaa/summer-research/). The funders had no role in study design, data collection and analysis, decision to publish, or preparation of the manuscript.

Competing Interests: The authors have declared that no competing interests exist.

* Email: ksheppar@skidmore.edu

Introduction

Bdellovibrio bacteriovorus HD100 preys on other Gram-negative bacteria by inserting into the host's periplasm where *B. bacteriovorus* grows and replicates, taking advantage of the nutrient rich environment of the host cell [1]. Because this predatory process kills the host, *B. bacteriovorus* is being studied as a living antibiotic for therapeutic, agriculture, and waste treatment purposes [2–6]. Based on its genome, the bacterium was predicted to be missing biosynthetic pathways for nine of the proteinogenic amino acids including Asn, likely making *B. bacteriovorus* protein synthesis dependent on host degradation products [7].

For translation, *B. bacteriovorus* encodes all twenty aminoacyl-tRNA synthetases (aaRSs) [7], typical of δ-proteobacteria but unlike most bacteria which usually encode 18–19 aaRSs [8,9]. The aaRSs missing are either glutaminyl-tRNA synthetase (GlnRS) and or asparaginyl-tRNA synthetase (AsnRS) [9]. In bacteria missing an AsnRS to directly ligate Asn to $tRNA^{Asn}$, Asn is synthesized on $tRNA^{Asn}$ using an indirect two-step pathway by taking advantage of an aspartyl-tRNA synthetase (AspRS) with relaxed tRNA specificity, a non-discriminating AspRS (ND-AspRS) [10,11]. The ND-AspRS forms Asp-$tRNA^{Asn}$, which is then amidated by the amidotransferase GatCAB to Asn-$tRNA^{Asn}$ [10,12,13]. GatCAB can also be used for Gln-$tRNA^{Gln}$ formation in bacteria lacking a GlnRS, about two-thirds of all known bacteria [8,14–18].

Despite coding for both AsnRS and GlnRS, *B. bacteriovorus*, like other δ-proteobacteria, also encodes GatCAB [7,8]. In most γ-proteobacteria like *Escherichia coli* that encode AsnRS and GlnRS, GatCAB is typically absent [8]. It has been hypothesized that *B. bacteriovorus* could use GatCAB for tRNA-dependent Asn synthesis given that it lacks both asparagine synthetases (AsnA and AsnB) [19]. For *B. bacteriovorus* to use the two-step Asn-$tRNA^{Asn}$ synthetic pathway, it also must encode a ND-AspRS despite having an AsnRS. Previously, two bacteria were known to encode GlnRS and both routes for Asn-$tRNA^{Asn}$ formation: *Deinococcus radiodurans* and *Thermus thermophilus* [10,12,20–22]. They encode both routes by acquiring an additional AspRS from archaea. The additional AspRS serves as the ND-AspRS required for tRNA-dependent Asn synthesis [10,12,20–22].

However, *B. bacteriovorus* codes for only one AspRS [7]. We therefore predicted the lone *B. bacteriovorus* AspRS is non-discriminating in order to facilitate GatCAB synthesis of Asn on $tRNA^{Asn}$. We demonstrate that the *B. bacteriovorus* AspRS can readily form Asp-$tRNA^{Asn}$ as the first step in tRNA-dependent Asn biosynthesis. By analyzing bacterial genomes, we found a significant number of bacteria may also encode both routes for Asn-$tRNA^{Asn}$ synthesis including additional species with a second

AspRS. However, only a limited number of bacteria encode AsnRS, GlnRS, GatCAB, and only one AspRS but neither Asn synthetase like *B. bacteriovorus*.

Materials and Methods

General

Oligonucleotides were from Integrated DNA Technologies (San Diego, California). *B. bacteriovorus* HD100 genomic DNA was a gift from Dr. John Tudor (Saint Joseph's University). Samples were sequenced at the Yale DNA Analysis Facility on Science Hill (New Haven, CT). Nuclease P1 and amino acids were from Sigma-Aldrich (St. Louis, MO). Phenol, ATP, and chloroform were from Fisher Scientific (Pittsburg, PA). [α-^{32}P]ATP (10 mmol/μCi) was from Perkin Elmer (Shelton, CT). Polyethylenimine (PEI)-cellulose thin layer chromatography (TLC) glass plates were from EMD Millipore (Billerica, MA). Restriction enzymes, *Escherichia coli* BL21(DE3) and NEB10β strains, OneTaq DNA Polymerase, and T4 DNA ligase were from New England Biolabs (Ipswich, MA). *E. coli* JF448 was from the Yale Coli Genetic Stock Center (New Haven, CT). *E. coli trpA*34 was a gift from the Söll Laboratory at Yale University (New Haven, CT).

Over-production and purification aaRSs

B. bacteriovorus aspS (Bd3311) was cloned between the *Nde*I and *Bam*HI restriction sites in pET28a to be N-terminally His_6-tagged. *B. bacteriovorus* AspRS was overproduced using the autoinduction method [23] and purified by nickel-affinity chromatography in the same manner as the *S. aureus* AspRS following manufacturer's protocols (Qiagen) [24]. The purified enzyme was dialyzed, concentrated, and stored as described [24]. The enzyme preparation was determined>95% pure by Coomassie-stained polyacrylamide gel [24]. The *Legionella pneumophila aspS* was chemically synthesized (Life Technologies, GeneArt) and then subcloned between the *Nde*I and *Bam*HI sites in pET28a and overproduced and purified as described for the *S. aureus* AspRS [24].

The *B. bacteriovorus asnC* (Bd1054) was chemically synthesized with optimized codons for overproduction in *E. coli* (Life Technologies GeneArt). The optimization increased the number of codons, 52% to 90%, in the GeneArt's top codon class (90–100) based on frequency of codon usage in *E. coli* [25]. The gene was then subcloned into pET28a between the *Nde*I and *Bam*HI sites to be N-terminally His_6-tagged. AsnRS was over-produced as described previously for the *S. aureus* homolog [24] using the autoinduction method [23] and purified by nickel-affinity chromatography following manufacturer's protocols (Qiagen) with a buffer of 50 mM Tris-HCl, pH 7.6 with 10 mM $MgCl_2$ and 300 mM NaCl. The purified AsnRS was dialyzed in 50 mM Tris-HCl, pH 7.6 with 10 mM $MgCl_2$, 30 mM NaCl, and 50% glycerol, and then concentrated and stored as described [24].

In vitro transcription, tRNA folding, and ^{32}P labeling

The tRNA genes were *in vitro* transcribed and the resultant tRNA was purified by chromatography as described [22]. The tRNAs were heated to 95°C for 5 min and slowly cooled to room temperature to refold with $MgCl_2$ added to a final concentration of 5 mM at 65°C. Samples were stored at −20°C and ^{32}P-labeled as described previously using the *E. coli* CCA-adding enzyme [8]. *Methanothermobacter thermautotrophicus* $tRNA^{Gln}$ was *in vitro* transcribed, purified, and folded as described previously [26] and ^{32}P-labeled as described previously using the *E. coli* CCA-adding enzyme [8].

^{32}P-based tRNA aminoacylation assay

The aminoacylation activities of the aaRSs were monitored using the established ^{32}P-based assay [8,27–30]. The AspRS reactions contained 50 mM HEPES-KOH, pH 7.2, 30 mM KCl, 15 mM $MgCl_2$, 5 mM DTT, 4 mM L-Asp, and 4 mM ATP. The AsnRS reactions contained 50 mM HEPES-KOH, pH 7.5, 30 mM KCl, 15 mM $MgCl_2$, 5 mM DTT, 4 mM L-Asn, and 4 mM ATP. For plateau aminoacylation of tRNA, reactions were carried out at 37°C with 1.0 μM ^{32}P-labeled tRNA, 11.0 μM tRNA, and 3.0 μM enzyme. Steady-state kinetic studies with 5 nM AspRS were carried out at 37°C with 0.055–1.0 μM ^{32}P-labeled tRNA, and 0–10.0 μM tRNA over 6 min. Steady-state kinetic studies with 5 nM AsnRS were carried out at 37°C with 0.055–1.0 μM ^{32}P-labeled $tRNA^{Asn}$, and 0–12.0 μM $tRNA^{Asn}$ over 6 min. Reaction mixtures and enzymes were pre-incubated for 30 sec at 37°C. Reactions were started by the addition of enzyme and repeated three to four times. Time points were quenched, digested, separated by TLC, processed and analyzed as described previously [8,24,29,30]. The activity of the *L. pneumophila* AspRS was measured in the presence of 0.1 μM ^{32}P-labelled tRNA, 50 mM HEPES-KOH, pH 7.2, 30 mM KCl, 15 mM $MgCl_2$, 5 mM DTT, 4 mM L-Asp, and 4 mM ATP over 5 minutes. Reactions were started by the addition of *L. pneumophila* AspRS to a final concentration of 10 nM at 37°C. Time points were quenched, digested, separated by TLC, processed and analyzed as described previously [8,29].

E. coli trpA34 in vivo assay

The *B. bacteriovorus aspS* was cloned into pCBS2 between the *Nde*I and *Bgl*II restriction sites (pCBS2-*Bb-aspS*) [31]. In a similar fashion the *L. pneumophila aspS* was subcloned into pCBS2 (pCBS2-*Lp-aspS*). Following transformation into *E. coli trpA*34 cells, the cultures were grown and assayed as described previously on M9 minimal media agar plates with or without Trp with minor adjustments [24]. Briefly, cultures were grown overnight at 37°C in LB in the presence of ampicillin (100 μg/ml). The overnight culture was used to inoculate 5 mL of M9 minimal media supplemented with ampicillin (100 μg/ml) and all twenty amino acids (20 μg/ml each). The cultures were then grown shaking at 37°C for four hours. The samples were adjusted to the same $O.D._{600}$ by diluting with M9 minimal media before 5 mL of adjusted culture was spun down at 1,500×g for 5 min, and washed three times with M9 minimal media supplemented with ampicillin (100 μg/ml). After washing, the samples were resuspended in 0.2 mL of M9 minimal media supplemented with ampicillin (100 μg/ml) before spotting 2 μL of culture on M9 minimal media agar ampicillin (100 μg/ml) plates with or without L-Trp (20 μg/ml), and supplemented with the other 19 amino acids (20 μg/ml each).

E. coli JF448 *in vivo* assay

The *B. bacteriovorus aspS* and *gatCAB* (operon of Bd0058, Bd0059, Bd0060) were fused into an artificial operon as described previously [31]. The artificial operon was subcloned into the pCBS2 plasmid between the *Nde*I and *Bgl*II restriction sites (pCBS2-*Bb-aspS-gatCAB*) and transformed into *E. coli* JF448 cells. The cells were grown and assayed as described previously on M9 minimal media agar plates with or without Asn [24]. Briefly, cultures were grown overnight at 37°C in LB in the presence of ampicillin (100 μg/ml). The overnight culture was used to inoculate 5 mL of M9 minimal media supplemented with ampicillin (100 μg/ml) and all twenty amino acids (20 μg/ml each). The cultures were then grown shaking at 37°C for four hours before being spun down at 1,500×g for 5 min, and washed

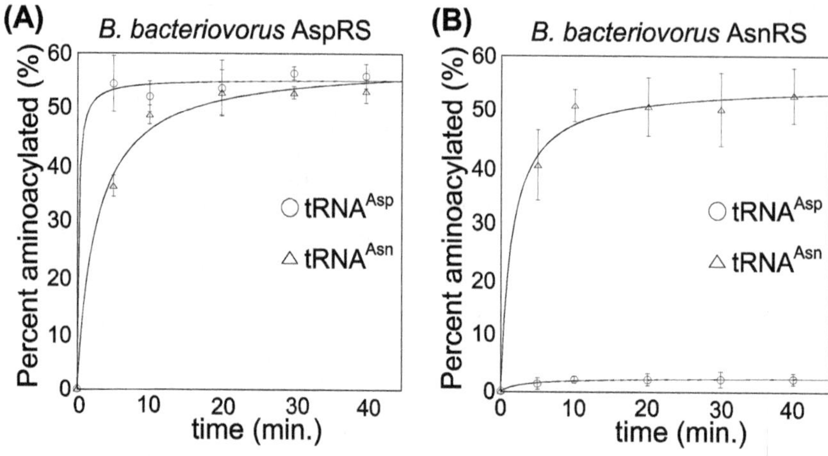

Figure 1. *B. bacteriovorus* AspRS aspartylates $tRNA^{Asn}$. Aminoacylation of *in vitro* transcribed $tRNA^{Asp}$ (○) and $tRNA^{Asn}$ (△) by either (A) *B. bacteriovorus* AspRS or (B) *B. bacteriovorus* AsnRS. Reactions were carried out at 37°C with 1.0 μM ^{32}P-labeled $tRNA^{Asp\ or\ Asn}$, 11.0 μM $tRNA^{Asp\ or\ Asn}$, 4.0 mM ATP, 4.0 mM relevant amino acid (L-Asp or L-Asn) and 3.0 μM enzyme. Experiments were repeated three times and error bars represent standard deviations.

three times with M9 minimal media supplemented with ampicillin (100 μg/ml). After washing, the samples were resuspended in 1 mL of M9 minimal media supplemented with ampicillin (100 μg/ml) and then diluted to an $O.D._{600}$ of 0.45. The samples were then diluted 100-fold in M9 minimal media before spotting 2 μL of culture on M9 minimal media agar ampicillin (100 μg/ml) plates with or without L-Asn (20 μg/ml), and supplemented with the other 19 amino acids (20 μg/ml each).

Bioinformatic survey of bacterial genomes

Bacterial genomes representing 547 different genera were analyzed for genes encoding AsnRS (*asnS*), GlnRS (*glnS*), GatCAB (*gatC, gatA, gatB*), AsnA (*asnA*), AsnB (*asnB*), and AspRS (*aspS*). Genes were searched either in the UniProt (http://www.uniprot.org) or KEGG: Kyoto Encyclopedia of Genes and Genomes (http://www.genome.jp/kegg/) databases. Sequences were then compared to known relevant enzymes by BLAST to validate the presence of the relevant active sites and domain architecture. The *E. coli* CFT AsnRS (AAN79540), GlnRS (AAN79239), AsnA (AAN83104), and AsnB (AAN79222), the *H. pylori* J99 GatA (AAD06348) and GatB (AAD06184), and the *Deinococcus radiodurans* discriminating AspRS (AAF10918) and ND-AspRS (AAF10623) sequences were used for the analysis. This was of particular importance to distinguish AsnA, which lacks an anticodon-binding domain, from its orthologs, AspRS and AsnRS [32]. Bacterial-type AspRS sequences were distinguished from archaeal-type AspRS sequences by the presence of a GAD insertion domain specific to bacterial AspRSs [33]. When no gene was initially identified for an enzyme, a tBLASTn search was performed with an enzyme sequence from a related organism. In addition, the $tRNA^{Asn}$ isoacceptors of these bacteria were analyzed for the presence of a U1-A72 base pair. The tRNA isoacceptors sequences studied were either from the Genomic tRNA Database (http://lowelab.ucsc.edu/GtRNAdb/) or the KEGG database (http://www.genome.jp/kegg/). The results of the survey are detailed in Table S1.

Results

In vitro aminoacylation of $tRNA^{Asp}$ and $tRNA^{Asn}$

Bacterial GatCAB recognizes the U1-A72 base pair present in many bacterial $tRNA^{Asn}$ isoacceptors [34,35]. *B. bacteriovorus* $tRNA^{Asn}$ has a U1-A72 base pair (Table S1) meaning if aspartylated, the tRNA could serve as a substrate for GatCAB. However, the presence of $tRNA^{Asn}$ with a U1-A72 base pair and GatCAB does not necessarily mean *B. bacteriovorus* encodes the two-step pathway for Asn-$tRNA^{Asn}$ formation. For example, *Lactobacillus delbruekii bulgaricus* encodes $tRNA^{Asn}$ with a U1-A72 pair along with GatCAB (Table S1) but does not synthesize Asn on $tRNA^{Asn}$ as it lacks a ND-AspRS and uses only AsnRS to form Asn-$tRNA^{Asn}$ [36].

Table 1. Aminoacylation kinetics of *B. bacteriovorus* AspRS and AsnRS at 37°C.

	k_{cat} (s^{-1})	K_M (μM)	k_{cat}/K_M (s^{-1} μM^{-1})	L*
AspRS				
$tRNA^{Asp}$	0.52±0.09	1.4±0.7	(40±20)×10^{-2}	3
$tRNA^{Asn}$	0.17±0.04	1.4±0.8	(12±7)×10^{-2}	1
AsnRS				
$tRNA^{Asn}$	0.70±0.07	2.1±0.5	(33±9)×10^{-2}	3

*L = Specificity relative to the catalytic efficiency of AspRS with $tRNA^{Asn}$ as a substrate, $(k_{cat}/K_M)/(k_{cat}/K_M)$ of AspRS for $tRNA^{Asn}$. Experiments were repeated three to four times and standard deviations are reported.

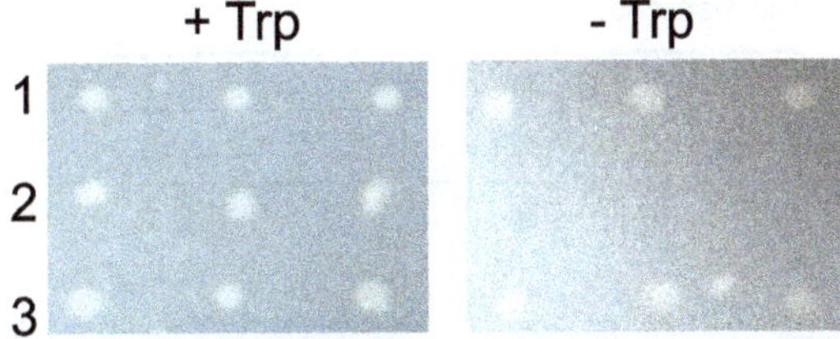

Figure 2. The *B. bacteriovorus aspS* rescues the Trp auxotrophy of *E. coli trpA34*. *E. coli trpA34* was grown with pCBS2 containing either 1) the *ND-aspS* from *D. radiodurans* as a positive control, 2) the *discriminating(D)-aspS* from *D. radiodurans* as a negative control, or 3) the *B. bacteriovorus aspS*. The cultures were grown in triplicate on M9 minimal media agar plates with 100 µg/ml of ampicillin in the presence (+ Trp, 20 µg/ml) or absence (- Trp) of Trp at 37°C for three days. Representative results are shown from three separate trials.

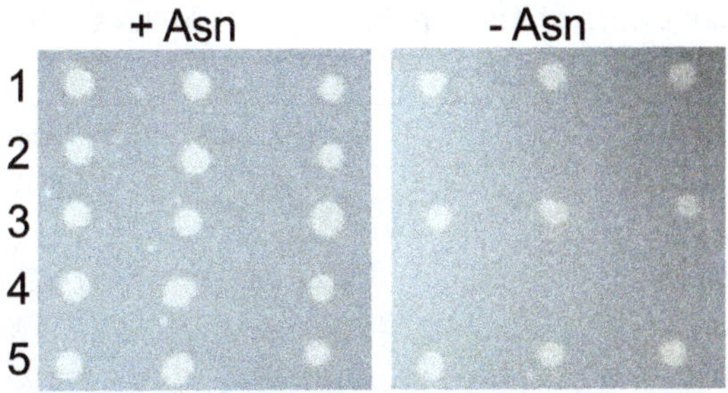

Figure 3. Co-production of *B. bacteriovorus* AspRS and GatCAB results in an Asn prototroph. *E. coli* JF448 was grown with pCBS2 containing either 1) the *D. radiodurans ND-aspS* and *gatCAB* or 2) the *D. radiodurans D-aspS* and *gatCAB* as positive and negative controls, respectively. 3) To control for possible toxic effects of *B. bacteriovorus aspS* expression, *E. coli* NEB10β, an Asn prototroph, was grown with pCBS2-*B. bacteriovorus aspS*. *E. coli* JF448 was also grown with pCBS2 containing either 4) the *B. bacteriovorus aspS* alone or 5) the *B. bacteriovorus aspS* and *gatCAB* in an operon together. The resultant strains were grown in triplicate on M9 minimal media agar plates with 100 µg/ml of ampicillin in the presence (+Asn, 20 µg/ml) or absence (-Asn) of Asn at 37°C for two days. Representative results are shown from three separate trials.

For *B. bacteriovorus* to encode the two-step pathway for Asn-$tRNA^{Asn}$ formation, the organism must code for a ND-AspRS along with GatCAB and $tRNA^{Asn}$ with a U1-A72 base pair. Given the presence of GatCAB in *B. bacteriovorus* despite encoding GlnRS and AsnRS, and the absence of both asparagine synthetases (AsnA and AsnB) to synthesize Asn [7], we predicted the lone *B. bacteriovorus* AspRS aspartylates $tRNA^{Asn}$ to enable the bacterium to synthesize Asn in a tRNA-dependent manner.

To determine if the *B. bacteriovorus* AspRS is non-discriminating, we overproduced the enzyme in *E. coli* and purified it to homogeneity [24]. The recombinant enzyme was readily able to aspartylate both $tRNA^{Asp}$ and $tRNA^{Asn}$ to similar levels (Figure 1A). Discriminating AspRS enzymes typically prefer $tRNA^{Asp}$ to $tRNA^{Asn}$ by a factor of 500–2,250 [22,37]. In contrast and similar to other ND-AspRS enzymes [22,24,37,38], the *B. bacteriovorus* AspRS preferred $tRNA^{Asp}$ as a substrate by only 3-fold (Table 1). The difference in catalytic efficiency by AspRS was attributed to an increased k_{cat} with $tRNA^{Asp}$ as a substrate (Table 1). The *B. bacteriovorus* AsnRS also readily uses $tRNA^{Asn}$ as a substrate, reaching a similar aminoacylation plateau (Figure 1B). The $tRNA^{Asn}$ was a better substrate for AsnRS by 3-fold with a higher k_{cat} compensating for an increased K_M relative to AspRS (Table 1).

B. bacteriovorus aspS rescues *E. coli* Trp auxotroph

To establish whether *B. bacteriovorus* AspRS also uses $tRNA^{Asn}$ as a substrate in a cellular context where there exists competition from other aaRSs and modified tRNA isoacceptors, we used the established *E. coli trpA*34 complementation assay [24,31,39]. Tryptophan synthetase alpha subunit (TrpA) is required for Trp synthesis in *E. coli*. The *trpA*34 strain is a Trp auxotroph due to mutation of codon 60 from an essential Asp codon to an Asn codon [40]. Production of a ND-AspRS in the strain rescues the phenotype, because the missense suppressor Asp-$tRNA^{Asn}$ formed by the ND-AspRS allows decoding of the mutant Asn codon with Asp and production of active TrpA [24,31,39]. Consistent with our *in vitro* results demonstrating the *B. bacteriovorus* AspRS readily uses $tRNA^{Asn}$ as a substrate, the *trpA*34 strain with the *B. bacteriovorus aspS* was able to grow in the absence of Trp (Figure 2).

B. bacteriovorus aspS with *gatCAB* rescues *E. coli* Asn auxotroph

We predicted *B. bacteriovorus* encodes a ND-AspRS so the bacterium could synthesize Asn in a tRNA-dependent manner using GatCAB. Bacterial GatCABs readily amidate Asp-$tRNA^{Asn}$ to Asn-$tRNA^{Asn}$ *in vitro* [8,12,15–17,21]. To verify co-production of *B. bacteriovorus* AspRS and GatCAB *in vivo* leads to Asn synthesis, we used the established *E. coli* JF448 system [22,41]. The JF448 strain is an Asn auxotroph due to mutation of both Asn synthetase genes in *E. coli* [41] and the phenotype can be rescued by introducing the tRNA-dependent route for Asn biosynthesis [22,24]. Consistent with our hypothesis, co-production of the *B. bacteriovorus* AspRS and GatCAB enabled the JF448 strain to grow in the absence of Asn in the media (Figure 3).

Bioinformatic analysis

To determine how common it is for a bacterium to encode AsnRS, GlnRS, and GatCAB, we surveyed genomes from 547 different bacterial genera (Table S1). The three enzymes are encoded together in 68 different genera (Table 2). Like *B. bacteriovorus*, only 18 genera coded for all three while not encoding an Asn synthetase in their genomes. They represent a diverse range of bacteria from the δ-proteobacteria, the Deinococcus-Thermus, Bacteroidetes, and Verrucomicrobiae clades. All these bacteria have a $tRNA^{Asn}$ with a U1-A72 base pair required for recognition by bacterial GatCAB, consistent with these bacteria possibly synthesizing Asn on $tRNA^{Asn}$. A second AspRS is encoded in five of the 18 genomes, all from the Deinococcus-Thermus phylum (Tables S1 and S2). This second AspRS in this clade is of the archaeal type and may stabilize GatCAB at higher growth temperatures [42]. Similar to *B. bacteriovorus*, only 13 bacterial genera encode GlnRS, AsnRS, GatCAB and one AspRS but neither Asn synthetase (Table S1).

The γ-proteobacteria *L. pneumophila* belongs to the group encoding GlnRS, AsnRS, one AspRS, and GatCAB but neither Asn synthetase. The lack of an Asn synthetase suggested its AspRS recognizes $tRNA^{Asn}$ as the first step in tRNA-dependent Asn biosynthesis as we hypothesized for *B. bacteriovorus*. We therefore tested whether the *L. pneumophila* AspRS could use $tRNA^{Asn}$ as a substrate both *in vivo* (Fig. 4A), using the *E. coli trpA*34 assay, and *in vitro* (Fig. 4B). Like the *B. bacteriovorus* AspRS, the *L. pneumophila* was able to aspartylate $tRNA^{Asn}$ suggesting this γ-proteobacteria potentially encodes the two-step pathway for Asn-$tRNA^{Asn}$ formation. The *L. pneumophila* AspRS has about a 1.6-

Table 2. Presence of Asn, Asn-tRNA^{Asn}, and Gln-tRNA^{Gln} biosynthetic pathways in bacteria.[1]

	No AsnA/B	AsnA Only	AsnB Only	AsnA & B	Total
GatCAB Only	81 (81)	0 (0)	105 (105)	0 (0)	186 (186)
AsnRS, GlnRS	9 (8)	11 (8)	43 (35)	26 (26)	89 (77)
GlnRS, GatCAB	41 (41)	0 (0)	57 (57)	0 (0)	98 (98)
AsnRS, GatCAB	31 (28)	15 (9)	56 (55)	5 (4)	107 (96)
AsnRS, GlnRS, GatCAB	18 (18)	2 (2)	41 (41)	6 (6)	67 (67)

[1]Representative genomes from 547 different bacterial genera were analyzed for the presence of genes coding for AsnA, AsnB, AsnRS, GlnRS, and GatCAB. The results are detailed in Table S1. In parentheses is the number of bacterial genera with a tRNA^{Asn} isoacceptor containing a U1-A72 base pair.

fold preference for tRNA^{Asp} over tRNA^{Asn}, similar to other ND-AspRS enzymes [22,24,37,38].

Beyond the Deinococcus-Thermus phylum, an additional 15 bacterial genera encode an extra AspRS (Table S2). *C. acetobutylicum* is unique in encoding three AspRSs and an additional GatCAB [18]. All 15 genomes encoded GatCAB and a tRNA^{Asn} with a U1-A72 base pair. The majority of the bacteria in this group (10 out of 15) lack an AsnRS and the indirect route with GatCAB is the only means for Asn-tRNA^{Asn} synthesis. Primarily, these bacteria are Actinobacteria in the order Actinomycetales. Of those with GatCAB and AsnRS, four are Firmicutes in the Clostridiaceae family and all encode at least one Asn synthetase. However, in the case of *C. acetobutylicum*, the AsnB is split into two halves and does not appear to be functional under normal physiological conditions with Asn synthesis being tRNA-dependent [18]. One Actinomycetales, *Amycolatopsis mediterranei* RB, also encoded an additional AspRS with AsnRS and GatCAB. It also in its genome has four *asnB* genes encoding the glutamine-dependent Asn synthetase (AsnB).

We also examined how many other bacteria potentially use both Asn-tRNA^{Asn} biosynthetic pathways. AsnRS and GatCAB are encoded in 174 bacterial genera with 163 also coding for a tRNA^{Asn} with a U1-A72 base pair (Table 2). In 46 of these bacteria, Asn synthesis could only be tRNA-dependent as they lack AsnA and AsnB, as was found in *Staphylococcus aureus* [24]. The other 117 genera encode at least one Asn synthetase. Of these, 49 also have a GlnRS suggesting GatCAB may be used for tRNA-dependent Asn synthesis including other δ-proteobacteria. *Bacteriovorax marinus* was an exception among the δ-proteobacteria as it coded for AsnRS and GlnRS but not GatCAB. It is also possible that instead the GatCAB is for Gln-tRNA^{Gln} formation but to date no bacteria are known to encode both routes for Gln-tRNA^{Gln} synthesis. The other 68 genera lack a GlnRS and retain GatCAB for Gln-tRNA^{Gln} though that does not exclude GatCAB from also being used for Asn-tRNA^{Asn} formation in these bacteria. As noted previously, AsnRS is present in all prokaryotes with AsnA, the ammonia-dependent Asn synthetase, [8,32].

Discussion

The *B. bacteriovorus* AspRS is non-discriminating readily able to form Asp-tRNA^{Asn}. The ND-AspRS may provide the bacterium the ability synthesize Asn in a tRNA-dependent manner using GatCAB, providing a potential functional role for the amidotransferase in an organism with both AsnRS and GlnRS [19]. Thus, unlike the initial prediction after the *B. bacteriovorus* genome was sequenced [7], the bacterium does potentially encode an Asn biosynthetic pathway. Also, *B. bacteriovorus* with a ND-AspRS, GatCAB, and AsnRS could encode both routes for Asn-tRNA^{Asn} formation in addition to the direct route for Gln-tRNA^{Gln} synthesis similar to *Deinococcus radiodurans* and *Thermus thermophilus* [10,12,18,20–22]. However, unlike those

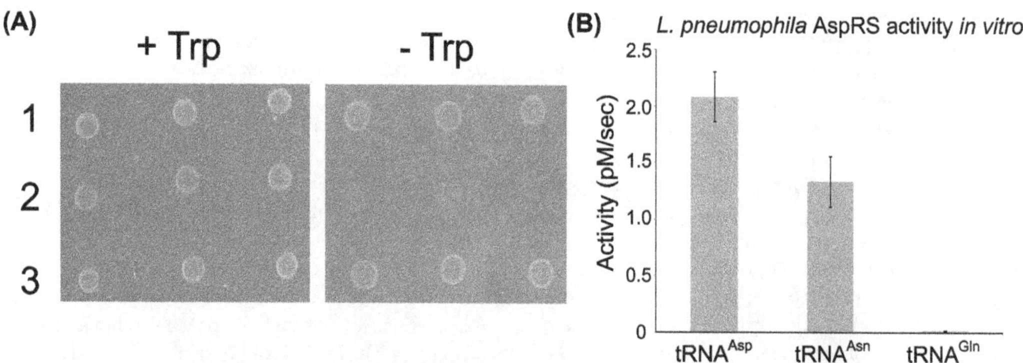

Figure 4. The *L. pneumophila* AspRS aspartylates tRNA^{Asn}. (A) *E. coli trpA34* was grown with pCBS2 containing either 1) the *ND-aspS* from *D. radiodurans* as a positive control, 2) the *discriminating(D)-aspS* from *D. radiodurans* as a negative control, or 3) the *L. penumophila aspS*. The cultures were grown in triplicate on M9 minimal media agar plates with 100 μg/ml of ampicillin in the presence (+ Trp, 20 μg/ml) or absence (- Trp) of Trp at 37°C for three days. Representative results are shown from three separate trials. (B) Asparylation of *in vitro* transcribed tRNA^{Asp}, tRNA^{Asn}, and tRNA^{Gln} by the *L. pneumophila* AspRS. Reactions were carried out at 37°C with 0.1 μM ^{32}P-labeled tRNA$^{Asp,\ Asn,\ or\ Gln}$, 4.0 mM ATP, 4.0 mM L-Asp and 10 nM AspRS. Experiments were repeated three times and error bars represent standard deviations.

two species that acquired an additional AspRS from archaea [10,12,20–22], *B. bacteriovorus* has only one AspRS.

Acquisition of an additional AspRS in bacteria is rare. The second AspRS may provide an advantage in particular environmental niches. In the case of *T. thermophilus*, GatCAB binding to the archaeal-type ND-AspRS stabilizes the amidotransferase at elevated temperatures and the second AspRS may be an adaptation to a thermophilic environment [42]. The second AspRS in other thermophiles may also serve the same purpose in addition to allowing the bacteria to synthesize Asn in a tRNA-dependent manner. For *C. acetobutylicum*, the additional AspRS enzymes and GatCAB have been linked to the organism switching from acidogenesis to solventogenesis [18,43].

Why *B. bacteriovorus* may still retain the tRNA-dependent route for Asn production despite coding for AsnRS is not clear. The organism is unable to synthesize eight other proteinogenic amino acids [7]. During growth away from a host cell, basal protein synthesis in *B. bacteriovorus* with those eight amino acids requires recycling them from protein degradation [7]. Recycling Asn residues may be problematic as Asn residues in polypeptides are susceptible to deamidation [44,45]. ND-AspRS and GatCAB could provide *B. bacteriovorus* the means to compensate for deamidation of Asn residues by synthesizing Asn from Asp on $tRNA^{Asn}$. The route would also provide direct coupling of Asn synthesis for use in translation [9]. In addition, the indirect pathway would allow *B. bacteriovorus* to convert Asp to Asn from hosts that underutilize Asn. Conversely, *B. bacteriovorus* may retain AsnRS to take advantage of the Asn provided by a host as well as to efficiently recycle non-deamidated Asn residues following protein degradation as has been hypothesized in *T. thermophilus*, *D. radiodurans*, and *S. aureus* [10,12,22,24].

It is also possible one of the *B. bacteriovorus* Asn-$tRNA^{Asn}$ pathways is non-functional. However, it should be noted *asnS* expression increased significantly during *B. bacteriovorus* growth in a host cell like the other aaRS genes including AspRS with only minimal increased expression of the GatCAB genes [46]. The increased expression of the aaRS genes during the growth phase was after initial predation by *B. bacteriovorus* as expression levels were unchanged 30 min. after host infection like the GatCAB genes [47]. Interestingly, host-independent *B. bacteriovorus* cultures grown in the presence of peptone and tryptone exhibited increases in the expression of not only *aspS* and *asnS* but also the genes for GatCAB [47]. The available gene expression results are consistent with a role for GatCAB in *B. bacteriovorus* when free-living and AsnRS during growth when Asn is present either from a host or in the media. Such a scenario would be similar to what is hypothesized in *S. aureus*, that encodes a ND-AspRS, GatCAB, and AsnRS but lacks a GlnRS and either Asn synthetase [24]. The *S. aureus* GatCAB is predicted for Gln-$tRNA^{Gln}$ formation and tRNA-dependent Asn biosynthesis, while AsnRS is predicted for growth in Asn-rich environments like the human body [24].

Interestingly, *L. pneumophila* like *B. bacteriovorus* is capable of free-living and growth in a host cell [48]. The presence of a ND-AspRS along with GlnRS, AsnRS and GatCAB in *L. pneumophila* raises the possibility it may also encode both routes for Asn-$tRNA^{Asn}$ formation. In *L. pneumophila*, the genes for AsnRS, ND-AspRS, and GatCAB are all up regulated during post-exponential and transmissive growth phases, suggesting a role in the organism's life cycle [49].

Two distinct routes for Asn-$tRNA^{Asn}$ formation in bacteria like *B. bacteriovorus* were likely acquired in a stepwise manner. Both AspRS and GatCAB were likely present in the last universal common ancestor (LUCA) while AsnRS evolved from a duplication of an archaeal AspRS [32,50,51]. Therefore, it has been hypothesized that Asn-$tRNA^{Asn}$ formation in LUCA was via the indirect pathway using a ND-AspRS and GatCAB [50,51]. Under this scenario, early bacteria likely also used a ND-AspRS and GatCAB for Asn-$tRNA^{Asn}$ formation [51]. Consistent with that hypothesis, the phylogenies of the GatCAB subunits and bacterial-type AspRSs suggest they were vertically inherited in bacteria [51,52]. AsnRS was later likely acquired in different bacterial lineages via horizontal gene transfer from archaea [32,50,52].

In some bacterial species like *L. delbruekii bulgaricus*, the presence of AsnRS to aminoacylate $tRNA^{Asn}$ appears to have lessened the selective pressure for AspRS to recognize $tRNA^{Asn}$ as a substrate, facilitating the AspRS to evolve specicity for just $tRNA^{Asp}$ [36]. Accordingly, the role of the *L. delbruekii bulgaricus* GatCAB is for only Gln-$tRNA^{Gln}$ synthesis [36]. In *B. bacteriovorus*, AspRS retained its relaxed tRNA specificity along with GatCAB after acquisition of AsnRS possibly for tRNA-dependent Asn biosynthesis as was found in *S. aureus* [24]. Given the above and the presence of GatCAB and AsnRS in most δ-proteobacteria, the ancestral δ-proteobacteria likely encoded GatCAB for Asn-$tRNA^{Asn}$ formation before acquiring AsnRS to directly attach the Asn to $tRNA^{Asn}$. As many δ-proteobacteria, like the predatory bacterium *Myxococcus xanthus* [53] and the metal-reducing anaerobic *Geobacter metallireducens* [54], encode one or more asparagine synthetase (AsnA and or AsnB) along with AsnRS and GlnRS, the retention of GatCAB in these bacteria maybe vestigial. However, it may be beneficial for these bacteria to retain both routes for Asn-$tRNA^{Asn}$ formation as hypothesized for *B. bacteriovorus*, *L. pneumophila*, and *S. aureus* [24].

It is unclear how many other bacteria encode both routes for Asn-$tRNA^{Asn}$ formation. Based on our genomic analysis, about 30% of bacterial genera surveyed could code for both routes as they encode GatCAB, AsnRS, and $tRNA^{Asn}$ isoacceptors with a U1-A72 base pair. In the few bacteria like *B. bacteriovorus* that also encode GlnRS but neither Asn synthetase, it is likely GatCAB was retained for tRNA-dependent Asn production. Similarly, those bacteria like *S. aureus* without a GlnRS and both Asn synthetases may also use GatCAB for tRNA-dependent Asn biosynthesis in addition to Gln-$tRNA^{Gln}$ formation [24]. However, the majority of bacteria that encode AsnRS and GatCAB also code for at least one Asn synthetase (AsnA and/or AsnB); in total, 21% of all bacterial genera analyzed. Since these organisms have alternate means to synthesize Asn, GatCAB may not be for Asn-$tRNA^{Asn}$ formation as their AspRS enzymes might be specific for $tRNA^{Asp}$. For example, *L. delbruekii bulgaricus* AspRS does not aspartylate $tRNA^{Asn}$ and uses GatCAB only for Gln-$tRNA^{Gln}$ formation [36]. Clarifying the tRNA specificity of the AspRSs in other bacteria encoding both AsnRS and GatCAB would establish how many other bacteria code for both routes for Asn-$tRNA^{Asn}$ formation, providing the foundation to better understand how Asn metabolism and protein synthesis are integrated into bacterial physiology and adaptation to certain environmental niches.

Supporting Information

Table S1 Genomic analysis of bacterial genomes from 547 different genera for genes related to Asn-$tRNA^{Asn}$, Gln-$tRNA^{Gln}$, and Asn synthesis.

Table S2 Genomic analysis of bacterial genomes encoding at least two AspRS enzymes for genes related to Asn-$tRNA^{Asn}$, Gln-$tRNA^{Gln}$, and Asn synthesis.

Acknowledgments

We thank Dr. John Tudor at Saint Joseph's University for *B. bacteriovorus* HD100 genomic DNA. We thank Mark Benhaim, Melanie Feen, Stefani Mladenova, and Nilendra Nair at Skidmore College for their support. We also thank at Skidmore College Drs. Sarita Lagalwar and Sylvia McDevitt for critical discussions, and Dr. Patricia Hilleren for use of her Storm 820 phosphorimager.

Author Contributions

Conceived and designed the experiments: AA KS. Performed the experiments: AA BU DMG RD HR KS. Analyzed the data: AA BU DMG RD HR KS. Contributed reagents/materials/analysis tools: AA BU DMG RD HR. Wrote the paper: AA BU DMG HR RD KS.

References

1. Sockett RE (2009) Predatory lifestyle of *Bdellovibrio bacteriovorus*. Annu Rev Microbiol 63: 523–539.
2. Dwidar M, Monnappa AK, Mitchell RJ (2012) The dual probiotic and antibiotic nature of *Bdellovibrio bacteriovorus*. BMB Rep 45: 71–78.
3. Monnappa AK, Dwidar M, Seo JK, Hur JH, Mitchell RJ (2014) *Bdellovibrio bacteriovorus* inhibits *Staphylococcus aureus* biofilm formation and invasion into human epithelial cells. Sci Rep 4: 3811.
4. Dashiff A, Junka RA, Libera M, Kadouri DE (2011) Predation of human pathogens by the predatory bacteria *Micavibrio aeruginosavorus* and *Bdellovibrio bacteriovorus*. J Appl Microbiol 110: 431–444.
5. Wand H, Vacca G, Kuschk P, Kruger M, Kastner M (2007) Removal of bacteria by filtration in planted and non-planted sand columns. Water Res 41: 159–167.
6. Atterbury RJ, Hobley L, Till R, Lambert C, Capeness MJ, et al. (2011) Effects of orally administered *Bdellovibrio bacteriovorus* on the well-being and *Salmonella* colonization of young chicks. Appl Environ Microbiol 77: 5794–5803.
7. Rendulic S, Jagtap P, Rosinus A, Eppinger M, Baar C, et al. (2004) A predator unmasked: life cycle of *Bdellovibrio bacteriovorus* from a genomic perspective. Science 303: 689–692.
8. Sheppard K, Akochy PM, Salazar JC, Söll D (2007) The *Helicobacter pylori* amidotransferase GatCAB is equally efficient in glutamine-dependent transamidation of Asp-tRNA^{Asn} and Glu-tRNA^{Gln}. J Biol Chem 282: 11866–11873.
9. Yuan J, Sheppard K, Söll D (2008) Amino acid modifications on tRNA. Acta Biochim Biophys Sinica 40: 539–553.
10. Becker HD, Kern D (1998) *Thermus thermophilus*: a link in evolution of the tRNA-dependent amino acid amidation pathways. Proc Natl Acad Sci USA 95: 12832–12837.
11. Cathopoulis T, Chuawong P, Hendrickson TL (2007) Novel tRNA aminoacylation mechanisms. Mol Biosyst 3: 408–418.
12. Becker HD, Min B, Jacobi C, Raczniak G, Pelaschier J, et al. (2000) The heterotrimeric *Thermus thermophilus* Asp-tRNA^{Asn} amidotransferase can also generate Gln-tRNA^{Gln}. FEBS Lett 476: 140–144.
13. Curnow AW, Ibba M, Söll D (1996) tRNA-dependent asparagine formation. Nature 382: 589–590.
14. Curnow AW, Hong K, Yuan R, Kim S, Martins O, et al. (1997) Glu-tRNA^{Gln} amidotransferase: a novel heterotrimeric enzyme required for correct decoding of glutamine codons during translation. Proc Natl Acad Sci USA 94: 11819–11826.
15. Raczniak G, Becker HD, Min B, Söll D (2001) A single amidotransferase forms asparaginyl-tRNA and glutaminyl-tRNA in *Chlamydia trachomatis*. J Biol Chem 276: 45862–45867.
16. Salazar JC, Zúñiga R, Raczniak G, Becker H, Söll D, et al. (2001) A dual-specific Glu-tRNA^{Gln} and Asp-tRNA^{Asn} amidotransferase is involved in decoding glutamine and asparagine codons in *Acidithiobacillus ferrooxidans*. FEBS Lett 500: 129–131.
17. Wu J, Bu W, Sheppard K, Kitabatake M, Kwon ST, et al. (2009) Insights into tRNA-dependent amidotransferase evolution and catalysis from the structure of the Aquifex aeolicus enzyme. J Mol Biol 391: 703–716.
18. Saad NY, Schiel B, Brayé Heap JT, Minton NP, Dürre P, et al. (2012) Riboswitch (T-Box)-mediated control of tRNA-dependent amidation in *Clostridium acetobutylicum* rationalizes gene and pathway redundancy for asparagine and asparaginyl- tRNA^{Asn} synthesis. J Biol Chem 287: 20382–20394.
19. Karlin S, Brocchieri L, Mrazek J, Kaiser D (2006) Distinguishing features of delta-proteobacterial genomes. Proc Natl Acad Sci USA 103: 11352–11357.
20. Becker HD, Reinbolt J, Kreutzer R, Giegé R, Kern D (1997) Existence of two distinct aspartyl-tRNA synthetases in *Thermus thermophilus*. Structural and biochemical properties of the two enzymes. Biochemistry 36: 8785–8797.
21. Curnow AW, Tumbula DL, Pelaschier JT, Min B, Söll D (1998) Glutamyl-tRNA^{Gln} amidotransferase in *Deinococcus radiodurans* may be confined to asparagine biosynthesis. Proc Natl Acad Sci USA 95: 12838–12843.
22. Min B, Pelaschier JT, Graham DE, Tumbula-Hansen D, Söll D (2002) Transfer RNA-dependent amino acid biosynthesis: an essential route to asparagine formation. Proc Natl Acad Sci USA 99: 2678–2683.
23. Studier FW (2005) Protein production by auto-induction in high density shaking cultures. Protein Expr Purif 41: 207–234.
24. Mladenova SR, Stein KR, Bartlett L, Sheppard K (2014) Relaxed tRNA specificity of the *Staphylococcus aureus* aspartyl-tRNA synthetase enables RNA-dependent asparagine biosynthesis. FEBS Lett 588: 1808–1812.
25. Sharp PM, Li WH (1987) The codon Adaptation Index–a measure of directional synonymous codon usage bias, and its potential applications. Nucleic Acids Res 15: 1281–1295.
26. Feng L, Sheppard K, Tumbula-Hansen D, Söll D (2005) Gln-tRNA^{Gln} formation from Glu-tRNA^{Gln} requires cooperation of an asparaginase and a Glu-tRNA^{Gln} kinase. J Biol Chem 280: 8150–8155.
27. Wolfson AD, Uhlenbeck OC (2002) Modulation of tRNA^{Ala} identity by inorganic pyrophosphatase. Proc Natl Acad Sci USA 99: 5965–5970.
28. Bullock TL, Uter N, Nissan TA, Perona JJ (2003) Amino acid discrimination by a class I aminoacyl-tRNA synthetase specified by negative determinants. J Mol Biol 328: 395–408.
29. Sheppard K, Akochy PM, Söll D (2008) Assays for transfer RNA-dependent amino acid biosynthesis. Methods 44: 139–145.
30. Silva GN, Fatma S, Floyd AM, Fischer F, Chuawong P, et al. (2013) A tRNA-independent mechanism for transamidosome assembly promotes aminoacyl-tRNA transamidation. J Biol Chem 288: 3816–3822.
31. Min B, Kitabatake M, Polycarpo C, Pelaschier J, Raczniak G, et al. (2003) Protein synthesis in Escherichia coli with mischarged tRNA. J Bacteriol 185: 3524–3526.
32. Roy H, Becker HD, Reinbolt J, Kern D (2003) When contemporary aminoacyl-tRNA synthetases invent their cognate amino acid metabolism. Proc Natl Acad Sci USA 100: 9837–9842.
33. Delarue M, Poterszman A, Nikonov S, Garber M, Moras D, et al. (1994) Crystal structure of a prokaryotic aspartyl tRNA-synthetase. EMBO J 13: 3219–3229.
34. Bailly M, Giannouli S, Blaise M, Stathopoulos C, Kern D, et al. (2006) A single tRNA base pair mediates bacterial tRNA-dependent biosynthesis of asparagine. Nucleic Acids Res 34: 6083–6094.
35. Nakamura A, Yao M, Chimnaronk S, Sakai N, Tanaka I (2006) Ammonia channel couples glutaminase with transamidase reactions in GatCAB. Science 312: 1954–1958.
36. Kim SI, Nalaskowska M, Germond JE, Pridmore D, Söll D (1996) Asn-tRNA in *Lactobacillus bulgaricus* is formed by asparaginylation of tRNA and not by transamidation of Asp-tRNA. Nucleic Acids Res 24: 2648–2651.
37. Becker HD, Roy H, Moulinier L, Mazauric MH, Keith G, et al. (2000) *Thermus thermophilus* contains an eubacterial and an archaebacterial aspartyl-tRNA synthetase. Biochemistry 39: 3216–3230.
38. Chuawong P, Hendrickson TL (2006) The nondiscriminating aspartyl-tRNA synthetase from *Helicobacter pylori*: anticodon-binding domain mutations that impact tRNA specificity and heterologous toxicity. Biochemistry 45: 8079–8087.
39. Cathopoulis TJ, Chuawong P, Hendrickson TL (2008) Conserved discrimination against misacylated tRNAs by two mesophilic elongation factor Tu orthologs. Biochemistry 47: 7610–7616.
40. Shirvanee L, Horn V, Yanofsky C (1990) *Escherichia coli* mutant trpA34 has an Asp–Asn change at active site residue 60 of the tryptophan synthetase alpha chain. J Biol Chem 265: 6624–6625.
41. Felton J, Michaelis S, Wright A (1980) Mutations in two unlinked genes are required to produce asparagine auxotrophy in *Escherichia coli*. J Bacteriol 142: 221–228.
42. Blaise M, Bailly M, Frechin M, Behrens MA, Fischer F, et al. (2010) Crystal structure of a transfer-ribonucleoprotein particle that promotes asparagine formation. The EMBO J 29: 3118–3129.
43. Saad NY, Stamatopoulou V, Braye M, Drainas D, Stathopoulos C, et al. (2013) Two-codon T-box riboswitch binding two tRNAs. Proc Natl Acad Sci USA 110: 12756–12761.
44. Washington EJ, Banfield MJ, Dangl JL (2013) What a difference a Dalton makes: bacterial virulence factors modulate eukaryotic host cell signaling systems via deamidation. Microbiol Mol Biol Rev 77: 527–539.
45. Yang H, Zubarev RA (2010) Mass spectrometric analysis of asparagine deamidation and aspartate isomerization in polypeptides. Electrophoresis 31: 1764–1772.
46. Karunker I, Rotem O, Dori-Bachash M, Jurkevitch E, Sorek R (2013) A global transcriptional switch between the attack and growth forms of *Bdellovibrio bacteriovorus*. PLoS One 8: e61850.
47. Lambert C, Chang CY, Capeness MJ, Sockett RE (2010) The first bite–profiling the predatosome in the bacterial pathogen *Bdellovibrio*. PLoS One 5: e8599.
48. Molofsky AB, Swanson MS (2004) Differentiate to thrive: lessons from the *Legionella pneumophila* life cycle. Mol Microbiol 53: 29–40.
49. Weissenmayer BA, Prendergast JG, Lohan AJ, Loftus BJ (2011) Sequencing illustrates the transcriptional response of *Legionella pneumophila* during infection and identifies seventy novel small non-coding RNAs. PLoS One 6: e17570.
50. Woese CR, Olsen GJ, Ibba M, Söll D (2000) Aminoacyl-tRNA synthetases, the genetic code, and the evolutionary process. Microbiol Mol Biol Rev 64: 202–236.

51. Sheppard K, Söll D (2008) On the evolution of the tRNA-dependent amidotransferases, GatCAB and GatDE. J Mol Biol 377: 831–844.
52. Charrière F, O'Donoghue P, Helgadóttir S, Marechal-Drouard L, Cristodero M, et al. (2009) Dual targeting of a $tRNA^{Asp}$ requires two different aspartyl-tRNA synthetases in *Trypanosoma brucei*. J Biol Chem 284: 16210–16217.
53. Goldman BS, Nierman WC, Kaiser D, Slater SC, Durkin AS, et al. (2006) Evolution of sensory complexity recorded in a myxobacterial genome. Proc Natl Acad Sci USA 103: 15200–15205.
54. Aklujkar M, Krushkal J, DiBartolo G, Lapidus A, Land ML, et al. (2009) The genome sequence of *Geobacter metallireducens*: features of metabolism, physiology and regulation common and dissimilar to *Geobacter sulfurreducens*. BMC Microbiol 9: 109.

5

Role of Type IV Pili in Predation by *Bdellovibrio bacteriovorus*

Ryan M. Chanyi, Susan F. Koval*

Department of Microbiology and Immunology, University of Western Ontario, London, Ontario, Canada

Abstract

Bdellovibrio bacteriovorus, as an obligate predator of Gram-negative bacteria, requires contact with the surface of a prey cell in order to initiate the life cycle. After attachment, the predator penetrates the prey cell outer membrane and enters the periplasmic space. Attack phase cells of *B. bacteriovorus* have polar Type IV pili that are required for predation. In other bacteria, these pili have the ability to extend and retract via the PilT protein. *B. bacteriovorus* has two *pilT* genes, *pilT1* and *pilT2*, that have been implicated in the invasion process. Markerless in-frame deletion mutants were constructed in a prey-independent mutant to assess the role of PilT1 and PilT2 in the life cycle. When predation was assessed using liquid cocultures, all mutants produced bdelloplasts of *Escherichia coli*. These results demonstrated that PilT1 and PilT2 are not required for invasion of prey cells. Predation of the mutants on biofilms of *E. coli* was also assessed. Wild type *B. bacteriovorus* 109JA and the *pilT1* mutant decreased the mass of the biofilm to 35.4% and 27.9% respectively. The *pilT1pilT2* mutant was able to prey on the biofilm, albeit less efficiently with 50.2% of the biofilm remaining. The *pilT2* mutant was unable to disrupt the biofilm, leaving 92.5% of the original biofilm after predation. The lack of PilT2 function may impede the ability of *B. bacteriovorus* to move in the extracellular polymeric matrix and find a prey cell. The role of Type IV pili in the life cycle of *B. bacteriovorus* is thus for initial recognition of and attachment to a prey cell in liquid cocultures, and possibly for movement within the matrix of a biofilm.

Editor: Christophe Beloin, Institut Pasteur, France

Funding: Funding was provided to SFK by the Natural Sciences and Engineering Research Council of Canada (155492), www.nserc-crsng.gc.ca. The funders had no role in study design, data collection and analysis, decision to publish, or preparation of the manuscript.

Competing Interests: The authors have declared that no competing interests exist.

* Email: skoval@uwo.ca

Introduction

Bdellovibrio bacteriovorus is a Gram-negative obligate predator of other Gram-negative bacteria. The cells are small, vibroid in shape and highly motile via a single polar sheathed flagellum. Their life cycle consist of two stages, a motile attack phase and an intraperiplasmic growth phase. During the attack phase the cells will reversibly attach to potential prey cells for a short recognition period. If deemed suitable, *B. bacteriovorus* will irreversibly attach to the prey cell and begin to secrete hydrolytic enzymes to create a pore in the outer membrane and the peptidoglycan of the prey [1]. *B. bacteriovorus* will squeeze through the pore into the periplasmic space. The pore is resealed and an osmotically stable 'bdelloplast' is formed. This signifies the end of the attack phase and the beginning of the growth phase.

Evans *et al.* [2] showed the presence of polar pili on *B. bacteriovorus* and that the disruption of the *pilA* gene abolished predation. Mahmoud and Koval [3] showed that these fibers were in fact Type IV pili (TFP) and that *B. bacteriovorus* in a coculture containing anti-PilA antibody was not able to prey. These results indicate that a direct interaction between TFP and the prey cell is required for successful predation. However, it was not determined if TFP are required for the attachment to or invasion of a prey cell. These small (8 nm in width) polar fibers are incredibly strong and, as has been shown in *Pseudomonas aeruginosa* [4] and *Myxococcus xanthus* [5], have the ability to extend, adhere to a surface and retract. Successive cycles of extension, adherence and retraction is called twitching motility. The PilT protein, via ATP hydrolysis, powers the retraction of the pilus and can generate forces exceeding 100 pN per single fiber, making it one of the strongest biological motors [6]. Deletion of PilT in *P. aeruginosa* abolishes twitching motility and results in a hyperpiliated cell, unable to retract the pilus but still able to extend it. *B. bacteriovorus* contains two annotated *pilT* genes, *pilT1* and *pilT2* [7]. In a study of host-independent mutants of *B. bacteriovorus* 109J, Medina *et al.* [8] reported that transposon insertion into the *pilT* gene which encodes the motor which retracts the pilus (*pilT2* Bd3852) produced a mutant that had an impaired ability to prey on a preformed biofilm of *Escherichia coli* cells. Although predation in liquid cocultures was not included in this study, this result with biofilms suggested that *B. bacteriovorus* uses the retraction of TFP to pull itself into the periplasmic space of the prey.

Koval *et al.* [9] described an epibiotic predator, *Bdellovibrio exovorus*, which does not invade the periplasmic space of the prey but remains attached to the outer surface. Polar pili were demonstrated on the type strain *B. exovorus* JSS^{T} [3] and the genome of this strain contains a full set of genes encoding TFP including *pilT1* and *pilT2* [10]. *B. exovorus* does not require these two genes to invade the prey cell, and thus they must have some other function in the epibiotic life cycle. If *pilT1* and *pilT2* are essential for the epibiotic life cycle, mutants would need to be constructed in a prey-independent mutant. However, so far isolation of such mutants in *B. exovorus* JSS has not been successful [10]. Therefore, we undertook a study of the function of the *pilT1* and *pilT2* genes during the periplasmic life cycle of *B. bacteriovorus* 109J.

Markerless in-frame deletion mutants of *pilT1*, *pilT2* as well as a *pilT1pilT2* double deletion mutant were constructed in the prey-independent strain 109JA. All mutants produced bdelloplasts in cocultures with *E. coli* prey cells and thus the retraction of TFP is not required for successful invasion of a prey cell. These results, combined with previous studies on *pilA* mutants and the presence of TFP on the epibiotic predator *B. exovorus*, suggest that TFP are required for initial attachment to prey cells. This study also demonstrated that *pilT2* is required for efficient predation on a biofilm.

Materials and Methods

Bacterial strains, media and culture conditions

E. coli strains (Table 1) were grown routinely in LB medium at 30°C overnight. When required, cells were grown in LB containing kanamycin (50 µg/ml) or chloramphenicol (25 µg/ml). *B. bacteriovorus* 109J was maintained in coculture with *E. coli* ML35. Cocultures were prepared by mixing a 1:3 ratio of predator to prey in HM buffer and incubated at 30°C overnight [11]. The facultative predator (prey-independent strain) *B. bacteriovorus* 109JA was grown in PY medium overnight at 30°C [3].

Construction of *B. bacteriovorus* 109JA markerless in-frame deletion mutants

Construction of the *pilT* mutants was performed in the facultative predator *B. bacteriovorus* 109JA because these gene deletions have the potential to be lethal. The *pilT1* gene (locus tag Bd1510) and the *pilT2* gene (locus tag Bd3852) were knocked out by allelic exchange using an in-frame deletion cloned on the suicide plasmid pSSK10. Deletion constructs were prepared as described in Steyert and Piñeiro [12] using primers listed in Table 2 with modification of the restriction sites used. A *Spe*I site was used to ligate the upstream and downstream fragments together. *Nde*I and *Xho*I were used for the upstream and downstream fragments respectively, to directionally ligate into pSSK10.

The correct construction of the plasmids pRCT1 and pRCT2 was confirmed by sequencing. *E. coli* SM10*λpir* harboring either pRCT1 or pRCT2 was used as a donor for conjugation into the recipient strain 109JA using the method of Cotter and Thomashow [13]. Dilutions were plated onto 1% PY agar containing streptomycin to select for *B. bacteriovorus* 109JA and chloramphenicol to select for the presence of plasmid. The presence of merodiploid strain 109JA in the cultures was confirmed by PCR using the same outer flanking primers that had been used for cloning. An aliquot of a merodiploid culture was added to fresh PY with no antibiotic and grown for 24 h at 30°C. This was repeated 3 times to allow for excision of the suicide plasmid by a second homologous recombination event. Counterselection was performed using PY containing 5% sucrose to select for excisants.

Table 1. Bacterial strains and plasmids used in this study.

Bacterial Strains	Source/Reference
E. coli ML35	Laboratory Strain
E. coli SY327λpir	Laboratory Strain
E. coli SM10 λpir	Laboratory Strain
E. coli CO1	Peter Cadieux, University of Western Ontario, London, Canada
B. bacteriovorus 109J	Laboratory Strain
B. bacteriovorus 109JA	John Tudor, St. Joseph's University, Philadelphia, USA
B. bacteriovorus 109JA::Δ*pilT1*	This study
B. bacteriovorus 109JA::Δ*pilT2*	This study
B. bacteriovorus 109JA::Δ*pilT1pilT2*	This study
P. aeruginosa PAK	Lori Burrows, McMaster University, Hamilton, Canada
P. aeruginosa PAK + pBADGr	This study
P. aeruginosa PAK + pBADGr::*pilT1*	This study
P. aeruginosa PAK + pBADGr::*pilT2*	This study
P. aeruginosa PAK::Δ*pilT*	Lori Burrows, McMaster University, Hamilton, Canada
P. aeruginosa PAK::Δ*pilT* + pBADGr	This study
P. aeruginosa PAK::Δ*pilT* + pBADGr::*pilT1*	This study
P. aeruginosa PAK::Δ*pilT* + pBADGr::*pilT2*	This study
Plasmids	
pSSK10	Silvia Piñeiro, University of Maryland, Baltimore, USA
pRCT1	This study
pRCT2	This study
pBADGr	Lori Burrows, McMaster University, Hamilton, Canada
pBADGr::*pilT1*	This study
pBADGr::*pilT2*	This study

Table 2. PCR oligonucleotide primers used for amplification of *pilT1* and *pilT2*.

Primer	Sequence (5′-3′)	Target Gene
PilT1-F1	ATAGTAACATATGACGTGAACATCTCCACCG	*pilT1*
PilT1-R1	AATAGCATGCATTCCGTCATCAGCGCTG	*pilT1*
PilT1-F2	ACATGCATGCAACATCTGATTCGTCGTCG	*pilT1*
PilT1-R2	ATACTCGAGACCGGAACCAGCGAAGAAT	*pilT1*
PilT2F1	CACACATATGGTCTTCGATACGACGGGAAA	*pilT2*
PilT2R1	AATAGCATGCGTTCAATGGAACCATGTTTC	*pilT2*
PilT2F2	AATCGCATGCACCAAAGTGGGCTAA	*pilT2*
PilT2R2	ATACTCGAGAAGTTCCGCGCAGGTCTT	*pilT2*

*restriction sites are underlined.

Dilutions were plated and colony PCR was performed to identify individual mutants.

Liquid predation assay

Liquid coculture predation assays were performed in triplicate using a 100-well BioScreen plate incubated for 48 h at 30°C. Samples were prepared by growing *B. bacteriovorus* 109JA and mutants axenically in PY medium (5% v/v inoculum) with shaking at 30°C. An aliquot of cells from each culture was washed and resuspended in 1 ml of HM buffer. The optical density at 600 nm (OD_{600}) was used to normalize the number of cells used for individual experiments.

An aliquot of each predator sample was normalized to have an OD_{600} of 0.2. One millilitre was centrifuged at 2500 *g* for 5 min, resuspended in HM buffer and added to a 20 ml coculture. A 16 h culture of *E. coli* ML35 was concentrated in HM buffer and added to obtain a coculture with an OD_{600} reading of 1.0. Two hundred microlitres were added to each well and readings were taken every 30 min for a total of 48 h.

Biofilm predation assay

Using a protocol modified from Kadouri and O'Toole [14], biofilms were formed in round-bottom microtiter dishes (BD Falcon). Microtiter wells were inoculated (200 µl per well) from 18 h *E. coli* LB-grown cultures diluted 1:100 in LB. Cells were grown for 48 h at 30°C to allow the biofilm to form. Quantification of biofilm bacteria was performed as follows. The wells were washed 3 times with HM buffer in order to remove any planktonic cells and 200 µl of crystal violet was then added for 15 min. Wells were washed with ddH_20 4 times and allowed to dry. Crystal violet was solubilized by adding 200 µl of 30% acetic acid for 20 min. An aliquot (125 µl) of each well was transferred to a flat-bottom microtiter dish (BD Falcon) and analyzed using a microplate reader at 600 nm. Preliminary studies confirmed the use of *E. coli* CO1 as the best candidate for biofilm studies (Fig. S2).

To assess predation of *B. bacteriovorus* on *E. coli* biofilms, the preformed biofilms were grown as described above and washed three times with HM buffer to remove planktonic cells. *B. bacteriovorus* cultures grown in PY overnight were washed and diluted 1:100 in HM buffer and 200 µl added to each well. As a control, 200 µl of a filtered sterilized lysate was prepared by passing the *B. bacteriovorus*-containing lysate through a 0.22 µm pore size filter. After filtering, no predator could be detected as judged by setting up a coculture as described earlier. The microtiter dish was incubated at 30°C for 24 h. Quantification of biofilm was performed as described above. Data presented were an average of 12-wells per replicate repeated three times. Statistical significance was measured using a 1-way ANOVA with a Bonferonni corrected post-hoc Students T-test, *$p<0.005$.

Immunofluorescence Microscopy

Immunofluorescence microscopy was used to count the number of cells which had TFP. Cells of strain 109JA and the *pilT* mutants were grown as described earlier. An aliquot (200 µl) was pre-fixed with 0.25% paraformaldehyde in Dulbecco's buffered saline (DPBS, 2.7 mM KCl, 1.5 mM KH_2PO_4, 136.9 mM NaCl and 8.9 mM $Na_2HPO_4 \cdot 7H_2O$) and centrifuged at 5000 *g* for 5 min. Samples were resuspended in full strength fixative (2.5% paraformaldehyde in DPBS) and incubated at 37°C for 10 min before being washed twice in 2.5% bovine serum albumin in DPBS (BSA-DPBS). Samples were resuspended in 1:100 anti-pilA antibody [3] in BSA-DPBS, incubated for 1 h at 37°C then washed twice in BSA-DPBS. Antibody detection was performed by resuspending samples in 1:100 sheep anti-rabbit IgG conjugated to Cy3 in BSA-DPBS, incubated for 1 h at 37°C and washed three times in BSA-DPBS. Samples were visualized by placing 10 µl onto CELLSTAR microscope slides and viewed using an Axioskop II epifluorescence microscope (Zeiss) equipped with a QImaging Retiga 1300 cooled monochrome 12-bit camera and an HBO100/2 mercury lamp for epifluorescence illumination. Imaging was performed using Northern Eclipse software version 6.0 (Empix Imaging Inc.).

Electron Microscopy

To visualize TFP, cells were negatively stained and viewed by transmission electron microscopy (TEM). Cells were negatively stained on a 400-mesh Formvar-carbon-coated copper grid that was inverted over a drop of cells for 1 min. The grid was then washed on 2 drops of water and the cells were stained with 1% uranyl acetate containing bacitracin (50 µg/mL) as a wetting agent. Negative stains were visualized using a Philips EM 410 transmission electron microscope operating at 60 kV.

To visualize biofilms, scanning electron microscopy (SEM) was used. *E. coli* CO1 biofilms were developed on a 12-by-22-mm PVC plastic coverslip (Fisher Scientific, Pittsburgh, PA). The coverslips were placed in a 6-well polystyrene cell culture plate (Corning, Inc., Corning, NY) and inoculated with a 1:100 dilution of a culture of *E. coli* CO1 in LB. Plates were incubated at 30°C for 48 h with shaking. Preformed biofilms were rinsed three times in a buffer consisting of equal parts 0.07 M sodium phosphate dibasic and 0.07 M potassium phosphate monobasic (SEM buffer, pH 6.8) to remove any planktonic cells. *B. bacteriovorus* 109JA and the *pilT* mutants were prepared as described above for previous biofilm experiments and added to an appropriate well. The biofilms were incubated at 30°C with shaking (120 rpm) for 24 h before being rinsed three times in SEM buffer. Residual biofilms were fixed for 30 min in SEM buffer containing 2% glutaraldehyde, followed by another three washes in SEM buffer and a 30 min secondary osmium fixation (2% osmium tetroxide). The samples were washed in sterile water and dehydrated step-wise in an ethanol series (70%, 80%, 90%, 3×100%). Each step was left to sit for 10 min. The samples were critically point dried at the Biotron Integrated Microscopy Facility (University of Western Ontario, London, ON, Canada). The samples were platinum coated and visualized on a Hitachi S-4500 field emission scanning electron microscope operating in high vacuum using Quartz XOne version 9.50 imaging software at the Surface Science Facility (University of Western Ontario, London, ON, Canada). Scanning electron microscopy was repeated on two separate occasions.

Results

Comparative analysis of active domains in *pilT* genes of *Bdellovibrio bacteriovorus*

The number and chromosomal arrangement of TFP genes is the same in *B. bacteriovorus* strains HD100 and 109J [3]. As described by Mahmoud and Koval [3], *pilT1* (Bd1510) is located within an operon containing *pilB* upstream and *pilC*, *pilS* and *pilR* directly downstream. It does not contain a Walker A box necessary for binding ATP. In contrast, *pilT2* (Bd3852) contains the Walker A (GPTGSGKS) box and the two turn amphipathic α-helical AIRNLIRE sequence important for its function [15]. The *pilT2* gene is not found within an operon and is not close to any other genes involved in TFP assembly, regulation or function.

In this study, a closer analysis of *pilT1* revealed that, although it does not contain a recognized nucleotide-binding site, it does contain a Walker B box. However this would be insufficient for binding ATP. It does not contain any other domains recognized in the PilT proteins of *P. aeruginosa* or *Neisseria gonorrhoeae* that are important for function. Because of this, PilT1 may not be involved in the retraction process but may function in regulation or stability of the pilus, similar to PilU of *P. aeruginosa* [16].

Analysis of PilT2 revealed that it contains both the Walker A and the Walker B domains. It also contains many of the active domains found in PilT of *P. aeruginosa* with minor differences in amino acid composition. There was a minor variation in the AIRNLIRE sequence, with two amino acid differences (AI<u>S</u>NL<u>V</u>RE). It is believed this sequence may be involved in the regulation of PilT function and does not contain catalytic activity. Although both isoleucine residues are essential for PilT function [15], an isoleucine to valine substitution is generally regarded as a homologous substitution and should not inhibit proper function of PilT2. The Asp Box (EDPIE) is identical to that found in *P. aeruginosa* while the His Box (H<u>L</u>VFGTVH) has a single leucine to isoleucine substitution. Structural studies of the PilT gene in *Aquifex aeolicus* and *Vibrio cholerae* identified the ASP and His boxes as forming the catalytic pocket [17]. Therefore, PilT2 contains the domains which bind ATP, form the catalytic pocket and is capable of regulating its own function. A mutant lacking this gene should not be able to retract its pilus and thus would help to determine if *B. bacteriovorus* uses PilT2 to generate the force necessary for the invasion process.

Growth and Predation Assay of *pilT* mutants

The facultative predator *B. bacteriovorus* 109JA was used to create *pilT* mutants because these genes were hypothesized to be essential for predation. The ability of the *pilT* mutant to grow axenically as well as a predator could then be tested.

In axenic growth mode, wild type strain 109JA and the *pilT2* mutant grew in PY medium at the same rate reaching a final OD_{600} of 0.4 after 36 h (Fig. 1). The *pilT1* mutant and the *pilT1pilT2* double mutant also grew at similar rates, reaching a final OD_{600} of 0.5 after 36 h. After 30 h, growth of all strains began to plateau and there was no further increase in optical density. Examination by light microscopy did not reveal any variation in cellular morphology which would account for the increased optical density of cultures of the two mutants, such as elongated, undivided or corkscrew shaped cells which are occasionally observed during axenic growth of these predators. Nor is the lower optical density of wild type and the *pilT2* mutant cultures due to aggregation of cells. In all three mutants and the wild type, cells became elongated and occasionally formed a corkscrew shape if left in medium for longer than 48 h.

In predatory mode, *B. bacteriovorus* 109JA preyed efficiently and produced bdelloplasts on *E. coli* ML35. Maximum growth was reached after 18 h as measured by a decrease in prey cell turbidity (Fig. 2). The *pilT1* mutant reached maximum growth after only 15 h. The *pilT2* mutant grew more slowly, reaching the maximal decrease in optical density after 30 h. The *pilT1pilT2* double mutant preyed within 24 h which was slower than the *pilT1* mutant but quicker than the *pilT2* mutant, restoring it to a near wild type predation pattern. All *pilT* mutants produced bdelloplasts of *E. coli* (as judged by phase contrast light microscopy; Fig. S1). Although minor variations in predation kinetics were observed (Fig. 2), overall neither PilT1 nor PilT2 were required for invasion of prey cells.

Presence of Type IV Pili on *pilT* mutants

Western blot analysis of whole cell lysates of axenically grown strain 109JA and the *pilT* mutants showed similar levels of the PilA protein present (Fig. S3). To examine the piliation of individual cells, electron microscopy and immunofluorescence light microscopy were used.

Electron microscopy evaluation of strain 109JA showed that 22.3% (n = 336) of cells expressed TFP on their surface (Table 3). This was significantly ($p<0.005$) different than the *pilT1* mutant, of which only 8.2% ($p<0.005$, n = 346) of cells were observed to have TFP on their surface. Only 16.5% (n = 250) of the *pilT2* mutant cells were piliated. Surprisingly, the *pilT1pilT2* double mutant had near wild type levels with 21.2% (n = 330) piliation.

None of the mutants were hyperpiliated as is observed with *P. aeruginosa* (Fig. S3B). In our study, the TFP were always located on the non-flagellated pole. In the rare instance that the cell did not have a flagellum, TFP were never observed on both poles. The number of TFP per cell (one) did not appear to change, however, this could only be assessed by electron microscopy. On occasion a cell was observed to have more than one pilus (Fig. S4).

Due to the short length of TFP on *B. bacteriovorus* (Fig. S4), staining artefacts and the laborious task of finding cells that were stained optimally for visualizing and thus counting TFP, electron microscopy was not an ideal method for determining the level of piliation. Therefore immunofluorescence microscopy with antibodies to the PilA protein was used for a more accurate measurement. In all cases (wild type and mutants) the number of cells with TFP increased. Both the wild type and the *pilT1pilT2* mutant had about 27% of cells with TFP (n = 4308, n = 3103). Surprisingly, the *pilT1* mutant had 33.4% (n = 5252) piliation, a

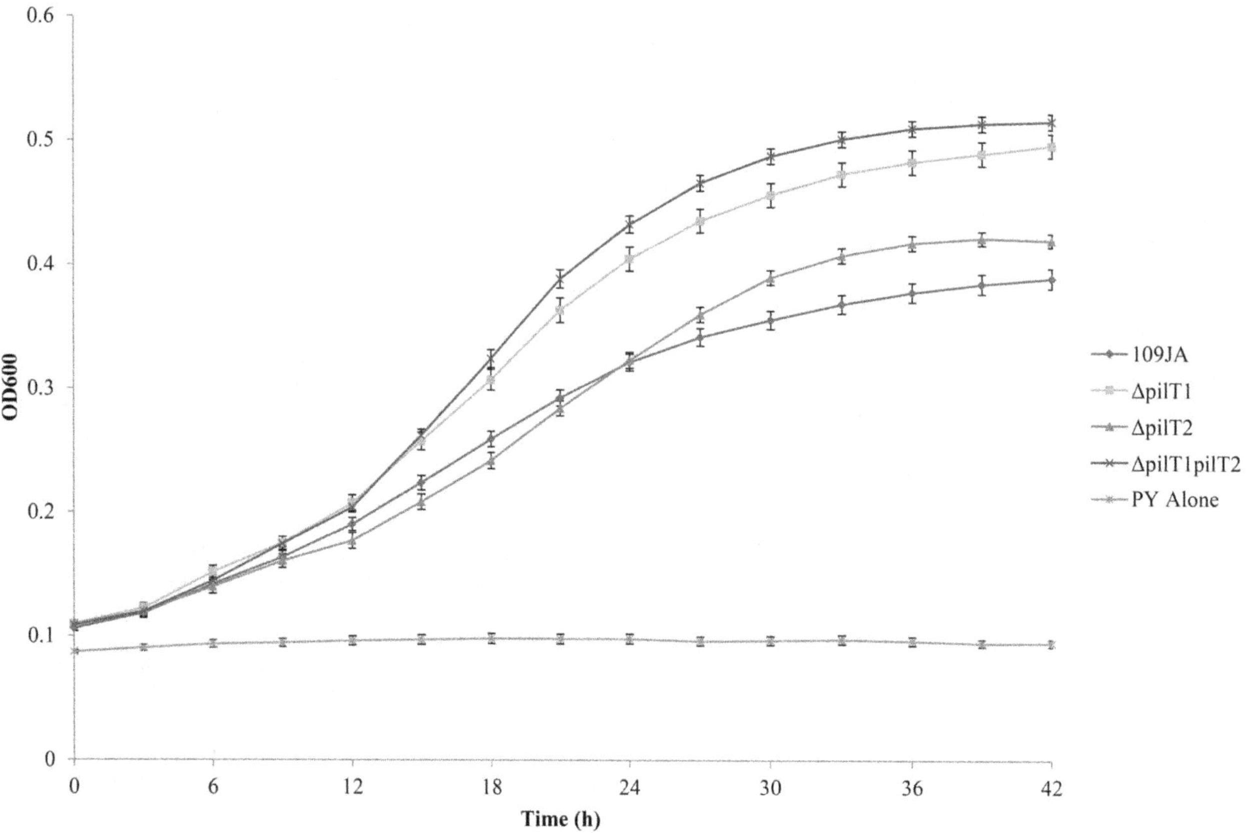

Figure 1. Axenic growth curve of *B. bacteriovorus* 109JA and *pilT* mutants. Growth in PY medium over a 42 h period. Results shown are an average of 20 replicates repeated in triplicate. Error bars represent standard error.

significant ($p<0.005$) increase from 8.2% measured by electron microscopy. The *pilT2* mutant had 29.2% of cells ($n = 3073$) with TFP, a small but significant increase above wild type levels ($p< 0.005$).

Biofilm Predation Assay

During this study, it was noted that the mutant with the markerless in-frame deletion of *pilT2* did not form plaques on lawns of *E. coli*, however the *pilT1* and *pilT1pilT2* mutants did (Data not shown). This observation confirmed the results of Medina *et al.* [8], who reported that a *pilT2* transposon mutant could not form lytic halos on lawns of *E. coli*. Therefore a robust biofilm model was used to assess predation of these predators. *E. coli* CO1 was used as the prey cell as it is a known biofilm-producing strain isolated from the stool of a healthy woman. This strain showed an increased ability to attach to uroepithelial cells of the elderly and agglutinate P-type erythrocytes, indicating it has potential to cause renal infections [18]. This strain was chosen over other known biofilm producers such as *P. aeruginosa* and *E. coli* strains GR12 and C1212 as it formed the best biofilm under the experimental conditions tested to allow for *Bdellovibrio* growth (Fig. S2). Wild type *B. bacteriovorus* 109JA reduced the mass of the biofilm by 65% while the *pilT2* mutant was unable to prey on the biofilm (Fig. 3). The *pilT1* mutant reduced the biofilm mass by 72%, slightly more than wild type which coincides with results observed in liquid cocultures where the *pilT1* mutant was slightly more aggressive in predation. Surprisingly, the *pilT1pilT2* double mutant was able to prey on the biofilm, but only able to reduce the mass by 49%. This indicates that although *pilT2* is required for predation on the biofilm, the *pilT1* deletion is able to partially restore function.

Scanning Electron Microscopy

Biofilm formation is a complex process which involves the secretion of many factors to create the extracellular polymeric substance (EPS). The EPS secreted by *E. coli* CO1 was seen as a dense, fibrous matrix which protects the cells beneath (Fig. 4A). Cells on the surface are not as protected as those buried beneath the EPS. *B. bacteriovorus* 109JA was able to penetrate the EPS layer and disrupt the biofilm. An overall loss in surface biofilm combined with the observation of bdelloplasts and many attack phase *B. bacteriovorus* (Fig. 4B) suggested that the EPS was not a barrier to predation. The *pilT1* mutant was able to disrupt the biofilm (Fig. 4C), but less predation was observed with the *pilT2* (Fig. 4D) and the double mutant (Fig. 4E). The biofilm exposed to the *pilT2* mutant had more EPS surrounding the prey cells than that seen when the biofilm was exposed to the *pilT1pilT2* mutant. These results suggest that the *pilT1pilT2* mutant can kill prey cells in a biofilm but the *pilT2* mutant has not been able to penetrate and disrupt the EPS.

Discussion

The mechanism which facilitates invasion of the prey cell periplasmic space by *B. bacteriovorus* is beginning to be understood. Publication of the genome sequence of *B. bacteriovorus* in 2004 allowed us to speculate which genes may be important for this process. Evans *et al.* [2] and Mahmoud and

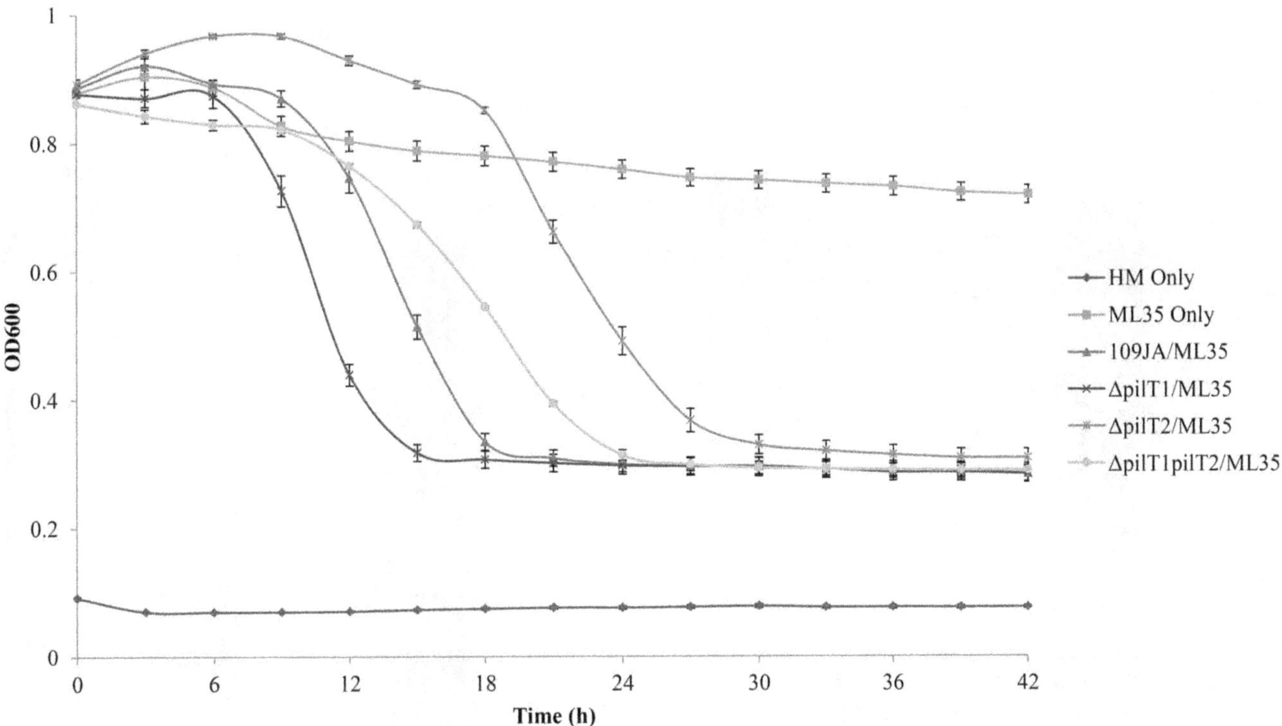

Figure 2. Effect of mutations in *pilT* genes on predation by *B. bacteriovorus* on *E. coli*. The decrease in prey cell optical density was used to assess predator growth. Results shown are an average of 20 replicates repeated in triplicate. Controls included: HM Buffer alone, *E. coli* ML35 prey cells alone and the wild type strain 109JA. Error bars represent standard error.

Koval [3] demonstrated that TFP are essential for predation to occur. Deletion of the gene for the major pilin protein, PilA, resulted in non-predacious *B. bacteriovorus*. Mahmoud and Koval [3] raised antibodies against the PilA protein and showed that the presence of anti-PilA in cocultures was enough to delay and inhibit predation. No attachment between *B. bacteriovorus* and the prey cell was observed for 16 h, suggesting that TFP are important for attachment to prey cells. However, because TFP in other bacteria are able to extend and retract, it was necessary to ask the question: Is retraction of TFP the driving force behind prey cell entry? Amino acid sequence analysis of PilT2 showed high sequence identity to other retraction ATPases, and thus it is suggested that this gene encodes an ATPase that could retract TFP in *B. bacteriovorus*. This study has demonstrated that PilT1 and PilT2 are not required for predation in liquid cocultures and, more specifically, they are not required for invasion of prey cells.

Bioinformatic analysis of the active domains present in PilT1 of *B. bacteriovorus* revealed that although it was annotated as a homologue of a PilT protein, it does not appear to function in this regard. It may be more similar to PilU in *P. aeruginosa*, which is thought to increase pilus strength or stability [16]. This may account for the decrease in number of TFP assessed by negative staining of the *pilT1* mutant. This staining method may possibly shear the pilus during the preparation process, giving the appearance of decreased surface associated TFP.

Bioinformatic analysis of PilT2 demonstrated high conservation within the active domains among other PilT proteins known to retract TFP, such as *P. aeruginosa* and *N. gonorrhoeae*. Twitching motility could not be demonstrated in *B. bacteriovorus* 109JA or the *pilT* mutants by use of a conventional sub-surface twitching motility assay designed for studies with *P. aeruginosa* [19]. It is possible that the prey-independent strain used in our study (strain 109JA) is unable to use twitching motility. The obligate predatory life cycle of wild type strains does not allow for twitching motility to be directly measured using this method. Expression of *B. bacteriovorus* PilT1 or PilT2 in a *P. aeruginosa* PilT deletion

Table 3. Presence of type IV pili on the surface of cells as assessed by electron microscopy and immunofluorescence microscopy.

Strain	% Number of cells with TFP	
	Electron Microscopy	**Immunofluorescence**
109JA	22.3	27.0
Δ*pilT1*	8.2*	33.4*
Δ*pilT2*	16.5*	29.2*
Δ*pilT1pilT2*	21.2	27.7

Statistical significance determined using the Chi-Square test, *p<0.005.

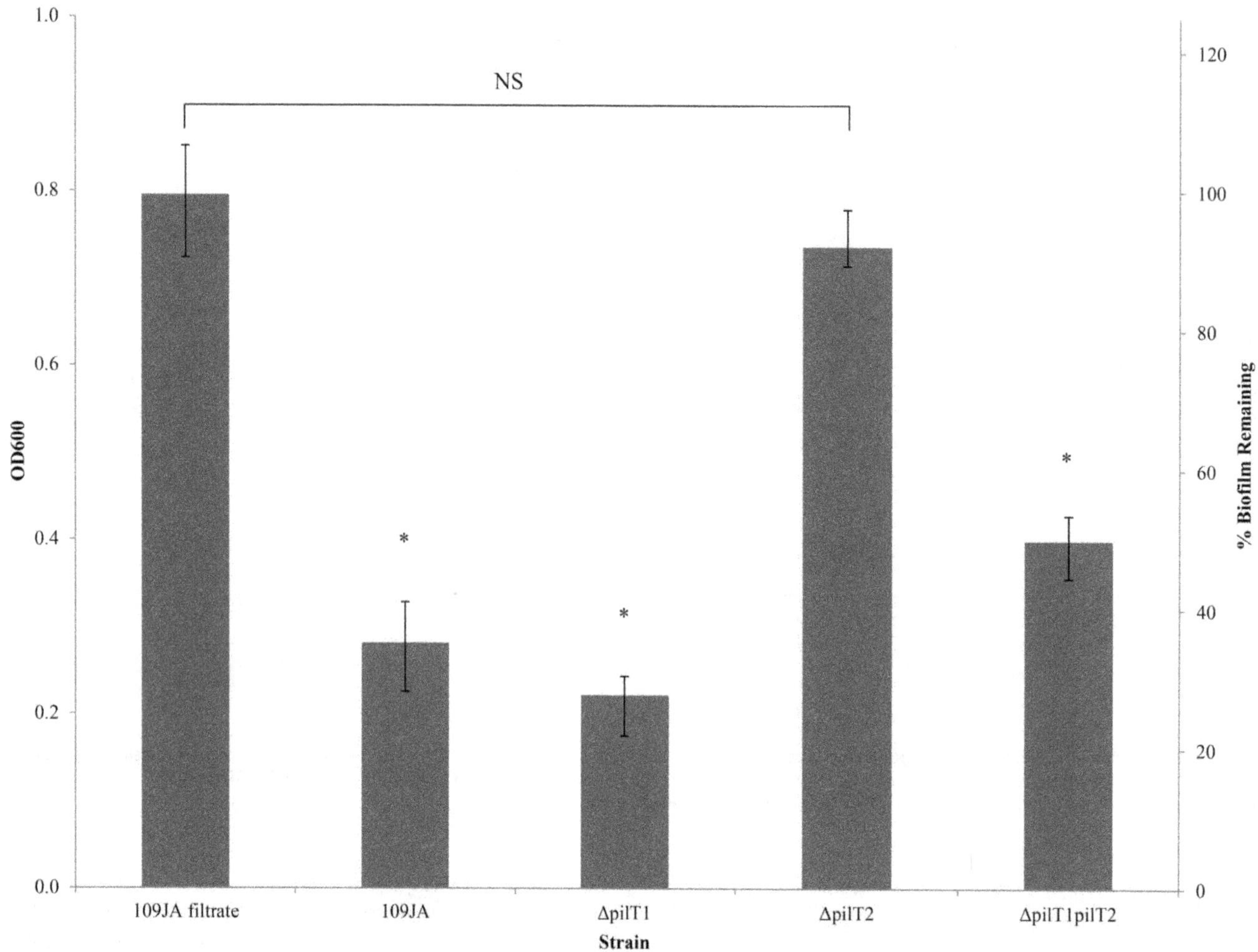

Figure 3. Biofilm predation assay. Biofilms of *E. coli* CO1 were pre-formed for 48 h in 96-well microtiter plates. Predator cultures containing either *B. bacteriovorus* 109JA or a *pilT* mutant were added and the plates incubated for a further 24 h. Residual biofilm cells were stained with crystal violet and the optical density at 600 nm (OD600) determined (*$p<0.005$). To exclude secreted factors contributing to the decrease in remaining biofilm, a 0.45 μm filtrate of a *B. bacteriovorus* 109JA culture used as a control. The percent biofilm remaining relative to the 109JA filtrate is shown on the secondary axis. Data presented were an average of 12-wells per replicate repeated in triplicate. Statistical significance was measured using a 1-way ANOVA with a Bonferonni corrected post-hoc Students T-test, *$p<0.005$.

mutant was unable to restore the twitching phenotype (Fig. S5). Lambert *et al.* [20] described *Bdellovibrio* gliding motility, a type of surface motility not flagellar or pilus mediated. In our study, the *pilT* mutants were still able to use gliding motility, confirming this is independent of TFP and PilT1 or PilT2 (Data not shown).

It is an unresolved question as to why so few cells of *B. bacteriovorus* express TFP on their surface in *in vitro* cocultures, yet are essential for survival. It is not known in a batch culture what percentage of *B. bacteriovorus* cells are capable of predation. It is possible that only a minority of the population are able to prey and therefore only this minority would express TFP. Both the *pilT1* and *pilT2* mutants showed a small increase in the number of cells with TFP as assessed by immunofluorescence microscopy. This result suggests that these genes may be involved in TFP function but does not elucidate what role each gene is playing, whether it be in regulation or direct involvement.

The ability of *B. bacteriovorus* to disrupt and kill cells within a biofilm has been studied previously [14,21,22]. We have shown that *pilT2* is not required for efficient predation in liquid coculture but is required for predation on a biofilm, confirming the results of Medina *et al.* [8]. SEM analysis revealed that an EPS layer is produced by *E. coli* cells during biofilm formation. This layer adds a level of protection from the outer environment but not to *B. bacteriovorus* predation, as predators were capable of penetrating and disrupting the biofilm. However, the *pilT2* and *pilT1pilT2* mutants were not able to do this as efficiently. It appears predation is able to occur on the cells at the surface of the biofilm. Eventually other cells would be preyed upon, but this would require a higher concentration of *Bdellovibrio* or more time. Taken together, the delay in predation on biofilms by *pilT2* and *pilT1pilT2* mutants is possibly due to the inability to retract their TFP via inactivation of the *pilT2* gene. The *pilT1* mutant still disrupted the biofilm and also partially restored the ability of the *pilT1pilT2* mutant to prey on the biofilm. It is known that *B. bacteriovorus* secretes hydrolytic enzymes such as proteases into the culture supernatant and that some of these enzymes have the ability to disrupt biofilms of Gram-positive pathogens such as *Staphylococcus aureus* [23]. The regulation of production of such enzymes is unknown and possibly could be signalled through the TFP system. Removal of the EPS layer in a biofilm by proteases could expose the underlying prey cells to predation. If PilT1 is a negative regulator of protease synthesis, then deletion of *pilT1* would result in an upregulation of

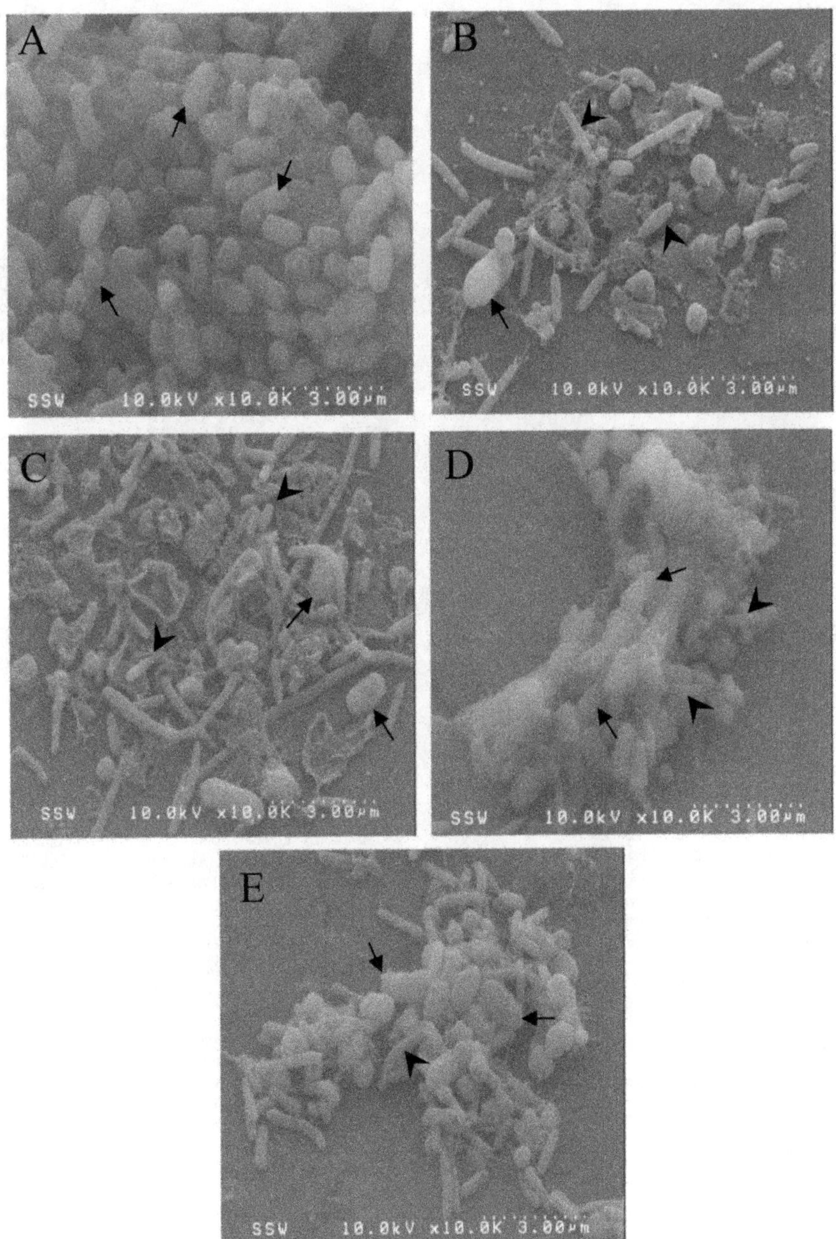

Figure 4. Scanning electron microscopy of predation on pre-formed biofilms of *E. coli.* (**A**) Biofilms of *E. coli* CO1 (arrows) were formed for 48 h on polyvinyl chloride plastic coverslips. Predator cultures (arrowheads) of (B) *B. bacteriovorus* 109JA, (C) Δ*pilT1*, (D) Δ*pilT2* (E) Δ*pilT1pilT2* were added for a further 24 h.

enzyme production. This may explain why, although not statistically significant, deletion of *pilT1* was able to prey slightly better in liquid cocultures (Fig. 2) and on biofilms (Fig. 3). Also, it partially restored predation of the *pilT1pilT2* mutant closer to wild type in both predation assays. An increase in secreted hydrolytic enzymes may prime *B. bacteriovorus* for predation and allow easier entry into the prey cell.

The genome sequence of the epibiotic predator, *B. exovorus* JSS, contains a fully annotated set of TFP genes, including both *pilT1* and *pilT2* [10]. In this species invasion does not occur, an observation which supports the view of this study that retraction of TFP is not the driving force behind the invasion process. A PilT2 deletion in *B. exovorus* could not be isolated as it would not form plaques on an agar surface. To avoid this problem prey-independent strains are generally used. However, our lab and that of Jurkevitch [10] have been unable to produce a prey independent derivative of *B. exovorus* JSS.

The ability of most BALOs to penetrate the outer membrane of a Gram-negative cell without membrane fusion, in direct opposition of the internal turgor pressure of the prey cell, still remains an enigma. Other mechanisms must be at play to allow this event to occur. Pasternak *et al.* [10] hypothesized that the epibiotic predator *B. exovorus* evolved from a periplasmic predator by gene loss, including the loss of the genes that enable prey invasion. The identification of such genes would be step forward in understanding the invasion process.

Supporting Information

Figure S1 Phase contrast micrograph of predation of *pilT2* mutant (white arrow) on *E. coli* ML35. Bdelloplasts (black arrow) and uninfected *E. coli* prey cells (black arrowhead).

Figure S2 Prey cell biofilm formation. Variation in biofilm formation of *E. coli* strains was assessed by staining residual cells with crystal violet. *E. coli* CO1 produced the best biofilm under the experimental conditions used.

Figure S3 PilA western blot analysis. The level of PilA (19.6 kDa) in whole cell lysates of *B. bacteriovorus* 109JA and the *pilT* mutants grown prey-dependently on *E. coli* ML35 was assessed.

Figure S4 Electron micrographs of type IV pili (arrows) of (A) *B. bacteriovorus* 109J and (B) *P. aeruginosa* PAK::*pilT*. Cells were negatively stained with uranyl acetate. Note the hyperpiliated phenotype of the *P. aeruginosa* mutant.

Figure S5 Subsurface twitching motility assay. *P. aeruginosa* PAK and a PilT mutant expressing PilT1 or PilT2 from *B. bacteriovorus* were used to assess twitching motility. PilT proteins were expressed on an arabinose inducible plasmid (pBADGr) and the *P. aeruginosa* cultures plated on LB agar containing 0.1% arabinose. As a control, 0.2% glucose was added to repress expression. The zone of motility was visualized using crystal violet and the diameter of each zone measured.

Acknowledgments

We would like to thank Lori Burrows, Department of Biochemistry and Biomedical Sciences, McMaster University for providing us with *Pseudomonas aeruginosa* PAK and the *pilT* deletion mutant; Peter Cadieux, Department of Microbiology and Immunology, University of Western Ontario for *Escherichia coli* strains CO1, C1212 and GR-21; Cezar Khursigara, Department of Molecular and Cellular Biology, University of Guelph for technical advice on biofilm assays; Ross Davidson at Surface Science Western, University of Western Ontario and Richard Gardner at the Biotron Integrated Imaging Facility, University of Western Ontario who both provided technical assistance for scanning electron microscopy.

Author Contributions

Conceived and designed the experiments: RMC SFK. Performed the experiments: RMC. Analyzed the data: RMC SFK. Contributed reagents/ materials/analysis tools: RMC SFK. Wrote the paper: RMC SFK.

References

1. Lambert C, Hobley L, Chang CY, Fenton A, Capeness M, et al. (2009) A predatory patchwork: Membrane and surface structures of *Bdellovibrio bacteriovorus*. Adv Microb Physiol 54: 313–361.
2. Evans KJ, Lambert C, Sockett RE (2007) Predation by *Bdellovibrio bacteriovorus* HD100 requires type IV pili. J Bacteriol 189: 4850–4859.
3. Mahmoud KK, Koval SF (2010) Characterization of type IV pili in the life cycle of the predator bacterium *Bdellovibrio*. Microbiology 156: 1040–1051.
4. Bradley DE (1980) A function of *Pseudomonas aeruginosa* PAO pili: twitching motility. Can J Microbiol 26: 146–154.
5. Wu SS, Kaiser D (1995) Genetic and functional evidence that type IV pili are required for social gliding motility in *Myxococcus xanthus*. Mol Microbiol 18: 547–558.
6. Maier B, Potter L, So M, Long CD, Seifert HS, et al. (2002) Single pilus motor forces exceed 100 pN. PNAS 99: 16012–16017.
7. Rendulic S, Jaqtap P, Rosinus A, Eppinger M, Baar C, et al. (2004) A predator unmasked: life cycle of *Bdellovibrio bacteriovorus* from a genomic perspective. Science 303: 689–692.
8. Medina AA, Shanks RM, Kadouri DE (2008) Development of a novel system for isolating genes involved in predator-prey interactions using host independent derivatives of *Bdellovibrio bacteriovorus* 109J. BMC Microbiol 8: 33.
9. Koval SF, Hynes SH, Flannagan RS, Pasternak Z, Davidov Y, et al. (2013) *Bdellovibrio exovorus* sp. nov., a novel predator of *Caulobacter crescentus*. IJSEM 63: 146–151.
10. Pasternak Z, Njagi M, Shani Y, Chanyi R, Rotem O, et al. (2013) In and out: an analysis of epibiotic vs periplasmic bacterial predators. ISME J 8: 625–635.
11. Chanyi RM, Ward C, Pechey A, Koval SF (2013) To invade or not to invade: two approaches to a prokaryotic predatory life cycle. Can J Microbiol 59: 273–279.
12. Steyert SR, Piñeiro SA (2007) Development of a novel genetic system to create markerless deletion mutants of *Bdellovibrio bacteriovorus*. Appl Environ Microbiol 73: 4717–4724.
13. Cotter TW, Thomashow MF (1992) A conjugation procedure for *Bdellovibrio bacteriovorus* and its use to identify DNA sequences that enhance the plaque-forming ability of a spontaneous host-independent mutant. J Bacteriol 174: 6011–6017.
14. Kadouri D, O'Toole GA (2005) Susceptibility of biofilms to *Bdellovibrio bacteriovorus* attack. Appl Environ Microbiol 71: 4044–4051.
15. Aukema KG, Kron EM, Herdendorf TJ, Forest KT (2004) Functional dissection of a conserved motif within the pilus retraction protein PilT. J Bacteriol 187: 611–618.
16. Chiang P, Burrows LL (2003) Biofilm formation by hyperpiliated mutants of *Pseudomonas aeruginosa*. J Bacteriol 185: 2374–2378.
17. Chiang P, Sampaleanu LM, Ayers M, Pahuta M, Howell PL, et al. (2008) Functional role of conserved residues in the characteristic secretion NTPase motifs of the *Pseudomonas aeruginosa* type IV pilus motor proteins PilB, PilT and PilU. Microbiology 154: 114–126.
18. Reid G, Zorzitto ML, Bruce AW, Jewett MAS, Chan RCY, et al. (1984) Pathogenesis of urinary tract infection in the elderly: the role of bacterial adherence to uroepithelial cells. Curr Microbiol 11: 67–72.
19. Asikyan ML, Kus JV, Burrows LL (2008) Novel proteins that modulate type IV pilus retraction dynamics in *Pseudomonas aeruginosa*. J Bacteriol 190: 7022–7034.
20. Lambert C, Fenton AK, Hobley L, Sockett RE (2011) Predatory *Bdellovibrio* bacteria use gliding motility to scout for prey on surfaces. J Bacteriol 193: 3139–3141.
21. Dwidar M, Hong S, Cha M, Jang J, Mitchell RJ (2012) Combined application of bacterial predation and carbon dioxide aerosols to effectively remove biofilms. Biofouling 28: 671–680.
22. Kadouri DE, Tran A (2013) Measurement of predation and biofilm formation under different ambient oxygen conditions using a simple gasbag-based system. Appl Environ Microbiol 79: 5264–5271.
23. Monnappa AK, Dwidar M, Seo JK, Hur JH, Mitchell RJ (2014) *Bdellovibrio bacteriovorus* inhibits *Staphylococcus aureus* biofilm formation and invasion into human epithelial cells. Sci Rep 4: 3811.

6

Overexpression of the Novel MATE Fluoroquinolone Efflux Pump FepA in *Listeria monocytogenes* Is Driven by Inactivation of Its Local Repressor FepR

François Guérin[1,2], Marc Galimand[3], Fabrice Tuambilangana[1], Patrice Courvalin[3], Vincent Cattoir[1,2]*

1 CHU de Caen, Service de Microbiologie, F-14033 Caen, France, 2 Université de Caen Basse-Normandie, EA4655 (équipe "Antibio-résistance"), F-14032 Caen, France, 3 Institut Pasteur, Unité des Agents Antibactériens, Paris, France

Abstract

Whereas fluoroquinolone resistance mainly results from target modifications in gram-positive bacteria, it is primarily due to active efflux in *Listeria monocytogenes*. The aim of this study was to dissect a novel molecular mechanism of fluoroquinolone resistance in this important human pathogen. Isogenic *L. monocytogenes* clinical isolates BM4715 and BM4716, respectively susceptible and resistant to fluoroquinolones, were studied. MICs of norfloxacin and ciprofloxacin were determined in the presence or in the absence of reserpine (10 mg/L). Strain BM4715 was susceptible to norfloxacin (MIC, 4 mg/L) and ciprofloxacin (MIC, 0.5 mg/L) whereas BM4716 was highly resistant to both drugs (MICs 128 and 32 mg/L, respectively). Reserpine was responsible for a 16-fold decrease in both norfloxacin and ciprofloxacin MICs against BM4716 suggesting efflux associated resistance. Whole-genome sequencing of the strains followed by comparative genomic analysis revealed a single point mutation in the gene for a transcriptional regulator, designated *fepR* (for fluoroquinolone efflux protein regulator) belonging to the TetR family. The frame-shift mutation was responsible for the introduction of a premature stop codon resulting in an inactive truncated protein. Just downstream from *fepR*, the structural gene for an efflux pump of the MATE family (named FepA) was identified. Gene expression was quantified by qRT-PCR and demonstrated that *fepA* expression was more than 64-fold higher in BM4716 than in BM4715. The clean deletion of the *fepR* gene from BM4715 was responsible for an overexpression of *fepA* with resistance to norfloxacin and ciprofloxacin, confirming the role of FepR as a local repressor of *fepA*. In conclusion, we demonstrated that overexpression of the new MATE efflux pump FepA is responsible for fluoroquinolone resistance in *L. monocytogenes* and secondary to inactivation of the FepR repressor.

Editor: Rajeev Misra, Arizona State University, United States of America

Funding: This research was funded by a grant from the Ministère de l'Enseignement Supérieur et de la Recherche (EA4655), Université Caen Basse-Normandie, France. The funders had no role in study design, data collection and analysis, decision to publish, or preparation of the manuscript.

Competing Interests: The authors have declared that no competing interests exist.

* Email: cattoir-v@chu-caen.fr

Introduction

Listeria monocytogenes is a gram-positive rod-shaped facultative intracellular food-borne pathogen widely distributed in the environment [1]. It is responsible for severe human infections (such as bacteremia and central nervous system [CNS] infections) primarily in neonates, elderly people and patients with impaired cellular immunity, as well as abortions in pregnant women [2]. The reference treatment of listeriosis currently relies on a synergistic combination of high-dose ampicillin or amoxicillin and gentamicin administered intravenously [3]. Nonetheless, despite an effective therapy, CNS infections are associated with a high mortality rate (ca. 30%) while sequels are common [4].

Except for intrinsic resistance to cephalosporins and fosfomycin, *L. monocytogenes* is susceptible to all antibiotics in vitro, including fluoroquinolones (FQs). However, FQs are not recommended in the treatment of listeriosis even if newer compounds (i.e. levofloxacin, moxifloxacin) might represent an efficient alternative in the treatment of CNS listeriosis, as previously reported in one human case report [5] and several animal models [6,7]. Due to the extensive use of FQs for the treatment of multiple infections, there is also an important selective pressure for recovery of in vivo FQ-resistant mutants in gram-positive bacteria including *L. monocytogenes*.

In gram-positive bacteria, FQ resistance is chromosomally encoded and most commonly results from the accumulation of mutations in molecular targets that are type II topoisomerases, DNA gyrase ($GyrA_2GyrB_2$) and topoisomerase IV ($ParC_2ParE_2$) [8]. Alterations predominantly occur within short conserved regions of the *gyrA*, *gyrB*, *parC*, and *parE* genes, the so-called quinolone-resistance determining regions (QRDRs). In contrast, FQ resistance in *L. monocytogenes* seems to be primarily due to active efflux, especially through overexpression of the *lde* gene coding for a transporter belonging to the major facilitator superfamily (MFS) [9–12].

In this study, we have elucidated a novel molecular mechanism of FQ resistance in a clinical isolate of *L. monocytogenes* (a preliminary report of this work was presented at the 53rd Interscience Conference on Antimicrobial Agents and Chemotherapy, Denver, CO, 10–13 September 2013 [abstract C1-1440]).

Materials and Methods

Bacterial strains and molecular typing

Strains BM4715 and BM4716 (both belonging to the serotype 1/2b) were collected in France from a unique blood sample of the same patient suffering from listeriosis. Strain BM4715 was fully susceptible to ciprofloxacin whereas BM4716 was resistant. Reference strain *L. monocytogenes* EGD-e (serotype 1/2a) [13] was included for molecular typing. *Staphylococcus aureus* ATCC 29213, *S. aureus* SA-1199B (NorA-overproducing) [14], and *Escherichia coli* EC1000 (Life Technologies) were used as controls for antimicrobial susceptibility testing, efflux, and subcloning, respectively.

Strains BM4715 and BM4716 were typed by pulsed-field gel electrophoresis (PFGE) using the SmaI endonuclease, as previously described [15]. The PFGE patterns were analyzed in accordance with Tenover et al. [16].

Antimicrobial susceptibility testing

MICs of antimicrobial agents (except for sparfloxacin and levofloxacin) were determined on Mueller-Hinton agar using E-test strips (bioMérieux, Marcy l'Etoile, France) with or without reserpine (10 mg/L). MICs of sparfloxacin and levofloxacin as well as those of antiseptics and dyes were determined on Muller-Hinton by the agar dilution method (tested range, from 0.06 to 256 mg/L) with or without reserpine (10 mg/L) with 10^4 CFU per spot after 24 h of incubation at 35°C [17]. Determination of MICs was performed in three independent experiments.

PCR amplification and sequencing

Genomic DNA from *L. monocytogenes* BM4715 and BM4716 was extracted using the QIAmp DNA Mini kit (Qiagen). QRDRs of *gyrA*, *gyrB*, *parC*, and *parE* genes were amplified by PCR with specific primers (Table S1) [10] and the purified PCR products were sequenced with the same sets of primers in both directions (GATC Biotech, Konstanz, Germany).

Whole-genome sequencing

Genomic DNA was extracted from mid-log phase cultures of *L. monocytogenes* BM4715 and BM4716 using the NucleoBond buffer set III and the NucleoBond AX-G 100 (Macherey-Nagel, Hoerdt, France) following the manufacturer's instructions. High-throughput sequencing was performed using an Illumina MiSeq Benchtop sequencer (ProfileXpert-LCMT, Lyon, France). The shotgun sequencing for *L. monocytogenes* BM4715 led to an assembly of 93 contigs sizing from 203 to 320,858 bp with an aggregate genome size of 2,984,196 bp and a 145.3X average coverage of the genome while data for BM4716 were as follows: 89 contigs sizing from 203 to 476,158 bp, aggregate genome of 2,984,010 bp, and a 150.6X average genomic coverage. Comparative genomic analysis was performed using the CLC Genomics Workbench software 6.5.1 (CLC bio, Aarhus, Denmark). The nucleotide and deduced amino acid sequences for each contig were analyzed with BlastN and BlastX programs available over the Internet at the National Center for Biotechnology Information website (http://blast.ncbi.nlm.nih.gov/Blast.cgi).

RNA manipulations

Total RNA was extracted from BM4715 and BM4716 using the ZR Fungal/Bacterial RNA Miniprep kit (Zymo Research, Irvine, CA). Residual chromosomal DNA was removed by treating samples with the TURBO DNA-*free* kit (Life Technologies, Saint Aubin, France). Samples were quantified using the Biospec-Nano spectrophotometer (Shimadzu, Noisiel, France) and the integrity was assessed using the Agilent 2100 bioanalyzer.

For RT-PCR experiments, cDNA was synthesized from total RNA (~1 μg) using the QuantiTect Reverse Transcription kit (Qiagen, Courtaboeuf, France) according to the manufacturer's instructions. For operon mapping, PCR reactions were then carried out according to standard conditions using specific primers synthesized by Sigma-Aldrich France (Table S1). Each PCR amplification was performed on cDNA and chromosomal DNA (used as positive control). Transcript levels of the *fepA* gene were determined by the DeltaDelta Ct method and the *bglA* gene [18] was used as a housekeeping control gene (Table S1). Each experiment was performed in triplicate.

The transcription start site (TSS) and promoter sequences were determined using the 5′RACE System kit (Life Technologies SAS, Saint Aubin, France) using specific primers (Table S1) according to the manufacturer's instructions.

Construction of a BM4715 *fepR* deletion mutant

A fepR deletion mutant was derived from L. monocytogenes BM4715 (named BM4715ΔfepR) by allelic exchange with a truncated copy of fepR using the pWS3 suicide vector as previously described [19]. Approximately 500-bp fragments upstream and downstream from fepR were amplified by PCR using BM4715 chromosome as template and primer pairs Lmo-fepR-F1-EcoRI/Lmo-fepR-R1 and Lmo-fepR-F2/Lmo-fepR-R2-EcoRI (Table S1). Following EcoRI restriction, ligation and amplification using Lmo-fepR-F1-EcoRI and Lmo-fepR-R2-EcoRI, the resulting fragment carrying the truncated fepR copy was cloned in the temperature-sensitive pG(+)host9-derived shuttle vector pWS3 to create plasmid pWS3ΩfepR-KO. The hybrid plasmid was then introduced into the chromosome of BM4715 by electro-transformation and homologous recombination followed by excision of the wild-type copy as described [19]. Deletion of the fepR gene was confirmed by PCR and sequencing.

Multiple alignment and phylogenetic analysis

Sequence comparison and phylogenetic analysis was performed by the neighbor-joining algorithm with the ClustalX software (version 1.83) and the resulting tree was displayed with TreeView software (version 1.6.6).

Nucleotide sequence accession numbers

The nucleotide sequences of the *fepR/fepA* locus from strains BM4715 and BM4716 have been deposited in the GenBank database under accession no. KJ000253 and KJ000254, respectively.

Ethics statement

Ethical approval was not required for the study since there was no direct patient involvement and only bacterial strains were retrospectively studied. In addition, clinical samples were de-identified and no identifiable patient information is available.

Results

Efflux-mediated FQ resistance in BM4716

Comparison of antibiotic susceptibility profiles showed that strain BM4715 was susceptible to norfloxacin and ciprofloxacin (MICs of 4 and 0.5 mg/L, respectively) whereas BM4716 was resistant (MICs of 128 and 32 mg/L, respectively) (Table 1). Besides these two FQs, no significant differences in MICs between

Table 1. MIC of antibiotics, antiseptics, and dyes against *L. monocytogenes* clinical isolates BM4715 and BM4716 as well as BM4715 *fepR* deletion mutant (BM4715Δ*fepR*).

Compound[a]	MIC (mg/L)[b]					
	BM4715		BM4716		BM4715Δ*fepR*	
	−R	+R	−R	+R	−R	+R
FQ antibiotic						
Norfloxacin	4	2 (2)	128	8 (**16**)	64	4 (**16**)
Ciprofloxacin	0.5	0.5 (1)	32	2 (**16**)	16	1 (**16**)
Sparfloxacin	1	1 (1)	2	2 (1)	2	2 (1)
Levofloxacin	1	1 (1)	2	1 (2)	2	1 (2)
Moxifloxacin	0.25	0.25 (1)	0.5	0.25 (2)	0.5	0.25 (2)
Non-FQ antibiotic						
Amoxicillin	0.25	0.25 (1)	0.25	0.25 (1)	0.25	0.25 (1)
Cephalothin	2	2 (1)	2	2 (1)	2	2 (1)
Gentamicin	0.5	0.5 (1)	1	1 (1)	1	1 (1)
Erythromycin	0.25	0.25 (1)	0.25	0.25 (1)	0.25	0.25 (1)
Clindamycin	2	2 (1)	2	2 (1)	2	2 (1)
Chloramphenicol	4	4 (1)	4	4 (1)	4	4 (1)
Tetracycline	0.5	0.25 (2)	0.5	0.25 (2)	0.25	0.25 (1)
Tigecycline	0.12	0.12 (1)	0.12	0.06 (2)	0.06	0.06 (1)
Vancomycin	1	1 (1)	1	1 (1)	1	1 (1)
Linezolid	2	2 (1)	2	2 (1)	2	2 (1)
Daptomycin	1	0.5 (2)	1	0.5 (2)	1	0.5 (2)
Cotrimoxazole	0.03	0.03 (1)	0.06	0.03 (2)	0.03	0.03 (1)
Rifampin	0.06	0.06 (1)	0.06	0.06 (1)	0.06	0.06 (1)
Fusidic acid	2	2 (1)	2	2 (1)	2	2 (1)
Fosfomycin	≥1,024	≥1,024 (1)	≥1,024	≥1,024 (1)	≥1,024	≥1,024 (1)
Antiseptic						
Benzalkonium chloride	4	2 (2)	8	8 (1)	8	8 (1)
Cetylperidinium chloride	2	2 (1)	8	4 (2)	8	4 (2)
Chlorhexidine	2	2 (1)	8	8 (1)	8	8 (1)
Tetraphenylphosphonium	128	64 (2)	256	128 (2)	128	64 (2)
Dye						
Acridine orange	256	128 (2)	256	128 (2)	256	128 (2)
Acriflavine	64	64 (1)	128	128 (1)	64	64 (1)
Crystal violet	4	4 (1)	8	8 (1)	4	4 (1)
Ethidium bromide	32	16 (2)	256	128 (2)	256	128 (2)
Rhodamine	16	8 (2)	16	8 (2)	16	8 (2)

[a]FQ, fluoroquinolone.
[b]MICs determined in the presence (+R) or absence (−R) of reserpine (10 mg/L). Values in parentheses indicate the *n*-fold decrease in MIC in the presence of reserpin compared to its absence. Values in bold indicate significant changes in MIC.

the two isolates were observed for other antibiotics (Table 1). However, few changes were noted for some antiseptics and dyes, such as cetylperidinium chloride (4-fold), chlorehexidine (4-fold), and ethidium bromide (8-fold) (Table 1). PFGE analysis confirmed that these strains (isolated in a unique clinical specimen from the same patient) were isogenic, revealing that BM4716 was a FQ-resistant mutant derived from BM4715 (Figure S1). QRDRs of *gyrA*, *gyrB*, *parC*, and *parE* genes were sequenced but no mutations were found, suggesting another mechanism of FQ resistance. Since norfloxacin and ciprofloxacin are hydrophilic FQs and well-known substrates of efflux pumps, we determined their MICs with or without reserpine. There was a significant change (16-fold decrease) in MICs against BM4716 in the presence of the efflux pump inhibitor, confirming efflux-related resistance (Table 1). Expression levels of *lde* and *mdrL* genes (both coding for efflux pumps known to be associated with fluoroquinolone resistance in *L. monocytogenes*) were not significantly different between BM4716 and BM4715 strains (Figure 1A), suggesting the implication of other(s) transporter(s).

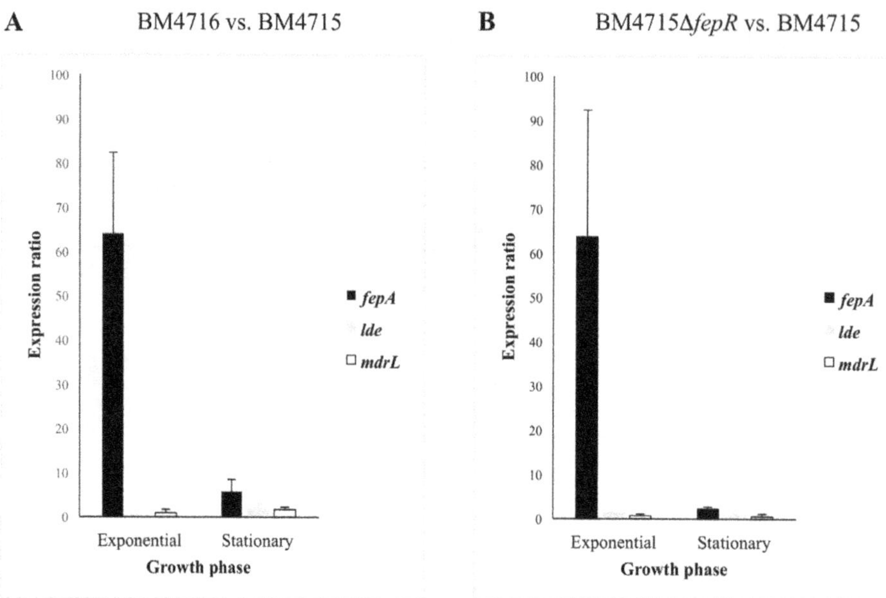

Figure 1. Expression ratios of the *fepA* gene in *L. monocytogenes* according to the bacterial growth phase. BM4716 vs. BM4715 (A), and *fepR* deletion mutant (BM4715Δ*fepR*) vs. BM4715 (B).

Single mutation within a *tetR*-like gene in BM4716

By comparing the entire genome *L. monocytogenes* BM4716 to that of BM4715, we found only three mutations, including two silent mutations in the same gene (corresponding to *lmo0460* in *L. monocytogenes* EGD-e) coding for a membrane-associated lipoprotein and, most importantly, a single mutation within a 594-bp gene (corresponding to *lmo2088* in *L. monocytogenes* EGD-e) coding for a 197-amino-acid TetR-like transcriptional regulator. This mutation (G61T) was responsible for the occurrence of a premature stop codon (E21*) leading to a nonfunctional truncated protein (Figure 1). Immediately downstream from this gene, a 1,332-bp gene (corresponding to *lmo2087* in *L. monocytogenes* EGD-e) was identified, which coded for a 443-amino-acid efflux pump of the MATE family (Figure 2). Interestingly, this protein only shared 12% to 27% identity with other bacterial MATE efflux pumps (Figure 3).

Overexpression of a MATE efflux pump in BM4716

Assuming that the efflux-mediated FQ resistance was likely due to this novel MATE efflux pump, we assessed the expression of its corresponding gene by qRT-PCR. The gene was highly overexpressed in BM4716 as compared to BM4715 in exponential phase (64-fold increase), while expression alteration was moderate in stationary phase (6-fold increase) (Figure 1A). The role of the *tetR*-like gene as repressor of *fepA* was confirmed by construction of the clean deletion mutant. Indeed, the strain BM4715Δ*fepR* was resistant to norfloxacin and ciprofloxacin (MICs of 64 and 16 mg/L, respectively) as observed with the BM4716 clinical isolate while there was a significant change (16-fold decrease) in MICs against

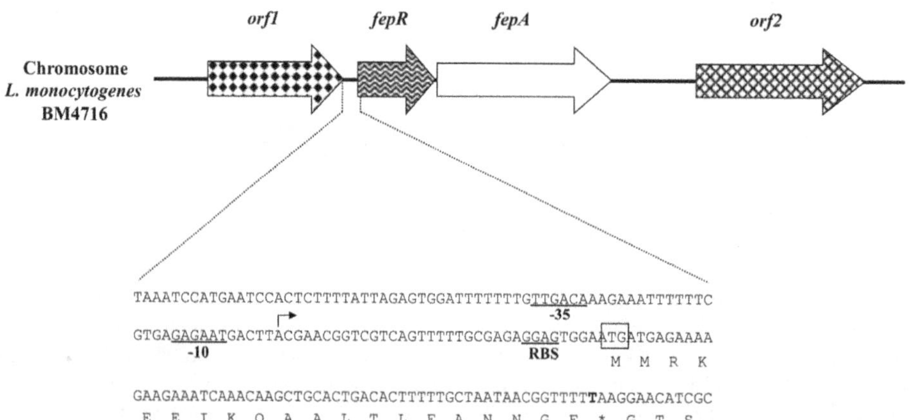

Figure 2. Schematic map of the genetic environment of *fepR*/*fepA* in *L. monocytogenes* BM4716 chromosome. Open reading frames (ORFs) are indicated by horizontal arrows. Genes *orf1* and *orf2* putatively encode a lipase and a DNA-binding protein, respectively. The sequence corresponding to the upstream region of *fepR*/*fepA* genes is presented in details. The −35 and −10 promoter boxes are underlined and the transcription start site (TSS) is represented by an arrow. The start codon of *fepR* and its putative ribosome-binding site (RBS) are indicated. The non-synonymous mutation G61T (leading to substitution E21*) is shown in bold.

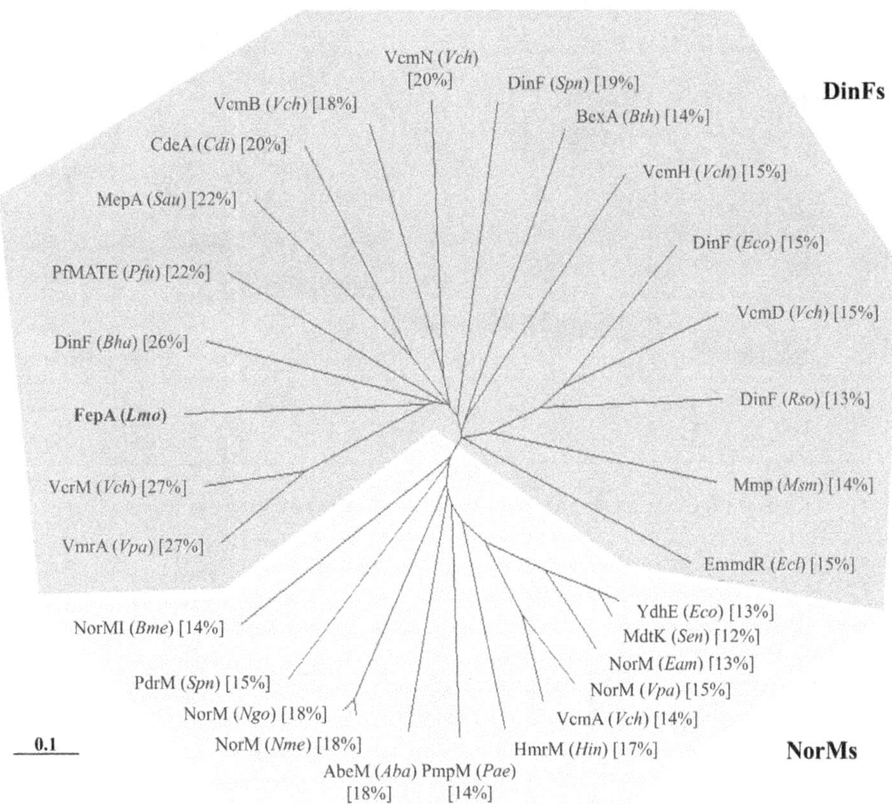

Figure 3. Phylogenetic tree based on neighbor-joining analysis of sequences of bacterial efflux proteins belonging to the MATE family. The various homologs were identified in: *Aba, Acinetobacter baumannii; Bha, Bacillus halodurans; Bme, Brucella melitensis; Bth, Bacteroides thetaiotaomicron; Cdi, Clostridium difficile; Eam, Erwinia amylovora; Ecl, Enterobacter cloacae; Eco, Escherichia coli; Hin, Haemophilus influenzae; Lmo, Listeria monocytogenes; Msm, Mycobacterium smegmatis; Ngo, Neisseria gonorrhoeae; Nme, Neisseria meningitidis; Pae, Pseudomonas aeruginosa; Pfu, Pyrococcus furiosus; Rso, Ralstonia solanacearum; Sen, Salmonella enterica* serovar Typhimurium; *Sau, Staphylococcus aureus; Spn, Streptococcus pneumoniae; Vch, Vibrio cholerae;* and *Vpa, Vibrio parahaemolyticus.* The scale bar represents 10% difference in amino acid sequences. Amino acid identities of each MATE protein as compared to FepA are indicated in square brackets. The two DinF and NorM subfamilies are highlighted.

BM4715Δ*fepR* in the presence of reserpine (Table 1). Also, there was an overexpression of *fepA* in BM4715Δ*fepR* as compared to BM4715 in exponential phase (64-fold increase) as observed in BM4716 (Figure 1B). The gene for the MATE efflux pump was named FepA (for fluoroquinolone efflux protein A) and that for the TetR-like transcriptional regulator designated FepR (for fluoroquinolone efflux protein regulator).

Description of the *fepRA* operon

Since an operon structure was bioinformatically predicted for the *fepA* and *fepR* genes (only 9 intervening bp), this was confirmed by RT-PCR (Figure 4). In addition, we experimentally determined a unique TSS 34 bp uptstream from the start codon of *fepR* (Figure 1). The *fepRA* locus was surrounded by two genes (according to *L. monocytogenes* EGD-e numbering): *lmo2089* (upstream) and *lmo2086* (downstream) coding for a lipase (347 amino acids) and a DNA-binding protein (423 amino acids), respectively (Figure 2).

Discussion

Whereas FQ resistance in gram-positive bacteria mainly occurs through point mutations in QRDRs in ParC and GyrA, it has so far only been related to active efflux in *L. monocytogenes* [9–12]. There are five families of drug efflux systems: the major facilitator superfamily (MFS), the resistance-nodulation-cell division (RND), the small multidrug resistance (SMR), the multidrug and toxic-compound extrusion (MATE), and the ATP-binding cassette (ABC) family [20]. The first four catalyze drug extrusion by exploiting the transmembrane electrochemical (H^+ or Na^+) gradient whereas ABC transporters are powered by ATP hydrolysis. In gram-positive bacteria, active efflux is mainly associated with overexpression of MFS pumps, such as NorA in *S. aureus* and PmrA in *Streptococcus pneumoniae*, which preferentially extrude hydrophilic FQs (i.e. norfloxacin, ciprofloxacin) [21].

In *L. monocytogenes*, only two chromosomal pumps, both belonging to the MFS, have been involved in antimicrobial resistance [10,22]. The first transporter, encoded by the *mdrL* gene, is related to the efflux of macrolides, cefotaxime, and heavy metals [22] while the second, encoded by the *lde* gene, is associated with resistance to hydrophilic FQs as well as with acriflavine and ethidium bromide resistance [10]. Hence, this is the characterization of the third antimicrobial efflux pump in *L. monocytogenes*. Even though the prevalence of FepA-mediated FQ resistance is not known, it might be not so rare since 13 out of 15 ciprofloxacin-resistant foodborne isolates of *L. monocytogenes* did not show overexpression of the *lde* gene, suggesting the existence of other resistance mechanisms [11].

Transporters of the MATE family have been rarely demonstrated to be involved in antimicrobial resistance [21]. However,

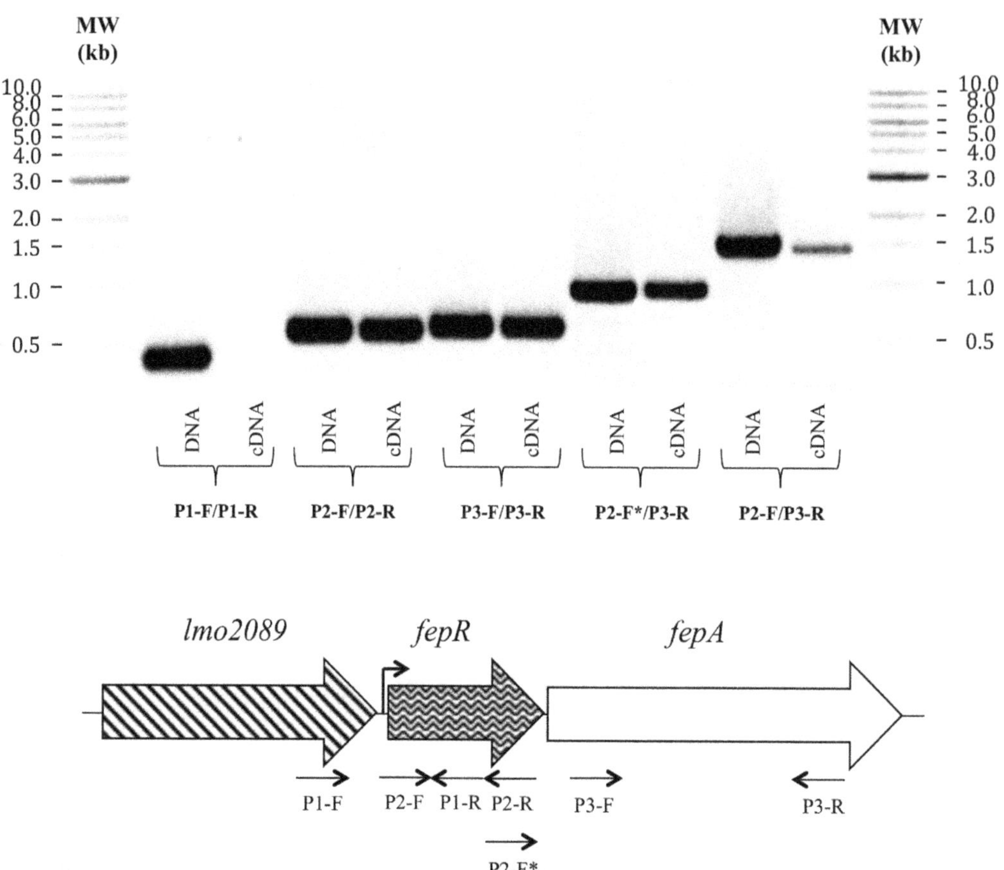

Figure 4. Agarose gel electrophoresis showing PCR products corresponding to transcripts of *fepR* and *fepA* genes. Different sets of primers were designed to amplify specific regions of *fepR* (P2-F/P2-R) or *fepA* (P3-F/P3-R), the intergenic region (P2-F*/P3-R), the long cotranscript (P2-F/P3-R) and a negative control (P1-F/P1-R) (Table S1). Each PCR amplification was carried out on chromosomal DNA (used as positive control) and on cDNA, as indicated. MW, 1-kb ladder (New England Biolabs, France).

almost all known MATE transporters can recognize FQs as substrates while acriflavine and ethidium bromide can also be pumped out [23]. The prototype of this family is NorM from *Vibrio parahaemolyticus* and its homolog in *E. coli* is YdhE [24,25]. All members of the MATE family possess 12 transmembrane domains and usually function as Na+/drug antiporters. By phylogenetic analysis, MATE transporters are divided in three clusters. The first and the third cluster include homologs of NorM and DinF, respectively, whereas the members of the second cluster are exclusively found in eukaryotes [25]. FepA appears to be related to DinF homologs even if the degrees of identity are low.

In gram-positive bacteria, only four MATE members have been described: CdeA in *Clostridium difficile*, MepA in *S. aureus*, and DinF and PdrM in *S. pneumoniae* [26–30]. CdeA is able to confer FQ resistance in *E. coli* when overexpressed and resistance to acriflavin and ethidium bromide is also observed [26]. MepA has a broad substrate profile including biocides, FQs (norfloxacin, ciprofloxacin), and tigecycline [27,28]. In *S. pneumoniae*, DinF is involved in FQ resistance while PdrM confers resistance to norfloxacin, acriflavine, and 4′,6-diamidino-2-phenylindole (DAPI) [29,30]. Taken together, all MATE-family proteins described so far are able to extrude FQ agents in gram-positive bacteria.

Little is known about regulation of genes encoding MATE-family proteins in gram-positive bacteria. Only the transcriptional regulation of the *mepA* gene has been extensively studied. More specifically, it has been shown that it was controlled by a local MarR-type repressor called MepR [27]. In this work, we have demonstrated that *fepA* was also negatively controlled, but by a TetR-type repressor. TetR proteins constitute a well-known family of transcriptional repressors [31]. They have been extensively studied in the regulation of several genes for drug efflux systems, such as TetR and *tetA* in *E. coli*, AcrR and *acrAB* in *E. coli*, AdeN and AdeIJK in *Acinetobacter baumannii*, or QacR and *qacA*/*qacB* in *S. aureus* [32,33]. As previously reported for other TetR-like repressors, FepR also autoregulates expression of its own gene [31]. As opposed to what was observed for *fepR*, TetR-like-encoding genes are usually divergently transcribed and are not part of an operon with the structural gene for the efflux pump [31]. Finally, no data are available about the expression of efflux pumps during the cell cycle. For *fepA*, it seems to be highly expressed during the exponential phase, like most of the genes controlled by σ^{70} factors, but further investigations are needed.

In conclusion, this is the first characterization of a MATE efflux pump involved in FQ resistance in *L. monocytogenes*. The substrate profile appears to be narrow, including only hydrophilic FQs. Finally, we also report transcriptional regulation of the expression of a MATE family efflux pump-encoding gene through a TetR-like repressor. Similar molecular mechanisms may be involved in FQ resistance within other important gram-positive pathogens in which FepA homologs are chromosomally encoded and for which FQ are indicated.

Supporting Information

Figure S1 PFGE patterns of SmaI-digested genomic DNA of *L. monocytogenes* strains. Lanes: 1, BM4715; 2, BM4716; 3, EGD-e.

Table S1 Deoxynucleotide primers used in the study.

Acknowledgments

The technical assistance of Michel Auzou, François-Xavier Toublet, and Marguerite Fines-Guyon is gratefully acknowledged.

Author Contributions

Conceived and designed the experiments: FG VC. Performed the experiments: FG FT. Analyzed the data: FG MG PC VC. Contributed reagents/materials/analysis tools: FG VC. Contributed to the writing of the manuscript: FG MG PC VC.

References

1. Farber JM, Peterkin PI (1991) *Listeria monocytogenes*, a food-borne pathogen. Microbiol Rev 55: 476–511.
2. Lorber B (1997) Listeriosis. Clin Infect Dis 24: 1–9; quiz 10–11.
3. Temple ME, Nahata MC (2000) Treatment of listeriosis. Ann Pharmacother 34: 656–661.
4. Hof H, Nichterlein T, Kretschmar M (1997) Management of listeriosis. Clin Microbiol Rev 10: 345–357.
5. Viale P, Furlanut M, Cristini F, Cadeo B, Pavan F, et al. (2007) Major role of levofloxacin in the treatment of a case of *Listeria monocytogenes* meningitis. Diagn Microbiol Infect Dis 58: 137–139.
6. Grayo S, Lott-Desroches MC, Dussurget O, Respaud R, Fontanet A, et al. (2008) Rapid eradication of *Listeria monocytogenes* by moxifloxacin in a murine model of central nervous system listeriosis. Antimicrob Agents Chemother 52: 3210–3215.
7. Sipahi OR, Turhan T, Pullukcu H, Calik S, Tasbakan M, et al. (2008) Moxifloxacin versus ampicillin + gentamicin in the therapy of experimental *Listeria monocytogenes* meningitis. J Antimicrob Chemother 61: 670–673.
8. Hooper DC (2002) Fluoroquinolone resistance among Gram-positive cocci. Lancet Infect Dis 2: 530–538.
9. Lampidis R, Kostrewa D, Hof H (2002) Molecular characterization of the genes encoding DNA gyrase and topoisomerase IV of *Listeria monocytogenes*. J Antimicrob Chemother 49: 917–924.
10. Godreuil S, Galimand M, Gerbaud G, Jacquet C, Courvalin P (2003) Efflux pump Lde is associated with fluoroquinolone resistance in *Listeria monocytogenes*. Antimicrob Agents Chemother 47: 704–708.
11. Jiang X, Zhou L, Gao D, Wang Y, Wang D, et al. (2012) Expression of efflux pump gene *lde* in ciprofloxacin-resistant foodborne isolates of *Listeria monocytogenes*. Microbiol Immunol 56: 843–846.
12. Morvan A, Moubareck C, Leclercq A, Herve-Bazin M, Bremont S, et al. (2010) Antimicrobial resistance of *Listeria monocytogenes* strains isolated from humans in France. Antimicrob Agents Chemother 54: 2728–2731.
13. Glaser P, Frangeul L, Buchrieser C, Rusniok C, Amend A, et al. (2001) Comparative genomics of *Listeria* species. Science 294: 849–852.
14. Kaatz GW, Seo SM (1995) Inducible NorA-mediated multidrug resistance in *Staphylococcus aureus*. Antimicrob Agents Chemother 39: 2650–2655.
15. Buchrieser C, Brosch R, Catimel B, Rocourt J (1993) Pulsed-field gel electrophoresis applied for comparing *Listeria monocytogenes* strains involved in outbreaks. Can J Microbiol 39: 395–401.
16. Tenover FC, Arbeit RD, Goering RV, Mickelsen PA, Murray BE, et al. (1995) Interpreting chromosomal DNA restriction patterns produced by pulsed-field gel electrophoresis: criteria for bacterial strain typing. J Clin Microbiol 33: 2233–2239.
17. Wiegand I, Hilpert K, Hancock RE (2008) Agar and broth dilution methods to determine the minimal inhibitory concentration (MIC) of antimicrobial substances. Nat Protoc 3: 163–175.
18. Tasara T, Stephan R (2007) Evaluation of housekeeping genes in *Listeria monocytogenes* as potential internal control references for normalizing mRNA expression levels in stress adaptation models using real-time PCR. FEMS Microbiol Lett 269: 265–272.
19. Zhang X, Vrijenhoek JE, Bonten MJ, Willems RJ, van Schaik W (2011) A genetic element present on megaplasmids allows Enterococcus faecium to use raffinose as carbon source. Environ Microbiol 13: 518–528.
20. Piddock LJ (2006) Multidrug-resistance efflux pumps - not just for resistance. Nat Rev Microbiol 4: 629–636.
21. Poole K (2005) Efflux-mediated antimicrobial resistance. J Antimicrob Chemother 56: 20–51.
22. Mata MT, Baquero F, Perez-Diaz JC (2000) A multidrug efflux transporter in *Listeria monocytogenes*. FEMS Microbiol Lett 187: 185–188.
23. Kuroda T, Tsuchiya T (2009) Multidrug efflux transporters in the MATE family. Biochim Biophys Acta 1794: 763–768.
24. Morita Y, Kodama K, Shiota S, Mine T, Kataoka A, et al. (1998) NorM, a putative multidrug efflux protein, of *Vibrio parahaemolyticus* and its homolog in *Escherichia coli*. Antimicrob Agents Chemother 42: 1778–1782.
25. Brown MH, Paulsen IT, Skurray RA (1999) The multidrug efflux protein NorM is a prototype of a new family of transporters. Mol Microbiol 31: 394–395.
26. Dridi L, Tankovic J, Petit JC (2004) CdeA of *Clostridium difficile*, a new multidrug efflux transporter of the MATE family. Microb Drug Resist 10: 191–196.
27. McAleese F, Petersen P, Ruzin A, Dunman PM, Murphy E, et al. (2005) A novel MATE family efflux pump contributes to the reduced susceptibility of laboratory-derived *Staphylococcus aureus* mutants to tigecycline. Antimicrob Agents Chemother 49: 1865–1871.
28. Kaatz GW, McAleese F, Seo SM (2005) Multidrug resistance in *Staphylococcus aureus* due to overexpression of a novel multidrug and toxin extrusion (MATE) transport protein. Antimicrob Agents Chemother 49: 1857–1864.
29. Tocci N, Iannelli F, Bidossi A, Ciusa ML, Decorosi F, et al. (2013) Functional analysis of pneumococcal drug efflux pumps associates the MATE DinF transporter with quinolone susceptibility. Antimicrob Agents Chemother 57: 248–253.
30. Hashimoto K, Ogawa W, Nishioka T, Tsuchiya T, Kuroda T (2013) Functionally cloned *pdrM* from *Streptococcus pneumoniae* encodes a Na(+) coupled multidrug efflux pump. PLoS One 8: e59525.
31. Ramos JL, Martinez-Bueno M, Molina-Henares AJ, Teran W, Watanabe K, et al. (2005) The TetR family of transcriptional repressors. Microbiol Mol Biol Rev 69: 326–356.
32. Grkovic S, Brown MH, Skurray RA (2002) Regulation of bacterial drug export systems. Microbiol Mol Biol Rev 66: 671–701, table of contents.
33. Rosenfeld N, Bouchier C, Courvalin P, Perichon B (2012) Expression of the resistance-nodulation-cell division pump AdeIJK in *Acinetobacter baumannii* is regulated by AdeN, a TetR-type regulator. Antimicrob Agents Chemother 56: 2504–2510.

7

Sorting through the Wealth of Options: Comparative Evaluation of Two Ultraviolet Disinfection Systems

Michelle M. Nerandzic[1]*, Christopher W. Fisher[3], Curtis J. Donskey[1,2]

1 Research Service, Louis Stokes Cleveland Veterans Affairs Medical Center, Cleveland, Ohio, United States of America, 2 Geriatric Research, Education and Clinical Center, Cleveland Veterans Affairs Medical Center, Cleveland, Ohio, United States of America, 3 STERIS Corporation, Healthcare Group, Mentor, Ohio, United States of America

Abstract

Background: Environmental surfaces play an important role in the transmission of healthcare-associated pathogens. Because environmental cleaning is often suboptimal, there is a growing demand for safe, rapid, and automated disinfection technologies, which has lead to a wealth of novel disinfection options available on the market. Specifically, automated ultraviolet-C (UV-C) devices have grown in number due to the documented efficacy of UV-C for reducing healthcare-acquired pathogens in hospital rooms. Here, we assessed and compared the impact of pathogen concentration, organic load, distance, and radiant dose on the killing efficacy of two analogous UV-C devices.

Principal Findings: The devices performed equivalently for each impact factor assessed. Irradiation delivered for 41 minutes at 4 feet from the devices consistently reduced *C. difficile* spores by $\sim 3\ log_{10}CFU/cm^2$, MRSA by$>4\ log_{10}CFU/cm^2$, and VRE by $>5\ log_{10}CFU/cm^2$. Pathogen concentration did not significantly impact the killing efficacy of the devices. However, both a light and heavy organic load had a significant negative impacted on the killing efficacy of the devices. Additionally, increasing the distance to 10 feet from the devices reduced the killing efficacy to $\leq 3\ log_{10}CFU/cm^2$ for MRSA and VRE and $<2\ log_{10}CFU/cm^2$ for *C.difficile* spores. Delivery of reduced timed doses of irradiation particularly impacted the ability of the devices to kill *C. difficile* spores. MRSA and VRE were reduced by $>3\ log_{10}CFU/cm^2$ after only 10 minutes of irradiation, while *C. difficile* spores required 40 minutes of irradiation to achieve a similar reduction.

Conclusions: The UV-C devices were equally effective for killing *C. difficile* spores, MRSA, and VRE. While neither device would be recommended as a stand-alone disinfection procedure, either device would be a useful adjunctive measure to routine cleaning in healthcare facilities.

Editor: Peter Setlow, University of Connecticut, United States of America

Funding: This work was supported by a Merit Review grant from the Department of Veterans Affairs to C.J.D. The funders had no role in study design, data collection and analysis, decision to publish, or preparation of the manuscript.

Competing Interests: C.W.F. is an employee of STERIS Corporation. The other authors declare that no competing interests exist.

* Email: michellenerandzi@aim.com

Introduction

Environmental surfaces may play an important role in transmission of healthcare-associated pathogens such as *Clostridium difficile*, methicillin-resistant *Staphylococcus aureus* (MRSA), and vancomycin-resistant *Enterococcus* (VRE) [1–6]. Pathogens are shed onto environmental surfaces and will remain for several days, or possibly months, if the surfaces are not effectively disinfected [1–6]. Unfortunately, several recent studies have demonstrated that environmental cleaning is often suboptimal in healthcare facilities [5–8]. Interventions such as education of housekeeping staff or use of fluorescent markers to provide feedback to housekeepers may result in improved cleaning [5–8]. Yet, despite the promise of improvement in routine cleaning, there remains a demand for novel, automated technologies that are effective against hard to kill *Clostridium difficile* spores, but are also safe and rapid. As a consequence, there has been an upsurge in automated disinfection technologies on the market, many of which have yet to be rigorously evaluated.

Novel ultraviolet disinfection devices are currently on the forefront of burgeoning automated technologies due to the well documented efficacy of ultraviolet-C (UV-C) irradiation for killing bacteria, viruses, and persistent spores [9–15]. The mechanism of killing of microorganisms by UV-C is primarily due to inactivation of DNA and RNA through absorption of photons resulting in formation of pyrimidine dimers from thymine and cytosine [9,12,15]. We previously demonstrated that an automated room disinfection device that utilizes low pressure mercury lamps for emitting UV-C radiation was effective for significantly reducing *C. difficile*, MRSA, and VRE contamination in hospital rooms (Tru-D Rapid Room Disinfection, Lumalier, Memphis, TN, USA) [16]. Similarly, Rutala et al. evaluated the Tru-D device and concluded that it was an efficacious and environmentally friendly method for disinfecting surfaces in healthcare facilities [17]. Here, we performed a side-by-side comparative evaluation of a homologous automated UV-C room disinfection device (Pathogon UV Disinfection System, Steris Corporation, Mentor, Ohio, USA) against the previously tested Tru-D device, in the laboratory setting. For each device, the impact of pathogen concentration, organic load, distance, and radiant dose on killing efficacy was assessed.

Materials and Methods

C. difficile, MRSA, and VRE Strains

Two clinical isolates of C. difficile, MRSA and VRE were studied. The MRSA strains were a pulsed-field gel electrophoresis (PFGE) type USA300 (community-associated) and USA800 (hospital-associated). The VRE strains were a VanA-type isolate (C37) and a VanB-type isolate (C68). The *C. difficile* strains were VA 17, a restriction endonuclease analysis (REA) type BI strain, and VA 11, an REA type J strain.

Preparation of *C. difficile* Spores

Spores were prepared by growth on brain-heart infusion agar (Becton Dickinson, Cockeysville, MD) supplemented with yeast extract (5 mg/ml) and L-cysteine (0.1%) at 37°C under anaerobic conditions as previously described [18]. Spores were stored at 4°C in sterile distilled water until use. Prior to testing, spore preps were confirmed by phase contrast microscopy and malachite green staining to be > 99% dormant, bright-phase spores.

Microbiology

For VRE, MRSA, and *C. difficile* cultures, media included Enterococcosel agar (Becton Dickinson, Cockeysville, MD) containing 20 µg/mL of vancomycin, CHROMagar (CHROMagar, Paris, France) containing 6 µg/mL of cefoxitin, and cycloserine-cefoxitin-brucella agar containing 0.1% taurocholic acid and lysozyme 5 mg/L (CDBA), respectively [19]. Plates containing MRSA or VRE were incubated aerobically at 37°C for 48 hours. *C. difficile* plates were incubated in a Whitley Workstation MG1000 anaerobic chamber (Microbiology International, Frederick, MD) at 37°C for 48 hours.

The UV-C Disinfection Devices

Figure 1A/B is a photograph of the devices. The Pathogon device (1A) is 28 inches wide, 31 inches long, and stands 67 inches tall. The system is a wheeled mobile unit that is controlled remotely by a Windows-based tablet controller. It is placed in the center of the room and commonly touched surfaces are arranged close to the device for optimal exposure to UV-C radiation (i.e. bedrails pulled up, call buttons placed on the bed, tables placed near the device). The device contains motion and heat sensor that are connected to a safety rated relay, aborting the UV-C cycle if someone enters the room during use. The unit has 24, 45 inch low pressure mercury bulbs. Once the operator has exited the room, a pre-programed germicidal dose is chosen based on the dimensions of the room. UV-C radiation penetrates all areas of the room that receive light, but the highest exposure occurs for areas that are in direct line of exposure to the output of the device; areas that are not in direct line of exposure to UV-C may receive radiation that is reflected from the walls and ceiling or from other surfaces in the room.

The Tru-D device (1B) is 72 inches tall and measures 24 inches at the widest portion of the base. It is a wheeled mobile device that is placed strategically in the center of the room just as described above for the Pathogon device. The operator exits the room, closes the door, and places a door sensor on the frame of the door. Continuous monitoring during operation is not required because the sensor triggers automatic discontinuation of the cycle if the door is opened. A handheld remote is used to select either a vegetative cycle that is effective for killing of non-spore forming organisms or a spore cycle that is effective in killing spores. The unit has 28, 36 inch low pressure mercury bulbs. The device contains eight sensors spaced at equal distances on a ring at the top of the device. The sensors measure the amount of UV-C light reflected back to the device. The device automatically ends the cycle when the area reflecting the lowest level of UV-C back to the sensors (i.e. shaded areas in the room) has received an adequate dose.

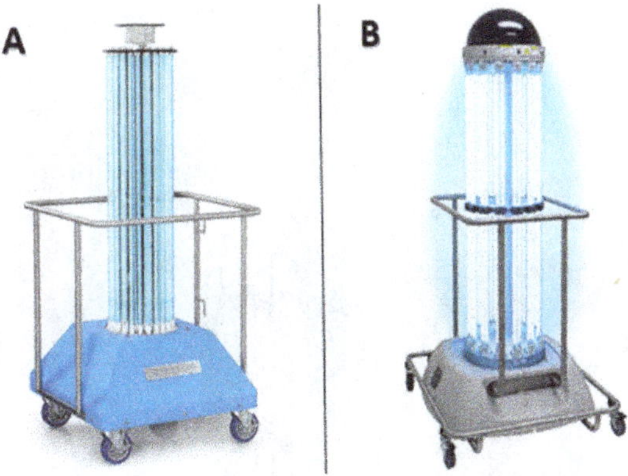

Figure 1. Photographs of the Pathogon (A) and Tru-D (B) devices.

Efficacy of the UV-C Devices for the Reduction of Pathogens on Carriers

The Effect of Pathogen Concentration and Organic Load. Initial experiments were conducted to determine whether pathogen concentration (i.e. concentration of organisms per cm^2) or organic load influenced the disinfection efficacy of the UV-C devices. For pathogen concentration experiments, ten µl aliquots of two strains of *C. difficile* spores, MRSA, and VRE were suspended in sterile phosphate buffered saline (PBS), inoculated onto stainless steel carriers, and then spread to cover a 1 cm^2 area. Organisms were allowed to desiccate onto the carriers under ambient room conditions. For *C. difficile* spores, the inoculum applied to the carriers ranged from 2 to >5 $log_{10}CFU/cm^2$. Previous experiments demonstrated that a reduction in vegetative organisms (MRSA and VRE) was observed after initial desiccation onto the carriers, however no further reduction was observed within the duration of treatment time (author's unpublished data). For each vegetative pathogen, the inoculum applied to the slide was adjusted such that 2 to >5 $log_{10}CFU/cm^2$ were recovered from the positive control specimens after desiccation. For organic load experiments, two strains of *C. difficile* spores, MRSA, and VRE were suspended in either sterile PBS, light organic load (5% fetal calf serum), or heavy organic load (5% fetal calf serum and 5% tryptone) and inoculated onto stainless steel carriers as described above. However, the inoculum was altered such that each carrier yielded 6 log_{10} CFU at baseline.

The carriers were placed on a laboratory bench top 4 feet from the UV-C device, within the direct field of radiation. Baseline slides were left untreated outside of the room (i.e., positive controls). The room dimensions were approximately 10×10 feet. Based on these dimensions, the UV-C devices were run for 41 minutes, as suggested by the manufacturer of the Pathogon device to deliver a spore-killing dose of UV-C within a 10×10 foot range.

To quantify viable organisms, the carriers were submersed in 10 mL of sterile PBS, vortexed vigorously, and dilutions of the

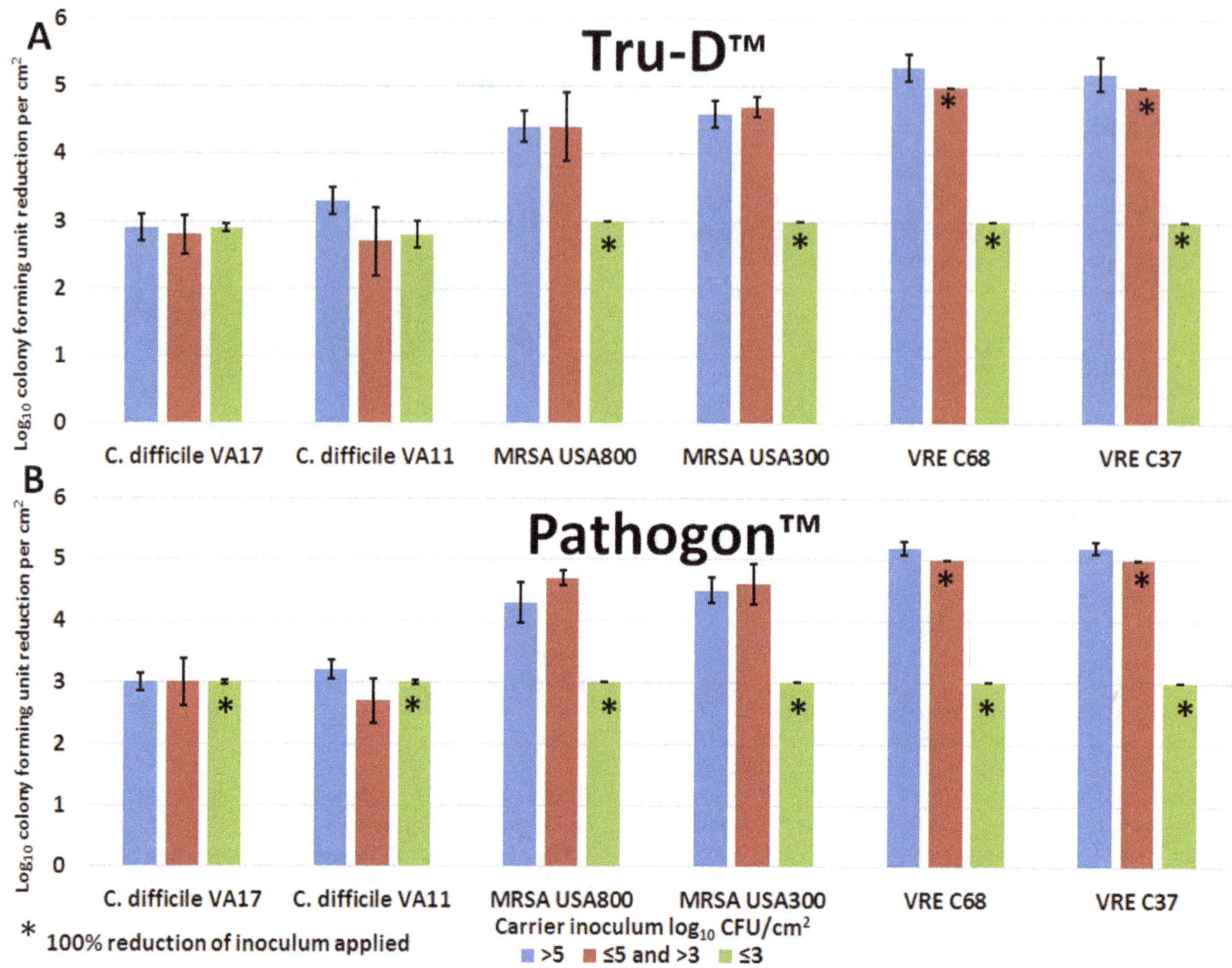

Figure 2. The effect of pathogen concentration on the efficacy of the UV-C devices. The $\log_{10}$CFU reduction/cm^2 of two strains of *C. difficile* spores, MRSA, and VRE inoculated onto carriers. Carriers contained either >5, ≤5 and >3, or ≤3 $\log_{10}$CFU of each pathogen. The carriers were irradiated for 41 minutes at a distance of 4 feet from the Tru-D (2A) or Pathogon (2B) device. The means of the data from experiments conducted in triplicate are presented. Error bars indicate standard error.

suspensions were plated onto selective media as described in *Microbiology*. Following 48 hours of incubation, $\log_{10}$ colony forming unit (CFU) reductions were calculated by comparing the $\log_{10}$CFU recovered from carriers post UV-C disinfection to untreated controls. All experiments were repeated three times.

The Effect of Distance and Indirect Irradiation. The killing efficacies of the UV-C devices were evaluated at increasing distances and shaded from the direct field of radiation. Carriers were prepared and processed as described above in *The Effect of Pathogen Concentration and Organic Load*, however, the organisms were suspended in PBS and altered such that each carrier yielded 6 $\log_{10}$ CFU at baseline. Additionally, carriers were placed 6 inches, 4 feet and 10 feet within the direct field of radiation, and also 4 feet shaded from direct radiation. The UV-C devices were run for 41 minutes.

The Effect of Radiant Dose. The effect of radiant dose on the killing efficacies of the UV-C devices was determined for vegetative organism (MRSA and VRE) and spores (*C. difficile*). Carriers were prepared and processed as described above in *The Effect of Pathogen Concentration and Organic Load*, however, the organism were suspended in PBS and altered such that each carrier yielded 6 $\log_{10}$ CFU at baseline. Carriers were placed 4 feet from the device in the direct field of UV-C and irradiated for either 10, 20, or 40 minutes.

Statistical Analysis

Data were analyzed using STATA 9.0 software (StataCorp, College Station, TX). Continuous data were analyzed using paired *t* tests.

Results

Efficacy of the UV-C Devices for the Reduction of Pathogens on Carriers

The Effect of Pathogen Concentration and Organic Load. Figures 2A and 2B show the mean $\log_{10}$CFU/cm^2 reductions of two strains of *C. difficile*, MRSA, and VRE on carriers after the use of the Tru-D and Pathogon devices, respectively. There was no significant differences between the $\log_{10}$CFU reductions of the two strains of each pathogen tested. Therefore, in subsequent experiments, the two strains were calculated collectively in the mean. The concentration of pathogens on a surface (≤3 to>5 $\log_{10}$CFU) did not have an impact on the killing efficacy of the UV-C devices. Furthermore,

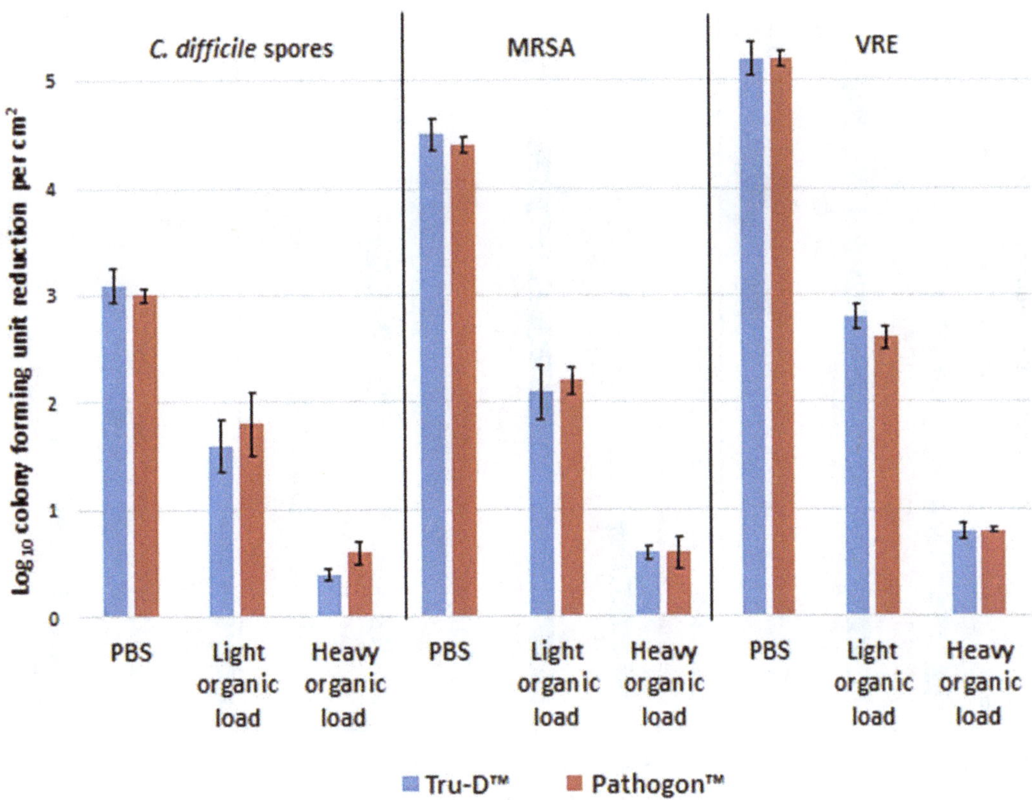

Figure 3. The effect of organic load on the efficacy of the UV-C devices. The log_{10}CFU reduction/cm^2 of *C. difficile* spores, MRSA, and VRE suspended in phosphate-buffered saline (PBS), light organic load (5% fetal calf serum), or heavy organic load (5% fetal calf serum, 5% tryptone). Carriers contained 6 log_{10}CFU of each pathogen. The carriers were irradiated for 41 minutes at a distance of 4 feet from the Tru-D or Pathogon device. The means of the data from experiments conducted in triplicate are presented. Error bars indicate standard error.

the UV-C devices were equally effective for reducing pathogens. There was no significant difference observed in the log_{10}CFU reductions achieved by the Tru-D or Pathogon device for the two strains of each pathogen assessed (Tru-D vs. Pathogon: $P = 0.57$ (*C.difficile*), $P = 1.0$ (MRSA) and $P = 0.97$ (VRE)). Irradiation delivered by the Tru-D and Pathogon devices for 41 minutes (spore killing dose) consistently reduced *C. difficile* spores by ~ 3 log_{10}CFU/cm^2, MRSA by >4 log_{10}CFU/cm^2, and VRE by > 5 log_{10}CFU/cm^2.

Figure 3 shows the effects of a light and heavy organic load on the killing efficacy of the Tru-D and Pathogon device. Both the light (5% fetal calf serum) and heavy organic load (5% fetal calf serum, 5% tryptone) had a significant deleterious impact on the efficacy of the devices. The light organic load decreased the log reductions achieved by the devices to <2 log_{10}CFU/cm^2 for *C.difficile* spores, <2.5 log_{10}CFU/cm^2 for MRSA, and < 3 log_{10}CFU/cm^2 for VRE. The heavy organic load had a more dramatic effect, decreasing the log reduction to <1 log_{10}CFU/cm^2 for each pathogen assessed.

The Effect of Distance and Indirect Irradiation. The germicidal efficacy of UV-C light as a function of distance follows an inverse relationship, as shown in Figure 4. The Tru-D and Pathogon devices achieved analogous log reductions for each distance assessed. Six inches away from the device, vegetative organisms (MRSA and VRE) were completely eliminated (≥ 6 log_{10}CFU/cm^2) and *C. difficile* spores were reduced by > 4 log_{10}CFU/cm^2. As the distance from the device was increased to 4 feet, the log reduction decreased to ≤5 log_{10}CFU/cm^2 for vegetative organisms and ≤3 log_{10}CFU/cm^2 for *C. difficile* spores. Shading the organisms from the direct field of radiation did not have a significant impact on the killing efficacy of the devices. Ten feet from the devices, the log reductions decreased further to ≤3 log_{10}CFU/cm^2 for vegetative organisms and < 2 log_{10}CFU/cm^2 for *C.difficile* spores.

The Effect of Radiant Dose. Figure 5 shows the effect of radiant dose on the killing efficacy of the UV-C devices. There was no significant difference between the killing efficacies of the Tru-D or Pathogon device for each of the timed doses of irradiation delivered. Killing achieved by the UV-C devices was directly proportional to the dose of irradiation delivered. MRSA and VRE were reduced by >3 log_{10}CFU/cm^2 after only 10 minutes of irradiation, while the hardier *C. difficile* spores required 40 minutes of irradiation to achieve a >3 log_{10}CFU/cm^2 reduction.

Discussion

We found that the Tru-D and Pathogon devices were equally effective for killing *C. difficile* spores, MRSA, and VRE in a laboratory setting. Surfaces in a real-world setting contain variable levels of contamination, and in our experience, yield between 4 to <1 log_{10}CFU when cultured (author's unpublished data). Here, we determined that the concentration of pathogens on a surface did not have a significant impact on the killing efficacy of the UV-C devices. Conversely, organic load did significantly reduce the killing efficacy of both devices. These findings are inconsistent with previously published data showing that organic load did not impact the killing efficacy of the Tru-D device [16]. However, the organic load used in the current study was much more

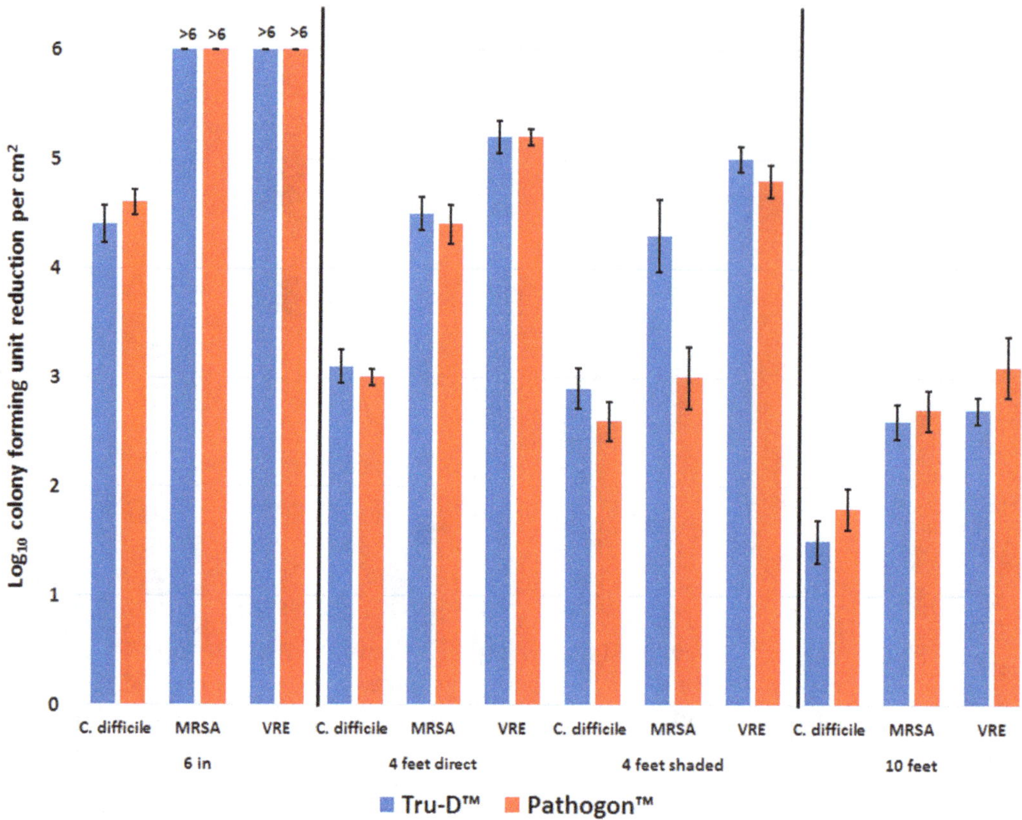

Figure 4. The effect of distance and indirect irradiation on the efficacy of the UV-C devices. The log_{10}CFU reduction/cm^2 of *C.difficile* spores, MRSA, and VRE at increasing distances and shaded from the direct field of radiation delivered by the UV-C device. Carriers contained 6 log_{10}CFU of each pathogen. The carriers were irradiated for 41 minutes at a distance of 6 in, 4 feet, 4 feet shaded, and 10 ft from the Tru-D or Pathogon device. The means of the data from experiments conducted in triplicate are presented. Error bars indicate standard error.

concentrated (5% fetal calf serum or 5% fetal calf serum plus 5% tryptone) than in the previously published study (1% bovine serum albumin). The current study demonstrates that as the matrix of organic load increased from light to heavy, the killing efficacy of the devices decreased, suggesting that UV-C light does not penetrate heavy soils, but may break through lighter organic loads.

Other factors known to impact the delivery of lethal doses of UV-C irradiation are distance from the device and time of radiant exposure [16–17]. The efficacy of the Tru-D and Pathogon devices significantly decreased as distance from the devices increased. We have previously demonstrated that shading from the direct field of irradiation inhibited the lethal effects of the Tru-D device (assessed at 10 feet from the device) [16]. Conversely, here we found that at 4 feet from the devices, shading did not have a significant impact on the killing efficacy of UV-C. For killing of *C. difficile* spores, time of UV-C exposure was of particular importance. While vegetative organisms were reduced by 3 log_{10}CFU after only 10 minutes of exposure, it took 40 minutes to achieve the same level of reduction for *C. difficile* spores. These results suggest that the Tru-D and Pathogon devices are similarly effective at delivering lethal doses of UV-C irradiation under analogous conditions. And as previously demonstrated for the Tru-D device, the Pathogon device may be a promising new environmental disinfection technology that could be a useful adjunct to routine cleaning measures in healthcare facilities.

UV-C devices have important advantages over other disinfection strategies that are effective against *C. difficile* spores. Sodium hypochlorite has corrosive effects on various materials, may irritate the eyes and respiratory tracts of cleaning staff and patients, and the efficacy is dependent on correct application by housekeeping staff [20]. Hydrogen peroxide vapor and hydrogen peroxide dry-mist have been shown to be highly effective in elimination of *C. difficile* spores [20–22]. However, these systems are relatively expensive to operate, a dedicated staff is required, and up to several hours may be required to complete room disinfection [20–22]. In contrast, after the initial purchase of the UV-C device, the cost of operating and maintaining them is minimal (i.e., electricity and annual bulb replacement of ~ $20 each), a dedicated staff is not essential, and a 3 log_{10}CFU reduction in *C. difficile* spores can be achieved in less than an hour. Additionally, UV-C may be less damaging to surfaces than bleach and does not produce emissions that are harmful or irritating to operators.

The Tru-D and Pathogon devices do have some potential limitations. First, because spores require a minimum of 40 minutes of irradiation to achieve significant reductions, it may not be feasible to use the devices in circumstances where rapid turn-over of rooms is required. Second, surface properties and organic debris may potentially inhibit lethal doses of UV-C from killing pathogens. For example, UV-C does not penetrate porous surfaces such as sheets, upholstery and curtains [15]. In our current study, lethal doses of UV-C irradiation were significantly or completely inhibited by organic matrices. Finally, the efficacy of the UV-C devices was reduced at sites further from the devices. Therefore, it is recommended that commonly touched surfaces (e.g., bedside table, call button, telephone) be arranged close to the device for optimal exposure to UV-C radiation.

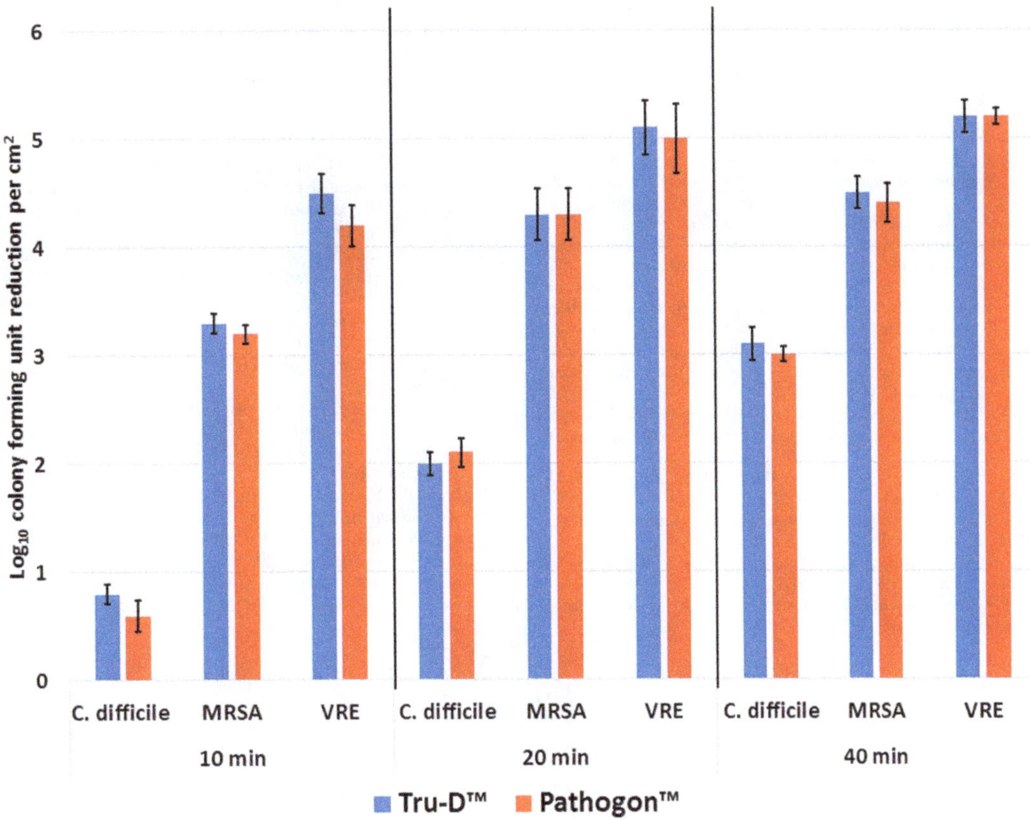

Figure 5. The effect of radiant dose on the efficacy of the UV-C devices. The $\log_{10}$CFU reduction/cm^2 of *C.difficile* spores, MRSA, and VRE after receiving increasing timed doses of irradiation delivered by the UV-C devices. Carriers contained 6 $\log_{10}$CFU of each pathogen. The carriers were irradiated for 10, 20, 40 minutes at a distance of 4 feet from the Tru-D or Pathogon device. The means of the data from experiments conducted in triplicate are presented. Error bars indicate standard error.

Both UV-C disinfection systems performed similarly in a laboratory setting, however, each system has certain advantages and limitations. The Tru-D device is unique in that it uses UV-C light reflected from the walls, ceilings, floors, and items in the room to calculate the amount of irradiation required to deliver a programmed lethal dose for either vegetative or spore-forming pathogens. This feature is advantageous because it delivers a customized dose of irradiation to each room based on the areas of the room that are hardest for light to penetrate. However, this advantage translates into longer cycle times for rooms that inhibit the reflection of light, often increasing the run time to greater than 50 minutes for a 10×10 foot room. On the other hand, the Pathogon device delivers a pre-programmed dose of irradiation that is configured based on the size of the room and type of pathogen contamination suspected, therefore, cycle time does not fluctuate. However, shaded areas of the room may not receive sufficient lethal doses of irradiation. In our opinion, the Pathogon control interface is more user-friendly, however, the next generation of the Tru-D device has been updated with an iPad interface. Last, the Pathogon device is significantly less expensive than the Tru-D device, however after the initial cost, both units require equivalent care and maintenance.

Our study does have some limitations. This study was not designed to address the impact of the UV-C devices on native pathogens found on surfaces in hospital rooms. Nevertheless, the two devices performed equivalently in a laboratory setting, and because the Tru-D device has been previously reported as effective for significantly reducing *C. difficile*, MRSA, and VRE contamination in hospital rooms it can be inferred that the Pathogon device may perform similarly. Further studies are needed to determine whether reductions achieved by these devices translates to reduced rates of infection.

Acknowledgments

We thank Steris Corporation, Mentor, OH for providing the Pathogon unit used in this study free of charge. Steris Corporation did not provide funding for the study, however C.W.F contributed to the study design and editing of the manuscript.

Author Contributions

Conceived and designed the experiments: MMN CWF CJD. Performed the experiments: MMN CWF. Analyzed the data: MMN CWF CJD. Contributed reagents/materials/analysis tools: CWF CJD. Wrote the paper: MMN CJD.

References

1. Goodman ER, Platt R, Bass R, Onderdonk AB, Yokoe DS, et al. (2008) Impact of an environmental cleaning intervention on the presence of methicillin-resistant *Staphylococcus aureus* and vancomycin-resistant enterococci on surfaces in intensive care unit rooms. Infect Control Hosp Epidemiol 29: 593–99.
2. Weber DJ, Anderson D, Rutala WA (2013) The role of the surface environment in healthcare associated infections. Curr Opin Infect Dis 26(4): 338–44.
3. Sitzlar B, Deshpande A, Fertelli D, Kundrapu S, Sethi AK, et al. (2013) An environmental disinfection odyssey: evaluation of sequential interventions to

improve disinfection of Clostridium difficile isolation rooms. Infect Control Hosp Epidemiol 34(5): 459–65.

4. Boyce JM, Potter-Bynoe G, Chenevert C, King T (1997) Environmental contamination due to methicillin-resistant *Staphylococcus aureus* (MRSA): possible infection control implications. Infect Control Hosp Epidemiol 18: 622–7.
5. Eckstein BC, Adams DA, Eckstein EC, Rao A, Sethi AK, et al. (2007) Reduction of *Clostridium difficile* and vancomycin-resistant Enterococcus contamination of environmental surfaces after an intervention to improve cleaning methods. BMC Infect. Dis 7: 61.
6. Hayden MK, Bonten JM, Blom DW, Lyle EA, van de Vijver D, et al. (2006) Reduction in acquisition of vancomycin-resistant *Enterococcus* after enforcement of routine environmental cleaning measures. Clin Infect Dis 42: 1552–60.
7. Hacek DM, Ogle AM, Fisher A, Robicsek A, Peterson LR (2010) Significant impact of terminal room cleaning with bleach on reducing nosocomial *Clostridium difficile*. Am J Infect Control 38(5): 350–3.
8. Carling PC, Parry MF, Bruno-Murtha LA, Dick B (2010) Improving environmental hygiene in 27 intensive care units to decrease multidrug-resistant bacterial transmission. Crit Care Med 38: 1212–4.
9. Conner-Kerr TA, Sullivan PK, Gaillard J, Jones RM (1998) The effects of ultraviolet radiation on antibiotic-resistant bacteria *in vitro*. Ostomy Wound Manage 44: 50–6.
10. Griego VM, Spence KD (1977) Inactivation of *Bacillus thuringiensis* spores by ultraviolet and visible light. Appl Env Microbiol 35: 906–10.
11. Hercik F (1936) Action of ultraviolet light on spores and vegetative forms of *Bacillus megatherium* sp. J Gen Physiol 20: 589–94.
12. Setlow P (2006) Spores of *Bacillus subtilis*: their resistance to and killing by radiation, heat and chemicals. J Appl Microbiol 101: 514–25.
13. Setlow P (2001) Resistance of spores of *Bacillus* species to ultraviolet light. Environ Mol Mutagen 38: 97–104.
14. Thai TP, Keast DH, Campbell KE, Woodbury MG, Houghton PE (2005) Effect of ultraviolet light C on bacterial colonization in chronic wounds. Ostomy Wound Manage 51: 32–45.
15. Owens MU, Deal DR, Shoemaker MO, Knudson GB, Meszaros JE, et al. (2005) High-dose ultraviolet C light inactivates spores of *Bacillus subtilis* var. niger and *Bacillus anthracis* Sterne on non-reflective surfaces. Appl Biosafety 10: 240–7.
16. Nerandzic MM, Cadnum JL, Pultz MJ, Donskey CJ (2010) Evaluation of an automated ultraviolet radiation device for decontamination of *Clostridium difficile* and other healthcare-associated pathogens in hospital rooms. BMC Infect Dis 10: 197.
17. Rutala WA, Gergen MF, Weber DJ (2010) Room decontamination with UV radiation. Infect Control Hosp Epidemiol 31(10): 1025–9.
18. Sorg JA, Sonenshein AL (2008) Bile salts and glycine as cogerminants for *Clostridium difficile* spores. J Bacteriol 190(7): 2505–2512.
19. Nerandzic MM, Donskey CJ (2009) Effective and reduced-cost modified selective medium for isolation of *Clostridium difficile*. J Clin Microbiol 47(2): 397–400.
20. Barbut F, Menuet D, Verachten M, Girou E (2009) Comparison of the efficacy of a hydrogen peroxide dry-mist disinfection system and sodium hypochlorite solution for reduction of *Clostridium difficile* spores. Infect Control Hosp Epidemiol 30: 507–14.
21. Boyce JM, Havill NL, Otter JA, McDonald LC, Adams NM, et al. (2008) Impact of hydrogen peroxide vapor room decontamination on *Clostridium difficile* environmental contamination and transmission in a healthcare setting. Infect Control Hosp Epidemiol 29: 723–9.
22. Boyce JM (2009) New approaches to decontamination of rooms after patients are discharged. Infect Control Hosp Epidemiol 30: 515–7.

8

Contribution of S-Layer Proteins to the Mosquitocidal Activity of *Lysinibacillus sphaericus*

Mariana Claudia Allievi, María Mercedes Palomino, Mariano Prado Acosta, Leonardo Lanati, Sandra Mónica Ruzal*, Carmen Sánchez-Rivas

Departamento de Química Biológica, Facultad de Ciencias Exactas y Naturales, Universidad de Buenos Aires, IQUIBICEN-CONICET, Buenos Aires, Argentina

Abstract

Lysinibacillus sphaericus strains belonging the antigenic group H5a5b produce spores with larvicidal activity against larvae of *Culex* mosquitoes. C7, a new isolated strain, which presents similar biochemical characteristics and Bin toxins in their spores as the reference strain 2362, was, however, more active against larvae of *Culex* mosquitoes. The contribution of the surface layer protein (S-layer) to this behaviour was envisaged since this envelope protein has been implicated in the pathogenicity of several bacilli, and we had previously reported its association to spores. Microscopic observation by immunofluorescence detection with anti S-layer antibody in the spores confirms their attachment. S-layers and BinA and BinB toxins formed high molecular weight multimers in spores as shown by SDS-PAGE and western blot detection. Purified S-layer from both *L. sphaericus* C7 and 2362 strain cultures was by itself toxic against *Culex sp* larvae, however, that from C7 strain was also toxic against *Aedes aegypti*. Synergistic effect between purified S-layer and spore-crystal preparations was observed against *Culex* sp. and *Aedes aegypti* larvae. This effect was more evident with the C7 strain. *In silico* analyses of the S-layer sequence suggest the presence of chitin-binding and hemolytic domains. Both biochemical characteristics were detected for both S-layers strains that must justify their contribution to pathogenicity.

Editor: Guido Favia, University of Camerino, Italy

Funding: Support was provided by Consejo Nacional de Investigaciones Científicas y Técnicas CONICET PIP0229 and Agencia Nacional de Promoción Científica y Tecnológica ANPCyT PICT-2012-0789. The funders had no role in study design, data collection and analysis, decision to publish, or preparation of the manuscript.

Competing Interests: The authors have declared that no competing interests exist.

* Email: sandra@qb.fcen.uba.ar

Introduction

Lysinibacillus sphaericus, together with *Bacillus thuringiensis* var. *israelensis*, represents the best ecological insecticide against mosquitoes and an environmental friendly alternative to chemical insecticides. These Gram-positive bacteria synthesize spores together with crystal-proteins, Cry and/or Cyt toxins, both being a very stable bioinsecticide [1]. In *B. thuringiensis* var. *israelensis*, crystal Cry and Cyt proteins are involved in the recognition of the insect target, the disruption of the membrane and finally the hemolysis. This last activity has been shown to be related to the low appearance of resistant mosquitoes. Moreover, other less stable toxins are produced during the cells vegetative growth: Vip for *Bacillus thuringiensis* [2] [3] or Mtx for *L. sphaericus* [4].

Lysinibacillus sphaericus, formerly *Bacillus sphaericus,* was renamed due to the presence of lysine and aspartic acid in the composition of their peptidoglycan [5]. These are a heterogeneous group of gram positive sporulating *Bacillus* some of which are entomopathogenic against mosquito larvae [4]. Hybridization studies of their DNA lead to classify them in 5 groups (I to V), but the most toxic strains belong to the homology group IIA [6] [7] and flagellar serotype H5a5b being 2362 the reference strain. These bacteria present particular metabolic traits: although they do not use hexoses or pentoses as carbon sources [8], they are able to use the amino-sugar N-acetylglucosamine, the monomer of chitin, and posses an active PTS transporter (Phosphoenolpyruvate phosphotransferase system) essentially implicated in its utilization [9] [10]. However, in contrast to *B. thuringiensis* strains [11], no chitinase activity has been detected in these bacteria.

L. sphaericus spores present an important exosporium allowing spore and crystals to remain firmly associated [12]. These crystalline inclusions are composed by two proteins named BinA and BinB, which can form dimer and/or associate in mixed proportions [13]. During the vegetative growth phase *L. sphaericus* strains produce several toxic proteins named Mtx1, 2, 3 [14] [15]. Besides being very efficient in synergic experiments with BinA-BinB, Mtx proteins are not synthesized during the sporulation phase and are degraded by proteases synthesized during this period. In fact, recombinants containing the cloned *mtx1* gene under a *bin* promoter allow Mtx1 synthesis during sporulation, but again the protein was rapidly degraded while sporulation proceeds [16].

While *B. thuringiensis* var. *israelensis* spore-crystal preparations are highly active against *Aedes*, *Culex*, *Anopheles* and *Simulium* [1] [4] those from *L. sphaericus* are essentially active against *Culex* and *Anopheles* species. This complementarity in behavior and targets has been exploited by using mixed preparations and recombinants containing the cloned toxic genes from *L. sphaericus* [17] [18] [19] into *B. thuringiensis*. However, *B. thuringiensis* is reported to be highly sensitive to the presence of chemical and metal contaminants, while *L. sphaericus* shows a better persistence in contaminated ponds [4]. Also several reports have shown the ability of this bacterium to survive [20] [21] and bioabsorb metal at concentrations otherwise toxic [22] and this property is linked to the presence of their S-layer envelope.

Moreover, it is worthwhile mentioning that the S-layer from several bacteria have also been implicated in their pathogenicity; this is especially so for *Bacillus* species as *B. cereus*, *B. anthracis* [23] [24], *Paenibacillus alvei* [25], different strain-variants from *B. thuringiensis* [26] [27] [28] and new isolated from *L. sphaericus* [29]. Altogether these properties have increased the interest in *L. sphaericus* strains and the drive to find new isolates of this species.

The reference strain 2362 is endowed with a high molecular weight S-layer (120 kDa) [30] that was also present in spore preparations [22]. Sequence based analysis performed for this protein allowed to predict the presence of hemolytic and chitin binding domains that might contribute to their entomopathogenicity. Such properties are associated with the pathogenicity of the Cry proteins from *B. thuringiensis* [31]. This led us to investigate the insecticidal properties of the S-layer from two *L. sphaericus* strains: the reference strain 2362 and C7 (a new isolate for which a higher insecticidal activity of the mixture of spore-crystals preparations has been observed). The mosquitocidal activity against *Culex* and *Aedes* larvae of spore-crystals and purified S-layers was assayed. Hemolytic activity and chitin binding properties of the S-layer were analyzed to characterize their influence on the mosquitocidal activity of *Lysinibacillus sphaericus*.

Materials and Methods

Bacterial strains, antibodies and media

L. sphaericus 2362 was obtained from Institut Pasteur. C7 strain was isolated from a crocodile lagoon in Cuba. Flagellar antisera (provided by A. Delécluse, Institut Pasteur) and phages typing (provided by A. Yousten) characterizations allowed us to classify them in the same group as 2362 (H5a, H5b). In addition, the antibody against the S-layer from 2362 (provided by L. Lewis) [30] recognizes the S-layer from the C7 strain. The anti-BinA or BinB were kindly provided by JF. Charles (Institut Pasteur). Strains were grown in LB (Luria Bertrani) or NYSM (Nutrient Yeast Extract Salt Medium) medium [32] as indicated in the text in aerated conditions at 32°C. For spores preparations growth was on solid NYSM medium at 32°C and spore-crystals were collected after 3 day incubation.

Phage typing assay

The phage-typing of the isolated strain was performed as described previously and the phages used were provided by A. A. Yousten [33]. Bacteria were grown on NYSM agar at 30°C. Phages 1A and 2 were propagated on strain *Bacillus sphaericus* SSII-1; phages 12 and SST were propagated on strain Kellen K; and phages 4 and 5 were propagated on strain 1593. For typing, bacteria were seeded into NYSM soft agar overlays and 20 μl of diluted phage suspension was spotted onto the surface. Plates were incubated at 30°C for 18 h. The production of individual plaques at the same dilution or at a 10-fold more concentrated dilution than those producing individual plaques on the normal phage propagating strain was noted as a positive result. Patterns of test results were compared with those for standard strains 1593 and 2362 (phage group 3), SSII-1 (phage group 2), and Kellen K (phage group 1).

Spores-crystal preparations

Strains were grown on solid NYSM medium and incubated 2–3 days at 32°C. Plates were then washed with 10 ml of 1 M NaCl, washed four times with milliQ H_2O and suspended in 10 mM sodium acetate pH 4.5 that prevents germination [34]. Dry-pellets containing a mixture of spores-crystals were stored at −20°C.

Cleaning off S-layer protein from spores

Spores-crystals from 3-day cultures were scrapped and washed with 10 ml of 1 M NaCl, then in deionized water, in 0.5 mM EDTA which eliminates the S-layer, and finally four times in deionized water [22]. These spores were used to evaluate synergy with purified S-layer proteins.

S-layer purification from cultures

Lysinibacillus sphaericus 2362 and C7 strains were grown in LB medium. Cells from 100 ml exponential cultures ($OD_{600} = 1$) were harvested and washed once with PBS. The S-layer proteins were extracted from the cells by cationic substitution by resuspending in 10 ml of 6 M LiCl, vortexing and incubating for 30 min at room temperature. After centrifugation at 15,000×g for 15 min, the supernatant was collected and dialyzed against 10 mM $CaCl_2$ overnight at 4°C. After centrifugation (10,000×g for 20 min), the pellets containing the S-layers were resuspended in deionized water and stored at −20°C [22].

Protein analysis

Spores proteins were alkali-solubilized (30 min at 37°C in 0.05 N NaOH) [17]. After alkaline extraction the preparations were centrifuged at 12,000×g 5 min to produce Pellet (P) and supernatant (S). The spores and S-layers preparations were heated at 90°C 5 min in loading buffer (10% glycerol, 4% SDS, 4 M urea, 2% β-mercaptoethanol, and 0.05% bromophenol blue) and subjected to electrophoresis in 12.5% SDS-PAGE. Gels were stained with Coomassie Brilliant Blue. The same amount of material was loaded in each well (10^7 CFU of spores and 0.5 μg of S-layer protein).

For western blot analysis, gels were electrotransferred with a Semi-dry Blotter (Amersham Biosciences) to PVDF membranes (Macherey-Nagel, Germany) soaked with polyclonal anti-rabbit antibodies against S-layer or BinA or BinB (diluted 1:2000 for S-layer and 1:50000 for BinA and BinB), and visualized with biotin-conjugated anti-rabbit followed by streptavidin-HRP conjugate (Pierce). Chemiluminescence was detected with luminol substrate (ECL from Sigma). Images were obtained with a Fuji LAS1000 digitalizer.

Detection of S-layer protein in spores by immunofluorescence

An immunofluorescence assay was performed in order to detect the presence of S-layer protein probably associated to the exosporium as described in *B. anthracis* [35] [36]. Spores from 3-day cultures were scrapped, washed with 10 ml of 1 M NaCl and resuspended in 10 mM sodium acetate pH 4.5 that prevents germination [34]. After centrifugation and double washing with PBS, the pellets were resuspended in PBS. A 20 μl (10^8 CFU) spore sample was then mixed with 0.5 μl antibody against S-layer and incubated 1 h at 25°C at a low shaking speed. The sample was then centrifuged, washed twice with PBS, resuspended in 100 μl PBS and subjected to the second antibody. 1.5 μl of antibody [Alexa fluor 647 goat anti-rabbit IgG (H+L) (Invitrogen)] was added and the sample was incubated for 30 min at 25°C. Spores were then pelleted, washed twice with PBS, and resuspended in

50 μl PBS. A drop was applied to a glass slide and overlaid with a coverslip. Immunofluorescence was observed with fluorescence microscopy on an Epifluorescense LED Axio Scope A1 model Microphot microscope (Carl Zeiss), N-Achroplan 100x/1.25 Oil PH3, filters mCherry FS64HE. The images were taken with a EOS T3 1100D digital camera (Canon) (10,1 Mega Pixel CMOS, 3.888×2.592 pixel eff, lens 18–55 mm).

Larvicidal assays

Experiments were carried with autochthonous seasonal species of *Culex sp* and *Aedes aegypti* larvae supplied by "Grupo de estudios de Mosquito" of the University of Buenos Aires [37]. Eggs were spread in dechlorinated water, fed with yeast extract, maintained at 28°C with daily illumination. After 4 days the majority of larvae reached 2–3rd instar development and were ready to use.

Petri dishes (10 cm diameter) containing 20 ml dechlorinated water and 20 larvae (2nd and 3rd instar) were inoculated with different concentrations of spores, S-layer protein or both as indicated in Tables 1 and 2 and were incubated at room temperature [38]. Larval mortality was evaluated 24 and 48 hours after exposure and compared to controls. The concentrations giving 50% mortality (LC50) was expressed as number of spores (CFU/ml) or S-layer proteins (μg/ml) after subtracting the mortality of controls. Assays were performed in duplicate and repeated three times. A representative experiment is shown.

Binding of S-layers to chitin compounds

To determine if the S-layer proteins recognize the chitin present in the cuticulum of insects, the binding specificity of these proteins for insoluble polysaccharides was used as follows: 50 μg of purified proteins were mixed with 10 μg of polysaccharide in a total volume of 500 μl PBS and incubated for 16 h at 32°C. The samples were centrifuged at 3,000×g for 3 min to separate free protein from substrate bound protein. Protein concentrations in the supernatant were determined using Bradford (BioRad). The percentage of adsorbed protein was calculated by substracting the free protein amount from the total protein in the sample. Polysaccharides were also incubated with bovine serum albumin, BSA, as negative control.

The polysaccharides used for this experiment were powder chitin, crab chitosan and fish chitosan resuspended in H_2O deionised and sonicated (to solubilise particles). These assays were performed at least six or more times as independent experiments in duplicate each time.

Hemolytic assays

Hemolysis is a way to evaluate membrane perturbation, a feature present in entomopathogenic toxins [31]. % Hemolysis was determined as previously described [39]. Briefly, sheep red blood cells were separated by centrifugation (1,000×g for 5 min) and resuspended to 1% with PBS. The final volume of the reaction mixture was 1 ml containing 0.5 ml of washed blood cells and various concentrations of S-layer protein (0.5–150 μg) in the same buffer and incubated at 37°C for 30 min. After centrifugation at 1,200×g for 5 min hemolysis was quantified measuring the absorbance of the supernatant at 405 nm. Positive control was 100% hemolysis after incubation of the same volume of sheep red blood cells with dechlorinated H_2O. Negative controls were red blood cells incubated with buffer. As a specificity control, antibodies (1:200 dilution) against S-layer were incubated for one hour along with the S-layer protein. Subsequently, the mixture was tested for hemolysis. These assays were performed five or more times as independent experiments in duplicate each time.

Statistical Analysis

Statistical significance was evaluated by the Mann Whitney-U test for nonparametric data by Infostat software [40]. $P<0.05$ was considered to be statistically significant.

Results

Spore-crystals preparations contain S-layers

In a search for *Lysinibacillus sphaericus* strains with enhanced insecticidal activity, one particular isolate C7 was obtained. The C7 strain belongs to the same phage group 3 as the 2362 control. Its spore-crystals preparations were at least 100-fold more active than 2362 against *Culex pipiens* 2nd and 3rd instars larvae (LC_{50} as μg total protein/assay were 1500±340 and 15±2 for 2362 and C7 respectively). Moreover, the strain had the same metabolic features as the reference strain 2362: growth with N-acetylglucosamine but not with glucose as carbon source, antigenic and phage-typing characteristics, presence of Mtx and Bin in spores. Since no clear difference in the yield of Bin toxins present in spores was observed, we undertook the analysis of the S-layer content of spores. We had previously observed that S-layers remain associated to spores [22], but these proteins could be a remnant of the vegetative envelope, or a constituent of the spore itself. For this purpose two approaches were used: Western Blot and immunofluorescence. Spore-crystal preparations were analyzed and detected using Western Blot. In this condition several high molecular weight bands (>120 kDa) are usually observed. This is generally attributed to associations between the toxic components BinA and BinB [13]. Western blot analyses with antibodies against Bin toxins and S-layer protein reveal the latters presence in the high molecular weight band (HMW up to 130 kDa), confirming that the S-layer and Bin components remain closely associated (Fig. 1A and B). Furthermore, the more toxic strain, C7, produced spores containing higher amounts of S-layer protein, which was also recognized by the specific antibody from 2362.

An immunofluorescence assay was used to directly detect the presence of S-layer protein in spores in association with the exosporium as has been described in *B. antracis* [35] [36] S-layer protein outside the spores, probably in association with the exosporium, was detected through direct observation of S-layer protein in non permeabilized spores (Fig. 2).

Larvicidal activity of S-Layer proteins from vegetative cultures and spores

S-layer extracted by LiCl and analyzed by SDS-PAGE and Western Blot showed a band running as 120–130 kDa (Fig. 1C and D). These purified S-layers proteins from C7 and 2362 strains were used to feed *Culex sp.* larvae as described in Materials and Methods. The dose response mortalities were obtained. The S-layer from C7 was as efficient as that of 2362 against *Culex* larvae (LC_{50} 2.2 and 2.0 μg/ml, respectively) (Table 1). Surprisingly, when assayed against *Aedes aegypti*, only that of C7 showed a substantial activity (10-fold more toxic C7 than 2362) (Table 1). This led us to investigate possible interactions of the S-layers and spores in the larvicidal activity.

Synergy of spores and S-layers in the larvicidal activity

For this purpose individual batches of spores-preparations from 2362 and C7 strains were analyzed in order to determine their activity in the same batch of *Culex sp.* and *Aedes aegypti* larvae.

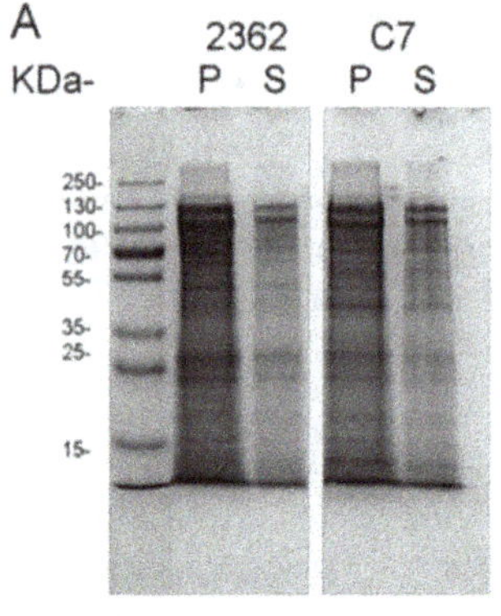

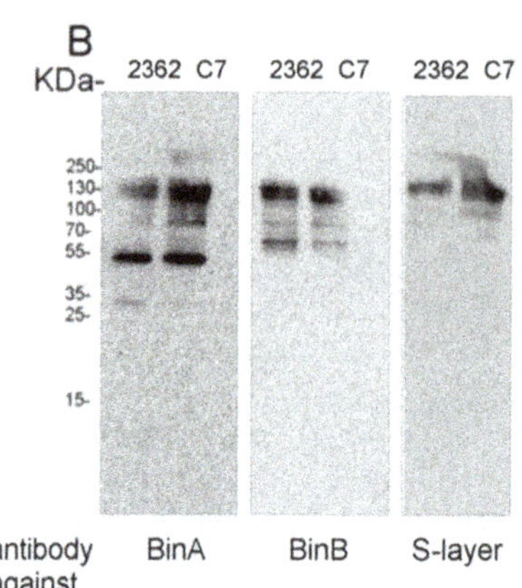

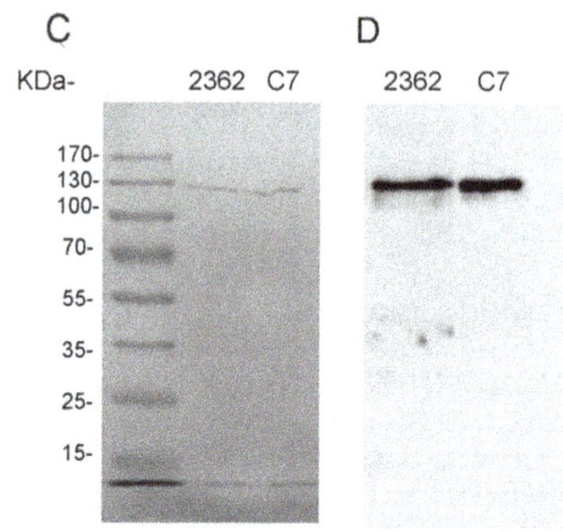

Figure 1. Detection of S-layers in spore-preparations and vegetative cultures. Pellet (P) and supernatant (S) fractions of spore-preparations from 2362 and C7 strains, obtained from the alkaline treatment described in Materials and Methods, were subjected to SDS-PAGE 12.5% (A) and Western Blot analysis for detection with specific antibodies against BinA or BinB or S-layer proteins (B). Purified S-layers from 2362 and C7 strains were obtained from vegetative cultures as described in Materials and Methods. 6 µg of each preparation were subjected to SDS-PAGE 12.5% electrophoresis (C) and analyses by Western Blot with specific antibody against the 2362 strain's S-layer (D).

This caution is necessary since batches of mosquitoes vary between seasons and origin.

In this condition we determined the LC_{50} for each preparation (spores and S-layers) and strain (2362 and C7). In this assay (Table 1), spores from C7 were 2 and 25 times more toxic against *Culex* and *Aedes* respectively than those from strain 2362.

In order to check for synergistic effects S-layer was removed from spores with EDTA [22]. Using concentrations of spores and S-layers giving mortality below 50%, a synergistic evaluation between them was assayed (Table 2). As shown, the mixtures (spores and S-layers), at sub-lethal concentrations, had 2 to 3 times higher activity than each component on its own, indicating that a synergistic effect between spores and S-layers took place.

These findings demonstrate the importance of the S-layer protein present in spores preparations in the larvicidal activity of these bacteria.

Sequence based analysis for putative functionalities that could support larvicidal activity in the S-layer protein

In order to characterize if a biochemical support for *Lysinibacillus sphaericus* pathogenicity is provided by it S-layer, a search of possible domains present in this protein was performed. Using different programs that are described in the Material S1 (SMART, EMBOSS Matcher Pairwise Sequence Alignment, and Clustal-O) the sequences of SlpC proteins were analyzed. AAA50256.1, surface layer protein of *Lysinibacillus sphaericus* 2362, was used. When we analyzed the SMART database to predict functionality and physical distribution, we found a similar arrangement of surface layer homology domains (SLH) and Internal Repeats as that observed for the chitinase ChiW (BAM67143.1) of *Paenibacillus sp.* FPU-7 (Fig. S1A) [41]. Homologue architecture was observed between the two proteins: three SLH domains and two Internal Repeats. A local similarity to GH 18 chitinase domains was found for the internal repeats (Fig. S1B and S1C) using EMBOSS Matcher program, which identifies local similarities in two input sequences using a rigorous algorithm and allows for the modification of default substitution scoring matrices (BLOSUM) for sequence alignment between distantly related proteins. Using Clustal-O, a global alignment tool for multiple proteins sequences, homology with ACA38715.1 hemolysin-type calcium-binding domain-containing protein (874 aa, from the complete sequence of *Lysinibacillus sphaericus* C3–41 accession number NC_010382), was obtained with a high score alignment (Fig. S1D). The predicted functionalities were then investigated using *in vitro* assays for chitin binding and hemolytic activity in these S-layers preparations.

Chitin binding assays of S-layers

An *in vitro* assay was performed in order to check whether the pathogenicity of these S-layers involved the recognition of the

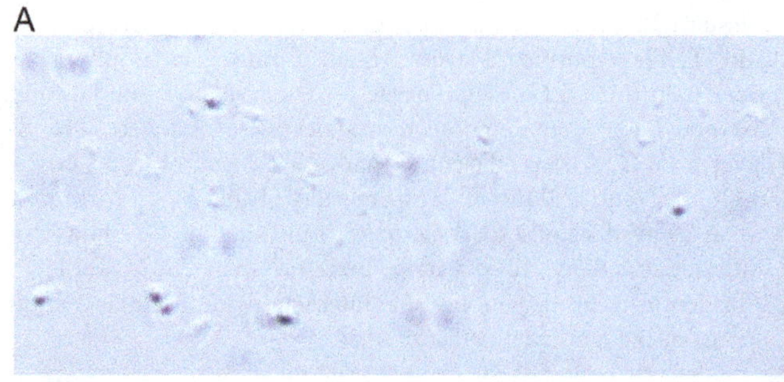

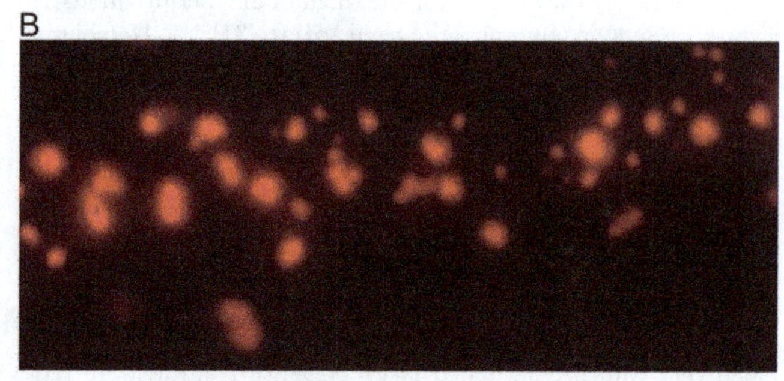

Figure 2. Detection of S-layers in spores by immunofluorescence. Spores preparations from C7 (A and B) were cleaned as described in Materials and Methods. A: Microscopic white light observation. B: Fluorescent (652 nm excitation and 668 nm emission) observation of the S-layer in

Table 1. Larvicidal activity of S-layers and spores from 2362 and C7 strains against *Culex sp* and *Aedes aegypti* larvae.

	Culex sp.	*Aedes aegypti*
S-layers from	**LC50 (μg/ml)**	**LC50 (μg/ml)**
2362	2.2 (0.2)	>80
C7	2.0 (0.2)	8.0 (0.5)
Spores from	**LC50 (cfu/ml)**	**LC50 (cfu/ml)**
2362	1.00×10^3 (0.02×10^3)	1.20×10^6 (0.06×10^6)
C7	0.50×10^3 (0.02×10^3)	5.00×10^4 (0.25×10^4)

S-layer and spores preparations were added with 20 *Culex sp.* or *Aedes aegypti* larvae as indicated in Material and Methods. After 48 h lethality was evaluated. Assays were performed with duplicate and repeated three times and a representative experiment is shown. In parenthesis, standard deviation.

chitin present in the cuticulum of their target. In a first assay, SDS-PAGE analysis of S-layers was performed in the presence or absence of colloidal chitin. Mobility retardation was observed for the S-layer but not for the bovine serum albumin (BSA) protein used as control (data not shown). To quantify the chitin binding property, a more sensitive assay was performed mixing purified S-layer proteins with insoluble chitosan, crab chitosan or colloidal chitin as described in Materials and Methods. As shown in Figure 3, the binding of S-layer proteins by these compounds featured similar efficacy, while this was not the case for BSA. Besides, the presence of N-acetylglucosamine (NAG), the monomer of chitin, inhibited the binding thus ensuring the specificity of the association between S-layers and chitin. Although binding to chitin was observed, we could not detect chitinase activity in the S-layer for these substrates.

Hemolytic activity of S-layers

We assayed hemolytic activity with purified S-layer preparation to determine the amount of hemoglobin release by the lyses of sheep red blood cells to obtain a direct quantitative measurement (see Materials and Methods). As shown in Figure 4, unlike BSA, the protein used as control, both S-layer proteins had hemolytic activity. We verified the specificity of the hemolysis by neutralizing the S-layer with specific antibodies that decreased the hemolytic activity to a 50%. We found that hemolytic activity was inhibited when N-acetylglucosamine was added, pointing to sugar recognition (data not shown).

Discussion

The pathogenic role of the S-layer has already been shown for several *Bacillus* species like *B. cereus* and *B. anthracis* [23] [24], *Paenibacillus alvei* [25], different strains from *B. thuringiensis* [26] [27] [28] and *Lysinibacillus sphaericus* [29]. However, no function has been assigned to it other than its role as an adhesion factor for *B. anthracis* [23].

In the present work we confirm our previous observation [22] that spores from *L. sphaericus* retain S-layer proteins. Western Blot analyses of spore preparations from 2362 and C7 strains revealed associations of S-layer and Bin proteins (Fig. 1B). Immunofluorescence was used to visualize the S-layer (Fig. 2) surrounding spores. Since Bin proteins are deposited on the exosporium, we suspect that S-layer proteins are also associated to this structure. In fact, in *B. anthracis* the analysis of the exosporium shows the presence of several envelope proteins, including S-layers [35] [36] [42]. On the other hand, unlike the Mtx toxins present during the exponential growth, the S-layers were not degraded [16] by the proteolytic activity of the sporulation process, suggesting that their presence in spores was not fortuitous and that their location on the exosporium must contribute to this.

Moreover, the S-layer protein present in vegetative cultures (in the absence of spores), was mosquitocidal by itself as was the S-layer protein from new isolates reported by Lozano *et al* [29]. Concerning the larvicidal activity, it can be remarked that the S-layer from C7 was more active than that from 2362 against mosquitoes, in particular against *Aedes aegypti*, a species poorly sensitive to *L. sphaericus* toxins (LC50 was 10-times lower as shown in Table 1). We also reported a synergistic effect between cleaned spores (devoid of S-layer by EDTA treatment) and S-layers. When both are present, their activity is higher than that of the individual preparations (Tables 2).

The *slpC* gene from 2362 is 100% homologous to several entomopathogenic strains from the same antigenic group and especially to C3-41, a completely sequenced strain [43]. This led us to investigate functionalities within the protein sequence that could account for pathogenicity. Sequence based analysis reported in Material S1 revealed two possible biochemical domains: a chitin

Table 2. Synergy between spores and S-layers against *Culex* sp. Larvae.

	2362	**C7**
Mortality with Spores	32%	20%
Mortality with S-layer	2%	5%
Mortality with both	62%	85%

S-layer and spores preparations were added separately or mixed at sub-lethal concentrations with 20 *Culex sp.* larvae as indicated in Material and Methods. After 24 h lethality was evaluated. Only the % of died larvae are reported. The concentration of spores or S-layer protein individually or together per experiment were 500 CFU/ml and 0.25 μg for 2362 strain and 50 CFU/ml and 0.6 μg for C7 strain.

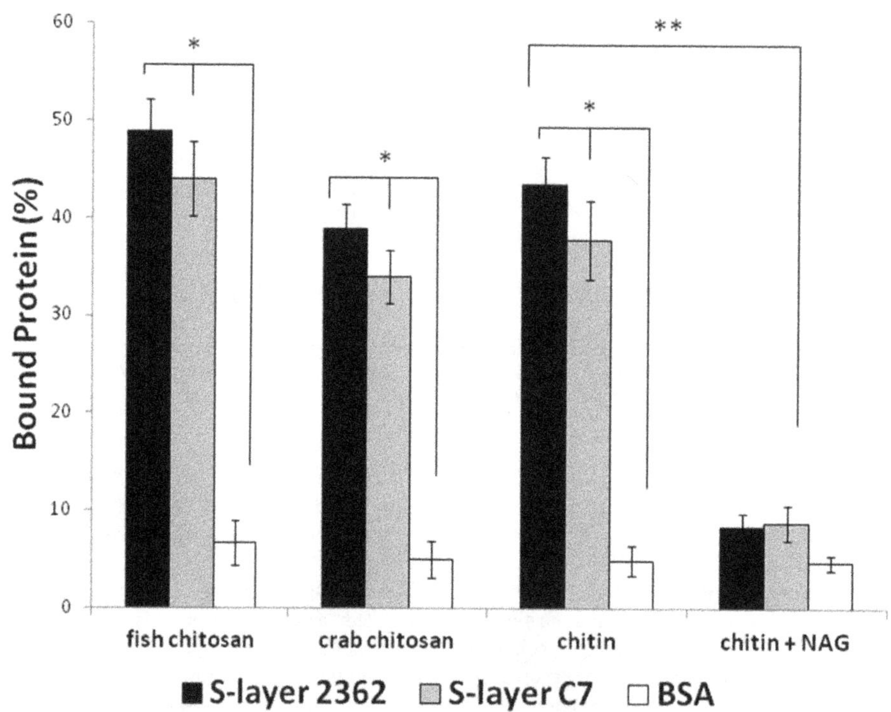

Figure 3. Binding assays of S-layer to chitin derivatives. Insoluble preparations of chitosan, crab chitosan, chitin (10 µg) were vigorously mixed with 50 µg of the different proteins and allow standing for 16 h at 32°C: S-layers from either 2362 or C7 strains, Bovine serum albumin (BSA). Free protein was determined and % bound calculated. Six independent experiments with duplicate samples were performed. Bars show the mean ± SD. Mann Whitney-U test was used to determine statistically significant differences between S-layers proteins and BSA control protein. *, $P<0.05$. With colloidal chitin, N-acetylglucosamine (NAG) (25 mM) was also added to verify binding inhibition. Five independent experiments with duplicate samples were performed. Mann Whitney-U test was used to determine statistically significant differences with and without NAG. **, $P<0.05$.

binding domain and a hemolytic domain that may contribute to the entomopathogenicity.

The chitin binding capacity (Fig. 3) and hemolytic activity (Fig. 4) were found to be present in the S-layer protein preparations and are both indicative of pathogenicity. Proteins with chitin-binding domains and chitinase activity have been described as pathogenicity factors in other species [44] [45]. In fact, the introduction of a chitinase gene from *B. thuringiensis* in a

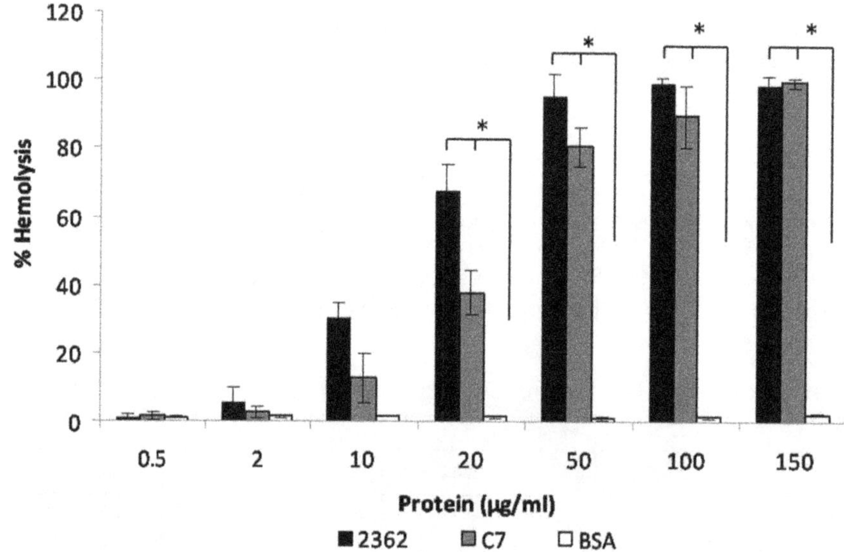

Figure 4. Hemolytic activity from S-layer of *L. sphaericus* 2362 and C7. Hemolysis was analyzed with 1% sheep red blood cells suspension in PBS. Bovine serum albumin (BSA) protein was used for unspecific effect. Six independent experiments with duplicate samples were performed. Bars show the mean ± SD. Mann Whitney-U test was used to determine statistically significant differences between S-layers proteins and BSA control protein. *, $P<0.05$.

L. sphaericus strain increases its insecticidal activity [46]. The presence of a ubiquitous chitin-binding domain in an external protein would be a welcome feature that would facilitate its binding to chitin containing substrates such as insects, thus expanding the number of susceptible mosquitoes species. A chitin binding domain in the S-layer protein would favor the interaction and attachment to insects as an anchor for the bacterium to its potential host. It is worthwhile to remark that chitinases and chitin-binding proteins have been implicated as virulence factors in several species [45]. Since chitinases are enzymes with a large spectrum of substrates and not easy to characterize, we have not been able to detect any chitinase activity in *L. sphaericus* cultures nor in their S-layer using the substrates reported in Fig. 3 (insoluble chitosan, crab chitosan or colloidal chitin). Although the S-layer protein sequence analysis showed a putative chitinase active site within the Glyco_18 domain, the carboxi-terminal cleft responsible for the catalytic activity seems to be absent (Fig. S1). However, it is worthwhile stressing that *L. sphaericus*, a species known for its failure to transport and use glucose (hexoses) as carbon source, has an active PTS transport system and functions for the use of N-acetylglucosamine (NAG), the monomer of chitin [9] [10]. Since *L. sphaericus* share the same habitat as fungi able to degrade chitin and deliver NAG, we may speculate that an efficient PTS-NAG would ensure the survival of these bacteria in the same environment while the presence of the S-layer protein with its chitin binding capacity would favor their interaction and also the attachment to insects, a step necessary for the development of pathogenicity. In fact, the presence of NAG interferes with chitin binding as shown in Figure 3.

Furthermore, the S-layer protein was shown to have hemolytic activity. Also as for chitin binding NAG was found to interfere with hemolytic activity (not shown). Hemolysis is attributed to membrane distortions caused by proteins interacting with lipids. This capacity should help in the pathogenic effect within the larvae, probably contributing to bring down the number of resistant mosquitoes, in the same way the Cyt proteins act in *B. thuringiensis* strains.

B. thuringiensis Cyt and Cry toxins showed a synergistic pattern against *Aedes* larvae [47] [31] similar to our observations with S-layers and spores containing Bins (Table 2).

The presence of S-layer in spore preparations has been reported for different *B. thuringiensis* strains [26] [27] [28] but their contribution to pathogenicity has not been confirmed. In such reports, the S-layer proteins were shown to be present as an inclusion inside the spore-crystals cytoplasm and were highly unstable, thus very difficult to isolate and characterize. Since we suspect that in *L. sphaericus* the S-layer is stabilized probably due to its location in the exosporium, as is the case in *B. anthracis*, it would be worthwhile to analyze it by cryo-EM [42].

Several strains belonging to the same antigenic group as 2362 presented S-layer proteins which cross-react with the antibody used in this work [30]; we also observed this cross-reactivity with that of the C7 strain, which implies they share a great homology. The reported sequences of the *slpC* gene from 2362 and C3-41 were identical. Moreover, the analysis of peptides obtained by trypsin digestion and MALDI-TOF analysis showed that the peptides sequences of S-layer from 2362 and C7 were identical [48]. We wonder why the larvicidal activity was so variable among these strains presenting similar SlpC proteins. We cannot account differences between strains to chitin binding or hemolysis activities since the *in vitro* assay using mimetic substrates (chitin compounds and sheep red blood cells) might not be exactly identical to their target in mosquitoes larvae gut.

The absence of larvicidal activity of the *slpC* gene from C3–41 cloned into *E. coli* [49], would suggest that post-translational modifications of the S-layer taking place in the original host, might be necessary to ensure pathogenicity. A relationship between pathogenicity and S-layer post-translational modifications has been reported for other species [50] [51] [52].

In conclusion, we observed that the S-layer of *L. sphaericus* play several functions. In fact their presence during an osmotic stress is essential [48]. Besides, the S-layers, either from vegetative cultures or associated to spores, were larvicidal against mosquitoes; however that of C7 strain was more active and presented a wider spectrum of activity. The S-layer proteins high molecular weight could indicate that they might result from an assembly of several functional domains in a new protein. Also, multiple *slp* genes have been annotated for the C3-41 strain, suggesting the importance of this protein for this species.

Concerning the bioinsecticide activity, this is the first report of the presence of both toxins (BinA-B) and S-layer protein in spores of *Lysinibacillus sphaericus* strains, where both contribute to its pathogenicity. Together with the high capacity for metal biosorption of these spores [22], which must explain their higher survival in contaminated ponds [20], strains with a wider bioinsecticide spectrum such as C7, would be an interesting alternative to *B. thuringiensis* formulations.

The analysis of the post-translational modifications of the S-layer proteins, which may contribute to the proteins' mosquitocidal activity and specificity, would be the aim of our future investigation.

Supporting Information

Figure S1 Sequence-based analysis. S1A: Comparison of the structural disposition obtained with SMART SEARCH (Simple Modular Architecture Research Tool) for gb|AAA50256.1| surface layer protein [*Lysinibacillus sphaericus*] and dbj|BAM67143.1| chitinase [*Paenibacillus* sp. FPU-7]. **S1B and S1C:** EMBOSS Matcher Pairwise Sequence Alignment between gb|AAA50256.1| surface layer protein [*Lysinibacillus sphaericus*] and dbj|BAM67143.1| chitinase [*Paenibacillus* sp. FPU-7]. B) for the predicted Pfam SLH domains, C) for both predicted Internal Repeat with Glyco_18_1 domains. Symbols: "*" identical aminoacid, ":" indicates group similarity, "." indicates low group similarity. Amino Acid Notation according to IUPAC-IUB-CBN. N° is for number of residues and position within proteins. Grey boxes are SLH domains. **S1D:** CLUSTAL O (1.2.0) sequence alignment between AAA50256.1 surface layer protein [*Lysinibacillus sphaericus* 2362, 1176 aa] and ACA38715.1 hemolysin-type calcium-binding domain-containing protein, *Lysinibacillus sphaericus*, C3-41, 874 aa): score 494.1 bits. Symbols: "*" identical aminoacid, ":" indicates group similarity, "." indicates low group similarity. Amino Acid Notation according to IUPAC-IUB-CBN. N° is for number of residues and position within proteins. Grey boxes are SLH domains in AAA50256.1 only.

Material S1 Sequence-based analysis. Free access sites were used to predict protein structure and function for the Surface layer protein AAA50256 [*Lysinibacillus sphaericus* 2362] and compared to possible orthologous proteins. URL links: **SMART**, (http://smart.embl-heidelberg.de/) Simple modular architecture research tool for Comparison of the structural disposition obtained with SMART SEARCH (Simple Modular Architecture Research Tool). **EMBOSS** Matcher Pairwise Sequence Alignment (http://www.ebi.ac.uk/Tools/services/web/toolform.

ebi?tool=emboss_matcher&context=protein) identifies local similarities in two input sequences using a rigorous algorithm based on Bill Pearson's lalign application. It enables to modify the default substitution scoring matrices (BLOSUM) for sequence alignment between distantly related proteins. **Clustal-O**, Global alignment tool (http://www.ebi.ac.uk/Tools/msa/clustalo/), Clustal Omega is a multiple sequence alignment program for proteins. It produces biologically meaningful multiple sequence alignments of divergent sequences.

Acknowledgments

We would like to specially thank: Linn Lewis, Jean François Charles, Armelle Delécscluse, Allan Yousten for providing us with the antibodies and phages necessary for identification of strains and proteins analyzed in this work. We also wish to thank Liliana Rondón for her help in the immunofluorescence microscopy examinations of spores with their S-layer and Silvia Strauss for careful language revision.

Author Contributions

Conceived and designed the experiments: MCA MMP MPA SMR CSR. Performed the experiments: MCA MMP MPA LL SMR. Analyzed the data: MCA MMP MPA SMR CSR. Contributed reagents/materials/analysis tools: MCA MMP MPA LL SMR CSR. Contributed to the writing of the manuscript: MCA MMP MPA SMR CSR.

References

1. Lacey LA (2007) *Bacillus thuringiensis* serovariety *israelensis* and *Bacillus sphaericus* for mosquito control. J Am Mosq Control Assoc. 23(2 Suppl): 133–63.
2. Estruch JJ, Warren GW, Mullins MA, Nye GJ, Craig JA, et al. (1996) Vip3A, a novel *Bacillus thuringiensis* vegetative insecticidal protein with a wide spectrum of activities against lepidopteran insects. Proc Natl Acad Sci U S A. 93(11): 5389–94.
3. van Frankenhuyzen KJ (2013) Cross-order and cross-phylum activity of *Bacillus thuringiensis* pesticidal proteins. Invertebr Pathol. 114(1): 76–85.
4. Berry C (2012). The bacterium, *Lysinibacillus sphaericus*, as an insect pathogen. J Invertebr Pathol 109: 1–10.
5. Ahmed I, Yokota A, Yamazoe A, Fujiwara T (2007) Proposal of *Lysinibacillus boronitolerans* gen. nov. sp. nov., and transfer of *Bacillus fusiformis* to *Lysinibacillus fusiformis* comb. nov. and *Bacillus sphaericus* to *Lysinibacillus sphaericus* comb. nov. Int. J. Syst. Evol. Microbiol. 57, 1117–1125.
6. Krych VK, Johnson JL, Yousten AA (1980) Deoxyribonucleic acid homologies among strains of *Bacillus sphaericus*. Int J Syst Bacteriol 30: 476–484.
7. Rippere KE, Johnson JL, Yousten AA (1997) DNA similarities among mosquito-pathogenic and nonpathogenic strains of *Bacillus sphaericus*. Int J Syst Bacteriol 47: 214–216.
8. Russell BL, Jelley SA, Yousten AA (1989) Carbohydrate metabolism in the mosquito pathogen *Bacillus sphaericus* 2362. Appl Environ Microbiol. 55(2): 294–7.
9. Alice AF, Pérez-Martínez G, Sánchez-Rivas C (2002) Existence of a true phosphofructokinase in *Bacillus sphaericus*: cloning and sequencing of the *pfk* gene. Appl Environ Microbiol. 68: 6410–5.
10. Alice AF, Pérez-Martínez G, Sánchez-Rivas C (2003) Phosphoenolpyruvate phosphotransferase system and N-acetylglucosamine metabolism in *Bacillus sphaericus*. Microbiol. 149: 1687–98.
11. Gomez-Ramírez M, Rojas-Avelizapa LI, Cruz-Camarillo R (2001) The chitinase of *Bacillus thuringiensis*, p. 273–282. In R. A. A. Muzzarelli ed.), Chitin enzymology. Atec Edizioni, Atec, Italy.
12. Yousten AA, Davidson EW (1982) Ultrastructural Analysis of Spores and Parasporal Crystals Formed by *Bacillus sphaericus* 2297. Environ Microbiol. 44(6): 1449–55.
13. Smith AW, Cámara-Artigas A, Brune DC, Allen JP (2005) Implications of high-molecular-weight oligomers of the binary toxin from *Bacillus sphaericus*. J Invertebr Pathol. 88: 27–33.
14. Thanabalu T, Hindley J, Jackson-Yap J, Berry C (1991) Cloning, sequencing and expression of a gene encoding a 100-kilodalton mosquitocidal toxin from *Bacillus sphaericus* SSII-1. J. Bacteriol. 173: 2776–2785.
15. Charles JF, Nielson-LeRoux C, Delécluse A (1996) *Bacillus sphaericus* toxins: molecular biology and mode of action. Annu Rev Entomol.41: 451–72.
16. Yang Y, Wang L, Gaviria A, Yuan Z, Berry C (2007) Proteolytic stability of insecticidal toxins expressed in recombinant bacilli. Appl Environ Microbiol 73: 218–25.
17. Thiéry I, Hamon S, Delécluse A, Orduz S (1998) The introduction into *Bacillus sphaericus* of the *Bacillus thuringiensis* subsp. *medellin* Cyt1Ab1 gene results in higher susceptibility of resistant mosquito larva populations to *B. sphaericus*. Appl Environ Microbiol. 64: 3910–6.
18. Wirth MC, Yang Y, Walton WE, Federici BA, Berry C (2007) Mtx toxins synergize *Bacillus sphaericus* and Cry11Aa against susceptible and insecticide-resistant *Culex quinquefasciatus* larvae. Appl Environ Microbiol. 73: 6066–71.
19. Chenniappan K, Ayyadurai N (2012) Synergistic activity of Cyt1A from *Bacillus thuringiensis* subsp. *israelensis* with *Bacillus sphaericus* B101 H5a5b against *Bacillus sphaericus* B101 H5a5b-resistant strains of *Anopheles stephensi Liston* (Diptera: Culicidae). Parasitol Res 110: 381–388.
20. Merroun ML, Raff J, Rossberg A, Hennig C, Reich T, et al (2005) Complexation of uranium by cells and S-layer sheets of *Bacillus sphaericus* JG-A12. Appl. Environ. Microbiol. 71: 5532–5543.
21. Pollmann K, Raff J, Merroun M, Fahmy K, Selenska-Pobell S (2006) Metal binding by bacteria from uranium mining waste piles and its technological applications. Biotechnology Advances 24: 58–68.
22. Allievi MC, Sabbione F, Prado-Acosta M, Palomino MM, Ruzal SM, et al. (2011) Metal Biosorption by Surface-Layer Proteins from *Bacillus* Species. Microbiol. Biotechnol. 21: 147–153.
23. Kern J, Schneewind O (2010) BslA, the S-layer adhesin of *B. anthracis*, is a virulence factor for anthrax pathogenesis. Mol Microbiol. 75: 324–32.
24. Wang YT, Oh SY, Hendrickx AP, Lunderberg JM, Schneewind O (2013) *Bacillus cereus* G9241 S-layer assembly contributes to the pathogenesis of anthrax-like disease in mice. J. Bacteriol. 195: 596–605.
25. Janesch B, Messner P, Schäffer C (2013) Are the SLH-Domains Essential for Cell Surface display and glycosylation of the S-Layer protein from *Paenibacillus alvei* CCM 2051T? J. Bacteriol. 195: 596–605.
26. Peña G, Miranda-Rios J, de la Riva G, Pardo-López L, Soberón M, et al. (2006) A *Bacillus thuringiensis* S-layer protein involved in toxicity against *Epilachna varivestis* (Coleoptera: Coccinellidae). Appl Environ Microbiol 72: 353–360.
27. Guo G, Zhang L, Zhou Z, Ma Q, Liu J, et al. (2008) A new group of parasporal inclusions encoded by the S-layer gene of *Bacillus thuringiensis*. FEMS Microbiol Lett 282: 1–7.
28. Zhou Z, Peng D, Zheng J, Guo G, Tian L, et al. (2011) Two groups of S-layer proteins, SLP1s and SLP2s, in *Bacillus thuringiensis* co-exist in the S-layer and in parasporal inclusions. BMB Rep. 44: 323–8.
29. Lozano LC, Ayala JA, Dussán J (2011) *Lysinibacillus sphaericus* S-layer protein toxicity against *Culex quinquefasciatus*. Biotechnol Lett 33: 2037–2041.
30. Lewis LO, Yousten AA, Murray RGE (1987) Characterization of the surface protein layers of the mosquito-pathogenic strains of *Bacillus sphaericus* J. Bacteriol 169: 72–79.
31. López-Diaz JA, Cantón PE, Gill SS, Soberón M, Bravo A (2013) Oligomerization is a key step in Cyt1Aa membrane insertion and toxicity but not necessary to synergize Cry11Aa toxicity in *Aedes aegypti* larvae. Environ Microbiol. 15: 3030–3039.
32. Myers PS, Yousten AA (1980) Localization of a mosquito-larval toxin of *Bacillus sphaericus* 1593. Appl. Environ. Microbiol. 39: 1205–1211.
33. Yousten AA (1984) *Bacillus sphaericus*: microbiological factors related to its potential as a mosquito larvicide. Adv Biotechnol Processes. 1984; 3: 315–43.
34. Travers RS, Martin PA, Reichelderfer CF (1987) Selective process for efficient isolation of soil *Bacillus* spp. Appl Environ Microbiol. 53(6): 1263–1266.
35. Redmond C, Baillie LW, Hibbs S, Moir AJ, Moir A (2004) Identification of proteins in the exosporium of *Bacillus anthracis*. Microbiology. 150(Pt 2): 355–63.
36. Kailas L, Terry C, Abbott N, Taylor R, Mullin N, et al. (2011) Surface architecture of endospores of the *Bacillus cereus/anthracis/thuringiensis* family at the subnanometer scale. Proc Natl Acad Sci U S A.108(38): 16014–9.
37. Fischer S, Schweigmann N (2004) *Culex mosquitoes* in temporary urban rain pools: seasonal dynamics and relation to environmental variables. J Vector Ecol. 29(2): 365–73.
38. Thiery I, Delécluse A, Tamayo MC, Orduz S (1997) Identification of a gene for Cyt1A-like hemolysin from *Bacillus thuringiensis* subsp. *medellin* and expression in a crystal-negative *B. thuringiensis* strain. Appl Environ Microbiol. 63(2): 468–73.
39. Rodriguez-Almazan C, Ruiz de Escudero I, Cantón PE, Muñoz-Garay C, Pérez C, et al. (2011) The amino- and carboxyl-terminal fragments of the *Bacillus thuringensis* Cyt1Aa toxin have differential roles in toxin oligomerization and pore formation. Biochem. 50: 388–96.
40. Balzarini MG, Gonzalez L, Tablada M, Casanoves F, Di Rienzo JA, et al. (2008) Infostat. Manual del Usuario, Editorial Brujas, Córdoba, Argentina.
41. Itoh T, Hibi T, Fujii Y, Sugimoto I, Fujiwara A, et al. (2013) Cooperative degradation of chitin by extracellular and cell surface-expressed chitinases from *Paenibacillus* sp. strain FPU-7. Appl Environ Microbiol. 79: 7482–90.
42. Rodenburg CM, McPherson SA, Turnbough CL Jr, Dokland T (2014) Cryo-EM analysis of the organization of BclA and BxpB in the *Bacillus anthracis* exosporium. J Struct Biol. 186(1): 181–7.

43. Hu X, Fan W, Han B, Liu H, Zheng D, et al. (2008) Complete genome sequence of the mosquitocidal bacterium *Bacillus sphaericus* C3–41 and comparison with those of closely related *Bacillus* species. J. Bacteriol 190: 2892–2902.

44. Sampson MN, Gooday GW (1998) Involvement of chitinases of *Bacillus thuringiensis* during pathogenesis in insects. Microbiology 144: 2189–2194.

45. Frederiksen RF, Paspaliari DK, Larsen T, Storgaard BG, Larsen MH, et al. (2013) Bacterial chitinases and chitin-binding proteins as virulence factors. Microbiol. 159: 833–47.

46. Cai Y, Ya J, Hu X, Han B, Yuan Z (2007) Improving the insecticidal activity against resistant *Culex quinquefasciatus* mosquitoes by expression of chitinase gene *chiAC* in *Bacillus sphaericus*. Appl Environ Microbiol. 73: 7744–7746.

47. Wirth MC, Delécluse A, Walton WE (2001) Cyt1Ab1 and Cyt2Ba1 from *Bacillus thuringiensis* subsp. *medellin* and *B. thuringiensis* subsp. *israelensis* synergize *Bacillus sphaericus* against *Aedes aegypti* and resistant *Culex quinquefasciatus* (Diptera: Culicidae). Appl Environ Microbiol. 67: 3280–4.

48. Allievi MC (2012) S-layer de *Bacillus sphaericus:* caracterización, regulación, análisis funcional y aplicaciones. Thesis. University of Buenos Aires. Available: http://bldigital.bl.fcen.uba.ar/Download/Tesis/Tesis_5143_Allievi.pdf Accessed 28 June 2012.

49. Hu X, Li J, Hansen BM, Yuan Z (2008) Phylogenetic analysis and heterologous expression of surface layer protein SlpC of *Bacillus sphaericus* C3–41. Biosci Biotechnol Biochem 72: 1257–1263.

50. Forsberg LS, Choudhury B, Leoff C, Marston CK, Hoffmaster AR, et al. (2011) Secondary cell wall polysaccharides from *Bacillus cereus* strains G9241, 03BB87 and 03BB102 causing fatal pneumonia share similar glycosyl structures with the polysaccharides from *Bacillus anthracis*. Glycobiol. 21: 934–948.

51. Balomenou S, Fouet A, Tzanodaskalaki M, Couture-Tosi E, Bouriotis V, et al. (2013) Distinct functions of polysaccharide deacetylases in cell shape, neutral polysaccharide synthesis and virulence of *Bacillus anthracis*. Mol Microbiol. 87: 867–83.

52. Friedlander A, Quinn CP, Kannenberg EL, Carlson RW (2012) Localization and structural analysis of a conserved pyruvylated epitope in *Bacillus anthracis* secondary cell wall polysaccharides and characterization of the galactose-deficient wall polysaccharide from avirulent *B. anthracis* CDC 684. Glycobiol. 22: 1103–1117.

9

Early Staphylococcal Biofilm Formation on Solid Orthopaedic Implant Materials: *In Vitro* Study

Hironobu Koseki[1]*, Akihiko Yonekura[1], Takayuki Shida[1], Itaru Yoda[1], Hidehiko Horiuchi[1], Yoshitomo Morinaga[2], Katsunori Yanagihara[2], Hideyuki Sakoda[3], Makoto Osaki[1], Masato Tomita[1]

1 Department of Orthopedic Surgery, Graduate School of Biomedical Sciences, Nagasaki University, Nagasaki, Japan, **2** Department of Laboratory Medicine, Graduate School of Biomedical Sciences, Nagasaki University, Nagasaki, Japan, **3** Division of Medical Devices, National Institute of Health Sciences, Tokyo, Japan

Abstract

Biofilms forming on the surface of biomaterials can cause intractable implant-related infections. Bacterial adherence and early biofilm formation are influenced by the type of biomaterial used and the physical characteristics of implant surface. In this *in vitro* research, we evaluated the ability of *Staphylococcus epidermidis*, the main pathogen in implant-related infections, to form biofilms on the surface of the solid orthopaedic biomaterials, oxidized zirconium-niobium alloy, cobalt-chromium-molybdenum alloy (Co-Cr-Mo), titanium alloy (Ti-6Al-4V), commercially pure titanium (cp-Ti) and stainless steel. A bacterial suspension of *Staphylococcus epidermidis* strain RP62A (ATCC35984) was added to the surface of specimens and incubated. The stained biofilms were imaged with a digital optical microscope and the biofilm coverage rate (BCR) was calculated. The total amount of biofilm was determined with the crystal violet assay and the number of viable cells in the biofilm was counted using the plate count method. The BCR of all the biomaterials rose in proportion to culture duration. After culturing for 2–4 hours, the BCR was similar for all materials. However, after culturing for 6 hours, the BCR for Co-Cr-Mo alloy was significantly lower than for Ti-6Al-4V, cp-Ti and stainless steel ($P<0.05$). The absorbance value determined in the crystal violet assay and the number of viable cells on Co-Cr-Mo were not significantly lower than for the other materials ($P>0.05$). These results suggest that surface properties, such as hydrophobicity or the low surface free energy of Co-Cr-Mo, may have some influence in inhibiting or delaying the two-dimensional expansion of biofilm on surfaces with a similar degree of smoothness.

Editor: Zezhang Wen, LSU Health Sciences Center School of Dentistry, United States of America

Funding: This work was supported by JSPS KAKENHI Grant Number 24592236 (http://www.jsps.go.jp/j-grantsinaid/16_rule/rule.html). The funders had no role in study design, data collection and analysis, decision to publish, or preparation of the manuscript.

Competing Interests: The authors have declared that no competing interests exist.

* Email: koseki@nagasaki-u.ac.jp

Introduction

Solid biomaterials with particular characteristics, such as high biocompatibility or corrosion resistance, are now being implanted in the human body more frequently for a wide range of purposes. However, implant-related infection is generally the most common serious complication and the risk of surgical site infection (SSI) increases when a foreign material is present [1]. When bacteria adhere to and proliferate on the biomaterial surface, the bacteria produce extracellular polymeric substances, primarily polysaccharides, which mediate cell-to-cell adhesion and form a biofilm. The biofilm enveloping the bacteria can protect them from the immune system. Moreover, the presence of biofilm changes gene expression, alters growth rate, and decreases susceptibility to antibiotics [2–6], so implant-related infection is extremely difficult to treat [7–9]. Various methods have been devised to prevent implant-related infections, including techniques to sterilize the surgical site and instruments, and the use of highly sterile operating rooms. However, these infections still occur today in 0.2–17.3% of orthopaedic surgery [10–12]. Therefore, research investigating the formation of biofilms on biomaterials is critically important from the clinical perspective.

The process of biofilm formation is generally thought to be a two-step model. Firstly, bacteria rapidly adhere to the biomaterial surface by means of physicochemical interactions (van der Waals forces, gravitational forces, electrostatic repulsion, and ionic and dipole interactions). Secondly, the bacteria proliferate and accumulate to form multilayered cell clusters on the surface through molecular and cellular interactions [13,14]. Most implant-related infections are caused by *Staphylococcus* species [15–17]. The skin commensal organism *Staphylococcus epidermidis* (*S. epidermidis*) has been recognized as the preeminent and important medical pathogen in orthopaedic implant-related infections. It is particularly capable of adhering to and aggregating on biomaterial surfaces and can form biofilms on many biomaterials [18,19]. Arciola et al demonstrated that multiple instances of resistance to antibiotics were more frequent among polysaccharides producing the *S. epidermidis* strain [3]. Research studies have shown that polysaccharide intercellular adhesin (PIA) plays an important role in biofilm formation and development along with genetic factors such as *ica ADBC* [19–23]. However, the detailed mechanism of this process has yet to be determined because of the complex combination of numerous other factors related to the bacteria, the *in vivo* environment and the use of artificial materials.

The solid biomaterials used for clinical purposes must be biocompatible and have a high resistance to wear, fracture and corrosion. Depending on their application, biomaterials can be made of just a few kinds of materials standardized by the International Organization for Standardization (ISO) and the American Society for Testing and Materials (ASTM). Oxidized Zirconium-Niobium alloy (Oxinium) was commercialized as a new biomaterial in Japan in 2008. This alloy forged from zirconium and niobium is permeated with oxygen at a high temperature, with only 5 µm of the surface changed to zirconium ceramic. As a result, Oxinium exhibits a low level of abrasion on sliding surfaces characteristic of a ceramic and has the strength of a metal. Oxinium also contains almost no toxic metals [24].

Recently, numerous factors related to the artificial solid biomaterials themselves, such as chemical structure, surface roughness, hydrophilicity, Z potential and surface free energy, have been identified as influencing bacterial adherence and early biofilm formation [25–33]. Although the evidence about the relationship between biomaterial and early phase of biofilm formation in previous studies is inconsistent, some previous reports have highlighted a relatively strong relationship between biofilm formation and surface roughness [30–32]. The rougher surface provides a wider area for bacterial adherence, multiplication and biofilm formation [33,34]. However, there have been no studies into the effects of surface characteristics on bacterial biofilm formation apart from roughness. Therefore, in order to accurately compare the biofilm formation ability of the various biomaterials, we must eliminate this bias.

Several investigative methods have been established to evaluate the development of biofilms on the surface of biomaterials. Methods to directly examine biofilm formation include fluorescence microscopy (FM) [35], scanning electron microscopy (SEM) [36]. With these forms of image analysis, we can directly observe and enumerate the number of bacteria. Confocal laser scanning microscopy (CLSM) is a newly developed, valuable method for morphological observation of biofilm [37–39]. Indirect methods applicable for estimating biofilm density include viable cell count (VCC) after sonication [40], ATP-bioluminescence (ATP) [41], trypsin treatment [42] and crystal violet (CV) assay [43]. The VCC method is the most basic and conventional method for counting viable bacterial and the CV assay assesses the total amount of biofilm, including dead cells and extracellular polymeric substances. The percentage of surface covered by a biofilm is calculated as the biofilm coverage rate (BCR) [44]. The method for measuring BCR can estimate the growth of the biofilm using the time course, as well as assessing its two-dimensional expansion on non-translucent biomaterials without disrupting it.

In this *in vitro* study, we used BCR, CV assay and VCC to quantify the amount of biofilm formed by *S. epidermidis* and to compare its ability to form such biofilms on the surfaces of five types of solid biomaterials with a similar degree of smoothness. We have discovered no previous research focusing on the biofilm formation ability of different biomaterials, including Oxinium, which eliminates the influence of surface roughness.

Materials and Method

Specimen preparation

We prepared circular specimens (12 mm in diameter, 6 mm thick) from Oxinium (ASTM F2384), cobalt-chromium-molybdenum alloy (Co-Cr-Mo) (ASTM F75 high carbon), titanium alloy (Ti-6Al-4V) (ASTM F136), pure titanium (cp-Ti) (ASTM F67) and stainless steel (SUS316L) (ASTM F138) that are actually used in clinical practice. Original materials were obtained from Smith & Nephew Orthopaedics Inc. (Memphis, TM, USA) and Kakushin Surgical Instruments Co.Ltd. (Shizuoka, Japan). The five kinds of test specimen were progressively polished using a basic lapping machine (Doctorlap ML-180SL, Maruto Co.Ltd., Tokyo, Japan) with polishing compounds, a polishing cloth and a diamond slurry (Maruto Instrument Co. Ltd., Tokyo, Japan; 1 µm particle diameter).

Surface characterization

Micrographs of the specimen disk surfaces were obtained using a field emission scanning electron microscope (SEM: JSM 6610LV, JEOL, Tokyo, Japan). The surface morphology and roughness of the specimens were measured by means of a 3D measuring laser microscope (OLS4000, Shimadzu, Tokyo, Japan) with a cut-off value (λc) of 80 µm at room temperature. Three readings were made of each surface on three random samples, and the average roughness (Ra) and mean roughness profile depth (Rz) were used to determine the roughness of the specimens. The initial contact angles of the surface of each specimen to deionized water (Milli-Q, EMD Millipore, Billerica, MA, USA) and diiodomethane (Wako Pure Chemical Industries Ltd. Osaka, Japan) were measured by the drop method using an automated contact angle measurement device (DSA30, Krüss GmbH, Hamburg, Germany) on each of three randomly selected specimens at room temperature (25°C). Prior to contact angle determination, all specimens were equilibrated with ethanol. The total surface free energy (γ^t) and its polar (γ^P) and disperse (γ^d) components were calculated from the contact angles of deionized water and diiodomethane according to the Owen's (1) and Young's equation (2) [45].

$$\gamma_L(1+\cos\theta_L)=2(\gamma_S{}^d\times\gamma_L{}^d)^{1/2}+2(\gamma_S{}^P\times\gamma_L{}^P)^{1/2} \quad (1)$$

$$\gamma_S=\gamma_S{}^d+\gamma_S{}^P \quad (2)$$

Where θ is the measured contact angle, γ_L is the surface free energy of the reference liquid, $\gamma_L=\gamma_L{}^d+\gamma_L{}^P$. $\gamma_L{}^d$ and $\gamma_L{}^P$ are the dispersive and polar components of surface free energy of the reference liquids, respectively. $\gamma_S{}^d$ and $\gamma_S{}^P$ are the dispersive and polar components of surface free energy of the solid surface, respectively. The contact angle θ is a measurable parameter. When two liquids with known $\gamma_L{}^d$ and $\gamma_L{}^P$ are used, $\gamma_S{}^d$ and $\gamma_S{}^P$ can be obtained by solving the two simultaneous equations. The total surface free energy of the solid (γ_S) is the sum of $\gamma_S{}^d$ and $\gamma_S{}^P$. Deionized water ($\gamma_L{}^d=21.8$ mJ/m^2, and $\gamma_L{}^P=51.0$ mJ/m^2) and diiodomethane ($\gamma_L{}^d=37.0$ mJ/m^2, and $\gamma_L{}^P=26.4$ mJ/m^2) were used as the reference liquids [46–48].

Experimental design

PIA-producing *S. epidermidis* strain RP62A (American Type Culture Collection [ATCC] 35984, American Type Culture Collection, Manassas, VA, USA) was grown overnight in Trypticase Soy Broth (TSB: Becton Dickinson Biosciences, Franklin Lakes, NJ, USA) at 37°C. The culture was diluted into TSB the following day at a ratio of 1: 10 and incubated for 3 hours to create a bacterial suspension of 1×10^5 CFU/mL (logarithmic growth: Optical Density [OD] $_{600}=0.2$; pH 7.0). Olson et al. investigated the superior adherence ability of PIA-producing *S. epidermidis* on biomaterial surfaces [23]. The test specimens were subjected to ultrasonic cleaning and autoclaving and then 200 µL of the bacterial suspension was dropped onto the specimens at room temperature and incubated for 60 minutes. The specimens

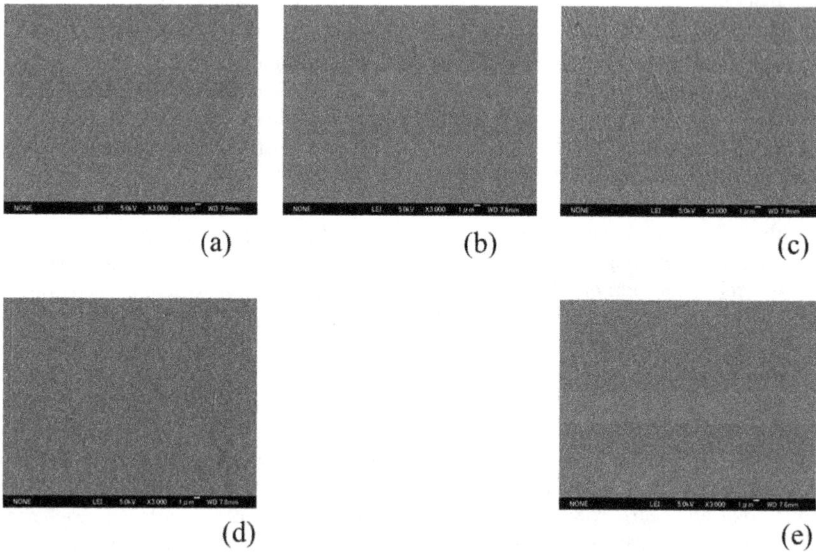

Figure 1. SEM micrographs. Although a few polishing micro-traces and marks were observed, all specimens had a generally featureless and smooth surface. Oxinium (a), Co-Cr-Mo (b), Ti-6Al-4V (c), cp-Ti (d), stainless steel (e) Original magnification ×3000 (Scale bar = 1 μm)

were rinsed twice with phosphate-buffered saline (PBS: Sigma-Aldrich St Louis, MO, USA; pH 7.0) to remove non-adherent and loosely adherent cells, and transferred into fresh TSB medium for culturing (culture duration: 2 hours, 4 hours and 6 hours).

The morphology of the biofilms on the different specimens after 6 hours culture was assessed by SEM. The biofilm was fixed with glutaraldehyde (2.5% v/v) in a 0.1 M cacodylate buffer (0.1 M Na-cacodylate trihydrate in H_2O, pH 7.4) for 4 hours at 4°C. The specimens were washed twice in the cacodylate buffer for 20 minutes followed by rinsing with H_2O for 1 minute and biofilm was then post-fixed in 1% OsO_4 for 2 hours at 4°C. The specimens were dehydrated with graded ethanol (50%, 70%, 80%, 90%, 95% and 99.5% v/v) for 10 minutes at each interval and dried using a freeze-dryer (ID-2, Engineering Co. Ltd., Mito, Japan). Finally, the biofilms were sputter-coated with platinum palladium using an ion-sputter (JFC-1600, JEOL, Tokyo, Japan), and viewed with a SEM at an accelerating voltage of 15 kV.

BCR measurements were performed as described previously [44]. The specimen surfaces were fixed with ethanol for 1 minute after which they were air dried and then stained with 0.5% crystal violet (Sigma-Aldrich, MD, USA) for 5 minutes. In order to remove the excess unbound dye, the specimens were then washed with distilled water and dried. The growth formation of the biofilm in the horizontal direction was observed using a digital optical microscope (VHX-100; Keyence, Osaka, Japan) and the percentage of the surface covered by bacteria was calculated as the biofilm coverage rate (BCR) [44,49]. Images with ×450 full color photographs of a random nine locations on each specimen were obtained and converted to gray-scale images with Paint Shop Pro 8 (Corel Co., Ltd., Eden Prairie, MN, USA). The BCRs were measured using the Scion Image software package (Scion Co., Ltd., Frederick, MD) [50] and the BCR value of nine areas were averaged for each specimen.

The total amount of biofilm was assessed using a CV assay. The biofilm formed after 6 hours of culturing was fixed with ethanol

Table 1. Surface roughness.

	Roughness (nm)	
	Ra	**Rz**
Oxinium	6.3(1.9)^{b,e}	47.6 (10.1)^{b,e}
Co-Cr-Mo	1.9 (0.8)^{a,c,d}	14.2 (7.4)^{a,c,d}
Ti-6Al-4V	6.7 (1.1)^{b,e}	50.8 (14.6)^{b,d,e}
cp-Ti	5.3 (1.2)^{b,e}	34.9 (7.2)^{b,c,e}
stainless steel	1.2 (0.4)^{a,c,d}	9.2 (0.8)^{a,c,d}

Data were expressed as a mean (standard deviation (SD)).
Ra: arithmetic mean of the departures of the roughness profile from the profile center-line.
Rz: average distance between the highest peak and the lowest valley.
^{a}: $P<0.01$ compared to Oxinium.
^{b}: $P<0.01$ compared to Co-Cr-Mo.
^{c}: $P<0.01$ compared to Ti-6Al-4.
^{d}: $P<0.01$ compared to cp-Ti.
^{e}: $P<0.01$ compared to stainless steel.

Table 2. Contact angles and Surface free energies.

	Contact angle (degree)				
	Water	**Diiodomethane**	$\gamma_s{}^d$ (mJ/m^2)	$\gamma_s{}^p$ (mJ/m^2)	γ_s (mJ/m^2)
Oxinium	69.0 (3.9)^{b,d,e}	37.8 (1.1)^{b,d,e}	34.3	9.6	43.9
Co-Cr-Mo	107.3 (5.2)^{a,c,d,e}	49.7 (1.2)^{a,c,d,e}	28.4	9.7	38.1
Ti-6Al-4V	71.7 (0.3)^{b,d,e}	36.5 (0.8)^{b,d,e}	35.6	7.8	43.4
Cp-Ti	96.9 (6.6)^{a,b,c}	41.1 (1.0)^{a,b,c}	39.9	0.1	40.0
stainless steel	90.4 (2.3)^{a,b,c}	38.7 (1.7)^{a,b,c}	39.5	0.9	40.4

Contact angle data were expressed as a mean (standard deviation (SD)). Surface free energies (γ_S) were calculated from the mean value of the contact angles of water and diiodomethane.
[a]: $P<0.01$ compared to Oxinium.
[b]: $P<0.01$ compared to Co-Cr-Mo.
[c]: $P<0.01$ compared to Ti-6Al-4.
[d]: $P<0.01$ compared to cp-Ti.
[e]: $P<0.01$ compared to stainless steel.

and dried for 5 minutes. The fixed biofilms were stained with 0.5% crystal violet (Sigma-Aldrich, MD, USA) for 5 minutes. The excess unbound dye was removed by washing the specimens with distilled water. After thorough air drying, the specimens were transferred into sterile conical tubes (Falcon, BD Biosciences, Franklin Lakes, NJ, USA) filled with 5 mL of PBS. The tubes were vortexed at full speed for 3 minutes and then placed in an ultrasonic bath and sonicated for 5 minutes at 120 W to release the biofilm attached to the biomaterial. After an additional vortex step, the specimens were removed. The remaining suspensions were plated in triplicate in 96-well microtiter plate and the absorbance values were measured at an optical density of 570 nm using a microplate reader (Infinite F200 PRO, Tecan, Männedorf, Switzerland).

The VCCs in the suspension were counted using the standard plate count method. The specimens with biofilm incubated for the same 6 hours were each placed in sterile conical tubes containing 5 ml PBS. In order to remove the biofilm from the specimen, these tubes were vortexed for 3 minutes, sonicated for 5 minutes, and vortexed again for 3minutes. The solution containing the biofilm was transferred into another sterile conical tube and diluted with PBS. The number of viable bacteria in the biofilm was determined by counting the colony-forming units (CFUs) with a Compact Dry TC culture kit (Nissui Pharmaceutical Co., Ltd., Tokyo, Japan), after which the bacterial density (CFU/ml) was calculated. As well as using uniform conditions for the bacteria, the five kinds of specimens were treated at the same time, and the experiments themselves were repeated using a uniform procedure to eliminate the effect of environmental factors.

Statistical analysis

The mean and standard deviation of the topographic parameters of the specimens (n = 6), contact angles (n = 12), BCR (n = 7), OD values (n = 10), and the VCC (n = 12) were analyzed for the different materials using SPSS 10.0 statistical software (SPSS Inc., Chicago, IL, USA). Statistical analysis was performed using one-way analysis of variance (one-way ANOVA), multiple comparison tests and the Tukey-Kramere and Bonferroni/Dunn multiple

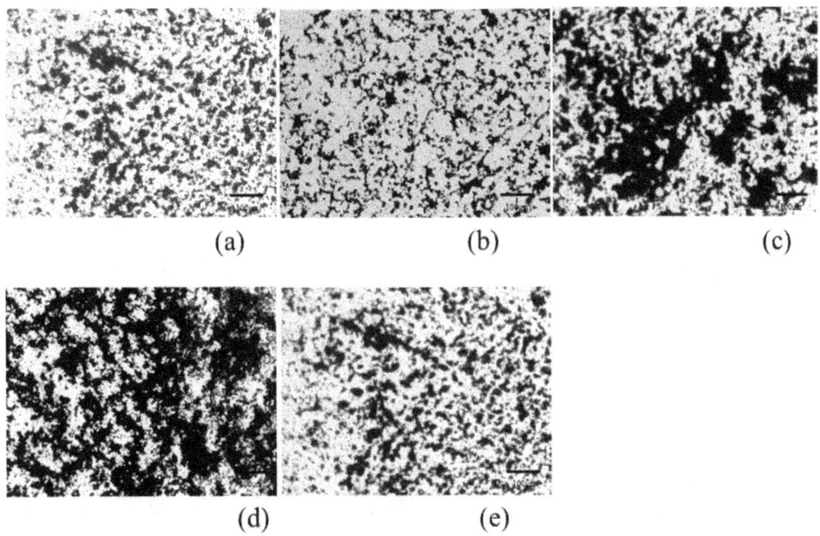

Figure 2. Digital optical micrographs. Biofilm incubated for 6 hours were stained with 0.5% crystal violet (black area). Oxinium (a), Co-Cr-Mo (b), Ti-6Al-4V (c), cp-Ti (d), stainless steel (e) Original magnification ×450 (Scale bar = 100 μm)

Figure 3. SEM images of biofilm. Oxinium (a), Co-Cr-Mo (b), Ti-6Al-4V (c), cp-Ti (d), stainless steel (e) Original magnification ×1000 (Scale bar = 10 μm)

comparison test for *post hoc* analysis. The value of statistical significance was set at $P<0.05$.

Results

Figure 1 shows SEM images of the prepared specimen surface. Although there are some fine polishing micro-traces and marks homogeneously distributed over the samples, all specimens were observed to be generally featureless with a smooth surface topography. The mean surface roughness parameters for each type of specimen are shown in Table 1. All specimens had comparatively smooth surfaces and recorded low average roughness (Ra<10 nm). The contact angles and surface free energies were shown in Table 2. The surface of Co-Cr-Mo had the highest water contact angle, followed by cp-Ti, stainless steel and Ti-6Al-4V. Oxinium yielded the lowest water contact angle. A greater water contact angle means a more hydrophobic surface. The total surface free energy of Co-Cr-Mo, which is composed of a low dispersive component, is relatively lower than that of the other biomaterials.

Biofilms formed on all of the specimens. Digital microscopic images of the biofilm after 6 hours culture is shown in Figure 2(a)~(e). These images demonstrate that a wide area was covered by the stained biofilm on the surface of Ti-6Al-4V and cp-Ti. Observation using SEM revealed that the bacteria on Co-Cr-Mo was aggregated and more tightly colonized than for Oxinium, Ti-6Al-4V and cp-Ti (Figure 3(a)~(e)). Biofilm colonies on the surface of Oxinium, Ti-6Al-4V and cp-Ti tended to be scattered and horizontally spread. The BCR rose as the culture duration increased (Figure 4). After culturing for 2 hours, the BCR was an

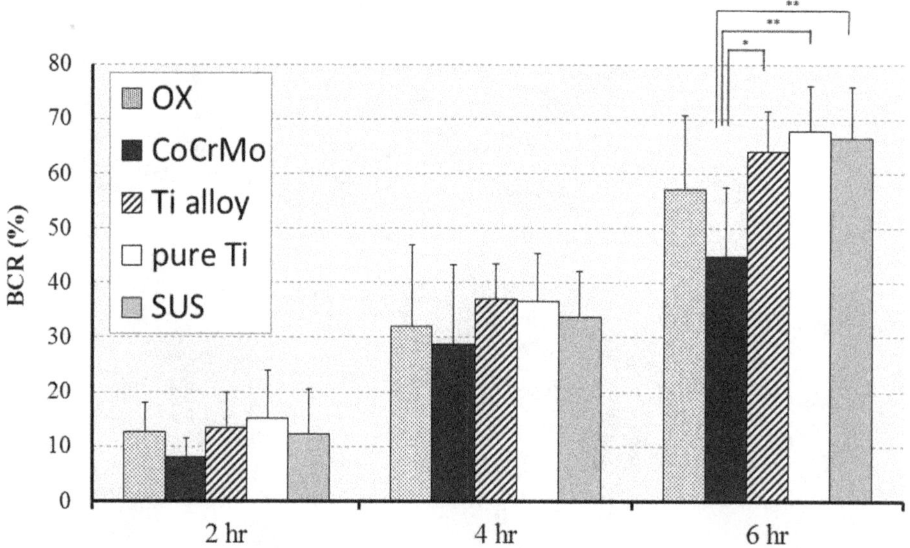

Figure 4. Biofilm coverage rate (BCR). Mean and standard deviation are shown. $^{*}p<0.05$, $^{**}p<0.01$

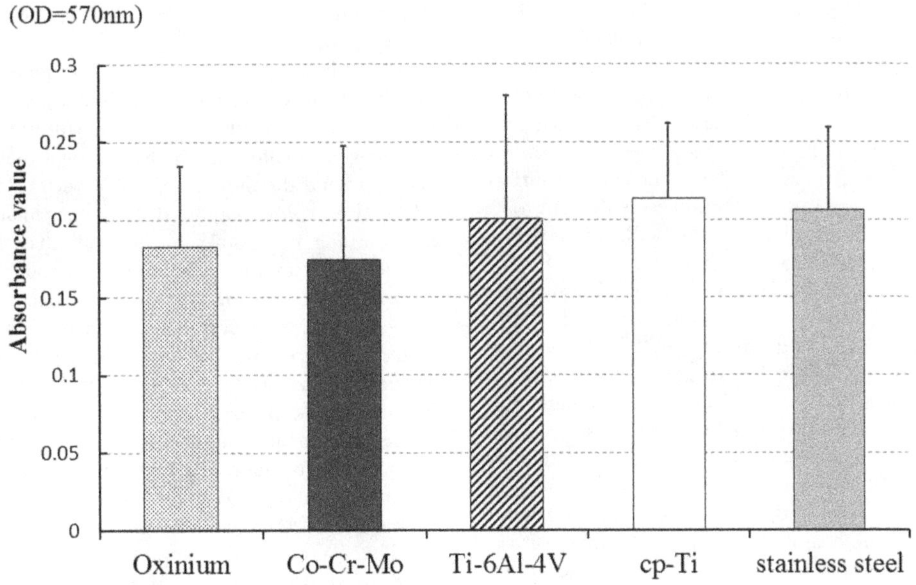

Figure 5. Absorbance value of crystal violet (CV) assay after 6 hours incubation. Mean and standard deviation are shown.

average of 12.6±5.4% for Oxinium, 8.0±3.6% for Co-Cr-Mo, 13.4±3.3% for Ti-6Al-4V, 15.2±6.1% for cp-Ti and 12.2±4.4% for stainless steel. Therefore, there was no significant difference in BCR between the materials. After culturing for 4 hours, the BCR was higher than after 2 hours for all test specimens. Similarly to the findings after 2 hours culturing, no statistically significant differences were observed. After culturing for 6 hours, Co-Cr-Mo had the lowest BCR and there were statistically significant differences between the BCR of Co-Cr-Mo (44.8±12.6%) and that of Ti-6Al-4V (64.1±7.3%), cp-Ti (67.8±8.3%) and stainless steel (66.6±9.4%) ($P<0.05$). During 2–4 hours of culturing, the biofilms on all the biomaterials grew at a similar pace to the increase in BCR. Conversely, between 4–6 hours of culturing, the BCR developed gradually on Co-Cr-Mo compared to the pattern for the other biomaterials. Figure 5 shows the total biofilm mass determined by CV staining. Although the absorbance value for Co-Cr-Mo tended to be lower than for the other materials, there was no significant difference between the materials ($P>0.05$). Correspondingly, the VCC values did not show any significant difference between the five materials (Figure 6) ($P>0.05$).

Discussion

In this research, we evaluated the difference in early biofilm formation of PIA-positive *S. epidermidis* on five types of biomaterials - including Oxinium, which is now being used as a new material for prosthetic joints. We also investigated the correlation between the physical characteristics of the various biomaterials and their ability to form biofilms at an early stage.

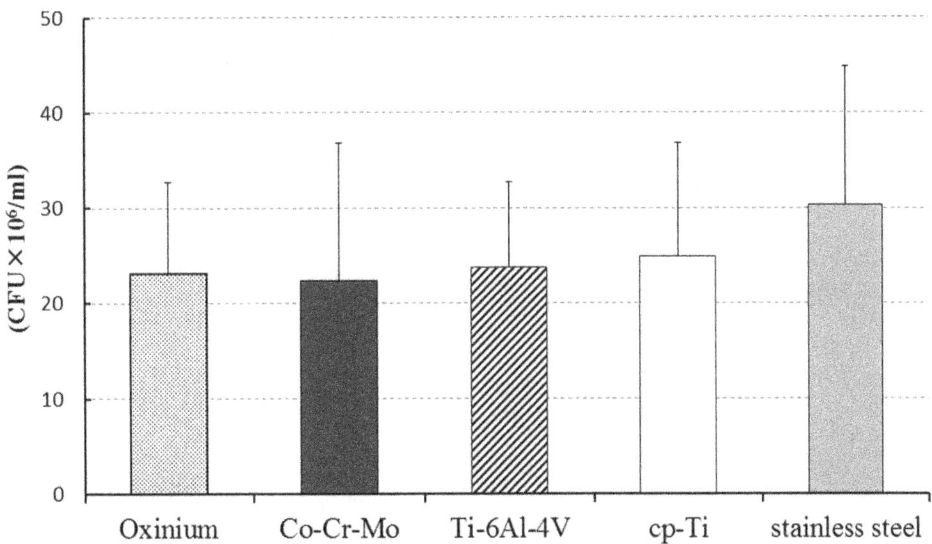

Figure 6. Colony forming units as determined by viable cell count (VCC) after 6 hours incubation. Mean and standard deviation are shown.

After culturing for 2 and 4 hours, biofilms had formed on all the test materials and the BCR values were similar for all of them ($P>0.05$). Previous reports have shown that bacterial adhesion is primarily determined by a threshold surface roughness of Ra more than 0.2 μm (200 nm) [33,51]. Therefore, we polished the specimen surfaces to similar degree of smoothness (Ra<10 nm) in order to eliminate any discrepancies due to the effect of surface roughness. This high level of surface smoothness is thought to be the reason that no significant difference in BCR was observed between specimens until culturing exceeds 4 hours.

After culturing for 6 hours, Co-Cr-Mo had a significantly lower BCR than Ti-6Al-4V, cp-Ti and stainless steel ($P<0.05$). However, the total biofilm mass determined by CV staining and the viable cell counts did not differ significantly between the materials ($P>0.05$). Boks et al reported that bond strengthening for four strains of *S. epidermidis* on a hydrophobic surface was limited to a minor increase [52]. Tang et al showed that more bacteria adhered to a hydrophilic surface than a hydrophobic surface [53]. As water molecules adjacent to a hydrophobic surface are not able to form hydrogen bonds with that surface (hydrophobic effect), bacterial adhesion to a hydrophobic surface is brought about by an entropically favorable release of water molecules. With regards to surface free energy, numerous studies in the dental field agree that surfaces with high surface free energy foster microbial adherence *in vitro* and *in vivo* [31–33,54–56]. Glantz et al reported that when analyzed gravimetrically, there was less dental plaque on low surface free energy hydrophobic substrata than on hydrophilic substrata due to the effect of interfacial thermodynamics [57]. On the other hand, Van Pelt et al suggested that surface free energy is presumably more directly related to the binding force rather than to the number of bacteria on the surface area [54]. Therefore, it can be speculated that bacteria on the relatively hydrophobic Co-Cr-Mo surface, which has the lowest surface free energy, binds cell-to-cell more tightly with polysaccharides than to a cell-to-material surface (bacteriophobic effect), and that it is difficult for bacteria to develop a biofilm on the horizontal plane on the Co-Cr-Mo surface. However, the ability of bacteria to adhere and form a biofilm, as described by Cerca et al, varies to a wide degree depending on the strain of *S. epidermidis* [58]. Schildhauer et al also reported that *S epidermidis* varied in its adherence to various metallic implants and there was no significant difference between them [59]. Thus, the literature does not agree on how the physical characteristics of a biomaterial influence early biofilm formation. It is also possible that additional physico-chemical characteristics, such as released metal ions and chemical structure, may have some influence that inhibits or delays biofilm development. Poortinga et al showed that the change in substratum potential as a function of the number of adhering bacteria is a measure of the amount of electric charge transferred between the substratum and the bacteria during adhesion [60]. Thus, early biofilm formation is a multi-factorial process that is unlikely to be explained by a single surface characteristic. Further study is needed to refine these results of this study.

Several limitations must be noted in interpreting the data. We established an *in vitro* model that imitates early biofilm growth on the surface of a biomaterial and measured the amount of undamaged biofilm with BCR, the total biofilm mass using the CV assay and the number of viable bacteria in the biofilm using VCC. However, we cannot deny the possibility that the polishing and washing processes may have affected the surface physical characteristics of the biomaterials. Although the complex phenomena that occur *in vivo* were not accurately reproduced, a simple comparison of biofilm formation capability on various material surfaces can be made. To our knowledge, studies that evaluate the bacteriological characteristics of biofilms on Oxinium have not yet been carried out. This study allowed greater control of the experimental variables and produced fewer artifacts in the results. The ultimate research goal is to identify how the pathogens causing implant-related infections interact with biomaterial surface characteristics to affect the process of biofilm formation. We consider that our study has provided valuable results in the early stages of assessment of biofilm formation. These simple configurations are particularly encouraging as tests for use and have demonstrated that surface wettability and surface free energy have an effect on horizontal expansion in the development of biofilm.

Conclusions

We compared early biofilm formation ability on the surface of five types of solid biomaterials, eliminating the effect of surface roughness on the process. After culturing for 2 and 4 hours, there was no significant difference in the BCR of the five materials. After culturing for 6 hours, the BCR for Co-Cr-Mo alloy was significantly lower than that for Ti-6Al-4V, cp-Ti and stainless steel. However, while the absorbance value determined by the crystal violet assay and the number of colony forming units calculated by a viable cell count tended to be low for Co-Cr-Mo alloy, there was no actual significant difference. These results suggest that surface characteristics, mainly wettability and surface free energy, may have some effect on horizontal expansion in biofilm development.

Acknowledgments

The authors gratefully acknowledge Smith & Nephew Richards Ltd. (Cirencester, UK) for kindly contributing the test components.

Author Contributions

Conceived and designed the experiments: HK. Performed the experiments: HK IY TS HH HS MT YM MO. Analyzed the data: HK IY MT YM MO. Contributed reagents/materials/analysis tools: HK AY KY MO. Wrote the paper: HK AY.

References

1. Mangram AJ, Horan TC, Pearson ML, Silver LC, Jarvis WR (1999) Guideline for Prevention of Surgical Site Infection, 1999. Centers for Disease Control and Prevention (CDC) Hospital Infection Control Practices Advisory Committee. Am J Infect Control 27: 97–132; quiz 133–134; discussion 196.
2. An YH, Friedman RJ (1997) Laboratory methods for studies of bacterial adhesion. Journal of Microbiological Methods 30: 141–152.
3. Arciola CR, Campoccia D, Gamberini S, Donati ME, Pirini V, et al. (2005) Antibiotic resistance in exopolysaccharide-forming Staphylococcus epidermidis clinical isolates from orthopaedic implant infections. Biomaterials 26: 6530–6535.
4. Donlan RM (2001) Biofilms and device-associated infections. Emerg Infect Dis 7: 277–281.
5. Costerton JW, Stewart PS, Greenberg EP (1999) Bacterial biofilms: a common cause of persistent infections. Science 284: 1318–1322.
6. Fux CA, Costerton JW, Stewart PS, Stoodley P (2005) Survival strategies of infectious biofilms. Trends in Microbiology 13: 34–40.
7. Gristina AG (1987) Biomaterial-centered infection: microbial adhesion versus tissue integration. Science 237: 1588–1595.
8. Stewart PS, Costerton JW (2001) Antibiotic resistance of bacteria in biofilms. Lancet 358: 135–138.
9. Hoyle BD, Costerton JW (1991) Bacterial resistance to antibiotics: the role of biofilms. Prog Drug Res 37: 91–105.
10. Phillips CB, Barrett JA, Losina E, Mahomed NN, Lingard EA, et al. (2003) Incidence rates of dislocation, pulmonary embolism, and deep infection during

the first six months after elective total hip replacement. J Bone Joint Surg Am 85-A: 20–26.

11. Spangehl MJ, Masri BA, O'Connell JX, Duncan CP (1999) Prospective analysis of preoperative and intraoperative investigations for the diagnosis of infection at the sites of two hundred and two revision total hip arthroplasties. J Bone Joint Surg Am 81: 672–683.
12. Wymenga AB, van Horn JR, Theeuwes A, Muytjens HL, Slooff TJ (1992) Perioperative factors associated with septic arthritis after arthroplasty. Prospective multicenter study of 362 knee and 2,651 hip operations. Acta Orthop Scand 63: 665–671.
13. Hori K, Matsumoto S (2010) Bacterial adhesion: From mechanism to control. Biochem Eng J 48(3): 424–434.
14. An YH, Friedman RJ (1998) Concise review of mechanisms of bacterial adhesion to biomaterial surfaces. J Biomed Mater Res 43: 338–348.
15. Chu VH, Crosslin DR, Friedman JY, Reed SD, Cabell CH, et al. (2005) Staphylococcus aureus bacteremia in patients with prosthetic devices: costs and outcomes. Am J Med 118: 1416.
16. Tsukayama DT, Estrada R, Gustilo RB (1996) Infection after total hip arthroplasty. A study of the treatment of one hundred and six infections. J Bone Joint Surg Am 78: 512–523.
17. Zimmerli W, Ochsner PE (2003) Management of infection associated with prosthetic joints. Infection 31: 99–108.
18. Mack D, Davies AP, Harris LG, Rohde H, Horstkotte MA, et al. (2007) Microbial interactions in Staphylococcus epidermidis biofilms. Anal Bioanal Chem 387: 399–408.
19. Gotz F (2002) Staphylococcus and biofilms. Mol Microbiol 43: 1367–1378.
20. Fluckiger U, Ulrich M, Steinhuber A, Doring G, Mack D, et al. (2005) Biofilm formation, icaADBC transcription, and polysaccharide intercellular adhesin synthesis by staphylococci in a device-related infection model. Infect Immun 73: 1811–1819.
21. Heilmann C, Schweitzer O, Gerke C, Vanittanakom N, Mack D, et al. (1996) Molecular basis of intercellular adhesion in the biofilm-forming Staphylococcus epidermidis. Mol Microbiol 20: 1083–1091.
22. O'Gara JP (2007) ica and beyond: biofilm mechanisms and regulation in Staphylococcus epidermidis and Staphylococcus aureus. FEMS Microbiol Lett 270: 179–188.
23. Olson ME, Garvin KL, Fey PD, Rupp ME (2006) Adherence of Staphylococcus epidermidis to biomaterials is augmented by PIA. Clin Orthop Relat Res 451: 21–24.
24. Hunter G, Dickinson J, Herb B, Graham R (2005) Creation of Oxidized Zirconium Orthopaedic Implants. J ASTM Int 2(7).
25. Katsikogianni M, Missirlis YF (2004) Concise review of mechanisms of bacterial adhesion to biomaterials and of techniques used in estimating bacteria-material interactions. Eur Cell Mater 8: 37–57.
26. Busscher HJ, van der Mei HC (1997) Physico-chemical interactions in initial microbial adhesion and relevance for biofilm formation. Adv Dent Res 11: 24–32.
27. Gottenbos B, Van Der Mei HC, Busscher HJ, Grijpma DW, Feijen J (1999) Initial adhesion and surface growth of Pseudomonas aeruginosa on negatively and positively charged poly(methacrylates). J Mater Sci Mater Med 10: 853–855.
28. Balazs DJ, Triandafillu K, Chevolot Y, Aronsson BO, Harms H, et al. (2003) Surface modification of PVC endotracheal tubes by oxygen glow discharge to reduce bacterial adhesion. Surf Interf Anal 35(3): 301–309.
29. Henrique MA, J.; Oliver, R. (2004) Adhesion of Candidaalbicans and Candida dubliniensis to acrylic and hydroxyapatite. Col Surf B Biointerf 33: 7.
30. Scheuerman TR, Camper AK, Hamilton MA (1998) Effects of Substratum Topography on Bacterial Adhesion. J Colloid Interface Sci 208: 23–33.
31. Teughels W, Van Assche N, Sliepen I, Quirynen M (2006) Effect of material characteristics and/or surface topography on biofilm development. Clin Oral Implants Res 17 Suppl 2: 68–81.
32. Subramani K, Jung RE, Molenberg A, Hammerle CH (2009) Biofilm on dental implants: a review of the literature. Int J Oral Maxillofac Implants 24: 616–626.
33. Quirynen M, Bollen CM (1995) The influence of surface roughness and surface-free energy on supra- and subgingival plaque formation in man. A review of the literature. J Clin Periodontol 22: 1–14.
34. Chang CC, Merritt K (1991) Effect of Staphylococcus epidermidis on adherence of Pseudomonas aeruginosa and Proteus mirabilis to polymethyl methacrylate (PMMA) and gentamicin-containing PMMA. J Orthop Res 9: 284–288.
35. Donlan RM (2005) New approaches for the characterization of prosthetic joint biofilms. Clin Orthop Relat Res: 12–19.
36. Gristina AG, Costerton JW (1985) Bacterial adherence to biomaterials and tissue. The significance of its role in clinical sepsis. J Bone Joint Surg Am 67: 264–273.
37. Patel JD, Ebert M, Ward R, Anderson JM (2007) S. epidermidis biofilm formation: effects of biomaterial surface chemistry and serum proteins. J Biomed Mater Res A 80: 742–751.
38. Braem A, Van Mellaert L, Mattheys T, Hofmans D, De Waelheyns E, et al. (2013) Staphylococcal biofilm growth on smooth and porous titanium coatings for biomedical applications. J Biomed Mater Res A.
39. Heydorn A, Nielsen AT, Hentzer M, Sternberg C, Givskov M, et al. (2000) Quantification of biofilm structures by the novel computer program COMSTAT. Microbiology 146 (Pt 10): 2395–2407.
40. Sheehan E, McKenna J, Mulhall KJ, Marks P, McCormack D (2004) Adhesion of Staphylococcus to orthopaedic metals, an in vivo study. J Orthop Res 22: 39–43.
41. Gracia E, Fernandez A, Conchello P, Lacleriga A, Paniagua L, et al. (1997) Adherence of Staphylococcus aureus slime-producing strain variants to biomaterials used in orthopaedic surgery. Int Orthop 21: 46–51.
42. Ha KY, Chung YG, Ryoo SJ (2005) Adherence and biofilm formation of Staphylococcus epidermidis and Mycobacterium tuberculosis on various spinal implants. Spine (Phila Pa 1976) 30: 38–43.
43. Lee BC, Jung GY, Kim DJ, Han JS (2011) Initial bacterial adhesion on resin, titanium and zirconia in vitro. J Adv Prosthodont 3: 81–84.
44. Kajiyama S, Tsurumoto T, Osaki M, Yanagihara K, Shindo H (2009) Quantitative analysis of Staphylococcus epidermidis biofilm on the surface of biomaterial. J Orthop Sci 14: 769–775.
45. Owens DK, Wendt RC (1969) Estimation of the surface free energy of polymers. Journal of Applied Polymer Science 13: 1741–1747.
46. Kwok SCH, Wang J, Chu PK (2005) Surface energy, wettability, and blood compatibility phosphorus doped diamond-like carbon films. Diamond and Related Materials 14: 78–85.
47. Clint JH (2001) Adhesion and components of solid surface energies. Current Opinion in Colloid & Interface Science 6: 28–33.
48. Chibowski E (2003) Surface free energy of a solid from contact angle hysteresis. Advances in Colloid and Interface Science 103: 149–172.
49. Scarano A, Piattelli M, Vrespa G, Caputi S, Piattelli A (2003) Bacterial adhesion on titanium nitride-coated and uncoated implants: an in vivo human study. J Oral Implantol 29: 80–85.
50. Wijekoon CP, Goodwin PH, Hsiang T (2008) Quantifying fungal infection of plant leaves by digital image analysis using Scion Image software. Journal of Microbiological Methods 74: 94–101.
51. Bollen CM, Lambrechts P, Quirynen M (1997) Comparison of surface roughness of oral hard materials to the threshold surface roughness for bacterial plaque retention: a review of the literature. Dent Mater 13: 258–269.
52. Boks NP, Busscher HJ, van der Mei HC, Norde W (2008) Bond-strengthening in staphylococcal adhesion to hydrophilic and hydrophobic surfaces using atomic force microscopy. Langmuir 24: 12990–12994.
53. Tang P, Zhang W, Wang Y, Zhang B, Wang H, et al. (2011) Effect of superhydrophobic surface of titanium on *staphylococcus aureus* adhesion. J Nanomaterials 2011: 1–8.
54. van Pelt AW, Weerkamp AH, Uyen MH, Busscher HJ, de Jong HP, et al. (1985) Adhesion of Streptococcus sanguis CH3 to polymers with different surface free energies. Applied and Environmental Microbiology 49: 1270–1275.
55. Busscher HJ, Uyen MH, van Pelt AW, Weerkamp AH, Arends J (1986) Kinetics of adhesion of the oral bacterium Streptococcus sanguis CH3 to polymers with different surface free energies. Applied and Environmental Microbiology 51: 910–914.
56. Puckett SD, Taylor E, Raimondo T, Webster TJ (2010) The relationship between the nanostructure of titanium surfaces and bacterial attachment. Biomaterials 31: 706–713.
57. Glantz P-O (1971) The adhesiveness of teeth. Journal of Colloid and Interface Science 37: 281–290.
58. Cerca N, Pier GB, Vilanova M, Oliveira R, Azeredo J (2005) Quantitative analysis of adhesion and biofilm formation on hydrophilic and hydrophobic surfaces of clinical isolates of Staphylococcus epidermidis. Res Microbiol 156: 506–514.
59. Schildhauer TA, Robie B, Muhr G, Koller M (2006) Bacterial adherence to tantalum versus commonly used orthopedic metallic implant materials. J Orthop Trauma 20: 476–484.
60. Poortinga AT, Bos R, Busscher HJ (1999) Measurement of charge transfer during bacterial adhesion to an indium tin oxide surface in a parallel plate flow chamber. J Microbiol Methods 38: 183–189.

10

Stable Heterogeneity for the Production of Diffusible Factors in Cell Populations

Marco Archetti*

School of Biological Sciences, University of East Anglia, Norwich, United Kingdom

Abstract

The production of diffusible molecules that promote survival and growth is common in bacterial and eukaryotic cell populations, and can be considered a form of cooperation between cells. While evolutionary game theory shows that producers and non-producers can coexist in well-mixed populations, there is no consensus on the possibility of a stable polymorphism in spatially structured populations where the effect of the diffusible molecule extends beyond one-step neighbours. I study the dynamics of biological public goods using an evolutionary game on a lattice, taking into account two assumptions that have not been considered simultaneously in existing models: that the benefit of the diffusible molecule is a non-linear function of its concentration, and that the molecule diffuses according to a decreasing gradient. Stable coexistence of producers and non-producers is observed when the benefit of the molecule is a sigmoid function of its concentration, while strictly diminishing returns lead to coexistence only for very specific parameters and linear benefits never lead to coexistence. The shape of the diffusion gradient is largely irrelevant and can be approximated by a step function. Since the effect of a biological molecule is generally a sigmoid function of its concentration (as described by the Hill equation), linear benefits or strictly diminishing returns are not an appropriate approximations for the study of biological public goods. A stable polymorphism of producers and non-producers is in line with the predictions of evolutionary game theory and likely to be common in cell populations.

Editor: James A.R. Marshall, University of Sheffield, United Kingdom

Funding: This work was supported by the Natural Environment Research Council grant NE/H015701/1 (www.nerc.ac.uk). The funder had no role in study design, data collection and analysis, decision to publish, or preparation of the manuscript.

Competing Interests: The author has declared that no competing interests exist.

* Email: m.archetti@uea.ac.uk

Introduction

Cooperation for the production of diffusible molecules is commonly observed in cell populations, from bacteria to eukaryotes [1]: bacteria, for example, produce molecules that contribute to population growth (like pyocyanin [2] and pyoverdine [3]), that enable the buildup of biofilms [4] or that confer resistance to antibiotics [5]; yeast cells produce invertase that catalyzes the hydrolysis of sucrose [6], and cancer cells produce growth factors that contribute to tumour expansion [7]. Because the effect of diffusible molecules is not limited to the producer cells, a mutant cell not producing the molecule can still benefit from the presence of its neighbour producers. The free-rider advantage enjoyed by non-producer cells may lead to an increase in their frequency in the population and drive producers to extinction, with a consequent reduction in average fitness for the population - similar to what is often referred to as "the tragedy of the commons" [8]. It is understood, however, that because this free-rider advantage is frequency-dependent, if the benefit conferred by the public good is non-linear, the dynamics is generally more complex and in well-mixed populations it can lead to a stable coexistence of producers and non-producers [9]. Whether this is also the case in spatially structured populations, however, is unclear.

In the study of public goods games in spatially structured populations it is usually assumed [10] that an individual's action affects only the fitness of individuals one node away and that an individual's fitness is the sum of all the payoffs accumulated in all the groups she belongs to (all the groups formed by the one-step neighbours of her one-step neighbours). This is reasonable for interactions in human social networks, but not for cellular networks, in which molecules typically diffuse beyond a cell's one-step neighbours, and in which the benefit for a cell is a function of the number of producer cells within the diffusion range of the molecule. In order to study diffusible public goods, therefore, one must decouple the interaction neighbourhood (the group playing the game, defined by the diffusion range of the molecule) and the update neighbourhood (the one-step neighbours). While such models have been used to study a simple two-person game with a linear benefit function (the prisoner's dilemma) on a regular lattice [11,12] only recently it has been used to study the dynamics of multi-player public goods games (which are appropriate for the study of biological molecules) and there seems to be no consensus on the conclusions of these studies.

Borenstein et al. [13] showed that in a 2-D model with diffusion and linear benefits producers and non-producers can never coexist. Scheuring [14] showed, instead, stable coexistence in a 1-D model with concave benefits (diminishing returns) and even

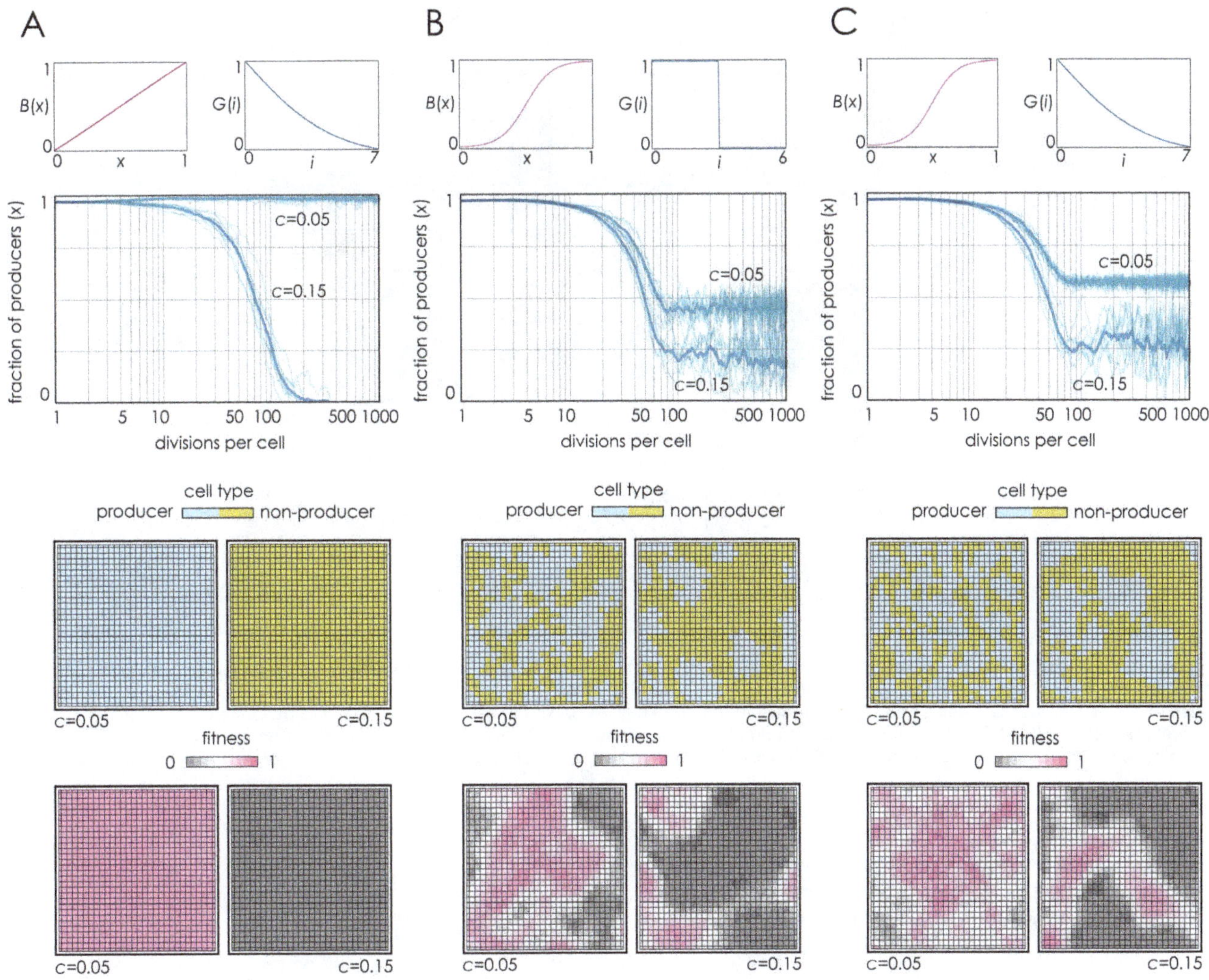

Figure 1. Realistic Hill coefficients lead to coexistence of producers and non-producers. For different benefit functions *B*(*x*) and gradients of diffusion *G*(*i*), the fraction of producers over time is show for $c = 0.05$ and $c = 0.15$. The lattices show the population after 1000 generations per cell. **A:** Linear benefit ($s = 1$, $h = 0.5$) with a diffusion gradient ($z = 3$, $d = 0$, $D = 7$). **B:** Sigmoid benefit ($s = 20$, $h = 0.5$) with no diffusion gradient ($z = 1000$, $d = 3$, $D = 6$). **C:** Sigmoid benefit ($s = 20$, $h = 0.5$) with a diffusion gradient ($z = 3$, $d = 0$, $D = 7$).

(although rarely) with linear benefits, depending on the initial conditions of the system. Archetti [15] showed that coexistence is the typical outcome of the dynamics in a 2-D model with diffusion, but did not take into account the fact that the efficacy of the diffusible molecule may decline with the distance form the source. Allen et al. [16] studied a model with diffusion and linear benefits, but did not investigate the possibility of coexistence of the two types, since in their finite stochastic population one of the strategies eventually goes to fixation (it is known, however, that in the presence of a stochastically stable polymorphism, coexistence in large populations is possible since fixation time increases exponentially with population size [17]).

A number of different assumptions in these studies [13–16] can account for the different conclusions about the possibility of a stable polymorphism. I will analyse two assumptions that seem the most prominent differences in the 2-D models described above: the shape of the diffusion gradient of the molecule (the efficacy of the molecule as a function of the distance from the producer cell) and the shape of the benefit function (the amount of public good produced as a function of the fraction of producers). In Borenstein et al. [13], Allen et al. [16] and Scheuring [14] the diffusion gradient is a smooth decreasing function, whereas in Archetti [15] the diffusion gradient is a step function. In Borenstein et al. [13] and Allen et al. [16] the benefit function is linear, whereas Scheuring [14] uses both linear and concave benefits, and Archetti [15] uses a variety of shapes, including linear, concave and sigmoid benefits. It is possible that the lack of a stable polymorphism reported by Borenstein et al. [13] is due to the fact that the benefit used in their model is linear, or it is possible that the stable polymorphism observed by Archetti [15] is due to the fact that the diffusion gradient in his model is a step function. Scheuring [14] showed that concave benefits can lead to coexistence, a result that is in contrast with both Borenstein et al. [13] and Archetti [15] but is not necessarily applicable to 2-D models and sigmoid benefit functions.

I will use a 2-D model that takes into account a variety of non-liner benefits (not analysed by Borenstein et al. [13]) and of smooth diffusion gradients (not used by Archetti [15]), extending therefore to 2-D and sigmoid benefits the results obtained by Scheuring [14] for linear and concave benefits in a 1-D model.

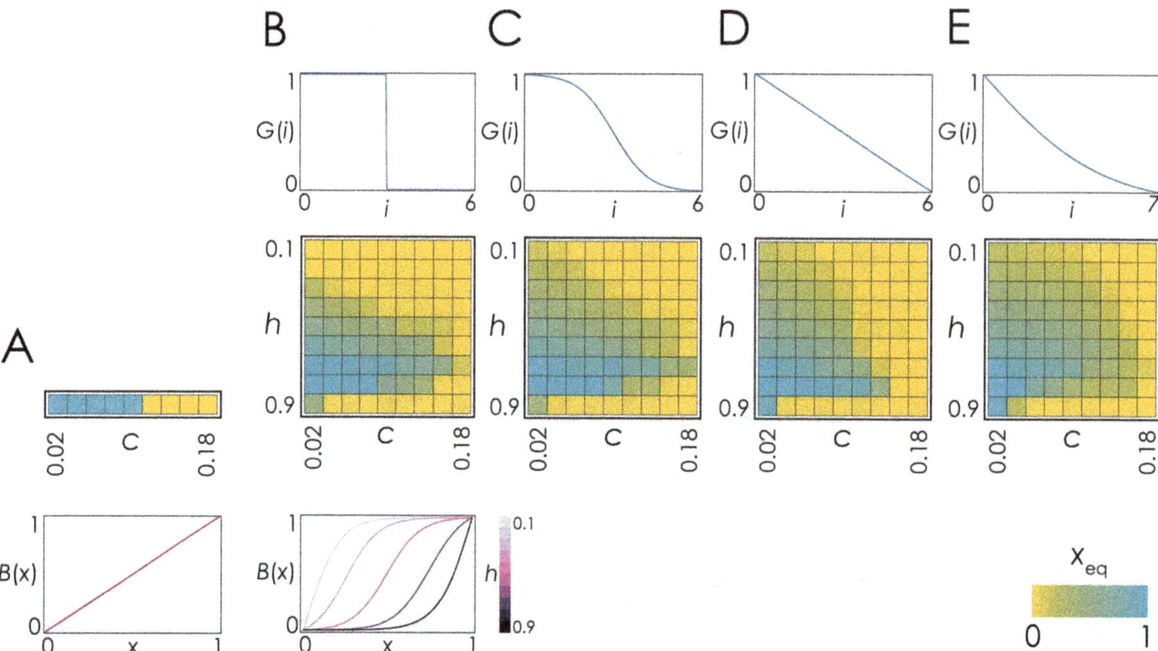

Figure 2. Different diffusion gradients allow coexistence of producers and non-producers. For different benefit functions *B*(*x*) and gradients of diffusion *G*(*i*), the contour plots show the fraction of producers at the stable mixed equilibrium (x_{eq}) as a function of *h* (the inflection point of the benefit function) and *c* (the cost of producing the molecule). **A**: Linear benefit (*s* = 1, *h* = 0.5; any of the diffusion gradients in **B-E**). **B-E**: Sigmoid benefit (*s* = 20). **B**: Fixed diffusion range with no diffusion gradient (*d* = 3, *D* = 6, *z* = 1000). **C**: Sigmoid diffusion gradient (*d* = 3, *D* = 6, *z* = 10). **D**: Linear diffusion gradient (*d* = 3, *D* = 6, *z* = 1). **E**: Convex diffusion gradient (*d* = 0, *D* = 7, *z* = 3).

The results of this extended model will help clarify what assumptions (diffusion gradient or non-linear benefits) are essential for an analysis of the problem, and whether a stable polymorphism is indeed possible with public goods involving diffusible biological molecules.

The model

Topology

Individual cells occupy all individual nodes of a planar graph, a regular 30×30 square lattice in which each node has four neighbours (Von Neumann neighbourhood) and the opposing edges are connected to form a toroidal network, in order to avoid edge effects. Differently from the standard approach (in which an individual's group is limited to her one-step neighbours and an individual plays multiple games centred on each of her neighbours [10]), here the interaction neighbourhood and the update neighbourhood are decoupled (the approach used by Archetti [15] and by Borenstein et al. [13]): a cell's group (of size n) is not limited to her one-step neighbours but is defined by the diffusion range D of the molecule. Group size for a cell is given by the number of cells within D nodes from that cell; in the Von Neumann neighbourhood, $n = 2D(D+1)+1$.

While in a model with a fixed diffusion range (a step function, in which the molecule is 100% effective up to a fixed distance from the source, and completely ineffective beyond that range) a cell's payoff can be determined simply by the number of producer cells within a given range from that cell, in a model with a diffusion gradient each cell receives contributions from other producer cells within a diffusion range D of the molecule, each contribution weighted by the distance i (the number of nodes) from that cell according to the following function:

$$G(i) = 1 - [g(i) - g(0)]/[g(D) - g(0)]$$

where

$$g(i) = 1/\left[1 + e^{-z(i-d)/D}\right]$$

The relative value of d and D determines the shape of the diffusion gradient, which is always decreasing but can be concave ($d = D$), convex ($d << D$) or sigmoid (intermediate values of $d < D$). For example, if $d = D/2$ and there are six producer cells all d nodes away from a focal cell ($i = d$ for all producers), $G(d) = 1/2$, that is, only half of each producer's contribution is available for the focal cell, hence the *weighted* number of producers (the sum of all the contributions) in that focal cell's group is three. The parameter z controls the steepness of the gradient at the inflection point: $z \to 0$ models a linear gradient; $z \to \infty$ models a step function equivalent to the one used by Archetti [15].

The game

We assume, as is standard, that there are two types of cells, producers and non-producers. Producers pay a cost c that non-producers do not pay ($0 < c < 1$). All cells (producers and non-producers) benefit from the public good produced by all the cells in their group. The benefit function is.

$$B(x) = [b(x) - b(0)]/[b(1) - b(0)]$$

where

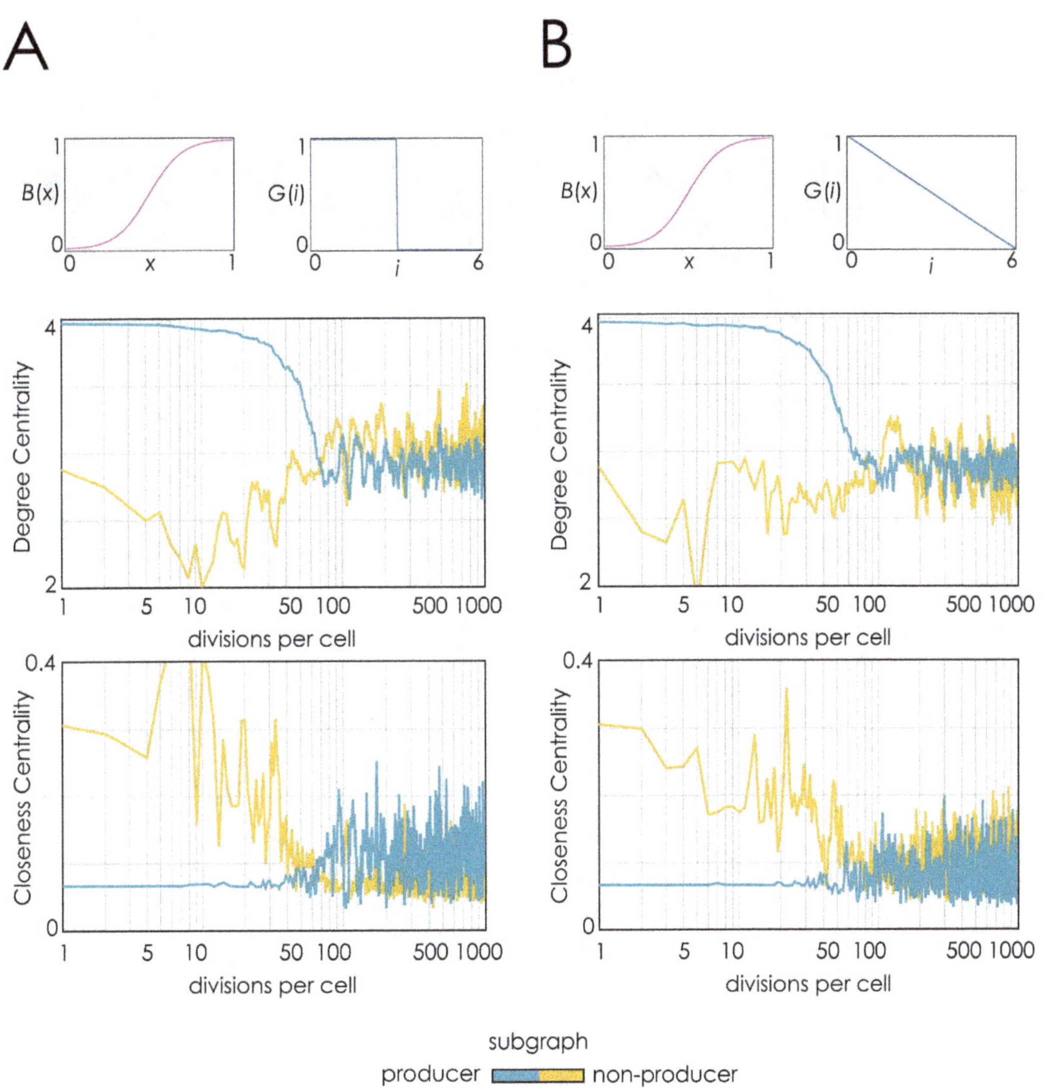

Figure 3. Diffusion gradients do not affect changes in population structure. Changes in degree centrality and closeness centrality over time are shown for the producer and non-producer subgraphs. **A**: Fixed diffusion range with no diffusion gradient ($d=3$, $D=6$, $z=1000$). **B**: Linear diffusion gradient ($d=3$, $D=6$, $z=1$).

$$b(x)=1/\left[1+e^{s(h-x)}\right]$$

is a function of the weighted fraction of producers x in the group that is, the weighted number of producers j in the group divided by group size n (see "*Topology*"). The parameter h controls the position of the inflection point ($h\to1$ gives strictly increasing returns and $h\to0$ strictly diminishing returns) and the parameter s controls the steepness of the function at the inflection point ($s\to\infty$ models a steep sigmoid function that is essentially an on/off switch; $s\to0$ models linear benefits) [18]. $B(j)$ is a simple normalisation of the logistic function $b(j)$ (without the normalisation, low values of s would yield constant benefits rather than linear increasing benefits). While input-output functions in biochemistry are often described by the Hill equation [19,20], we use this normalised logistic function because it enables to model not only sigmoid benefits but also linear and concave benefits, which are used by Scheuring [14] and Archetti [15], discussed by Borenstein [13], and used in models of public goods games in well-mixed populations [9].

A cell's payoff is a function of the amount of factor produced by the group she belongs to. I use a birth-death process equivalent to the one used by Archetti [15] and by Borenstein et al. [13]. The process starts with a number of non-producer cells placed on the graph; at each round a cell x with a payoff P_x is selected (at random) for update (death); a cell y (with a payoff P_y) is then chosen among x's neighbours. Two types of update are used: in the deterministic case, if $P_x>P_y$, no update occurs, while if $P_x<P_y$, x will adopt y's strategy (unconditional imitation); in the stochastic case, replacement occurs with a probability given by $(P_y-P_x)/M$, where M ensures the proper normalization and is given by the maximum possible difference between the payoffs of x and y [10]. Results are obtained averaging the final 200 of 1000 generations per cell, averaged over 10 different runs.

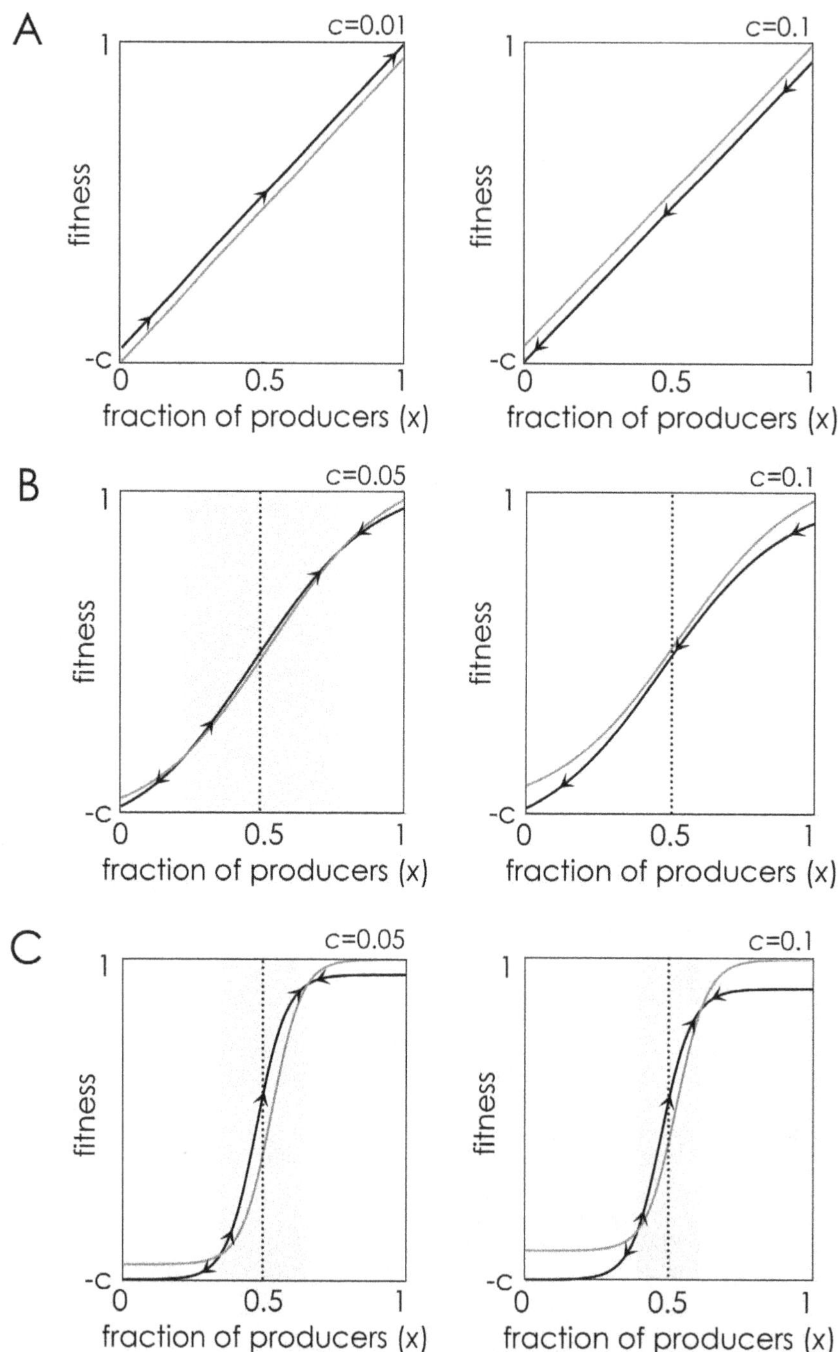

Figure 4. Why only non-linear benefits can lead to coexistence of producers and non-producers. The fitness of producers (black curve) and non producers (grey curve) as a function of the fraction (*x*) of producers within the diffusion range of the molecule, for different steepness coefficients (*s*). The arrows show the direction of the dynamics. The shaded area shows the basin of attraction of the internal stable equilibrium (if it exists). $n = 20$, $h = 0.5$. **A:** The benefit $B(x)$ of the molecule is an almost linear ($s = 0.001$) function of its concentration. **B:** The benefit $B(x)$ is a sigmoid ($s = 5$) function of its concentration. **C:** The benefit function is essentially a step function ($s = 20$).

Results

Sigmoid benefits lead to coexistence, with or without a diffusion gradient

First, I checked that the results of two degenerate versions of the model correspond to the ones reported previously: with a linear benefit and a smooth diffusion gradient [13], no coexistence of producers and non-producers is observed [**Figure 1A**]; with no diffusion gradient and a non-linear benefit [15], producers and non-producers can coexist [**Figure 1B**]. These results therefore are in line with the two simpler models [13,15] that led to opposing results. When benefits are non-linear, replacing the step function diffusion range [15] with a more realistic diffusion gradient [13] still leads to a coexistence of producers and non-producers [**Figure 1C**]. The shape of the benefit function,

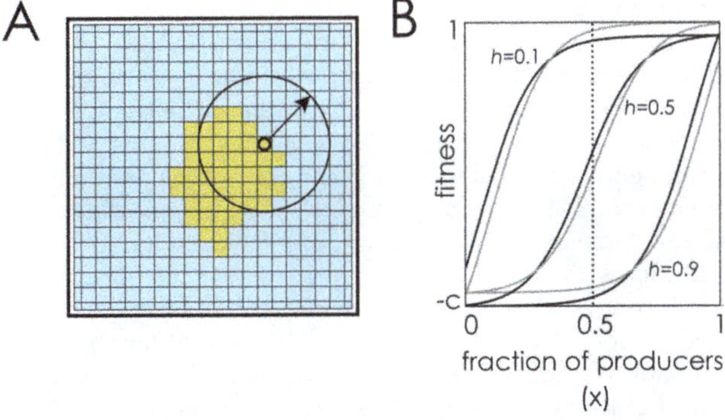

Figure 5. Why sigmoid benefits lead to different results from concave and convex benefits in spatially structured populations. The fraction of producers within the diffusion range (shown by an arrow) of molecules produced by cells at the producer/non-producer front remains approximately constant (~0.5) even as the front moves ahead. If the benefit function is concave ($h = 0.1$) or convex ($h = 0.9$), at this fraction of producers (~0.5) non-producers have an advantage, whereas producers have an advantage if the benefit function is sigmoid ($h = 0.5$). ($n = 20$, $c = 0.1$, $s = 10$).

therefore, seems crucial for the outcome of the dynamics, whereas the diffusion gradient seems irrelevant.

The result is robust and can hold even for concave benefits

To check the robustness of the previous conclusion I analyse the dynamics under a variety of diffusion gradients and benefit function. While linear benefits always lead to the extinction of one of the two types [**Figure 2A**], as shown by Borenstein et al. [13], non-linear benefits can lead to coexistence under a wide range of parameters if the cost is not too high [**Figure 2B-E**], in line with previous results in well-mixed populations [9], and as shown by Archetti [15] for diffusible molecules in spatially structured populations. The type of diffusion gradient is largely irrelevant, whereas the shape of the benefit function is the major determinant of the dynamics.

While Borenstein et al. [13] analyse linear benefits, they suggest that concave benefits, like linear benefits, would not lead to coexistence. With diffusion gradients, however, concave benefits (diminishing returns) can actually lead to a stable polymorphism, even though under a much more limited parameter range than with sigmoid benefits [**Figure 2B–E**]. Therefore diffusion gradients are even more conductive to polymorphic equilibria than the simple diffusion range (step function) used by Archetti [15].

No effect of diffusion gradients on population structure

Analysing changes in the topology of the subgraphs of producers and non-producers [**Figure 3**] reveals that the shape of the diffusion gradients does not affect significantly the spatial dynamics of the population. The degree centrality of the non-producer subgraph is only slightly lower (that is, non-producers have fewer neighbours in their clusters) when the molecule has a smooth decreasing diffusion gradient than when the diffusion is a step function; intermediate cases (not shown) lead to similar negligible differences. The closeness centrality (the inverse of the sum of the distance to all other vertices) is slightly higher for non producers and lower for producers (that is, non-producer clusters tend to be smaller and producer cluster bigger), but the differences are negligible. In short, whether diffusion is modelled as a gradient or not does not affect significantly the topology of the two subgraphs.

Rationale of the results

The logic of non-linear benefits on the dynamics of public goods is explained in **Figure 4** (see also previous discussions for well-mixed populations [9,18]: If the benefit of the molecule is a linear function of its concentration ($s \to 0$), either producers or non-producers have a higher fitness for any frequency of producers, depending only on the relative cost/benefit of producing the molecule (in sizeable groups, for reasonable costs producers will always have a disadvantage, this results in what we usually refer to as "N-person Prisoner's Dilemma", and hence to what is generally referred to as a "tragedy of the commons" [8]). If benefits are non-linear, however (larger s) and the cost c is not too high, a stable polymorphism is possible. Note that large s values (steep benefit functions) allow stable polymorphic equilibria for larger values of c, but they make the population less robust to random fluctuations in the fraction of producers, that is, to a smaller basin of attraction for the stable polymorphic equilibrium [**Figure 4**].

The above argument is valid for well-mixed populations. In spatially structured populations the logic requires one further step. The crucial point is that, while in a well-mixed population the fraction of producers in the group is approximately the same as the frequency in the population (because new groups are formed at each generation), this is not the case in a spatially structured population, where the local fraction of producers can be much lower or higher than in the rest of the population. Consider a cluster of non-producers in a population of producers [**Figure 5A**]: the group defined by the diffusion range of the molecules produced by a cell at the edge between producers and non-producers is made by approximately half producers and half non-producers (if the diffusion range and the clusters are large enough); if non-producers have a higher fitness than producers (which is always the case if benefits are linear and c is not too small), the non-producer cluster will expand and the producer/non-producer front will move ahead; the process will go on with new groups (new producer/non-producer edges) until the whole population is made of non-producers. The same will happen when benefits are concave or convex, but not when benefits are sigmoid [**Figure 5B**]. If population structure is more complex (that is, if there is more than one cluster of non-producers in a population of producers) the intuition provided by Figure 5 fails, although it is clear that with concave or convex benefits (h close 0 or close to 1 in **Figure 2**) producers will ultimately go to extinction, whereas with sigmoid benefits the position of producer and non-producer cells will fluctuate around a mixed stable equilibrium [**Figure 1, Movie S1**].

Discussion

Producers and non-producers coexist if benefits are sigmoid

In summary, in a public goods game with two types of cells, producers and non-producers of a diffusible molecule, a stable polymorphism of the two types is likely under a wide range of parameters if the effect of the molecule is a sigmoid function of its concentration. The shape of the diffusion gradient is largely irrelevant and even a simple step function (that is, a model in which the diffusible molecule is fully functional up to a certain range and completely ineffective beyond that range) is accurate enough. These results, therefore, are similar to results in well-mixed populations [9], where coexistence of producers and non-producers is commonly observed. In short, the difference between Borenstein et al. [13] and Archetti [15] is the shape of the benefit function, not the shape of the diffusion gradient. The conclusion of

Borenstein et al. [13] that coexistence is impossible depends on their assumption that benefits are linear.

Borenstein et al. [13] conclude that their assumption of linearity does not affect their result, and that non-linear benefits would lead to the same result (no coexistence of the two types). Their discussion of non-linearities, however, assumes a concave benefit function, rather than a sigmoid function. The extinction of one of the two types when benefits are concave is also observed in a model without diffusion gradient [15]. Scheuring [14] observes coexistence with concave benefits in a 1-D model. Here we have seen that concave benefits can actually lead to coexistence of the two types only under very restrictive conditions (the production cost must not be too high). Coexistence of producers and non-producers, however, is the typical result of the dynamics if benefits are sigmoid.

Sigmoid benefits are common in biological public goods

The relevance of these results depends on how common linear, sigmoid or concave benefits are in nature. Borenstein et al. [13] ask "could a saturating resource uptake curve, such as the Michaelis-Menten model facilitate coexistence?" (concluding that the answer is no). The Michaelis-Menten equation discussed by Borenstein et al. [13], however, is a concave function. Biological input-output systems, instead, generally follow the Hill equation [19,20], that is, biological input-output systems generally show a slow response at low inputs levels followed by a steep increase in response at intermediate levels and again a decreasing sensitivity as input levels increases. In other words, the effect of a biological molecule is often a sigmoid, not concave, function of its concentration [20–23]. Examples of sigmoid benefits have been reported for public goods in both microbes [24], and cancer cells [25].

There are various proximate explanations for the occurrence of sigmoid benefits in nature. The most basic explanation [20] is *cooperative binding*: transforming a single molecule to an active state may require simultaneous binding by multiple input signal molecules. Other explanations [23] include *titration of a repressor* (the initial reaction may inactivate the input signal molecule or reduce sensitivity to low intensity input signals), and *opposing saturated forward and back reactions* (a back reaction may return the active form produced by the initial reaction to the inactive state, and if the back reaction saturates at low signal input intensity, then a logarithmic output will result at low input intensity). More in general, the reason for the common occurrence of "Hill kinetics" is that the final physiological or behavioral response of a biological system is produced by a cascade of signal originating from cellular receptors and sensory systems. This series of reactions amplifies even the slightest departure from linearity of the underlying individual chemical reactions [21].

Since biological molecules generally follow the Hill kinetics (sigmoid benefits) rather than the Michaelis-Menten kinetics (concave benefits), and because linear benefits are unlikely to exist at all in nature, the coexistence of producers and non-producers is likely to be the typical outcome of cooperation for the production of diffusible molecules.

Further work on diffusible public goods

While, as we have seen, the exact shape of the diffusion gradient is not essential, the diffusion range of the molecule is important in determining the dynamics of the production of diffusible molecules. Here we have assumed that the diffusion range is small, thus group size (the interaction neighborhood) is always one order of magnitude smaller than population size. If the diffusion range is such that group size approaches population size, however, the extinction of one of the two types becomes possible [10,17]. Measuring the diffusion range of molecules that act as public goods is therefore important to understand their dynamics.

While Borenstein et al. [13] use a stochastic update rule, both Scheuring [14] and Archetti [15] show that whether update is deterministic or stochastic does not lead to substantially different results, except for the time required to reach a polymorphism. The details of the update rule, however, may be important. Scheuring [14] shows that, in his 1-D model, concave benefits lead to coexistence when using the birth-death update rule (the same rule used by both Borenstein et al. [13], Archetti [15] and here), but not with a death-birth rule. It is understood that very different results can be expected in spatially structured populations based on whether local and global selection are random or proportional to fitness (a more appropriate classification of update rules than the mere order of birth and death [26]).

The combined results of Borenstein et al. [13], Scheuring [14], Archetti [15], and the ones reported here help us understand the importance of non-linear benefits and of diffusion gradients; Scheuring [14] goes one step further and analyses a further update rule, although in a 1-D model and only with linear and concave benefits; Allen et al. [16] make a thorough analysis of diffusion and use different update rules, but their analysis is limited to linear benefits. It would be worth extending these studies of diffusible public goods to analyse the effect of different update rules on 2-D games with different update rules and benefits functions.

Conclusion

As noted recently by Borenstein et al. [13], Scheuring [14], Archetti [15] and Allen et al. [16], the dynamics of public goods production in biological systems must be analysed by games in which the interaction neighbourhood extends beyond the update neighbourhood. As we have seen here, however, the precise shape of the diffusion gradient has a relatively little impact on the results, and therefore an accurate treatment of diffusion is not essential; a simple diffusion range without decreasing efficacy is a good enough approximation. The crucial assumption of the models, instead, is the shape of the benefit function. Since the dynamics of non-linear public goods is so radically different from the dynamics with linear benefits, and since biological public goods are generally non-linear, the production of diffusible molecules in biological systems cannot be reliably approximated by models with linear benefits, even though the analysis of non-linear games is by far more complex and rarely amenable to analytical proofs. In a model with non-linear benefits, and therefore in a population of cells producing a diffusible molecule, stable polymorphism is not only possible, but the likely outcome of the dynamics.

Supporting Information

Movie S1 Long-term coexistence of producers and non-producers. Population structure (blue: producer cells; yellow: non-producer cells) and fitness, from $-c$ (grey) to 1 (pink) (as in Figure 1) during the first 200 divisions per cell, after a cluster of non-produce cells arises in a population of producers; sigmoid benefit ($s = 20$, $h = 0.5$), no diffusion gradient ($z = 1000$, $d = 3$, $D = 6$), $c = 0.05$.

Author Contributions

Conceived and designed the experiments: MA. Performed the experiments: MA. Analyzed the data: MA. Contributed reagents/materials/analysis tools: MA. Contributed to the writing of the manuscript: MA.

References

1. Crespi BJ (2001) The evolution of social behavior in microorganisms. Trends Ecol. Evol., 16: 178–183.
2. Wang Y, Kern SE, Newman DK (2010) Endogenous phenazine antibiotics promote anaerobic survival of Pseudomonas aeruginosa via extracellular electron transfer. Journal of Bacteriology 192: 365–369.
3. Meyer JM (2000) Pyoverdines: pigments, siderophores and potential taxonomic markers of fuorescent Pseudomonas species. Archives of Microbiology 174: 135–142.
4. Rainey PB, Rainey K (2003) Evolution of cooperation and conflict in experimental bacterial populations. Nature 425: 72–74.
5. Lee HH, Molla MN, Cantor CR, Collins JJ (2010) Bacterial charity work leads to population-wide resistance. Nature 467: 82–86.
6. Greig D, Travisano M (2004) The prisoner's dilemma and polymorphism in yeast SUC genes. Proc. Roy. Soc. B 271: S25–S26.
7. Axelrod R, Axelrod DE, Pienta KJ (2006) Evolution of cooperation among tumor cells. Proc. Natl. Acad. Sci. USA 103: 13474–13479.
8. Hardin J (1968) The tragedy of the commons. Science 162: 1243–1248.
9. Archetti M, Scheuring I (2012) Review: Game theory of public goods in one-shot social dilemmas without assortment. J. Theor. Biol. 299: 9–20.
10. Perc M, Gómez-Gardeñes J, Szolnoki A, Floria LM, Moreno Y (2013) Evolutionary dynamics of group interactions in structured populations: A review. J. R. Soc. Interface 10: 20120997.
11. Ifti M, Killingback T, Doebeli M (2004) Effects of neighbourhood size and connectivity on the spatial continuous prisoner's dilemma. J. Theor. Biol. 231: 97–106.
12. Ohtsuki H, Pacheco JM, Nowak MA (2007) Evolutionary graph theory: breaking the symmetry between interaction and replacement. J. Theor. Biol. 246: 681–694.
13. Borenstein DB, Meir Y, Shaevitz J, Wingreen NS (2013) Non-local interaction vie diffusible resource prevents coexistence of cooperators and cheaters in a lattice model. PLOS One, 8: e63304.
14. Scheuring I (2013) Diffusive public goods and coexistence of cooperators and cheaters on a 1D lattice. PLOS One, 9(7): e100769.
15. Archetti M (2013) Dynamics of growth factor production in monolayers of cancer cells and evolution of resistance to anticancer therapies. Evolutionary Applications 6: 1146–1159.
16. Allen B, Gore J, Nowak MA (2013) Spatial dilemmas of diffusible public goods. eLife 2: e01169.
17. Antal T, Scheuring I (2006) Fixation of strategies for an evolutionary game in finite populations. Bulletin of Mathematical Biology 68: 1923–1944.
18. Archetti M, Scheuring I (2011) Coexistence of cooperation and defection in public goods games. Evolution 65: 1140–1148.
19. Hill AV (1910). The possible effects of the aggregation of the molecules of hæmoglobin on its dissociation curves. Proceedings of the Physiological Society, Jan 1910.
20. Cornish-Bowden A (2012) *Fundamentals of Enzyme Kinetics*, 4th edition; Wiley Blackwel.
21. Frank, S. A. Input-output relations in biological systems: measurement, information and the Hill equation. Biology Direct 8: 31.
22. Tyson JJ, Chen KC, Novak B (2003) Sniffers, buzzers, toggles and blinkers: dynamics of regulatory and signaling pathways in the cell. Curr Opin Cell Biol 15: 221–231.
23. Zhang Q, Bhattacharya S, Andersen ME (2013) Ultrasensitive response motifs: basic amplifiers in molecular signalling networks. Open Biol, 3: 130031.
24. Chuang JS, Rivoire O, Leibler S (2010) Cooperation and Hamilton's rule in a simple synthetic microbial system. Mol Syst Biol. 6: 398.
25. Jourdan M, Mahtouk K, Veyrune J, Couderc G, Fiol G, et al. (2005) Delineation of the roles of paracrine and autocrine interleukin-6 (IL-6) in myeloma cell lines in survival versus cell cycle. A possible model for the cooperation of myeloma cell growth factors. Eur. Cytokine Netw. 16: 57–64.
26. Grafen A, Archetti M (2008) Natural selection of altruism in inelastic viscous homogeneous populations. J. Theor. Biol. 252: 694–710.

11

Comparison of the Bruker MALDI-TOF Mass Spectrometry System and Conventional Phenotypic Methods for Identification of Gram-Positive Rods

Claudia Barberis[1]*, Marisa Almuzara[1], Olivier Join-Lambert[2,3], María Soledad Ramírez[4], Angela Famiglietti[1], Carlos Vay[1]

1 Instituto de Fisiopatología y Bioquímica Clínica, Hospital de Clínicas José de San Martín, Facultad de Farmacia y Bioquímica, Universidad de Buenos Aires, Ciudad Autónoma de Buenos Aires, Argentina, **2** Université Paris Descartes, Paris, France, **3** Laboratoire de Microbiologie, Hopital Necker-Enfants malades, Paris, France, **4** Instituto de Microbiología y Parasitología Médica, Universidad de Buenos Aires-Consejo Nacional de Investigaciones Científicas y Tecnológicas, Ciudad Autónoma de Buenos Aires, Argentina

Abstract

In recent years, MALDI-TOF Mass Spectrometry (MS) method has emerged as a promising and a reliable tool for bacteria identification. In this study we compared Bruker MALDI-TOF MS and conventional phenotypic methods to identify a collection of 333 Gram-positive clinical isolates comprising 22 genera and 60 species. 16S rRNA sequencing was the reference molecular technique, and *rpo*B gene sequecing was used as a secondary gene target when 16Sr RNA did not allow species identification of *Corynebacterium* spp. We also investigate if score cut-offs values of ≥1,5 and ≥1,7 were accurate for genus and species-level identification using the Bruker system. Identification at species level was obtained for 92,49% of Gram-positive rods by MALDI-TOF MS compared to 85,89% by phenotypic method. Our data validates the score ≥1,5 for genus level and ≥1,7 for species-level identification in a large and diverse collection of Gram-positive rods. The present study has proved the accuracy of MALDI-TOF MS as an identification method in Gram-positive rods compared to currently used methods in routine laboratories.

Editor: Holger Rohde, Universitätsklinikum Hamburg-Eppendorf, Germany

Funding: This work was supported by grants from the Secretaría de Ciencia y Técnica de la Universidad de Buenos Aires (UBACyT) to C.V. M.S.R. is a member of the CONICET research career. The funders had no role in study design, data collection and analysis, decision to publish, or preparation of the manuscript.

Competing Interests: All the authors have declared that no competing interests exist.

* Email: claudiabar07@gmail.com

Introduction

Corynebacterium spp. and other Gram-positive rods (GPR) are widespread throughout nature, being most of the species part of the normal skin flora of humans and animals [1]. However, they are increasingly recognized as causes of human infection since immunosuppressive treatments, oncological diseases, antibiotic treatments, and multiple invasive procedures make patients vulnerable to opportunistic infections [2]. Correct identification of *Corynebacterium* spp. and other GPR species is a very challenging task because it will help to identify the real source of infection and install the appropriate treatment for the infection.

In routine laboratories the most used techniques to identify microorganisms are the conventional phenotypic tests [1,3]. However, these tests are time consuming and do not always give reliable identification at the species level.

The use of the rRNA gene sequence for GPR identification is one of the most commonly used molecular technique and is considered the gold standard approach for its identification in many cases [4]. Sequencing of the 16S rRNA gene discerns among most species in the genus *Corynebacterium* [5]. In order to arrive to a correct molecular identification, since the 16S rRNA gene of corynebacteria has very little polymorphism, sequencing the complete 16S rRNA gene (approximately 1,500 bp) is required [6]. The sequencing of a fragment of the *rpo*B gene resulted to be an alternative method for the identification of corynebacteria. This later method should be used to resolve ambiguous cases for definitive identification [6]. However, it has some limitation, as it is expensive for routine laboratories.

In the last years MALDI-TOF Mass Spectrometry (MS) has emerged as a promising technique for bacterial identification. MALDI-TOF MS is a rapid, reliable diagnostic tool for the identification of most microorganisms [7]. The technology is unique in clinical microbiology, allowing laboratories to definitively identify bacterial isolates within minutes. The rapid turn-around time and minimal cost for consumables per specimen compared with conventional identification methods have resulted in MALDI-TOF MS being increasingly used in clinical laboratories worldwide [8].

Using the Bruker Biotyper MALDITOF MS system, some organisms such as Gram-negative bacteria, are easily analyzed by directly smearing a colony onto the MS target [9] but other types of bacteria such as GPR generally require preparatory tube extraction or a direct on plate testing using formic acid [10].

Previous studies reported that the Bruker MALDI-TOF MS system accurately identified GPR species such as *Corynebacterium* spp. [7,11,12]. However, information of GPR other than *Corynebacterium* spp. is scarcely documented or poor identification results were obtained in the available studies [13].

In a recent study, Alatoom et al. [11] evaluated the MALDI-TOF MS method using Bruker Biotyper system (Bruker Daltonics, Billerica, MA) for the identification of 92 clinical isolates of *Corynebacterium* species and compared with the identification using 16S rRNA gene sequencing. The authors suggested a new lower score for genus and species identification level, being the proposed scores ≥ 1,7 and ≥1,5, respectively.

The aims of the present study were to compare the performance of MALDI-TOF MS and conventional methods for the identification of GRP and also validate the previous proposed score cut-off of ≥ 1,7 for species level identification in a greater collection of heterogeneous group of GPR. All the results were compared with molecular techniques as the gold standard.

Materials and Methods

Bacterial isolates

A total of 333 GPR clinical strains from a culture collection obtained during the period 2009–2013 at the University teaching hospital, Hospital de Clínicas "José de San Martín", Buenos Aires, Argentina, were used in this study. All isolates were recovered during clinical practice. We included every GPR species that were considered clinically significant as they were either recovered as unique pathogen from normally sterile body sites (e.g., blood culture or urines samples with a bacterial count of $>10^4$/ml) or from adequately collected clinical material where they are the predominant organisms.

Briefly, bacterial strains were grown on 5% sheep blood agar in 5% CO_2 at 35°C for 24 h (or later for slower growing species). The strains were stored at −70°C in brain heart infusion broth containing 20% glycerol until use.

All isolates were identified by conventional phenotypic methods as described before [1,3]. Phenotypic identification of colonies included morphology, Gram staining, catalase activity, lipophily for *Corynebacterium* spp. and biochemical methods using the algorithm previously described by Funke et al [3,14]. In parallel with conventional biochemical methods, molecular identification was performed.

Molecular identification was used as the gold standard method to compare the results obtained by MALDI TOF MS and conventional phenotypic methods. 16S rRNA gene sequencing was carried out for characterization of all isolates. In *Corynebacterium* spp. identification, *rpo*B gene was used as a secondary gene target when 16S rRNA gene did not allow correct identification to species level. PCR reactions were performed as previously described [15,16]. Sequencing of the PCR products were performed on both DNA strands using ABIPrism 3100 BioAnalyzer equipment at Macrogen Inc. sequencing facility, South Korea, sequencing facility. The sequences were analyzed using the BLAST v2.0 software (http://www.ncbi.nlm.nih.gov/BLAST/). A ≥99,0% (16S rRNA gene) and ≥95,0% (*rpo*B gene) similarity cut-off was required for species identification.

MALDI-TOF Mass Spectrometry

All isolates were retrospectively identified by MALDI-TOF MS. The clinical strains from the culture collection were subcultured on Columbia agar containing 5% sheep blood (Laboratorios Britania, Argentina) at 37°C with 5% CO_2 for 24 to 48 h for MALDI-TOF MS measurement. Bacterial isolates were identified by the direct colony on plate extraction method as previously described [10]. MALDI-TOF target plates were inoculated into the spots by picking a freshly grown overnight colony and overlaid with 1 µl of 70% formic acid (Sigma-Aldrich). Each spot was allowed to dry and subsequently overlaid with 1 µl of matrix (α-cyano-4-hydroxycinnamic acid).

Mass spectra were acquired using the MALDI-TOF MS spectrometer in a linear positive mode (Microflex, Bruker Daltonics). The bacterial test standard (BTS, Bruker) was used for instrument calibration. Mass spectra were analyzed in a m/z range of 2,000 to 20,000. The MALDI Biotyper library version 3.0 and MALDI Biotyper software version 3.1 were used for bacterial identification. Based on previous studies [11] cut-off scores for identification were: ≥1,5 for genus level, ≥1,7 for species-level. A score <1,5 was considered as resulting in no reliable identification. A minimum difference of 10% between the top and next closest score was required for a different genus or species [11].

Data analysis

Statistical analysis to calculate the efficiency of MALDI TOF MS and conventional phenotypic methods was carried out only in those genera comprising more than ≥ 20 isolates. Confidence intervals for identification were calculated with DAG Stat spreadsheet [17].

Results

Among the 333 clinical GPR included in our study, we identified 22 genera and 60 species, most of which are known to cause human infections. This shows the great diversity of genus and species that we have analyzed.

For all the isolates the rate of identification at the species level was higher for MALDI-TOF MS compared to conventional phenotypic method (92,49% vs 85,89%). All the obtained results were compared with the 16S RNA/*rpo*B sequence analysis as the gold standard.

Among 216 *Corynebacterium* isolates tested, identification by MALDI-TOF MS at species level was 93,52% (202/216) whereas identification by conventional methods was 92,13 (199/216) (Tables 1, 2).

Validating the proposed MALDI-TOF MS identification score for *Corynebacterium* at genus and species level [11], 153/202 isolates (75,7%) yielded a score ≥2,0, all of which were correctly identified to the species level. Forty-nine isolates (24,3%) yielded a score between <2,0-≥1,7. There were 4 isolates that could not be identified at species level because the difference between the top and the next score gave two different results (*Corynebacterium aurimucosum/Corynebacterium. minutissimum*) with less than 10%. Also, five isolates were misidentified (Table 3).

For *Actinomyces* and related genera (*Actinobaculum* spp., *Varibaculum* spp., *Arcanobacterium* spp., *Trueperella* spp., *Bifidobacterium* spp.), the results of MALDI-TOF MS at genus level was 95,77% (68/71) and to species level 92,96% (66/71). In these genera the phenotypic identification was less reliable to allow the correct species level identification (Table 1, 2). The MALDI-TOF MS identification scores were ≥2,0 in 41/65 isolates (63,1%), all of which were correctly identified to the species level. Twenty-four isolates (36,9%) yielded a score between <2,0-≥1,7 (Table 3).

For other GPR, identification by MALDI-TOF MS at genus level was 93,5% (43/46) and 86,9% (40/46) to species level. Thirty-two isolates (80%) generated identification scores >2,0 and 8 isolates (20%) scores of <2,0- ≥1,7 (Table 3). In the particular case of pigmented Gram-positive rods and aerobic actinomycetes

Table 1. Identification of 333 Gram- positive rods using MALDI-TOF MS.

Organism (No. of isolates tested)	**No. of isolates by level of identification**			
	Species level	**Genus level**	**No ID**	**Error**
	(≥1,7)	**(≥1,5)**	**(<1,5)**	
Corynebacterium spp.				
C. striatum (61)	61	61	0	0
C. amycolatum (38)	37	38	0	1[a]
C. urealyticum (14)	13	13	1	0
C. pseudodiphtheriticum (13)	12	13	0	1[b]
C. glucuronolyticum (12)	12	12	0	0
C. jeikeium (12)	12	12	0	0
C. tuberculostearicum (9)	7	7	2	0
C. aurimucosum (8)	4	8	0	0
C. propinquum (8)	8	8	0	0
C. accolens (7)	7	7	0	0
C. afermentans subsp *lipophilum* (4)	2	3	1	1[c]
C. group F (4)	4	4	0	0
C. mucifaciens (4)	4	4	0	0
C. diphtheriae (3)	3	3	0	0
C. simulans (3)	3	3	0	0
C. afermentans subsp *afermentans* (2)	1	1	1	0
C. coyleae (2)	1	2	0	1[d]
C. imitans (2)	2	2	0	0
C. kroppenstedtii (2)	2	2	0	0
C. bovis (1)	1	1	0	0
C. durum (1)	1	1	0	0
C. macginleyi (1)	1	1	0	0
C. minutissimum (1)	0	1	0	1[e]
C. pseudotuberculosis (1)	1	1	0	0
C. riegelli (1)	1	1	0	0
C. ulcerans (1)	1	1	0	0
C. xerosis (1)	1	1	0	0
Total (n: 216)	202 (93,52)	211(97,68)	5(2,3)	5(2,3)
Actinomyces turicensis (7)	6	7	0	1[f]
Actinomyces radingae (11)	9	10	1	1[g]
Actinomyces urogenitalis (6)	6	6	0	0
Actinomyces odontolyticus (5)	5	5	0	0
Actinomyces europaeus (3)	2	2	1	0
Actinomyces naeslundii/viscosus (5)	5	5	0	0
Actinomyces neuii (10)	10	10	0	0
Actinobaculum schaalii (10)	9	9	1	0
Actinomyces gravenitzii (1)	1	1	0	0
Varibaculum cambriense (1)	1	1	0	0
Arcanobacterium haemolyticum (6)	6	6	0	0
Trueperella bernardiae (2)	2	2	0	0
Bifidobacterium scardovii (1)	1	1	0	0
Propionibacterium acnes (1)	1	1	0	0
Propionibacterium avidum (2)	2	2	0	0
Total (n:71)	66(92,96)	68(95,77)	3(4,2)	2(2,8)
Leifsonia spp. (1)	0	0	1	0

Table 1. Cont.

Organism (No. of isolates tested)	No. of isolates by level of identification			
	Species level (≥1,7)	Genus level (≥1,5)	No ID (<1,5)	Error
Exiguobacterium aurantiacum. (2)	2	2	0	0
Exiguobacterium spp. (1)	0	0	1	0
Brevibacterium casei (2)	2	2	0	0
Brevibacterium ravenspurgense (1)	1	1	0	0
Microbacterium aerolatum (1)	1	1	0	0
Microbacterium aurum (1)	1	1	0	0
Microbacterium testaceum (1)	1	1	0	0
Microbacterium oxydans (1)	1	1	0	0
Microbacterium hominis (1)	0	1	0	0
Microbacterium spp. (1)	0	1	0	0
Arthrobacter protophormiae. (1)	1	1	0	0
Cellulosimicrobium cellulans (2)	2	2	0	0
Turicella otitidis (2)	2	2	0	0
Rothia aeria (2)	2	2	0	0
Dermabacter hominis (10)	10	10	0	0
Clostridium tertium (1)	1	1	0	0
Listeria monocytogenes (4)	4	4	0	0
Lactobacillus gasseri (1)	1	1	0	0
Lactobacillus rhamnosus (1)	1	1	0	0
Lactobacillus paracasei (1)	1	1	0	0
Rhodococcus equi (2)	2	2	0	0
Gordonia terrae (2)	2	2	0	0
Gordonia spp. (1)	0	1	0	0
Dietzia maris (1)	1	1	0	0
Dietzia natronolimnaea (1)	1	1	0	0
Dietzia spp. (1)	0	0	1	0
Total (n:46)	40(86,9)	43(93,5)	3(6,5)	0
TOTAL (n:333)	308 (92,49)	322 (96,69)	11(3,3)	7(2,1)

[a]Corynebabacterium aurimucosum; [b]Corynebacterium propinquum; [c]Corynebacterium jeikeium; [d]Corynebacterium afermentans, [e]Corynebacterium amycolatum, [f] Actinomyces radingae, [g]Actinomyces europaeus.

when we use the 16S RNA amplification to arrive to the correct identification, we observed that in some isolates such as *Leifsonia* spp., *Exiguobacterium* spp., *Dietzia* spp., *Gordonia* spp., we did not yield accurate results at species level (Table 1).

For the bacterial isolates in which we obtained the species level, we apply the DAG stat spreadsheet tool to visualize the efficiency of the MALDI-TOF MS and the conventional method compared with the molecular methods (Table 4).

Discussion

MALDI-TOF MS technology is now recognized as an efficient method for bacterial identification in routine laboratory [13].

Several authors have analyzed the differences between pre-analytical procedure, direct colony with and without formic acid and tube extraction method [11]. Direct on-plate testing and tube extraction of *Corynebacterium* spp. have yielded equivalent identification percentages at genus and species level using Andromas system [10]. Farfour et al. showed that the MALDI-TOF MS Andromas strategy was reliable to identify a set of 659 Gram-positive rods representing 16 bacterial genera and 72 species by the direct colony method [18]. In that report, Andromas MALDI-TOF MS system could not identify *Listeria* isolates to the species level because of very similar mass spectra [18]. In this study 4/4 *Listeria monocytogenes* isolates could be identified to the species level using Bruker Biotyper system. However, as no other *Listeria* species isolates were included in our collection of GPR, the real accuracy of this technique to identify *Listeria* remains to be established.

Species identification rates of *Corynebacterium* spp. isolates were similar by MALDI-TOF MS and conventional phenotypic methods. On the other hand in *Actinomyces* and related genera the efficiency was higher using MALDI-TOF MS (Table 4). However, MALDI-TOF MS is easy to perform, rapid and less costly than conventional phenotypic method for routine labora-

Tabla 2. Comparative identification rates of Gram-positive rods by conventional phenotypic method and MALDI-TOF MS.

Organism	No.	No.(%) ID conventional method		No.(%) ID MALDI-TOF method	
		Species	Genus/Group*	Species	Genus
Corynebacterium spp.	216	199 (92,13)	216 (100)	202 (93,52)	211 (97,68)
Actinomyces spp. and related genera	71	63 (88,73)	69 (97,18)	66 (92,96)	68 (95,78)
Other Gram positive rods					
Dermabacter hominis, Rothia aeria, Turicella otitidis	14	14 (100)	14 (100)	14 (100)	14 (100)
Pigmented Gram-positive rods (*Leifsonia* spp., *Microbacterium* spp., *Exiguobacterium* spp., *Cellulosimicrobium* spp., *Brevibacterium/Arthrobacter* spp.)	16	0 (0)	16 (100)*	12 (75)	14 (87,5)
Aerobic actinomycetes (*Gordonia* spp., *Rhodococcus equi*, *Dietzia* spp.)	8	2 (25)	6 (75)*	6 (75)	7 (87,5)
Listeria monocytogenes, Lactobacillus spp.,*Clostridium tertium*	8	8 (100)	8 (100)	8 (100)	8 (100)
Total	333	286 (85,89)		308 (92,49)	

* Correct identification was considered as a group.

tories. These features make this method a better option for bacterial identification.

Moreover, not only in pigmented GPR (*Exiguobacterium, Microbacterium, Leifsonia, Cellulosimicrobium*), but also in aerobic actinomycetes (*Dietzia, Gordonia*) the identification at species level was higher using the MALDI-TOF MS method. In these isolates the conventional phenotypic methods are difficult to perform because there are many similarity biochemical tests to correctly achieve the identification at species or genus level.

Finally, based on previous studies, a lower score cut-off is suitable for identification of GPR and equivalent in test accuracy than manufacturer recommended cut-off [11]. As the genus and species level identification cut-off score was reduced, correct identification increased more than 20% for all the studied isolates (Table 3). Our data suggests and expands upon results from Alatoom et al., since a great and diverse collection and other Gram-positive rods have been tested [11]. Other authors also showed that using the manufacturers cut-off, they obtained a lower level of identification at the species level. However, genus level identification remained equivalent [10].

In our experience the reduction in identification score required for species level and the direct on plate testing using formic acid, allows to identify the most commonly GPR isolated in the clinical microbiology laboratory today.

Table 3. % of Gram-positive rods identified to the species level according to cut-off values by MALDI-TOF MS.

Organism	No.(%) with cut-off scores ≥2,0 (species level)	No.(%) with cut-off scores <2,0-≥1,7 (species level)	No. (%) with cut-off scores ≥1,7-≥2,0 (species level)
Corynebacterium spp.	153 (75,7)	49 (24,3)	202 (100)
Actinomyces spp. and related genera	41 (63,1)	24 (36,9)	65 (100)
Other Gram positive rods			
Dermabacter hominis	10	0	10
Rothia aeria	2	0	2
Turicella otitidis	2	0	2
Microbacterium spp.	1	3	4
Exiguobacterium spp.	1	1	2
Brevibacterium/Arthrobacter spp.	4	0	4
Cellulosimicrobium spp.	2	0	2
Gordonia spp.	1	1	2
Rhodococcus equi	2	0	2
Dietzia spp.	0	2	2
Listeria monocytogenes	4	0	4
Lactobacillus spp.	2	1	3
Clostridium tertium	1	0	1
Total Other GPR	32 (80)	8 (20)	40 (100)

Table 4. Efficiency of MALDI-TOF MS and conventional method for isolates of *Corynebacterium* spp., and *Actinomyces* spp. and related genera.

Organism	No.	Efficiency (95% CI) conventional-molecular method	Efficiency (95% CI) MALDI-TOF- molecular method
Corynebacterium spp.	216	0,92 (CI: 0,88–0,95)	0,94 (CI:0,89–0,96)
Actinomyces spp. and related genera	71	0,89 (CI:0,79–0,95)	0,93 (CI:0,84–0,98)

To conclude, this study demonstrates the accuracy of Bruker MALDI-TOF MS system as an identification method for clinical GPR using direct colony extraction method. Our study clearly expose that the MALDI-TOF MS database has a wide range of different spectrum allowing the identification of most of the clinical genera and species recovered in the clinical settings. However, since new pathogens are emerging continuously, revision and addition of new spectra to the database must be considered.

Author Contributions

Conceived and designed the experiments: CB MA MSR CV. Performed the experiments: CB MA OJL MSR. Analyzed the data: CB MA MSR. Contributed reagents/materials/analysis tools: CB MA OJL MSR CV. Contributed to the writing of the manuscript: CB MSR. Revision of the manuscript: OJL AF CV.

References

1. Funke G, von Graevenitz A, Clarridge JE 3rd, Bernard KA (1997) Clinical microbiology of coryneform bacteria. Clin Microbiol Rev 10: 125–159.
2. Adderson EE, Boudreaux JW, Hayden RT (2008) Infections caused by coryneform bacteria in pediatric oncology patients. Pediatr Infect Dis J 27: 136–141.
3. Funke G BK (2011) Coryneform Gram-positive rods. In: Versalovic J CK, Funke G, Jorgensen JH, Landry ML, Warnock DW, editor. Manual of clinical microbiology. Washington, DC: ASM Press. pp. 413–422.
4. Bernard KA FG (2012) Genus Corynebacterium. In: WB W, editor. Bergey's manual of systematic bacteriology. New York: Springer. pp. 245–289.
5. Stackebrandt E EJ (2006) Taxonomic parameters revisited: tarnished gold standards. Microbiol Today 33: 152–155.
6. Khamis A, Raoult D, La Scola B (2005) Comparison between rpoB and 16S rRNA gene sequencing for molecular identification of 168 clinical isolates of *Corynebacterium*. J Clin Microbiol 43: 1934–1936.
7. Konrad R, Berger A, Huber I, Boschert V, Hormansdorfer S, et al. (2010) Matrix-assisted laser desorption/ionisation time-of-flight (MALDI-TOF) mass spectrometry as a tool for rapid diagnosis of potentially toxigenic *Corynebacterium* species in the laboratory management of diphtheria-associated bacteria. Euro Surveill 15.
8. Gaillot O, Blondiaux N, Loiez C, Wallet F, Lemaitre N, et al. (2011) Cost-effectiveness of switch to matrix-assisted laser desorption ionization-time of flight mass spectrometry for routine bacterial identification. J Clin Microbiol 49: 4412.
9. Saffert RT, Cunningham SA, Ihde SM, Jobe KE, Mandrekar J, et al. (2011) Comparison of Bruker Biotyper matrix-assisted laser desorption ionization-time of flight mass spectrometer to BD Phoenix automated microbiology system for identification of gram-negative bacilli. J Clin Microbiol 49: 887–892.
10. Theel ES, Schmitt BH, Hall L, Cunningham SA, Walchak RC, et al. (2012) Formic acid-based direct, on-plate testing of yeast and *Corynebacterium* species by Bruker Biotyper matrix-assisted laser desorption ionization-time of flight mass spectrometry. J Clin Microbiol 50: 3093–3095.
11. Alatoom AA, Cazanave CJ, Cunningham SA, Ihde SM, Patel R (2012) Identification of non-diphtheriae *Corynebacterium* by use of matrix-assisted laser desorption ionization-time of flight mass spectrometry. J Clin Microbiol 50: 160–163.
12. Vila J, Juiz P, Salas C, Almela M, de la Fuente CG, et al. (2012) Identification of clinically relevant *Corynebacterium* spp., *Arcanobacterium haemolyticum*, and *Rhodococcus equi* by matrix-assisted laser desorption ionization-time of flight mass spectrometry. J Clin Microbiol 50: 1745–1747.
13. Seng P, Drancourt M, Gouriet F, La Scola B, Fournier PE, et al. (2009) Ongoing revolution in bacteriology: routine identification of bacteria by matrix-assisted laser desorption ionization time-of-flight mass spectrometry. Clin Infect Dis 49: 543–551.
14. Bernard K (2012) The genus *Corynebacterium* and other medically relevant coryneform-like bacteria. J Clin Microbiol 50: 3152–3158.
15. Khamis A, Raoult D, La Scola B (2004) rpoB gene sequencing for identification of *Corynebacterium* species. J Clin Microbiol 42: 3925–3931.
16. Weisburg WG, Barns SM, Pelletier DA, Lane DJ (1991) 16S ribosomal DNA amplification for phylogenetic study. J Bacteriol 173: 697–703.
17. Mackinnon A (2000) A spreadsheet for the calculation of comprehensive statistics for the assessment of diagnostic tests and inter-rater agreement. Comput Biol Med 30: 127–134.
18. Farfour E, Leto J, Barritault M, Barberis C, Meyer J, et al. (2012) Evaluation of the Andromas matrix-assisted laser desorption ionization-time of flight mass spectrometry system for identification of aerobically growing Gram-positive bacilli. J Clin Microbiol 50: 2702–2707.

12

Protease-Dead Separase Is Dominant Negative in the *C. elegans* Embryo

Diana M. Mitchell[¤a], **Lindsey R. Uehlein-Klebanow**[¤b], **Joshua N. Bembenek***

Department of Biochemistry, Cellular and Molecular Biology, University of Tennessee, Knoxville, Tennessee, United States of America

Abstract

Separase is a protease that promotes chromosome segregation at anaphase by cleaving cohesin. Several non-proteolytic functions of separase have been identified in other organisms. We created a transgenic *C. elegans* line that expresses protease-dead separase in embryos to further characterize separase function. We find that expression of protease-dead separase is dominant-negative in *C. elegans* embryos, not previously reported in other systems. The *C. elegans* embryo is an ideal system to study developmental processes in a genetically tractable system. However, a major limitation is the lack of an inducible gene expression system for the embryo. We have developed two methods that allow for the propagation of lines carrying dominant-negative transgenes and have applied them to characterize expression of protease-dead separase in embryos. Using these methods, we show that protease-dead separase causes embryo lethality, and that protease-dead separase cannot rescue separase mutants. These data suggest that protease-dead separase interferes with endogenous separase function, possibly by binding substrates and protecting them from cleavage.

Editor: Anne C. Hart, Brown University/Harvard, United States of America

Funding: This work was supported by start-up funds from the University of Tennessee, Knoxville. The funders had no role in study design, data collection and analysis, decision to publish, or preparation of the manuscript.

Competing Interests: The authors have declared that no competing interests exist.

* Email: bembenek@utk.edu

¤a Current address: Department of Biological Sciences, University of Idaho, Moscow, Idaho, United States of America
¤b Current address: Department of Molecular Genetics, Lerner Research Institute, Cleveland, Ohio, United States of America

Introduction

Separase is a cysteine protease with multiple roles during cell division. In a number of these roles the protease activity of separase is required, including cohesin cleavage at the onset of anaphase [1–4], DNA repair [5], resolution of chiasmata in mouse oocytes [6], mitotic spindle elongation [7], and centriole duplication [8–10]. Additional non-proteolytic functions of separase have been identified, including anaphase exit [11] and Cdc14 early anaphase release (FEAR) pathway activation [12,13], and polar body extrusion in mouse oocytes [6]. Importantly, these studies examined protease-dead separase in separase mutant cells and concluded that separase can promote signaling events independent of its protease activity. There have been no studies to our knowledge that have examined any effects caused by the expression of protease-dead separase in a wild-type background, which could reveal more information about the activity of this mutant protein.

In *C. elegans*, separase has been shown to regulate chromosome segregation [14], centriole duplication ([10,15] and membrane trafficking [16–18]. However, the mechanism(s) by which separase controls these various cell division processes is not known. We created a transgenic worm line expressing protease-dead separase fused to GFP (SEP-1^{PD}::GFP) using standard methods to characterize its expression in *C. elegans* embryos [17]. As previously reported, strains expressing SEP-1^{PD}::GFP must be propagated on *gfp* RNAi, and removed from RNAi for several generations to examine expression [17]. In this report, we demonstrate that SEP-1^{PD}::GFP expression causes embryo lethality. Other researchers have encountered similar difficulties with other mutant proteins [19], highlighting the need for methods to control transgene expression. Here, we methodologically characterize the usefulness of *gfp* RNAi as a way to propagate toxic transgenes in the *C. elegans* embryo.

We find that *gfp* RNAi silences SEP-1^{PD}::GFP transgene expression and allows for maintenance of homozygous transgenic lines indefinitely. Upon removal from *gfp* RNAi, transgene re-expression takes several generations, with gradual reappearance of embryonic lethality. On average, we were able to propagate SEP-1^{PD}::GFP worm lines for 5 generations after removal from *gfp* RNAi. SEP-1^{PD}::GFP accumulates strongly at putative sites of separase activity, indicating that it could be substrate trapping. We also report that homozygous *sep-1* mutants expressing protease-dead separase are not viable.

We also describe a second method using male worms to propagate the transgene. The *pie-1* promoter is widely used to drive embryonic expression, and is expressed in the female germline [20]. Transgenic male worms carrying the SEP-1^{PD}::GFP transgene can be crossed to *unc-119* hermaphrodites for many generations without obvious deleterious effects. The resulting hermaphrodites, carrying a single copy of SEP-1^{PD}::GFP in a wild-type background, produce broods displaying high levels of embryonic lethality. Further, males can be used to propagate

SEP-1[PD]::GFP reliably to facilitate transgene characterization and to combine transgene expression with mutant alleles. Using this strategy, we find that protease-dead separase exacerbates phenotypes in heterozygous mutants.

We have successfully used these newly developed methods to provide the first characterization of SEP-1[PD]::GFP in the *C. elegans* embryo. These methods employ standard laboratory techniques used by *C. elegans* researchers, which will open new possibilities for analysis of gene function in the *C. elegans* embryo. SEP-1[PD]::GFP expression causes embryonic lethality in WT animals, and does not rescue mutant separase animals. We conclude that protease-dead separase is dominant negative and interferes with endogenous separase function, a finding that was not reported in other systems. Collectively, our results suggest that protease-dead separase may trap substrates, as has been found for other catalytically inactivated enzymes [21], which would interfere with substrate cleavage by endogenous separase.

Materials and Methods

Strains

C. elegans strains were maintained according to standard protocols (Brenner, 1974). Temperature sensitive strains were maintained at 16°C, unless otherwise indicated in the text, and shifted to non-permissive temperature as indicated. All other strains were maintained at 20°C. Strains containing the protease-dead *sep-1* transgene were maintained on lawns of *gfp* RNAi feeding bacteria as indicated in text and below, then transferred onto OP50 lawns as indicated. A full list of strains used in this study and genotypes are included in Table 1.

Some strains were obtained from the Caenorhabditis Genetics Center (CGC); see Table 1. Strain RQ372 was a kind gift from Dr. Risa Kitagawa. JAB18 was obtained by crossing WH520 males with OD56 hermaphrodites, and subsequent generations were maintained on *gfp* RNAi. At each generation following the cross, approximately half of the worms at L4 stage were moved to OP50 plates for 24 hours at 25°C and screened for the presence of both SEP-1[PD]::GFP and H2B::mCherry transgenes by microscopy. Worms were then singled from the original *gfp* RNAi feeding plate. This protocol was repeated until double homozygous transgenic lines were obtained, after which the line was maintained on *gfp* RNAi at 20°C. Feeding *gfp* RNAi did not silence expression of H2B::mCherry.

Molecular Biology

Cloning the protease-dead separase mutant DNA sequence into pjk#3 or pjk#7 vectors was performed as previously described [17]. Microparticle bombardment [22] was used to obtain transgenic worm lines as described in the text and Figure 1.

RNAi feeding

The *gfp* RNAi feeding construct in L4440 vector was obtained from Dr. Scott Kennedy [23]. To silence GFP fusion transgenes and maintain worm lines, worms were picked onto lawns of *gfp* RNAi feeding bacteria and L4 worms were picked at each generation onto fresh lawns. In order to provide the optimal RNAi effect for transgene silencing, worms were grown on *gfp* RNAi at 20°C (which is the semi-permissive temperature for the *sep-1(e2406)* allele), as we were unable to propagate some lines on *gfp* RNAi by feeding at 16°C. For transgene re-expression, L4 worms were removed to OP50 lawns and picked onto fresh OP50 feeding plates at each generation as indicated in the text and figures.

Microscopy

For imaging of mitotic embryos, young adult worms were dissected in M9 and embryos were mounted on agar pads as previously described [17]. For imaging of meiotic embryos, *in utero* imaging was performed using young adult worms immobilized in 1 mg/mL levamisole mounted on 2% agar pads and covered with a coverslip. Live cell imaging was performed using a Nikon Eclipse inverted microscope with a CSU-22 spinning disc imaging system equipped with a 60X 1.40NA objective from Visitech International, running metamorph software. Digital images were obtained with a Photometrics EM-CCD camera. Images were analyzed and time-lapse movies were made using FIJI (ImageJ) software using the Bio-Formats plugin from LOCI (www.loci.wisc.edu). Images were enhanced by adjusting minimum and maximum display levels in single color channels, then overlayed to display both channels.

Table 1. Strains used in this study.

Strain	Genotype	Reference and/or source
N2	Bristol (wild-type)	CGC
WH416	*unc-119(ed3)* III, *ojIs58[SEP-1::GFP unc119(+)]*	[16]
WH520	*unc-119(ed3)* III, *ojIs71[GFP::SEP-1(PD) unc119(+)]*	[17] and this study.
WH524	*unc-119(ed3)* III, *ojIs75[SEP-1(PD)::GFP unc119(+)]*	This study
WH408	*sep-1(e2406)* I/hT2*[bli-4(e937) let-? (q782) qIs48]* I	[18]
VC1279	*sep-1(ok1749)* I/hT2 I	CGC
WH458	*sep-1(e2406)* I/hT2 I; *unc-119(ed3)* III/hT2 III, *ojIs58* *[GFP::SEP-1 unc119(+)]*	This study
WH548	*sep-1(e2406)* I/hT2 I; *unc-119(ed3)* III/hT2 III, *ojIs71* *[GFP::SEP-1(PD) unc119(+)]*	This study
WH504	*sep-1(ok1749)* I/hT2 I; *unc-119(ed3)* III/hT2 III, *ojIs58* *[GFP::SEP-1 unc119(+)]*	This study
JAB7	*sep-1(ok1749)* I/hT2 I; *unc-119(ed3)* III/hT2 III, *ojIs71* *[GFP::SEP-1(PD) unc119(+)]*	This study
WH488	*sep-1(e2406)* I/hT2 I; *unc-119(ed3)* III/hT2 III	This study
JAB3	*sep-1(ok1749)* I/hT2 I; *unc-119(ed3)* III/hT2 III	This study
RQ372	*unc-119(ed3)* III, *ojIs58[SEP-1::GFP unc119(+)]* itIs37 *[Ppie-1::mCherry::his-58 (pAA64) + unc-119(+)]* IV	Dr. Risa Kitagawa
JAB18	*unc-119(ed3)* III, *ojIs71[GFP::SEP-1(PD) unc119(+)]* itIs37 *[Ppie-1::mCherry::his-58 (pAA64) + unc-119(+)]*	This study

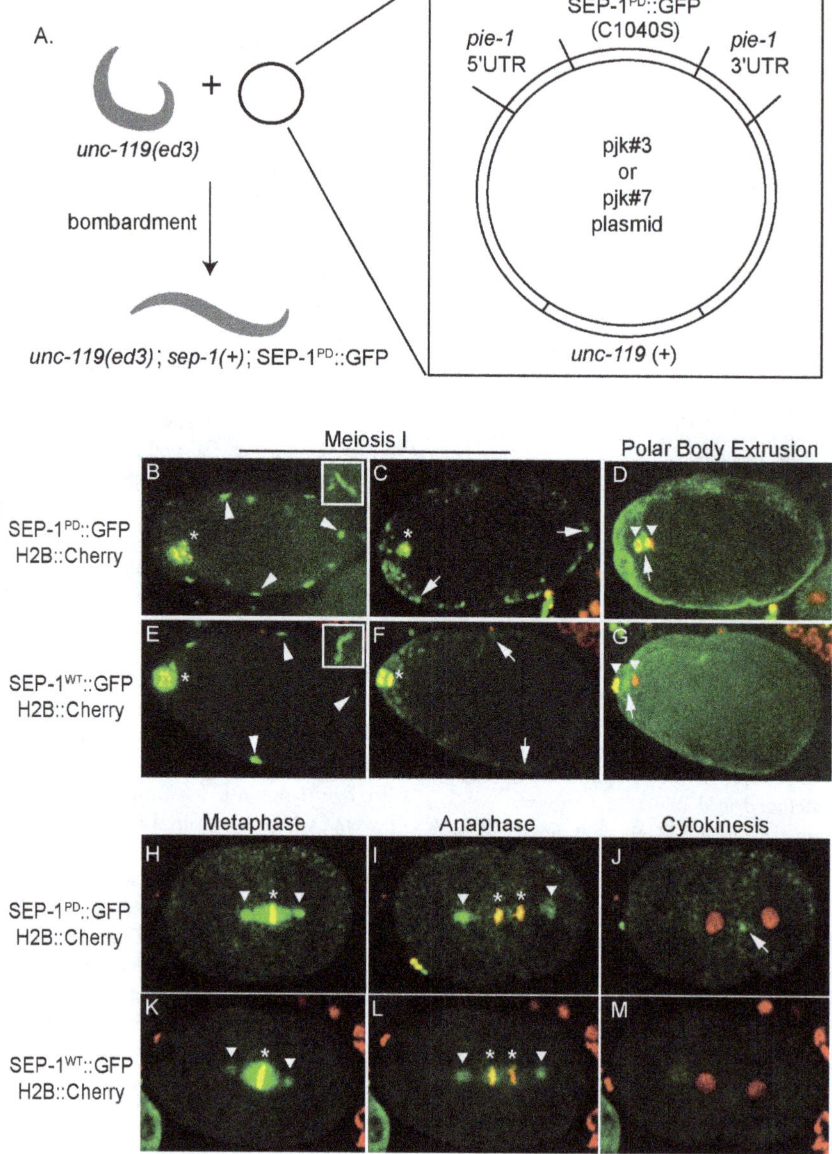

Figure 1. SEP-1^{PD}::GFP transgenic worm lines. A. Microparticle bombardment of homozygous *unc-119(ed3); sep-1(+)* worms with plasmid DNA bound to gold beads. The plasmid (enlarged panel) contains the *sep-1* coding sequence with mutation in the protease domain (C1040S) fused to GFP under control of the *pie-1* promoter and an *unc-119(+)* rescue sequence, allowing for identification of transformed worm lines. The designated alleles and transgene are homozygous in the resulting transgenic worm line. B-M: SEP-1^{PD}::GFP (top row) and SEP-1^{WT}::GFP (bottom row) localization in newly fertilized embryos with H2B::mCherry. Embryos were imaged after 5 generations removed from *gfp* RNAi feeding (see text and Figure 2). During meiosis I, SEP-1^{PD}::GFP and SEP-1^{WT}::GFP localize to chromosomes and the meiotic spindle (asterisk, B,C, E, F). During prometaphase, separase appears on cortical filaments that appear as puncta depending on whether they are oriented parallel to the focal plane (arrowheads, B and E, insets show examples of filaments oriented properly). Separase is localized to cortical granules by the onset of anaphase (arrows, C and F). During polar body extrusion, SEP-1^{PD}::GFP and SEP-1^{WT}::GFP accumulate at the base of the polar body (base of the polar body designated by arrows D and G, respectively) between the separating anaphase chromosomes (chromosomes designated by arrowheads, D and G). SEP-1^{PD}::GFP also accumulates strongly on the plasma membrane of the embryo after cortical granule exocytosis (D). During the indicated stages of mitosis (H-M), SEP-1^{WT}::GFP and SEP-1^{PD}::GFP localize to chromosomes (asterisk) and centrosomes (arrowhead). During cytokinesis, SEP-1^{PD}::GFP accumulates at the cleavage furrow (arrow, J).

Results

Creation of protease-dead separase transgenic worm lines

We created SEP-1^{PD}::GFP expressing transgenic worm lines using microparticle bombardment, using standard protocols [22], but GFP expressing lines could not be maintained for more than a few generations. The final construct contains several features that allowed us to propagate lines carrying dominant negative transgenes. The construct is designed to generate proteins fused to GFP driven by the *pie-1* promoter (Figure 1A). Importantly, the *pie-1* promoter is widely used to drive transgene expression in *C. elegans* oocytes and young embryos [20]. The construct also has an *unc-119(+)* selection marker allowing for identification of transformed worms. We cloned genomic *sep-1* sequence, with a point mutation that results in cysteine to serine substitution at amino acid 1040, located in the protease domain of SEP-1. We created multiple independent worm lines with integrated transgenes coding for both N- and C-terminal fusions of GFP to SEP-1^{PD} using this strategy, all of which had severe growth defects.

We generated lines expressing both SEP-1^{PD}::GFP and H2B::mCherry and examined localization of GFP tagged separase relative to chromosome segregation in the newly fertilized embryo (Figure 1B–M, Movies S1–S4). We found that SEP-1^{PD}::GFP localizes similarly to SEP-1^{WT}::GFP during meiosis I (Figure 1B-G, Movies S1 and S2) and mitosis (Figure 1H-M, Movies S3 and S4 and previously reported [17]). Both SEP-1^{WT}::GFP and SEP-1^{PD}::GFP localize to chromosomes and the meiotic spindle during meiosis (Figure 1B-C and E-F). SEP-1^{WT}::GFP and SEP-1^{PD}::GFP localize to filamentous structures (cortical filaments) in the oocyte prior to fertilization (not shown). We have previously reported the localization of separase and other proteins to cortical filaments [16], which are not well characterized. Following fertilization, during the progression of meiosis I, SEP-1^{PD}::GFP moves from cortical filaments to cortical granules (Figure 1B-C, Movie S1) as does SEP-1^{WT}::GFP (Figure 1E-F, Movie S2). Following cortical granule exocytosis and meiotic anaphase I, SEP-1^{PD}::GFP associates strongly with the embryo plasma membrane for an extended period of time and with the base of the polar body as compared to SEP-1^{WT}::GFP (Figure 1D and G, Movies S1 and S2). Interestingly, SEP-1^{PD}::GFP accumulates strongly compared to SEP-1^{WT}::GFP at several sites of putative action during mitosis, including centrosomes, mitotic spindle (compare Figure 1H-I and 1K-L) and the cleavage furrow during cytokinesis (compare Figure 1J and 1M, [17]). These sites of separase activity may contain substrates of separase, which may have stronger binding to the inactive protease leading to its accumulation relative to wild-type separase, suggesting that SEP-1^{PD}::GFP could be substrate-trapping.

Silencing of SEP-1^{PD}::GFP expression by *gfp* RNAi

If protease-dead separase remains bound to substrates, it could interfere with their cleavage by endogenous separase, therefore having a dominant-negative effect. Consistent with a dominant negative activity, SEP-1^{PD}::GFP expression caused embryo lethality (Figure 2 and Figure S1, see below) in the wild-type background with two copies of endogenous separase. Dominant negative activity of protease-dead separase has not been reported in other systems. High levels of embryonic lethality in SEP-1^{PD}::GFP expressing worm lines required us to develop methods to propagate this "toxic" transgene for further examination. RNAi provides a reliable system for targeted gene knock down in *C. elegans*. We took advantage of RNAi in order to silence expression of the SEP-1^{PD}::GFP transgene by maintaining transgenic worm lines on lawns of *gfp* RNAi feeding bacteria.

After bombardment following the standard protocol, Unc rescued animals were screened for GFP expression, and Unc rescued GFP positive lines had high lethality when grown under standard lab conditions. However, SEP-1^{PD}::GFP transgenic worm lines fed *gfp* RNAi showed no embryonic lethality and could be propagated indefinitely at 20°C and 25°C (Figure 2A). When worms were transferred from *gfp* RNAi onto OP50, embryonic lethality returned after several generations (Figure 2A), and higher levels of embryonic lethality correlated with higher GFP expression levels. Interestingly, broods from individual worms showed similar levels of embryonic lethality within group at each generation after removal from RNAi (note error bars for each data point, Figure 2A), in contrast to variability in individual offspring from the same brood following injection of RNAi [24]. This difference could be the result of uniform RNAi administration when feeding RNAi continually over multiple generations and selective pressure that would favor animals with more effective RNAi response. The return of embryonic lethality was dependent on temperature, as embryonic lethality occurred in 3–5 generations at 20°C and 2–3 generations at 25°C, which could be due to reduced generational *gfp*(RNAi) transmission, increased transgene expression at 25°C, or an increase in cell cycle timing leading to a decrease in fidelity of division. Picking a larger number of worms at each generation allows for propagation of the transgenic line on OP50 through one more generation (Figure 2B). This could be due to effects with picking animals of different penetrance of generational RNAi propagation, as seen previously [24]. Similar results were obtained for multiple independent worm lines expressing both N-terminal and C-terminal SEP-1^{PD} GFP fusion proteins, indicating that the position of GFP fusion is not a factor (Figure S1). SEP-1^{PD} expressing worms that survive hatching show abnormal developmental phenotypes including tail defects, slow growth, and sterility (Figure 2C), suggesting that protease-dead separase interferes with normal developmental processes in addition to causing embryonic lethality. Because transgene expression is *pie-1* driven, and should be most highly expressed in the germline and deposited in the egg, these results suggest that SEP-1^{PD} expressing worms show phenotypes that are a result of defects in the developing embryo.

Transgenic SEP-1^{PD}::GFP males can be used to propagate dominant negative SEP-1^{PD}::GFP to offspring

Expression of most transgenes in the *C. elegans* embryo, including our SEP-1^{PD}::GFP transgene, is under control of the maternal *pie-1* promoter [20]. We created transgenic SEP-1^{PD}::GFP male worms by heat shock and backcrossed to *unc-119* hermaphrodites to easily identify Unc-rescued *sep-1(+)/sep-1(+)*; SEP-1^{PD}::GFP/- males, which did not express significant levels of SEP-1^{PD}::GFP and gave rise to many cross progeny for our studies (Figure S2). Typically, expression of *pie-1* driven transgenes is not observed in sperm, although expression in the male germline has been previously observed [25], which may depend where the transgene is integrated. We therefore reasoned that we could use males with a single copy of SEP-1^{PD}::GFP to propagate the SEP-1^{PD}::GFP transgene. From the F1 progeny, males heterozygous for the transgene (in *unc-119(ed3); sep-1(+)* homozygous background) crossed to *unc-119(ed3)* homozygous hermaphrodites produce heterozygous SEP-1^{PD}::GFP males (Unc rescued), heterozygous SEP-1^{PD}::GFP hermaphrodites (Unc rescued), and both male and hermaphrodite *unc-119* offspring (Figure 3A). The heterozygous SEP-1^{PD}::GFP hermaphrodites, identified by Unc rescue (mobility), can be used for analysis of transgene expression, while the transgenic heterozygous males (also identified by Unc rescue) were continually backcrossed to *unc-119* hermaphrodites to maintain the line (Figure 3A), and were also used in crosses with other worm lines to test for genetic interactions. Single F1 heterozygous transgenic SEP-1^{PD}::GFP hermaphrodites were picked onto individual plates and their progeny were analyzed for embryonic lethality. We found that embryonic lethality in the F2 brood was consistently in the range of 40–60% (Figure 3B), which shows that even a single copy of SEP-1^{PD}::GFP has dominant-negative effects in a wild-type background. After backcrossing males to *unc-119* hermaphrodites more than 50 generations, embryonic lethality in the F2 remained within this range (Figure 3C), regardless of C-terminal or N-terminal GFP fusion (not shown). Propagation of SEP-1^{PD}::GFP in males bypasses lethality and can be done indefinitely, while also

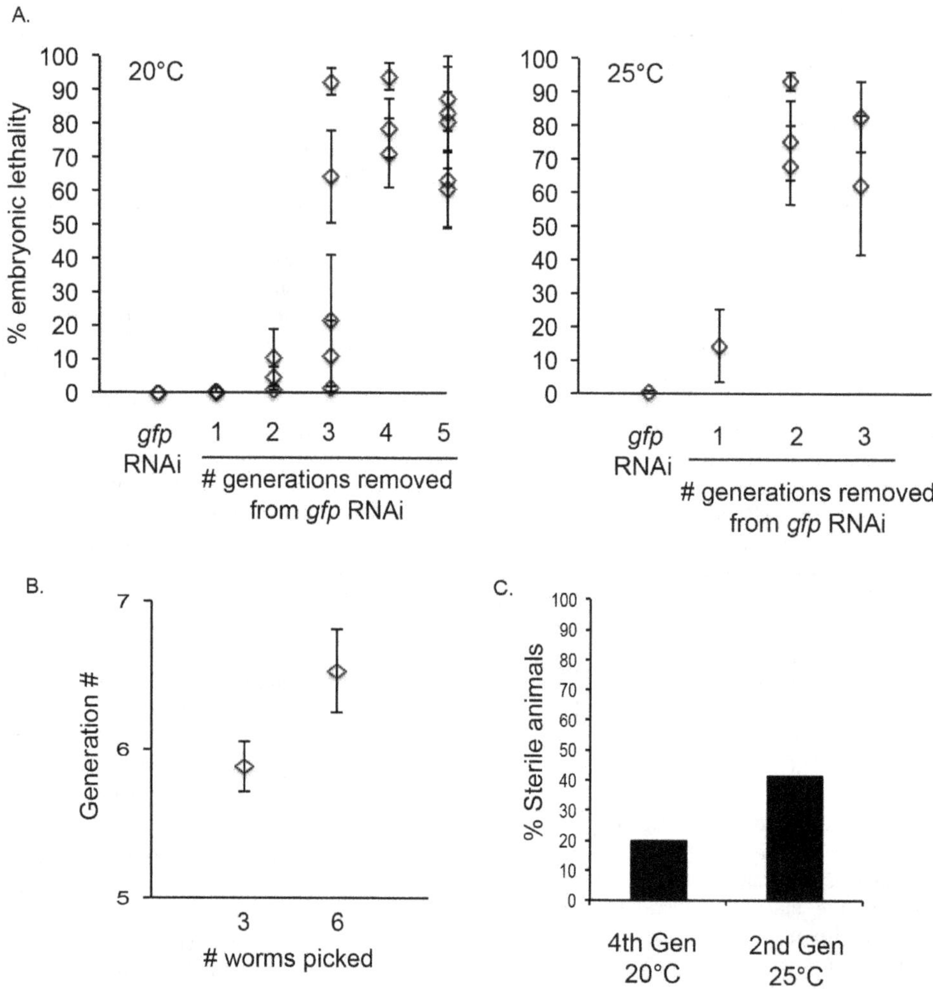

Figure 2. SEP-1[PD]::GFP worm lines can be maintained on *gfp* RNAi. A. Embryonic lethality of SEP-1[PD]::GFP line on *gfp* RNAi and after removal onto OP50 plates at 20°C (left) or 25°C (right). Each data point with error bars represents the average of embryonic lethality from 10 singled worms +/− SEM. B. Average generation +/− SEM that could be propagated for the SEP-1[PD]::GFP line after removal from *gfp* RNAi when the indicated number of worms are picked at each generation and kept at 20°C. C. Percentage of sterile animals in the SEP-1[PD]::GFP line after removal from *gfp* RNAi.

providing consistent transgene expression in the F1 generation. Therefore, this approach provides a convenient strategy to introduce transgenes into different mutant backgrounds to test for genetic interactions (see below) or to create worm lines in combination with other transgenes.

Genetic interactions of protease-dead separase with separase mutants

Previous studies demonstrated that protease-dead separase can rescue some loss of function separase phenotypes [11–13]. Therefore, we examined genetic interactions of SEP-1[PD]::GFP with mutant separase alleles: the hypomorphic *sep-1(e2406)* allele and the *sep-1(ok1749)* deletion allele. The *sep-1(e2406)* homozygous mutant is temperature sensitive and viable at 16°C, but 100% embryo lethal at the semi-permissive temperature, 20°C. The *sep-1(ok1749)* deletion mutant is likely a null allele since no protein can be detected by western blot [18]. At all temperatures, nearly all homozygous *sep-1(ok1749)* progeny die during embryogenesis, with very few surviving animals that arrest at early larval stages. Both *sep-1(e2406)* and *sep-1(ok1749)* are maintained as balanced heterozygotes with the *hT2[bli-4(e937) let-? (q782) qls48]* balancer chromosome which encodes GFP localized to the pharynx (hT2g). hT2g is a translocation balancer that can be used to balance mutations in LGI or LGIII. Scoring for GFP expression in the pharynx allows for identification of heterozygous (GFP+ pharynx) and homozygous (GFP- pharynx) mutants.

We examined *sep-1* mutant embryos expressing SEP-1[WT]::GFP or SEP-1[PD]::GFP, to determine if either of these transgenes can rescue *sep-1* mutants. We generated lines that were homozygous for either the SEP-1[WT]::GFP or SEP-1[PD]::GFP transgenes in these balanced separase mutant backgrounds. Both the balanced heterozygous *sep-1(e2406)* and *sep-1(ok1749)* deletion mutants with the SEP-1[PD]::GFP transgene could not be maintained on normal OP50 bacterial plates. Therefore, the balanced separase mutant lines with SEP-1[PD]::GFP were maintained on *gfp* RNAi at 20°C, because *gfp* RNAi feeding at 16°C did not allow for propagation of the strains (not shown). Furthermore, the balanced separase mutant lines homozygous for SEP-1[PD]::GFP could only be propagated for a maximum of 1-3 generations off of *gfp* RNAi at 20°C (compared to an average of 5 generations in the *sep-1(+)/*

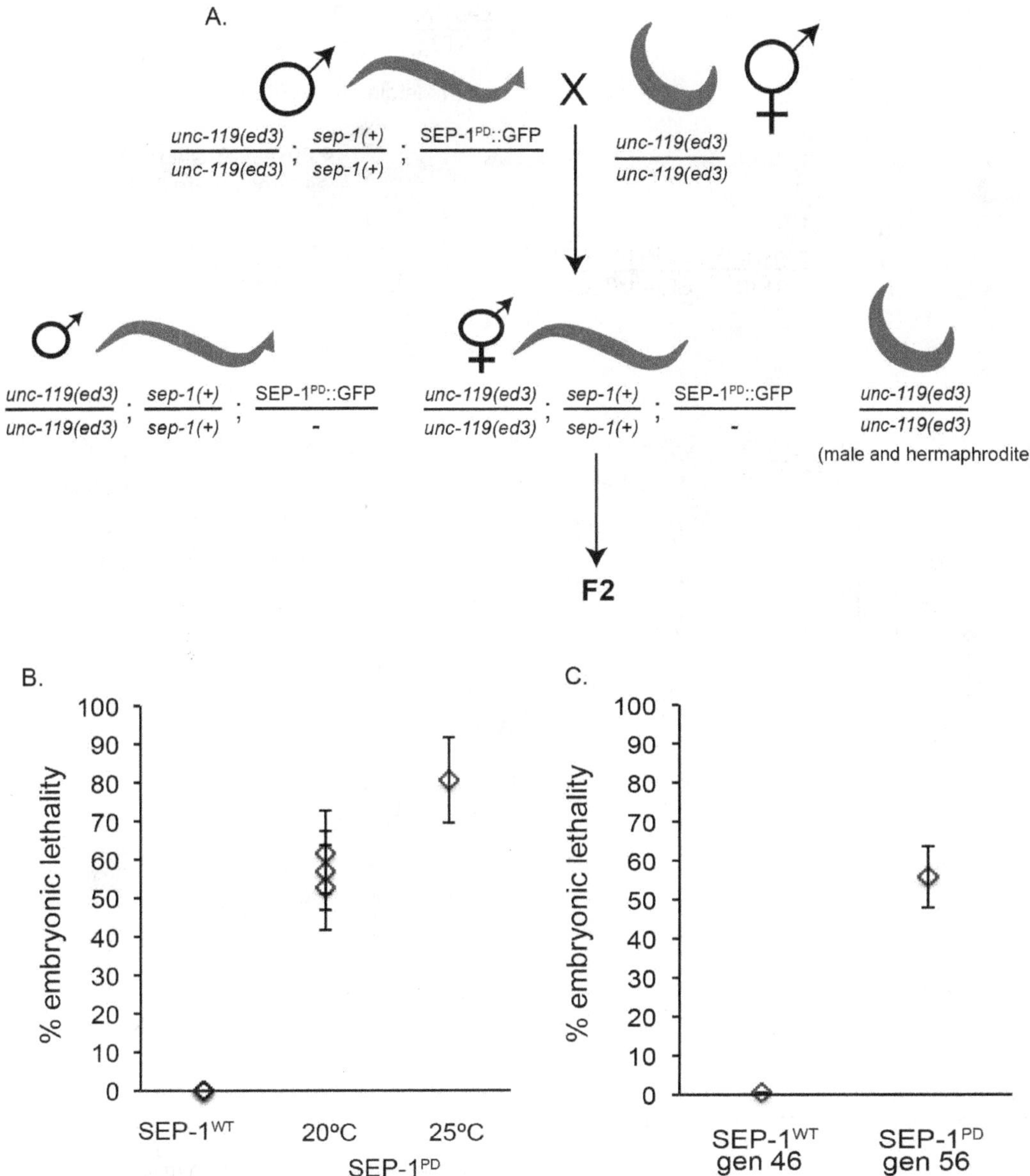

Figure 3. Propagation of the SEP-1^{PD}::GFP transgene by backcrossing. A. The diagram represents the strategy used to propagate the SEP-1^{PD}::GFP transgene using males. Transgenic *unc-119(ed3)/unc-119(ed3); sep-1(+)/sep-1(+);* SEP-1^{PD}::GFP/- males heterozygous for the transgene are continually crossed to *unc-119(ed3)* hermaphrodites to generate heterozygous *sep-1(+)/sep-1(+);* SEP-1^{PD}::GFP/- males and hermaphrodites in the *unc-119(ed3)* background. The resulting progeny (male and hermaphrodite) that carry the SEP-1^{PD}::GFP transgene are readily identified by mobility because they are Unc rescued due to presence of the transgene. B. Embryonic lethality in F2 broods from singled F1 *sep-1(+)/sep-1(+);* SEP-1^{WT}::GFP/- or *sep-1(+)/sep-1(+);* SEP-1^{PD}::GFP/- hermaphrodites at the indicated temperature. C. Embryonic lethality in the F2 after the indicated number of backcrosses of heterozygous *sep-1(+)/sep-1(+);* SEP-1^{WT}::GFP/- or *sep-1(+)/sep-1(+);* SEP-1^{PD}::GFP/- transgenic males to *unc-119* hermaphrodites. Data points represent the average of a group of 10 singled worms +/− SEM.

sep-1(+) background, Figure 2) before all progeny died, arrested prior to reaching adult, or were completely sterile.

We analyzed lethality in embryos from heterozygous, balanced mutant animals homozygous for SEP-1^{WT}::GFP or SEP-1^{PD}::GFP transgenes (Figure 4A). Consistent with the growth defects we observed, embryonic lethality in balanced mutant lines homozygous for SEP-1^{PD}::GFP was more severe than the mutant alone, and this effect was reduced when transgene expression was silenced with *gfp* RNAi (not shown). Further, SEP-1^{WT}::GFP expression was able to rescue both homozygous *sep-1(e2406)* and *sep-1(ok1749)* mutant progeny while SEP-1^{PD}::GFP could not (Figure 4B and 4C). Importantly, expression of SEP-1^{WT}::GFP can rescue both homozygous *sep-1(e2406)* hypomorphic and *sep-1(ok1749)* deletion mutants to produce a few gravid adult animals (not shown). These data indicate that protease-dead separase does not rescue viability in separase mutant embryos.

We employed our transgenic male propagation method to examine the phenotype of *sep-1(e2406)/+* embryos expressing SEP-1^{WT}::GFP or SEP-1^{PD}::GFP. Transgenic SEP-1^{WT}::GFP or SEP-1^{PD}::GFP males were crossed with *sep-1(e2406)* homozygous hermaphrodites at 16°C, and F1 progeny were grown to L4 at 16°C (Figure 5A). Unfortunately all other mutant alleles of

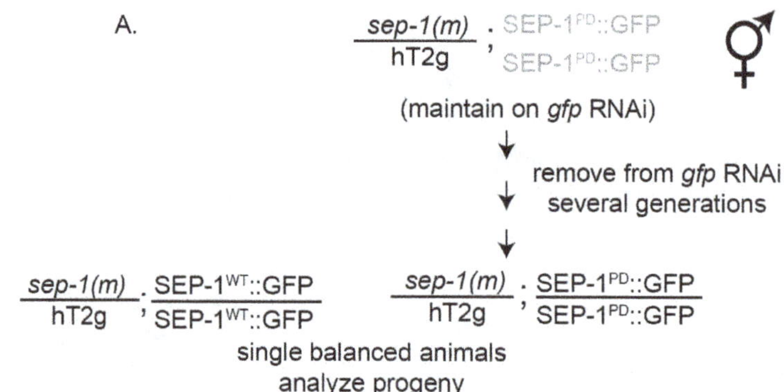

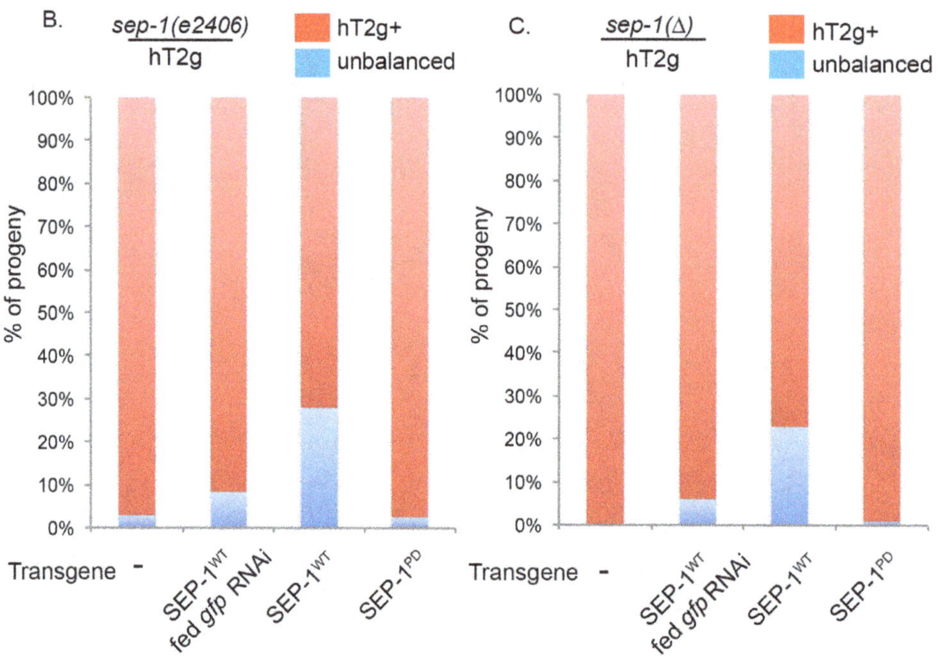

Figure 4. Wild-type SEP-1, but not protease-dead SEP-1, rescues *sep-1* mutants. A. Lines with SEP-1^{PD}::GFP transgene were maintained on *gfp* RNAi at 20°C and removed for several generations to allow transgene expression. Balanced worms were then singled and progeny were analyzed. B. Percentage of progeny that are balanced mutant or homozygous mutant from singled *sep-1(e2406)*/hT2g hermaphrodites homozygous for the indicated transgene. C. Percentage of progeny that are balanced mutant or homozygous mutant from singled *sep-1(ok1749)*/hT2g (referred to as Δ) hermaphrodites with indicated transgene at 20°C.

separase, including *sep-1(ok1749)*, are not viable as homozygotes and could not be tested this way. We attempted to cross *sep-1(ok1749)*/hT2g and *sep-1(e2406)*/hTg, both with the *unc-119(ed3)* background, hermaphrodites to transgenic SEP-1::GFP males to obtain the desired genotype, but had results inconsistent with the expected outcome in the F2 generation (not shown). Briefly, we observed that the SEP-1^{PD}::GFP transgene was not expressed in the Unc rescued F2 as expected, although the SEP-1^{WT}::GFP transgene was expressed. Given that the hT2g balancer breakdown has previously been reported [26], it is possible that there was a potential issue with hT2g. The F1 *sep-1(e2406)/+*; SEP-1^{WT}::GFP/- and *sep-1(e2406)/+*; SEP-1^{PD}::GFP/- worms were shifted to 20°C at the L4 stage and embryonic lethality was determined in the F2 brood. GFP expression was confirmed in the oocytes and embryos of the F1 hermaphrodites used in this analysis that give rise to F2 broods. Since the oocyte and early embryo is determined by the maternal genotype due to maternal deposition of cellular machinery, early F2 embryos reflect the *sep-1(e2406)/+*; SEP-1$^{WT \text{ or } PD}$::GFP maternal genotype. F2 embryos from *sep-1(e2406)/+*; SEP-1^{WT}::GFP/- animals were fully viable as expected, but *sep-1(e2406)/+*; SEP-1^{PD}::GFP/- F2 progeny showed 100% embryonic lethality (Figure 5B). This is consistent with a dominant negative activity of SEP-1^{PD}::GFP, because SEP-1^{PD}::GFP causes lethality in separase wild-type background (Figures 2 and 3), and enhances the phenotype of mutant separase alleles (Figure 4 and 5).

Discussion

We employed the methods described in this manuscript to examine the consequence of SEP-1^{PD} expression in the *C. elegans* embryo. We find that protease-dead separase is dominant negative and likely interferes with endogenous separase function in our system. A dominant negative activity of protease-dead separase has not been reported in other systems, and could alter the interpretation of phenotypes of cellular expression of protease-

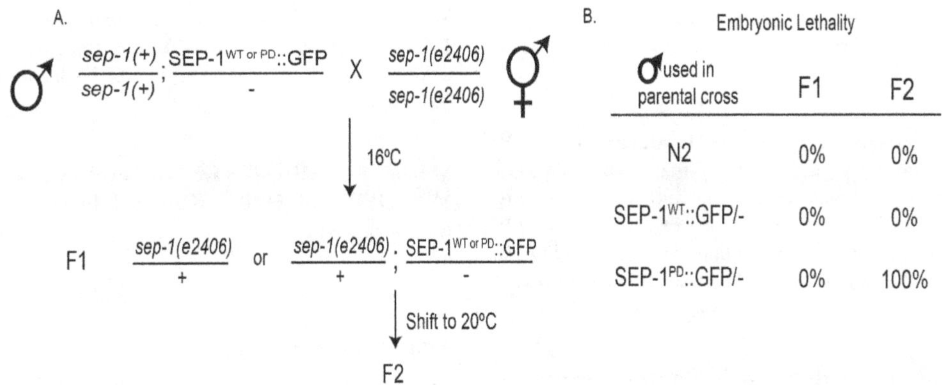

Figure 5. Genetic interactions of wild-type or protease-dead SEP-1 with *sep-1(e2406)*. A. Crossing scheme of heterozygous SEP-1^{WT}::GFP or SEP-1^{PD}::GFP transgenic males to *sep-1(e2406)* homozygous hermaphrodites. GFP transgene expression in the F1 was determined by microscopy after the shift to 20°C. Only progeny from animals expressing SEP-1^{WT}::GFP or SEP-1^{PD}::GFP were analyzed. B. The table shows embryonic lethality in the F1 and F2 progeny when males carrying the indicated transgene were used in the initial cross.

dead separase. If dominant negative activity of protease-dead separase arises due to substrate trapping by the mutant separase enzyme, substrate cleavage by endogenous separase could be prevented and cause embryo lethality. In support of this, we find accumulation of protease-dead separase at putative sites of separase activity where it could be preventing access to substrates by endogenous separase. However, protease-dead separase could still potentially rescue non-proteolytic functions of separase. Further, expression of protease-dead separase in mutant cells that are not true null alleles can further confound interpretations. This could explain discrepancies in findings and conclusions regarding the protease function of separase in anaphase spindle elongation. Using temperature sensitive separase alleles in budding yeast, two groups [27,28] concluded that separase proteolytic function was not required for anaphase spindle elongation. However, using more stringent alleles, [29] and [7] came to the opposite conclusion.

The lack of an inducible gene expression system in *C. elegans* makes it difficult to study dominant-negative or otherwise toxic mutant proteins. Here we discuss methods that facilitate studies involving worm lines with toxic *pie-1* driven transgenes. Although an inducible heat shock promoter has been suggested for inducing transgene expression [30], this approach does not lead to germline specific expression which could complicate phenotypic analysis. Combining soma or germline specific RNAi mutants (for example, *rrf-1* vs. *ppw-1* mutants) with *gfp* RNAi and inducible expression might be a way to circumvent this problem. While we were able to create SEP-1^{PD}::GFP transgenic lines by normal methods, recovery of bombarded animals directly on *gfp* RNAi could allow isolation of worm lines with transgenes that are more toxic than SEP-1^{PD} or allow for the isolation of overexpressing lines. These methods could also be combined with mutants defective in generational RNAi or temperature sensitive mutations in the RNAi machinery to more rapidly shut off the multigenerational RNAi silencing mechanism and more quickly induce transgene expression [31,32]. In addition, bombardment of *him;unc* lines could allow for immediate isolation of transgenic males, which can be maintained by backcrossing. The male propagation method bypasses the multigenerational propagation of *gfp* RNAi, but only introduces a single copy of the transgene, which may not lead to highest expression levels. On the other hand, backcrossing to *unc-119* each generation can help reduce selective pressure that might silence transgene expression or select for suppressor mutations.

The *gfp* RNAi feeding and male propagation methods allow for crossing schemes to study genetic interactions of mutant proteins. Mutant separase alleles are lethal when homozygous and must be maintained as heterozygotes with chromosomal balancers. We found that SEP-1^{PD}::GFP could not rescue hypomorphic or null separase mutants, and that SEP-1^{PD}::GFP expression exacerbated phenotypes in heterozygous hypomorphic separase mutants. We were unable to examine SEP-1^{PD}::GFP in separase null background because separase null worms are not viable. Further, male propagation and *gfp* RNAi feeding allow for the creation of double transgenic lines to study localization patterns and phenotypes by live imaging.

With the current advancements of CRISPR-Cas genome editing [33], the use of *gfp* RNAi and similar strategies may be advantageous. For example, *gfp* RNAi could be used to knock down expression of any endogenous gene of interest tagged with GFP. This strategy may also prove more effective than gene-specific RNAi for RNAi resistant genes. Further, one could temporally control mutant allele expression for multiple alleles by designing and utilizing *gfp* RNAi in combination with another, such as RNAi directed against mCherry.

Further analysis is required to determine which functions of separase require protease activity in the *C. elegans* embryo. Previous work indicated that polar body extrusion is independent of separase's proteolytic activity in mouse oocytes [6]. However, this analysis was performed in separase-null mouse embryos, which may have a different phenotype than seen in wild-type separase background. For example, separase could require autocleavage to efficiently bind substrates, which could be mediated in our SEP-1^{PD} transgenic lines by endogenous separase. Ultimately, detailed mechanistic understanding of separase function will require the identification of relevant substrates and characterization of how their cleavage works together with non-proteolytic signaling mechanisms to execute various cell division events.

Supporting Information

Figure S1 Embryonic lethality in lines with N-terminal or C-terminal GFP fusion to SEP-1^{PD}. Both WH520 (C-terminal fusion to SEP-1^{PD}) and WH524 (N-terminal fusion to SEP-1^{PD}) could be maintained on *gfp* RNAi (not shown). The graph shows embryonic lethality for WH520 and WH524 following removal from *gfp* RNAi at 20°C for the indicated number of generations. Each data point with error bars represents the average of a group of 10 singled worms +/- SEM examined in an individual experiment.

Figure S2 GFP expression in the male germline. GFP expression in N2 males (A) and WH520 males (B). Regions corresponding to sperm (box with dashed line) and testes (box with solid line) are outlined. C. Image of gonad of an F1 SEP-1^{PD}::GFP hermaphrodite derived from the cross outlined in Figure 3. The -1 oocyte and +1 and +2 embryos are designated by labels and the spermatheca is outlined by the box with dashed line.

Movie S1 Meiosis I, including cortical granule exocytosis and the first polar body extrusion, in an embryo expressing SEP-1^{PD}::GFP and H2B::mCherry. The movie shows a maximum projection of selected 1 µm z stacks to display cortical granules and chromosomes. Images were acquired every 20 seconds. Playback rate of the movie is 10 frames per second.

Movie S2 Meiosis I, including cortical granule exocytosis and the first polar body extrusion, in an embryo expressing SEP-1^{WT}::GFP and H2B::mCherry. The movie shows a maximum projection of selected 1 µm z stacks to display cortical granules and chromosomes. Images were acquired every 20 seconds. Playback rate of the movie is 10 frames per second.

Movie S3 Mitosis in an embryo expressing SEP-1^{PD}::GFP and H2B::mCherry. Images of a single z plane were acquired every 30 seconds. Playback rate of the movie is 10 frames per second.

Movie S4 Mitosis in an embryo expressing SEP-1^{WT}::GFP and H2B::mCherry. Images of a single z plane were acquired every 30 seconds. Playback rate of the movie is 10 frames per second.

Acknowledgments

We thank Judy Yanowitz, Gyorgyi Csankovski, Andrew Fire, Bruce McKee, Chris Turpin and Maitreyi Das for critical review, discussion, and/or feedback on the manuscript. Some *C. elegans* strains (N2, OD56 and VC1279) were obtained from the CGC (University of Minnesota), which is funded by the NIH Office of Research Infrastructure Programs (P40 OD010440).

Author Contributions

Conceived and designed the experiments: DMM JNB. Performed the experiments: DMM LRU JNB. Analyzed the data: DMM JNB. Contributed reagents/materials/analysis tools: DMM LRU JNB. Wrote the paper: DMM JNB.

References

1. Hauf S, Waizenegger IC, Peters JM (2001) Cohesin cleavage by separase required for anaphase and cytokinesis in human cells. Science 293: 1320–1323.
2. Uhlmann F, Lottspeich F, Nasmyth K (1999) Sister-chromatid separation at anaphase onset is promoted by cleavage of the cohesin subunit Scc1. Nature 400: 37–42.
3. Uhlmann F, Wernic D, Poupart MA, Koonin EV, Nasmyth K (2000) Cleavage of cohesin by the CD clan protease separin triggers anaphase in yeast. Cell 103: 375–386.
4. Waizenegger IC, Hauf S, Meinke A, Peters JM (2000) Two distinct pathways remove mammalian cohesin from chromosome arms in prophase and from centromeres in anaphase. Cell 103: 399–410.
5. Nagao K, Adachi Y, Yanagida M (2004) Separase-mediated cleavage of cohesin at interphase is required for DNA repair. Nature 430: 1044–1048.
6. Kudo NR, Wassmann K, Anger M, Schuh M, Wirth KG, et al. (2006) Resolution of chiasmata in oocytes requires separase-mediated proteolysis. Cell 126: 135–146.
7. Baskerville C, Segal M, Reed SI (2008) The protease activity of yeast separase (esp1) is required for anaphase spindle elongation independently of its role in cleavage of cohesin. Genetics 178: 2361–2372.
8. Matsuo K, Ohsumi K, Iwabuchi M, Kawamata T, Ono Y, et al. (2012) Kendrin is a novel substrate for separase involved in the licensing of centriole duplication. Curr Biol 22: 915–921.
9. Schockel L, Mockel M, Mayer B, Boos D, Stemmann O (2011) Cleavage of cohesin rings coordinates the separation of centrioles and chromatids. Nat Cell Biol 13: 966–972.
10. Cabral G, Sans SS, Cowan CR, Dammermann A (2013) Multiple mechanisms contribute to centriole separation in C. elegans. Curr Biol 23: 1380–1387.
11. Sullivan M, Uhlmann F (2003) A non-proteolytic function of separase links the onset of anaphase to mitotic exit. Nat Cell Biol 5: 249–254.
12. Lu Y, Cross F (2009) Mitotic exit in the absence of separase activity. Mol Biol Cell 20: 1576–1591.
13. Ross KE, Cohen-Fix O (2004) A role for the FEAR pathway in nuclear positioning during anaphase. Dev Cell 6: 729–735.
14. Siomos MF, Badrinath A, Pasierbek P, Livingstone D, White J, et al. (2001) Separase is required for chromosome segregation during meiosis I in Caenorhabditis elegans. Curr Biol 11: 1825–1835.
15. Schvarzstein M, Pattabiraman D, Bembenek JN, Villeneuve AM (2013) Meiotic HORMA domain proteins prevent untimely centriole disengagement during Caenorhabditis elegans spermatocyte meiosis. Proc Natl Acad Sci U S A 110: E898–907.
16. Bembenek JN, Richie CT, Squirrell JM, Campbell JM, Eliceiri KW, et al. (2007) Cortical granule exocytosis in C. elegans is regulated by cell cycle components including separase. Development 134: 3837–3848.
17. Bembenek JN, White JG, Zheng Y (2010) A role for separase in the regulation of RAB-11-positive vesicles at the cleavage furrow and midbody. Curr Biol 20: 259–264.
18. Richie CT, Bembenek JN, Chestnut B, Furuta T, Schumacher JM, et al. (2011) Protein phosphatase 5 is a negative regulator of separase function during cortical granule exocytosis in C. elegans. J Cell Sci 124: 2903–2913.
19. Hao Y, Boyd L, Seydoux G (2006) Stabilization of cell polarity by the C. elegans RING protein PAR-2. Dev Cell 10: 199–208.
20. Mello CC, Schubert C, Draper B, Zhang W, Lobel R, et al. (1996) The PIE-1 protein and germline specification in C. elegans embryos. Nature 382: 710–712.
21. Blanchetot C, Chagnon M, Dube N, Halle M, Tremblay ML (2005) Substrate-trapping techniques in the identification of cellular PTP targets. Methods 35: 44–53.
22. Praitis V, Casey E, Collar D, Austin J (2001) Creation of low-copy integrated transgenic lines in Caenorhabditis elegans. Genetics 157: 1217–1226.
23. Burton NO, Burkhart KB, Kennedy S (2011) Nuclear RNAi maintains heritable gene silencing in Caenorhabditis elegans. Proc Natl Acad Sci U S A 108: 19683–19688.
24. Alcazar RM, Lin R, Fire AZ (2008) Transmission dynamics of heritable silencing induced by double-stranded RNA in Caenorhabditis elegans. Genetics 180: 1275–1288.
25. Merritt C, Rasoloson D, Ko D, Seydoux G (2008) 3' UTRs are the primary regulators of gene expression in the C. elegans germline. Curr Biol 18: 1476–1482.
26. McKim KS, Rose AM (1990) Chromosome I duplications in Caenorhabditis elegans. Genetics 124: 115–132.
27. Severin F, Hyman AA, Piatti S (2001) Correct spindle elongation at the metaphase/anaphase transition is an APC-dependent event in budding yeast. J Cell Biol 155: 711–718.
28. Stegmeier F, Visintin R, Amon A (2002) Separase, polo kinase, the kinetochore protein Slk19, and Spo12 function in a network that controls Cdc14 localization during early anaphase. Cell 108: 207–220.
29. Jensen S, Segal M, Clarke DJ, Reed SI (2001) A novel role of the budding yeast separin Esp1 in anaphase spindle elongation: evidence that proper spindle association of Esp1 is regulated by Pds1. J Cell Biol 152: 27–40.
30. Zeiser E, Frokjaer-Jensen C, Jorgensen E, Ahringer J (2011) MosSCI and gateway compatible plasmid toolkit for constitutive and inducible expression of transgenes in the C. elegans germline. PLoS One 6: e20082.
31. Buckley BA, Burkhart KB, Gu SG, Spracklin G, Kershner A, et al. (2012) A nuclear Argonaute promotes multigenerational epigenetic inheritance and germline immortality. Nature 489: 447–451.
32. Calixto A, Ma C, Chalfie M (2010) Conditional gene expression and RNAi using MEC-8-dependent splicing in C. elegans. Nat Methods 7: 407–411.
33. Sander JD, Joung JK (2014) CRISPR-Cas systems for editing, regulating and targeting genomes. Nat Biotechnol.

Selection Pressure in Alternative Reading Frames

Katharina Mir*, Steffen Schober

Institute of Communications Engineering, Ulm University, Ulm, Germany

Abstract

Overlapping genes are two protein-coding sequences sharing a significant part of the same DNA locus in different reading frames. Although in recent times an increasing number of examples have been found in bacteria the underlying mechanisms of their evolution are unknown. In this work we explore how selective pressure in a protein-coding sequence influences its overlapping genes in alternative reading frames. We model evolution using a time-continuous Markov process and derive the corresponding model for the remaining frames to quantify selection pressure and genetic noise. Our findings lead to the presumption that, once information is embedded in the reverse reading frame −2 (relative to the mother gene in +1) purifying selection in the protein-coding reading frame automatically protects the sequences in both frames. We also found that this coincides with the fact that the genetic noise measured using the conditional entropy is minimal in frame −2 under selection in the coding frame.

Editor: Bryan A. White, University of Illinois, United States of America

Funding: Katharina Mir is funded by the Deutsche Forschungsgemeinschaft (DFG) under the grants BO867/23-3 in the priority program {SPP 1395}. Steffen Schober is supported by the DFG grant SCHO 1576/1. The funders had no role in study design, data collection and analysis, decision to publish, or preparation of the manuscript.

Competing Interests: The authors have declared that no competing interests exist.

* Email: katharina.mir@uni-ulm.de

Introduction

Overlapping genes are protein coding genes sharing the same DNA locus in different reading frames. As DNA consists of two strands and each amino acid is encoded by non-overlapping triplets (codons), up to six reading frames are possible at a given locus. Overlapping genes are a well known and accepted phenomenon in viruses, however this effect was explained from space limitations of the capsid volume [1]. Until lately most authors denied the existence of overlapping genes in bacterial genomes, consequently bacterial genome annotation programs excluded overlapping candidates in alternative reading frames deliberately [2–5]. Although an experimental verification of two protein-coding genes in the same DNA locus is extremely challenging, over the last years an increasing number of non trivially overlapping genes in prokayotes have been found [6–10].

This paper is concerned with the question how selection pressure in the protein-coding frame influences alternative reading frames. Is it possible to protect by selection two protein-coding sequences simultaneously? We explore this question using a stochastic model for the evolution of the protein-coding reading frame and predict the consequent behaviour in the alternative reading frames.

Sequence evolution can be described on nucleotide level [11–14], amino acid level, e.g. Dayhoff and Schwartz [15], or on codon level. Here we chose the latter approach using a time-continuous Markov process as suggested by Goldman and Yang [16] and Muse and Gaut [17]. We apply the model of Yang and Nielsen [18] which is based on [16]. An extended model was already used by Sabath *et al.* [19] to study the evolution of a random protein-coding sequence. In contrast to our approach, Sabath investigated the selection intensities of overlapping genes assuming that each gene of the overlapping pair faces selection independently.

Several studies analyzed selection intensities in virus genomes within overlapping gene regions investigating how nonsynonymous and synonymous mutations influence two reading frames simultaneously showing that a high rate of nonsynonymous mutations in one reading frame falls onto synonymous substitutions in an alternative frame at the same time, e.g. [20–22].

Our investigation reveals that selection pressure in the protein-coding reading frame +1 is correlated to the reverse reading frame −2, where in fact many examples of overlapping genes found so far are located e.g., [9,10]. Precisely there is a strong coupling of the nonsynonymous to synonymous substitutions rate ratios in these frames. In another approach following Yockey [23], we quantify the genetic noise using the conditional entropy and the mutual information as a measure of sequence similarity. The results obtained coincides with the former observations.

The outline of the paper is as follows: In Section *Methods* we introduce the evolutionary framework and the calculation of selection pressure. The biological and information theoretic measures are presented in Section *Results*, together with an application of the model to a bacterial genome and evidence on the robustness of our approach. Finally we discuss the results in the last section.

Methods

Framework of Evolutionary Model

This section introduces the evolutionary framework and the notations used. We denote a discrete random variable with X and their corresponding probability mass function with $p_X(x)$, where x

is the concrete realization of X. Throughout the paper the nucleotide alphabet is denoted with $\mathcal{N} = \{A, C, G, T\}$ and the codon alphabet is denoted with $\mathcal{C} = \{A, C, G, T\}^3$.

In the following we consider the well known Goldman and Yang model [16] in a simplified version as it was introduced by [18], where the following definitions can be found. (For more details on the derivation of the model see [24].) The model assumes a stationary codon distribution and independence of the evolving codon sites. The evolution of protein-coding DNA sequences is modelled by a time-continous Markov process described by the substitution rate matrix $Q = \{q_{xy}\}$, where q_{xy} is the rate from codon x to codon y with $x \neq y$

$$q_{xy}(\pi_y, \kappa, \omega) = \begin{cases} 0 & \text{if x and y differ at more than one position} \\ \pi_y & \text{if x and y differ by a synonymous transversion} \\ \kappa\pi_y & \text{if x and y differ by a synonymous transition} \\ \omega\pi_y & \text{if x and y differ by a non}-\text{synonymous transversion} \\ \kappa\omega\pi_y & \text{if x and y differ by a non}-\text{synonymous transition} \end{cases}, \quad (1)$$

where κ is the transition/transversion rate, ω is the nonsynonymous/synonymous rate ratio and π_y is the equilibrium frequency of codon y. Note that $\pi_y \in \mathcal{C}_{61} = \{\mathcal{C} \setminus (TAG, TGA, TAA)\}$ as transitions to stop codons are not allowed inside functional proteins. The row sums of the rate matrix $Q = \{q_{xy}\}$ have to be zero, which determines the main diagonal of the matrix. Further the rate matrix is multiplied by a scaling factor to normalize the expected number of nucleotide substitutions per codon to one. With every time-continuous Markov process, a discrete time Markov chain can be associated. This leads to a discrete evolution matrix $P(t) = P_{Y|X}(t)$ with conditional probabilities that describes a transition of an input $X \in \mathcal{C}_{61}$ to an output $Y \in \mathcal{C}_{61}$ for a fixed t. The evolutionary transition probability matrix is determined by

$$P(t) = \{p_{xy}\} = e^{Q\,t},$$

where p_{xy} is the probability that input codon x becomes y after time t. Note that

$$\pi P(t) = \pi \qquad \text{and} \qquad \pi Q = 0$$

holds, where row vector π is the stationary codon distribution.

We call $(X, Y, P_{Y|X})$ an evolutionary channel referring to the communication theoretic term [25]. Note that the rate matrix Q is also a channel matrix. Further the parameters of the rate matrix t, ω and κ are arbitrary but fixed.

Given a rate matrix for the protein-coding reading frame, we are interested in computing the resulting rate and evolutionary channel matrices in the other reading frames. We define the protein-coding reading frame as +1 and denote the shifted and reverse complement reading frames as non-coding reading frames $f = \{-1, \pm 2, \pm 3\}$. If we refer to a special reading frame, we use the index f. The setup we consider is as follows: In the protein-coding reading frame we assume that codons $x \in \mathcal{C}_{61}$ with codon usage π^{+1} from a bacterial organism are transmitted independently over the evolutionary channel $P(t, \kappa, \omega, \pi)$ to the output y. This is called a discrete memoryless channel. For convenience we write $P^{+1}_{Y|X}$ instead of $P^{+1}_{Y|X}(t,\kappa,\omega,\pi)$. Each codon in frame +1 consists of three random variables $c_j = (X_1^{(j)}, X_2^{(j)}, X_3^{(j)})$ with realizations $x_k^{(j)} \in \mathcal{N} = \{A, C, G, T\}$ and evolves to codon $\tilde{c}_j = (Y_1^{(j)}, Y_2^{(j)}, Y_3^{(j)})$ with $y_k^{(j)} \in \mathcal{N}$. In frame +1 we observe the scheme presented in Figure 1. Given $P^{+1}_{Y|X}$ and π^{+1} we want to determine the evolution matrix per reading frame $P^f_{Y|X}, f \in \{-1, \pm 2, \pm 3\}$. We solve this task directly via the rate matrix per reading frame Q^f given the rate matrix Q^{+1} and π^{+1} such that we are independent of the evolution time. For the alternative reading frames we combine two independent time-continuous Markov chains to the corresponding di-codon matrix in frame +1 by

$$Q^{+1}_{\text{di-codon}} = \left(Q^{+1} \otimes I_Q + I_Q \otimes Q^{+1}\right),$$

where $\otimes$ is the Kronecker product and I_Q is the identity matrix with the same dimension as the rate matrix [26]. The rates of the di-codon transitions are now combined to compute the rate matrices in the other frames. Without loss a generality, we consider frame +2 (black parts in Figure 1).

$$Q^{+2}(y_2^1 y_3^1 y_1^2 | x_2^1 x_3^1 x_1^2) = Q^{+2}(\tilde{y}|\tilde{x})$$
$$= \frac{\sum_{x_1^1, x_2^2, x_3^2 \in \mathcal{N}} \sum_{y_1^1, y_2^2, y_3^2 \in \mathcal{N}} \pi^{+1}_{\text{di-codon}}(x_1^1 \tilde{x} x_2^2 x_3^2) \cdot Q^{+1}_{\text{di-codon}}(y_1^1 \tilde{y} y_2^2 y_3^2 | x_1^1 \tilde{x} x_2^2 x_3^2)}{\sum_{x_1^1, x_2^2, x_3^2 \in \mathcal{N}} \pi^{+1}_{\text{di-codon}}(x_1^1 \tilde{x} x_2^2 x_3^2)},$$

where $\pi^{+1}_{\text{di-codon}}(c_j c_i) = \pi^{+1}(c_j) \cdot \pi^{+1}(c_i) \quad \forall\, c_i, c_j \in \mathcal{C}^{61}$. The rate matrices of the other frames can be determined accordingly. Note that Q^f is a 64×64 matrix for $f = \{\pm 2, \pm 3\}$ and a 61×61 matrix for $f = \{\pm 1\}$. Given the rate matrix Q^f of a time-continuous Markov chain the corresponding stationary distribution π^f in each reading frame as well as the transition matrix $P^f_{Y|X}$ for time t can be easily determined.

Selection pressure during evolution

An important parameter describing the selection pressure on the protein level is the ratio of nonsynonymous d_N to synonymous d_S substitution rates, denoted with $\omega = \frac{d_N}{d_S}$, see e.g., [16]. Three basic scenarios are distinguished e.g., [27]: Purifying selection when $\omega < 1$, adaptive selection for $\omega > 1$ and neutral mutation if $\omega = 1$. To determine the nonsynonymous/synonymous rate ratio ω in each reading frame, we apply the procedure presented in [18] and [24], that is based on the transition probability matrix $P_{Y|X}$, but can be easily adapted to the rate matrix Q.

Assume we determined, the rate matrix Q^f in each frame $f \in \{\pm 1, \pm 2, \pm 3\}$ as presented in *Framework of Evolutionary Model* as well as the stationary distributions π^f. The proportion of synonymous substitutions is the sum over all codon pairs x and y $(x \neq y)$ that code for the same amino acid

$$\rho^f_S = \sum_{x \neq y, aa_x = aa_y} \pi^f q^f_{xy},$$

where aa_x is the amino acid encoded by codon x. The proportion of nonsynonymous substitutions is calculated accordingly by

$$\rho^f_N = \sum_{x \neq y, aa_x \neq aa_y} \pi^f q^f_{xy}.$$

The transition/tranversion rate κ is the same in all reading frames. We assume that it is known from reading frame +1. To

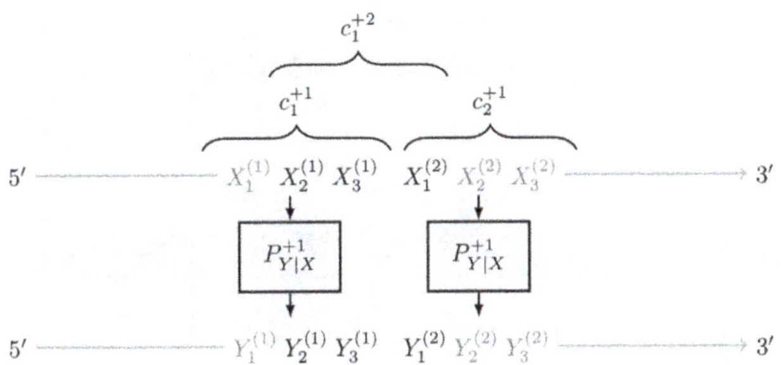

Figure 1. Transition Scheme. Scheme of transitions in sequence direction on forward strand and in time direction.

determine the proportion of synonymous sites, we calculate a new rate matrix following Eq. (1) for a fixed $\omega = 1.0$,

$$\rho_S^{1,f} = \sum_{x \neq y, aa_x = aa_y} \pi^f q_{xy}(\pi^f, \kappa, 1.0).$$

The proportion of nonsynonymous sites is calculated accordingly and denoted with $\rho_N^{1,f}$. The number of synonymous substitutions per synonymous site is

$$d_S^f = \frac{t\rho_S^f}{3\rho_S^{1,f}}.$$

The number of nonsynonymous substitutions per nonsynonymous site d_N^f is calculated accordingly. This results in

$$\omega^f = \frac{d_N^f}{d_S^f} = \frac{\rho_N^f \cdot \rho_S^{1,f}}{\rho_N^{1,f} \cdot \rho_S^f},$$

where the time as well as scaling factors of the rate matrices cancel out.

Results

Throughout the paper, we use the model genome *Escherichia coli* O157:H7 EDL933 (Accession number NC_002655, abbreviation EHEC), with a GC content of 50.4% and a length of 5528445 base pairs. In *File S1 Model verification* we present some simulation results to validate the calculations of the equilibrium frequencies per reading frame π^f.

To investigate the influence of selection pressure during evolution we chose two different input scenarios: The transition/transversion rates are $\kappa = 1.0$ or $\kappa = 5.0$ at time $t = 1.0$ and the nonsynonymous/synonymous rate ratio ω takes values between [0, 3]. The calculation of the nonsynonymous/synonymous rate ratio ω^f for $f \in \{\pm 1, \pm 2, \pm 3\}$ reveals the following results. Purifying selection refers to a selection against nonsynonymous substitutions on the DNA level, which protects the sequence. In Figure 2, we see that a protection of the coding frame +1 with $\omega < 1$, also protects the sequence in frame -2, the other alternative frames face adaptive selection. The opposite is observed for $\omega > 1$, where new information can be induced in frames +1 and -2, whereas the other frames are slightly below the neutral mutation line. The behaviour of ω^f is consistent for both scenarios. Note, there are numerous methods to determine the synonymous to nonsynonymous rate ratio. The *File S1 Selection pressure* shows a comparison of our approach with an alternative method.

Quantifying noise during evolution

In this part, we deal with the following question: An amino acid is transmitted over the evolutionary channel, how long is this information conserved in the different reading frames?

Evolution of a sequence can be considered as a communication process over time. In his book [23] proposed to use the conditional entropy to measure the amount of genetic information that can be transmitted over a noisy channel (based on [25]). We define the amino acid alphabet $\mathcal{A} = \mathcal{G}(\mathcal{C}_{61})$, where $\mathcal{G}$ is the genetic code, which results in a cardinality of $|\mathcal{A}| = 20$. The codon evolution matrix per frame $P_{Y|X}^f$ can be summarized to determine the amino acid evolution matrix $P_{Y|X}^{f,\mathcal{A}}$, based on the stationary distribution π^f. The stop codon probabilities are removed in all frames. The conditional entropy between two random variables X and Y over alphabets $\mathcal{X}, \mathcal{Y}$ is defined as, e.g., in [28]:

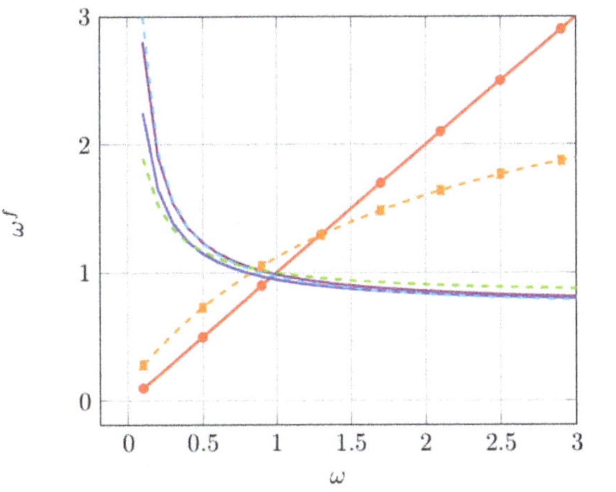

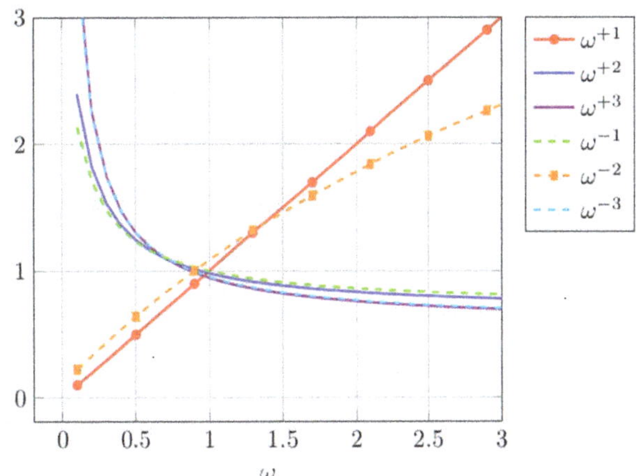

Figure 2. Selection Pressure. Estimation of nonsynonymous/synonymous rate ratio ω^f for different parameter settings. In the left panel $\kappa = 1.0$, $t = 1.0$ and on the right panel we set $\kappa = 5.0$, $t = 1.0$. Protection of protein-coding frame +1 for $\omega < 1$ is directly coupled with a protection of reading frame -2.

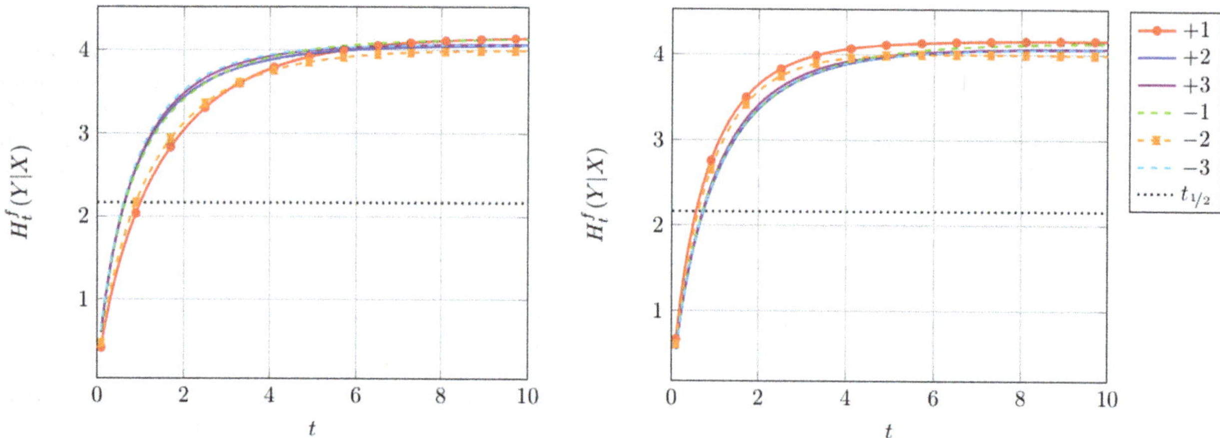

Figure 3. Information Loss. Conditional entropy for uniform input distribution over amino acids for different values of ω and $\kappa = 1.0$. On the left $\omega = 0.3$, the protein-coding frame as well as frame −2 are protected, which results in a slower information loss than for the other reading frames. On the right $\omega = 3.0$, we see the opposite scenario. At the black dotted line, half of the information is lost.

$$H(Y|X) = -\sum_{x\in\mathcal{X}} p_X(x) \sum_{y\in\mathcal{Y}} P_{Y|X}(y|x)\log_2 P_{Y|X}(y|x),$$

where $P_{Y|X}(y|x)$ is the conditional probability.

The conditional entropy between two randomly chosen amino acids X and Y in frame f conditioned on $X = a$ with $a \in \mathcal{A}$ is accordingly

$$H_t^f(Y|X=a) = -\sum_{y\in\mathcal{A}} P_{Y|X}^{f,\mathcal{A}}(y|a)\log_2 P_{Y|X}^{f,\mathcal{A}}(y|a),$$

where $P_{Y|X}^{f,\mathcal{A}}$ is the amino acid substitution matrix per reading frame. If we know that a specific amino acid was transmitted, how much of this knowledge is lost after time t? As the comparison of 20 values over time is inconvenient, we apply uniform weighting according to the amino acids $p^{f,\mathcal{A}}$, which results in

$$H_t^f(Y|X) = \sum_{x\in\mathcal{A}} p^{f,\mathcal{A}}(x) H_t^f(Y|X=x). \quad (2)$$

Note that Eq. (2) is bounded by

$$H_t^f(Y|X) \le H_t^f(Y) \le \log_2(20) = 4.32[\text{bit}],$$

where the entropy (or uncertainty) $H_t^f(Y) = -\sum_{y\in\mathcal{A}} p^{f,\mathcal{A}}(y) \log_2 p^{f,\mathcal{A}}(y)$ is maximal for a uniformly distributed random variable Y.

Yockey [23] additionally suggests the application of the mutual information as a measure of similarity between sequences. The mutual information between amino acid X and Y per frame f is defined as, e.g., in [28]:

$$I_t^f(X;Y) = H_t(Y) - H_t(Y|X). \quad (3)$$

Note, that the channel capacity, which is the maximal mutual information for all input distributions, can be determined numerically using the Blahut-Arimoto algorithm. But as there is no direct interpretation in our framework and the results match those of the mutual information, we abandoned the presentation.

Results at the example of EHEC. We chose two different input scenarios: Set $\omega = 0.3$ to model purifying selection and $\omega = 3.0$ to model adaptive selection. The transition/transversion rate is fixed to $\kappa = 1.0$ and the time t is changed during simulation. The same parameter setting was already used in [27]. We apply the conditional entropy introduced in Eq. (2) to answer the question how long the information which amino acid was transmitted is conserved in the different reading frames. The results are presented in Figure 3.

Evolution means loss of information over time or from a complementary point of view, an increase of uncertainty. To quantify this information loss, we determine the time needed to loose half of the information. As the conditional entropy in bounded by $\log_2(20)$, we determine for each frame $t_{1/2}$ such that

$$H_{t_{1/2}}^f(Y|X) = \frac{\log_2(20)}{2}.$$

The results are summarized in Table 1.

Table 1. Time for each frame where the conditional entropy is $\log_2{}^{(20)}/_2$.

ω	+1	+2	+3	−1	−2	−3
0.3	1.0	0.7	0.7	0.7	0.9	0.7
3.0	0.6	0.8	0.8	0.8	0.7	0.8

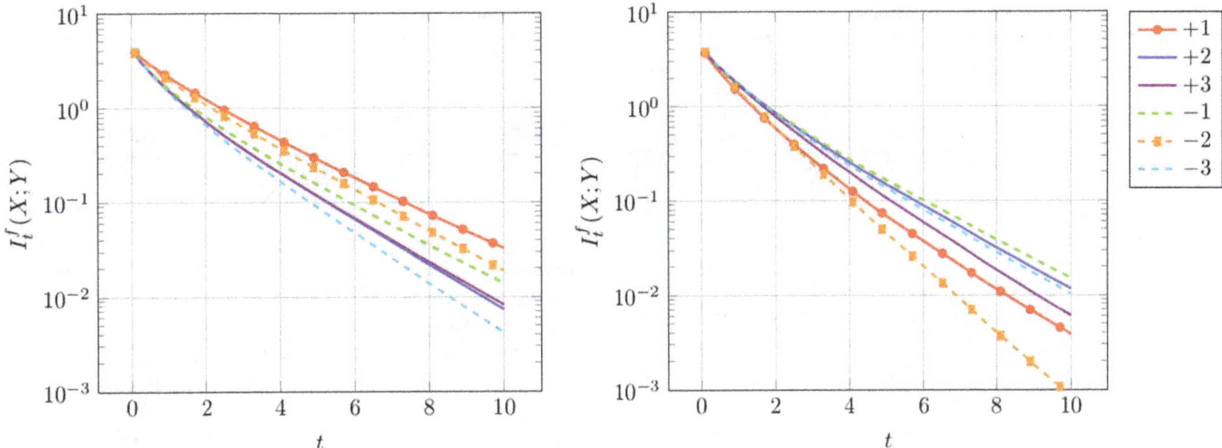

Figure 4. Sequence Similarity. Mutual information for uniform input distribution over amino acids for different values of ω and $\kappa = 1.0$. On the left $\omega = 0.3$, the amount of information transmitted over the channel is largest for the protected frames +1 and −2. On the right $\omega = 3.0$, where those frames are not protected, the opposite holds.

In accordance with the results of ω^f in Figure 2 we interpret Figure 3 and Table 1 as follows. Protected frames with $\omega<1$, store information longer than the unprotected frames with $\omega>1$. When frame +1 is protected, then the −2 frame is protected automatically, therefore those frames show a slower increasing uncertainty than the alternative frames.

Now, the mutual information $I_t^f(X;Y)$ per reading frame $f \in \{\pm1, \pm2, \pm3\}$ is investigated applying Eq. (3). The mutual information measures the similarity between X and Y, which is directly connected to the amount of information that can be transmitted over the channel [23]. We observe for the first scenario, where $\omega = 0.3$, presented in the left panel of Figure 4 that most information can be transmitted in reading frame +1 followed by reading frame −2. In general the proportion of information, that can be transmitted over the evolutionary channel decreases over time, but this information loss is faster in the frames, where $\omega^f>1$. In the right panel of Figure 4, where $\omega = 3.0$, we see that the mutual information is smallest, for the frames +1 and −2 which is also in accordance with Figure 2. This observation is confirmed in the *File S1 Conditional entropy and mutual information* for different values of ω.

Robustness of method. The question arises, how robust our method is, if we choose another codon substitution matrix. As we are able to determine the mutual information and the conditional entropy per reading frame, given only the evolution matrix in the coding reading frame $P_{Y|X}^{+1}$ and the stationary distribution of EHEC π^{+1} it is also possible to substitute the channel matrix $P_{Y|X}^{+1}$. In 2005 [29] published an empirical codon substitution matrix (P_{ECM}) obtained from an alignment of vertebrate DNA, which can also be applied to bacteria. Given a transition matrix we present in *File S1 Robustness of results* a method to estimate the transition matrices per reading frame based on the channel matrix $P_{Y|X}^{+1}$. For our investigations, the different time points t presented in Figure 5 are obtained by $P_{ECM}^t, t \in \mathbb{Z}$. The results confirm our findings, that most information can be transmitted in +1, followed by −2. That makes sense, as the matrix is based on genes with purifying selection, otherwise they would not have survived over time.

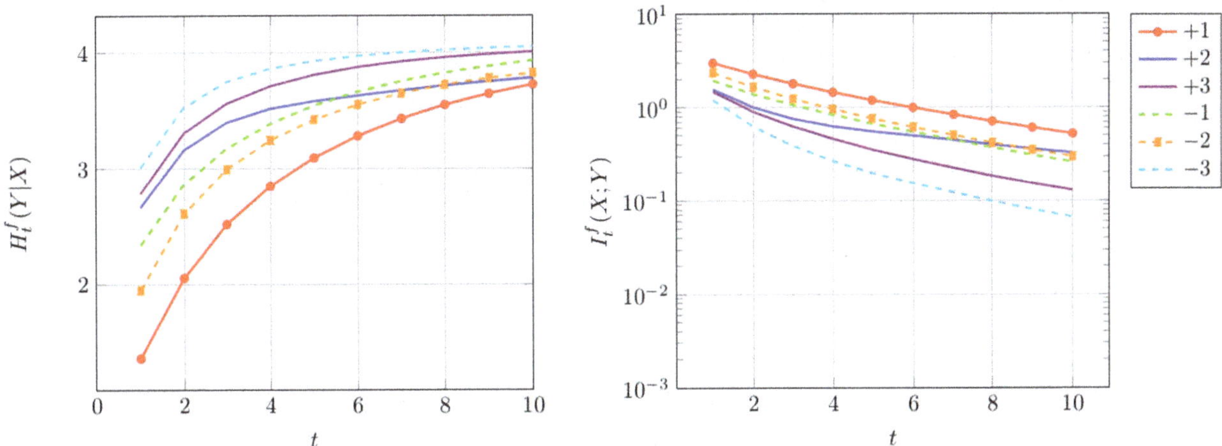

Figure 5. Empirical Substitution Matrix. Estimation of conditional entropy (left panel) and mutual information (right panel) for empirical codon substitution matrix P_{ECM}. A slower information loss for reading frames −2 is observed due to a protection of the protein-coding reading frame +1.

Table 2. Degree of freedom to choose amino acids according to the genetic code.

Frame	+2	+3	−1	−2	−3
Mean	2.94	2.93	2.67	1.59	3.12

Summary

In this paper we introduced a model to determine the codon evolution in different reading frames based on the protein-coding reading frame +1. The model is used to predict the selection pressure within different reading frames and reveals that a protection of the protein-coding reading frame also preserves the reverse reading frame −2. For the case of adaptive selection both frames are free to evolve. The remaining alternative frames show the reverse relation, i.e. they give preference to nonsynonymous substitutions while reading frame +1 is protected and are preserved when +1 is exposed to adaptive selection. These findings are further confirmed by the presented results on the conditional entropy. Namely, if $\omega<1$, the genetic noise is minimal in frames + 1 and −2, also the sequence similarity measured by the mutual information is largest. Conversely for $\omega>1$, where the genetic noise of +1 and −2 is largest and the sequence similarity accordingly smallest.

Discussion and Conclusion

At a first glance, understanding the evolution of overlapping protein-coding regions is extremely challenging, because one DNA segment codes for two proteins which are translated in different reading frames simultaneously, such that a mutation affects both proteins [30,31]. Biologist investigate evolutionary adaption of proteins for years now, assuming that adaption requires more nucleotide mutations at positions that change an amino acid than at positions that preserve a site [32]. The parameter of choice that measures the substitution rate at those sites is $\omega=\frac{d_N}{d_S}$ and is therefore used as an indicator of selective pressure within genes.

Meanwhile a large field emerged, investigating the evolutionary constraints within overlapping and non overlapping reading frames [30,33–36]. There exist empirical analyses describing, that a loss of a stop codon within a protein-coding gene by deletion, mutation or frame-shift, causes an elongation to the next stop codon, whereby an overlapping pair originates [37,38]. Other studies suggest, that the loss of a start codon is responsible for the development of an overlap [39–41]. From this point of view, a random formation can not be ruled out.

Our point of interest is slightly different, assuming we are given a protein-coding reading frame that evolves over time, we are interested in the evolutionary constraints implied within alternative reading frames. A biological interpretation of our findings is that during adaption many mutations occur that change amino acids in reading frames +1 and −2 simultaneously. Once a protein in reading frame +1 is fixed and adaptive selection is replaced by purifying selection, this process stops and the amount of synonymous substitutions increases, again in both reading frames. Note that we make no statement that both reading frames are already translated into proteins, since function of a sequence could also evolve later. As a matter of fact, over time the divergence of a sequence always increases even if it is *protected*, but we showed that this change happens slower in case of purifying selection for both, the +1 and −2 reading frame. No matter, how or if an overlapping gene pair evolved, our observations indicate the special role of the −2 reading frame. Interestingly, two recently experimentally verified examples of overlapping gene pairs in bacteria *yaaW/htga* by [10] and *dmdR1/adm* by [9] are in frame −2. We showed that it is possible to protect this frame by simply controlling the selection pressure within the protein-coding reading frame. This can be attributed to a property of the genetic code, as the most important codon positions are the first and second which fall onto the second respectively first position in the −2 frame. But this could also mean that a conserved sequence in −2 might be solely an artefact, providing not necessarily evidence for functionality.

Finally note that it is challenging to embed information in the overlapping reading frame −2, when the protein-coding reading frame +1 has a fixed amino acid sequence. Assume two amino acids $A_1,A_2 \in \mathcal{A}^*$, where $\mathcal{A}^*$ is $\mathcal{A}$ plus a stop label, should be encoded in the coding reading frame. Obviously each amino acid corresponds to an individual number of codons, hence it is possible to encode a certain number of different amino acids in the alternative reading frames without changing A_1 and A_2. The average taken over all possible pairs A_1, A_2 are shown in Table 2; it turns out that the degree of freedom is smallest in −2. It is worth noting that in general it is possible to embed information even in protein-coding sequences, see for example [42].

Supporting Information

File S1 Additional Data and Figures. Contains further information to verify the model predictions by comparison with simulation, another method to determine the selection pressure, different investigations on the conditional entropy and mutual information as well as a method to show the robustness of results.

Author Contributions

Analyzed the data: KM. Wrote the paper: KM SS. Developed the model in general: SS. Discussions and contributions in the mathematical part of the manuscript: SS. Developed the details of the model: KM. Implementation of the program: KM.

References

1. Chirico N, Vianelli A, Belshaw R (2010) Why genes overlap in viruses. Proceedings of the Royal Society B: Biological Sciences 277: 3809–3817.
2. Yooseph S, Sutton G, Rusch DB, Halpern AL, Williamson SJ, et al. (2007) The Sorcerer II Global Ocean Sampling expedition: expanding the universe of protein families. PLoS Biology 5: e16+.
3. Delcher AL, Bratke KA, Powers EC, Salzberg SL (2007) Identifying bacterial genes and endosymbiont DNA with Glimmer. Bioinformatics 23: 673–679.
4. Warren A, Archuleta J, Feng WC, Setubal J (2010) Missing genes in the annotation of prokaryotic genomes. BMC Bioinformatics 11: 131.
5. Johnson ZI, Chisholm SW (2004) Properties of overlapping genes are conserved across microbial genomes. Genome Research 14: 2268–2272.
6. McVeigh A, Fasano A, Scott D, Jelacic S, Moseley S, et al. (2000) IS1414, an *Escherichia coli* insertion sequence with a heat-stable enterotoxin gene embedded in a transposase-like gene. Infection and Immunity 68: 5710–5715.

7. Behrens M, Sheikh J, Nataro JP (2002) Regulation of the overlapping *pic/set* locus in *Shigella flexneri* and enteroaggregative *Escherichia coli*. Infection and Immunity 70: 2915–2925.
8. Silby MW, Levy SB (2008) Overlapping protein-encoding genes in *Pseudomonas fluorescens* Pf0-1. PLoS Genetics 4: e1000094.
9. Tunca S, Barreiro C, Coque JJR, Martin JF (2009) Two overlapping antiparallel genes encoding the iron regulator DmdR1 and the Adm proteins control siderophore and antibiotic biosynthesis in *Streptomyces coelicolor* A3(2). FEBS Journal 276: 4814–4827.
10. Fellner L, Bechtel N, Witting MA, Simon S, Schmitt-Kopplin P, et al. (2013) Phenotype of htga (mbia), a recently evolved orphan gene of escherichia coli and shigella, completely overlapping in antisense to yaaw. FEMS Microbiology Letters: 1–8.
11. Jukes TH, Cantor CR (1969) Evolution of Protein Molecules. Academy Press.
12. Kimura M (1980) A simple method for estimating evolutionary rates of base substitutions through comparative studies of nucleotide sequences. Journal of Molecular Evolution 16: 111–120.
13. Felsenstein J (1981) Evolutionary trees from dna sequences: a maximum likelihood approach. Journal of Molecular Evolution 17: 368–376.
14. Hasegawa M, Kishino H, aki Yano T (1985) Dating of the human-ape splitting by a molecular clock of mitochondrial dna. Journal of Molecular Evolution 22: 160–174.
15. Dayhoff MO, Schwartz RM (1978) Chapter 22: A model of evolutionary change in proteins. In: in Atlas of Protein Sequence and Structure.
16. Goldman N, Yang Z (1994) A codon-based model of nucleotide substitution for protein-coding DNA sequences. Molecular Biology and Evolution 11: 725–736.
17. Muse SV, Gaut BS (1994) A likelihood approach for comparing synonymous and nonsynonymous nucleotide substitution rates, with application to the chloroplast genome. Molecular Biology and Evolution 11: 715–724.
18. Yang Z, Nielsen R (2000) Estimating synonymous and nonsynonymous substitution rates under realistic evolutionary models. Molecular Biology and Evolution 17: 32–43.
19. Sabath N, Landan G, Graur D (2008) A method for the simultaneous estimation of selection intensities in overlapping genes. PLoS One 3.
20. Guyader S, Ducray D (2002) Sequence analysis of Potato leafroll virus isolates reveals genetic stability, major evolutionary events and differential selection pressure between overlapping reading frame products. Journal of General Virology 83: 1799–807.
21. Hughes AL, Westover K, da Silva J, O'Connor DH, Watkins DI (2001) Simultaneous positive and purifying selection on overlapping reading frames of the tat and vpr genes of simian immunodeficiency virus. Journal of Virology 75: 7966–72.
22. Hughes AL, Hughes MA (2005) Patterns of nucleotide difference in overlapping and non-overlapping reading frames of papillomavirus genomes. Virus Res 113: 81–88.
23. Yockey HP (1992) Information Theory in Molecular Biology. Cambridge: Cambridge University Press.
24. Yang Z (2006) Computational molecular evolution. Oxford: Oxford University Press.
25. Shannon CE (1948) A mathematical theory of communication. Bell system technical journal 27.
26. D'Argenio PR, Jeannet B, Jensen HE, Larsen KG (2001) Reachability analysis of probabilistic systems by successive refinements. In: APM-PROBMIV.
27. Zhang Z, Li J, Yu J (2006) Computing ka and ks with a consideration of unequal transitional substitutions. BMC Evolutionary Biology 6: 44.
28. Cover TM, Thomas JA (2006) Elements of Information Theory (Wiley Series in Telecommunications and Signal Processing). Wiley-Interscience.
29. Schneider A, Cannarozzi G, Gonnet G (2005) Empirical codon substitution matrix. BMC Bioinformatics 6.
30. Krakauer D (2000) Stability and Evolution of Overlapping Genes. Evolution; International Journal of Organic Evolution 54: 731–739.
31. Miyata T, Yasunaga T (1980) Molecular evolution of mRNA: a method for estimating evolutionary rates of synonymous and amino acid substitutions from homologous nucleotide sequences and its application. J Mol Evol 16: 23–36.
32. Kryazhimskiy S, Plotkin JB (2008) The population genetics of dN/dS. PLoS Genet 4: e1000304+.
33. Hein J, Stovlbaek J (1995) A maximum-likelihood approach to analyzing nonoverlapping and overlapping reading frames. Journal of Molecular Evolution 40: 181–9.
34. Pedersen AM, Jensen JL (2001) A dependent-rates model and an MCMC-based methodology for the maximum-likelihood analysis of sequences with overlapping reading frames. Mol Biol Evol 18: 763–76.
35. Fonseca M, Harris D, Posada D (2014) Origin and Length Distribution of Unidirectional Prokaryotic Overlapping Genes. G3 4: 19–27.
36. Rogozin I, Spiridonov A, Sorokin A, Wolf Y, Jordan I, et al. (2002) Purifying and directional selection in overlapping prokaryotic genes. Trends in Genetics 18: 228–232.
37. Fukuda Y, Washio T, Tomita M (1999) Comparative study of overlapping genes in the genomes of Mycoplasma genitalium and Mycoplasma pneumoniae. Nucleic Acids Research 27: 1847–1853+.
38. Fukuda Y, Nakayama Y, Tomita M (2003) On dynamics of overlapping genes in bacterial genomes. Gene 323: 181–187.
39. Cock P, Whitworth D (2007) Evolution of Gene Overlaps: Relative Reading Frame Bias in Prokaryotic Two-Component System Genes. Journal of Molecular Evolution 64: 457–462.
40. Cock PJA, Whitworth DE (2010) Evolution of relative reading frame bias in unidirectional prokaryotic gene overlaps. Mol Biol Evol 27: 753–6.
41. Sabath N, Graur D, Landan G (2008) Same-strand overlapping genes in bacteria: compositional determinants of phase bias. Biology Direct 3: 36+.
42. Haughton D, Balado F (2013) Biocode: Two biologically compatible algorithms for embedding data in non-coding and coding regions of dna. BMC Bioinformatics 14: 121+.

14

A Window of Opportunity to Control the Bacterial Pathogen *Pseudomonas aeruginosa* Combining Antibiotics and Phages

Clara Torres-Barceló[1]*, Flor I. Arias-Sánchez[1], Marie Vasse[1], Johan Ramsayer[1], Oliver Kaltz[1]❡, Michael E. Hochberg[1,2,3]*❡

1 Institut des Sciences de l'Evolution, CNRS-Université Montpellier 2, Montpellier, France, 2 Santa Fe Institute, Santa Fe, New Mexico, United States of America, 3 Wissenshaftskolleg zu Berlin, Berlin, Germany

Abstract

The evolution of antibiotic resistance in bacteria is a global concern and the use of bacteriophages alone or in combined therapies is attracting increasing attention as an alternative. Evolutionary theory predicts that the probability of bacterial resistance to both phages and antibiotics will be lower than to either separately, due for example to fitness costs or to trade-offs between phage resistance mechanisms and bacterial growth. In this study, we assess the population impacts of either individual or combined treatments of a bacteriophage and streptomycin on the nosocomial pathogen *Pseudomonas aeruginosa*. We show that combining phage and antibiotics substantially increases bacterial control compared to either separately, and that there is a specific time delay in antibiotic introduction independent of antibiotic dose, that minimizes both bacterial density and resistance to either antibiotics or phage. These results have implications for optimal combined therapeutic approaches.

Editor: Stefan Bereswill, Charité-University Medicine Berlin, Germany

Funding: This work was supported by the McDonnell Foundation (JSMF 220020294/SCS-Research Award), Agence National de la Recherche for funding ('EVORANGE' (ANR-09-PEXT-011) and 'EvolStress' (ANR-09-BLAN-099-01)), Erasmus Mundus Masters Programme and the CONACyT. The funders had no role in study design, data collection and analysis, decision to publish, or preparation of the manuscript.

Competing Interests: The authors have declared that no competing interests exist.

* Email: clara.torres@univ-montp2.fr (CTB); mhochber@univ-montp2.fr (MEH)

❡ These authors contributed equally to this work.

Introduction

Antibiotic resistant bacteria are a widespread problem that threatens human health. Due to the rapid adaptation of bacteria to old and new antibiotics there is an urgent need to develop alternative treatments [1] [2] [3]. Phage therapy, the use of parasitic viruses as antibacterial agents is attracting renewed attention due to their host specificity, innocuity for treated patients, and potential for evolution to outpace bacterial resistance [4]. Despite considerable research on single or combined therapies involving phage [5] [6], the underlying evolutionary processes remain poorly understood.

Evolutionary theory predicts that combined therapies can be more effective than a single component agent for preventing or limiting the evolution of antibiotic resistance [7], and this approach has gained attention in the control of pathogenic microbes [8] [9]. Specifically, adaptive trade-offs can emerge due to fitness costs associated with resistance to more than one antimicrobial agent, as shown in the evolution of resistance to multiple antibiotics [10]. Despite their potential, combined antimicrobial therapies are subject to the evolution of resistance due to convergent mechanisms of resistance if they target similar pathways, and the specific combination will determine the speed of resistance evolution [11]. Synergistic drug combinations, where joint antimicrobial effectiveness is greater than the individual effects, are more efficient and can be employed at lower doses, although selection for resistance can be substantial [11] [12]. Antagonistic drugs have a combined effect that is lower than predicted, and although they generally slow the evolution of resistance are rarely used in a clinical context [11] [12].

The actual implementation of antibiotic therapies also has important implications for the development of resistance [13] [14]. For instance, antibiotic dose can have an important effect on the evolution of resistance, but the mechanisms involved differ between low and high doses. In general, lower doses select for low cost resistance mutations that can be crucial to the stepwise acquisition of higher dose resistance, and higher doses impose stronger selection for resistant alleles [15] [16]. Another factor influencing the short and long-term efficiency of combined therapies is the timing of application, especially for antibiotics and phage, where phage replication and antibiotic effect are both density-dependent [17]. Phage population dynamics will be determined by the number of hosts in which they can replicate, with consequences for the amplification of phage densities and the therapeutic effectiveness [18]. If phages are administered at low bacterial densities or bacteria non-amenable physiologically, then the increase in phage densities will be lower and recurrent application of phages may be necessary [18].

We challenged the opportunistic pathogenic bacterium *Pseudomonas aeruginosa* PAO1 with a lytic bacteriophage and the antibiotic streptomycin, with the aim of uncovering the effects of independent and combined treatments. This nosocomial pathogen species represents a particular danger to cystic fibrosis patients, and is known to readily evolve antibiotic resistance [19]. The antibiotic streptomycin has been shown to act synergistically when used with other chemical antimicrobials and is commonly used to treat *P. aeruginosa* infections [20] [21]. By studying *in vitro* bacterial density dynamics, we show that phages and streptomycin have a synergistic negative effect against bacteria. We also find a specific window of opportunity in the addition time of the antibiotic, enhancing the suppression of populations already treated with phage. Antibiotic dose did not significantly affect bacterial density, contrary to conventional clinical practice of using high antibiotic doses [16]. Finally, we find no evidence that the synergistic effect of the combined treatments is driven by genetic trade-offs between resistances to the phage and to the antibiotic. A more likely explanation is a demographic feedback produced by phage addition, limiting the capacity of the bacteria to resist antibiotic exposure. Our study provides an evolutionary basis for the optimization of combined treatments.

Materials and Methods

Bacterium, phage and media

We used the bacterium *Pseudomonas aeruginosa* PAO1 and the phage LUZ7, from the *Podoviridae* family [22]. The experiment was carried out in 24-well plates, with bacteria growing in King's B (KB) medium at 37°C without agitation. M9 medium was used for dilutions. The antibiotic streptomycin (Sigma-Aldrich) was added to liquid medium at either 100 or 240 µg/mL, known to represent sub-lethal and MIC concentrations for PAO1, respectively [10]. The phage stock was prepared as described in [23]. Briefly, 10% vol/vol chloroform was added to phage-containing bacterial cultures, vortexed and centrifuged. Phage-containing supernatants were carefully recovered and stored at 4°C. This LUZ7 stock (10^7 PFU/mL) was used as the ancestral phages for all the experiments.

Experimental design

Six hours prior to the start of treatments, the 120 bacterial replicate populations were initiated from a *P. aeruginosa* PAO1 overnight culture, by adding 15 µL of culture to 1.5 mL of KB in 24-well plates. Phages were added (10^5 LUZ7 phages/mL) after 6 h (T_0), when bacterial populations were growing exponentially and therefore vulnerable to phage attack. We used a concentration of phages high enough to affect the bacterial population dramatically (decreasing density by 6 orders of magnitude), but without producing complete extinction. We established single treatments, with only phage or only antibiotic added, as well as combined phage-antibiotic treatments (Figure 1), named single-phage, single-strep and phage-strep, respectively. The antibiotic was added at one of three time points: simultaneously with the phage (+0 h), with a delay of +12 h, or with a delay of +24 h. Two antibiotic doses were tested: 100 or 240 µg/mL. For each treatment we established 9 replicate populations, 108 total: 2 phage (yes/no) x 2 antibiotic doses x 3 addition times x 9 replicates. Six control replicate lines were established for the single-phage treatment and for untreated control lines.

Density measurements and resistance assays

Bacterial density was measured at different time points (T_0, T_{14}, T_{45}, T_{70} = 0, 14, 45, and 70 hours post phage inoculation), by counting the number of growing colonies (colony-forming units, CFU) from samples plated on KB agar at appropriate dilutions.

At the end of the experiment (T_{70}), we assessed the surviving populations' resistance to streptomycin. 1 µL of the final populations were inoculated on to 250 µL of fresh KB containing streptomycin at different concentrations (12, 25, 50, 100, 200, 400 or 800 µg/mL). After 24 h, bacterial density was measured by means of optical density (OD) at 600 nm (Fluostar, BMG LABTECH). Resistance was taken as the Minimum Inhibitory Concentration (MIC), defined as the streptomycin concentration at which no bacterial growth was detected. For populations even resisting the highest concentration (800 µg/mL) the MIC was arbitrarily set to 1600 µg/mL.

To measure phage resistance, 1 µL of final bacteria was added to 250 µL of media containing ancestral phage (*c* 10^5 phages) and OD recorded after 24 h. Phage resistance was taken as a quantitative trait, calculated as the difference in OD obtained with and without phage added. The same assay was performed with evolved phage. To this end, the 9 replicates from the last time point of a given treatment were pooled and evolved phages extracted as described above. Thus, bacteria were confronted with a mix of phages from their own treatment. Bacteria from treatments without phage (single-strep, control, ancestral bacteria) were confronted with evolved phage from the +0 h phage addition time treatment with 100 µg/mL of streptomycin. All OD values were corrected for absorbance of blank wells; replicates for which positive control wells without phage showed zero growth were not used for analysis.

Statistical analysis

Using the JMP statistical package [24], we employed General Linear Model (GLM) techniques to analyze variation in bacterial density (CFU/mL, $\log_{10}$-transformed), antibiotic resistance (MIC, square-root-transformed) and phage resistance (OD difference between bacteria challenged with phage and not). In the main analyses, we tested fully factorial models, containing phage treatment (yes/no), antibiotic dose as explanatory factors and antibiotic addition time as a covariate. To test for non-linear effects of addition time, we also fitted its second-order polynomial term (addition time^2). Minimal adequate models were established through backward elimination of non-significant terms in the model. Where appropriate, analyses were carried out separately for single and combined treatments; additional tests compared evolved and ancestral bacteria.

To calculate expected final densities (70 h) in combined phage-antibiotic treatments, we paired single-phage with single-strep replicates. For both replicates in a pair, we calculated the reduction in bacterial density relative to the untreated controls (difference in CFU/mL). We then added together the two single density reductions to obtain the expected density in a hypothetical combined phage-antibiotic treatment. Specifically, for each combination of antibiotic dose and addition time, 36 of the possible 81 (9×9 replicates from single treatments) pairs were arbitrarily chosen and the density difference calculated relative to each of the two untreated control lines. This gave a total of 72 expected values that were to be compared with the corresponding observed values in the true combined phage-antibiotic treatment.

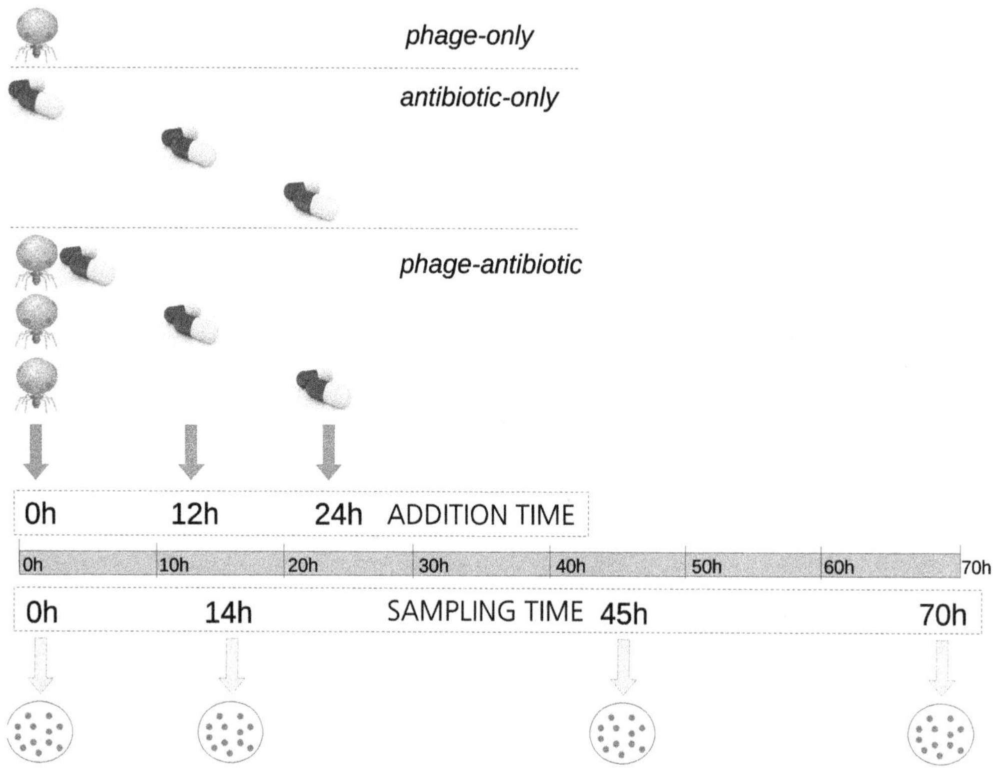

Figure 1. Overview of the experimental design. Exponentially growing bacteria were treated with (i) only phage (at 0 h), (ii) only antibiotic (1 dose at 0 h, 12 h, or 24 h) or (iii) first phage and then antibiotic (1 dose at 0 h, 12 h, or 24 h). Replicate populations of the bacteria were sampled at 0 h, 14 h, 45 h and 70 h, for density and resistance measurements. All antibiotic treatments were repeated for two streptomycin doses (100 and 240 µg/mL).

Results

1. Bacterial density

We challenged *P. aeruginosa* with either single or combined treatments of the phage LUZ7 and two doses of the antibiotic streptomycin (strep), administered at different time points. Bacterial population density was tracked over 70 h (Figure 1) to test the hypothesis that the use of phages can contribute to reduce antibiotic doses below the MIC, and that simultaneous or sequential administration of the two antimicrobials have different consequences on bacterial densities. Both single-phage and single-strep treatments strongly reduced bacterial density over the first 24 h, by up to 6 orders of magnitude (Figure 2). However, densities rebounded and nearly reached the levels of untreated controls by the end of the experiment (70 h) in all populations (Figure 2, 3A).

The combined phage-strep treatment caused a significantly stronger reduction in density compared to either single treatment (vs. single-phage: $t_{58} = 3.60$, $p = 0.0007$; vs. single-strep: $t_{105} = 9.43$, $p<0.0001$; Figure 3A). Unlike in the single treatments, almost 60% (30/54) of the populations did not recover from the combined treatment and showed strongly suppressed final densities ($<10^5$ CFU/mL). We then evaluated the relative effects of simultaneous (0 h) or delayed (+12 h, +24 h) addition of strep to populations containing phage (Figure 1). We found that bacterial density reduction at the end of the experiment was maximal when the antibiotic was added with a +12 h delay (phage x strep addition time^2 interaction: $F_{1, 95} = 5.03$, $p = 0.0272$). Streptomycin dose (100 vs. 240 µg/mL) had no significant effect on final density, nor were there significant interactions with other treatments (all $p>0.25$; Figure 2).

We further assessed whether the combined action of phage and antibiotic was additive or synergistic. To this end, we extrapolated outcomes in combined treatments from added effects on final bacterial density in the single treatments. Final densities were significantly lower than expected ($F_{1, 477} = 278.0$, $p<0.0001$; Figure 3B), indicating a positive synergistic action of phage and antibiotic. This positive synergy was most pronounced for the + 12 h antibiotic addition time (expected/observed x strep addition time interaction: $F_{2, 477} = 14.06$, $p<0.0001$; Figure 3B). Thus, an intermediate time delay in antibiotic addition in the combined treatment resulted in the strongest negative impact on bacterial population density.

2. Antibiotic and phage resistance

For the final bacterial populations (70 h) we analyzed variation in resistance to (ancestral) phage and to the antibiotic, the latter measured as the Minimum Inhibitory Concentration (MIC) of streptomycin. Bacteria from streptomycin treatments generally evolved very high levels of resistance (MIC≥800 µg/mL, Figure 4A). In the single-strep treatment, resistance reached maximum levels, whereas antibiotic resistance was lower in the combined phage-strep treatment ($F_{1, 79} = 27.6$, $p<0.0001$), but nonetheless higher than in the single-phage treatment ($F_{1, 40} = 47.41$, $p<0.0001$, Figure 4A). Resistance values were lower for populations where streptomycin was added to the phage with a +12 h delay (treatment x strep addition time^2 interaction: $F_{1, 79} = 4.48$, $p = 0.0375$) compared to the other treatments. Note that

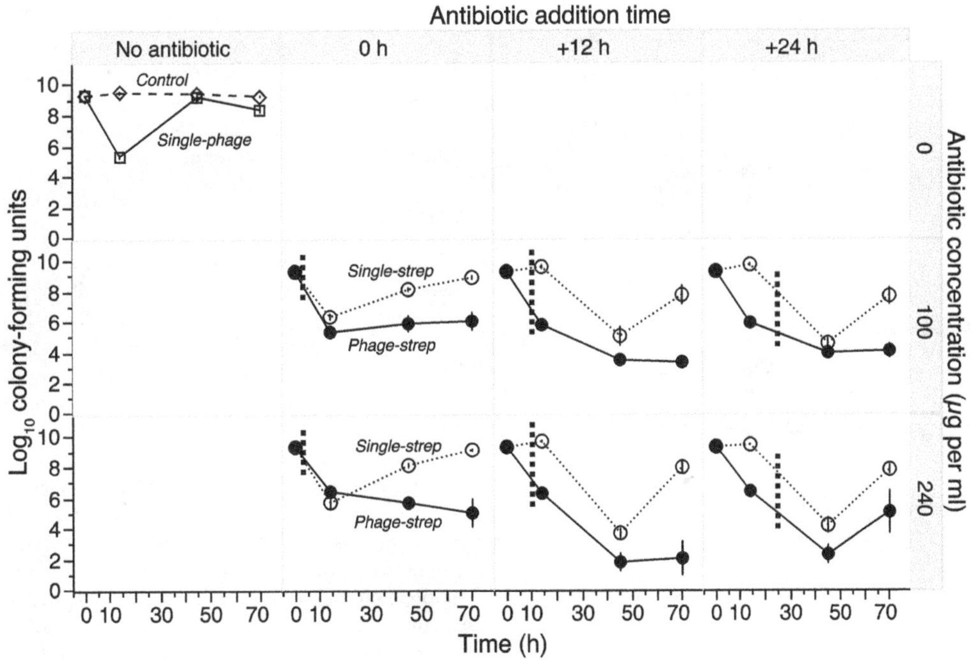

Figure 2. Changes in bacterial density over the course of the experiment. The six lower panels show the single antibiotic treatment ("single-strep") and the combined phage-strep treatment, for different addition times (red dotted lines) of the antibiotic streptomycin and for two antibiotic doses. The top-left panel shows the single-phage treatment and the unexposed control lines.

while ancestral bacteria were fully susceptible to streptomycin, moderate increases in resistance were detected for bacteria from the single-phage treatment and for totally unexposed controls (Figure 4A).

As expected, bacteria from phage treatments were more resistant to phage than bacteria from no-phage treatments ($F_{2, 110} = 5.90$, $p = 0.0037$; Figure 4B). Specifically, the absence of phage tended to produce a decrease in resistance, suggesting a possible cost of resistance. Similar to the finding for antibiotic resistance, increases in phage resistance in the combined treatment tended to be minimal when strep was added with a +12 h delay (strep addition time: $F_{1, 32} = 2.97$, $p = 0.0947$). In the single-strep treatments, we observed a loss of phage resistance significantly affected by the antibiotic addition time ($F_{1, 45} = 7.22$, $p = 0.01$). Very similar results were obtained when resistance was measured against evolved (from 70 h) rather than ancestral phage, with a clear minimum when adding the antibiotic at +12 h ($F_{1, 32} = 5.20$, $p = 0.0293$, Figure 4C). Resistance against ancestral phage was generally higher than against evolved phage (across all replicates: $t_{33} = 3.62$, $p = 0.001$), suggesting adaptation of phage to contemporary bacteria.

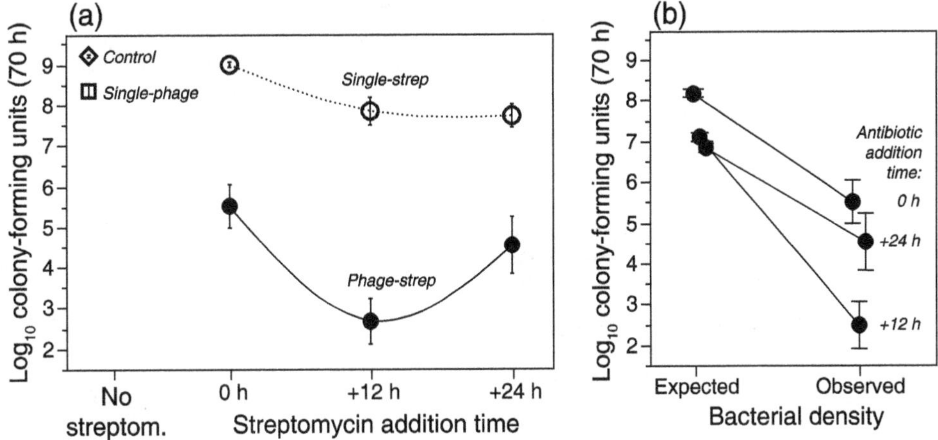

Figure 3. Final bacterial densities. (a) Effects of single treatments (single-strep, single-phage) and combined treatments (phage-strep), for different addition times of the antibiotic streptomycin in mean (± SE) final bacterial densities (70 h). Control lines were untreated, and ancestral bacteria regrown from frozen stocks for the assay. Note that the lines connect final densities for independently tested addition times, and do not represent time series of bacterial density. (b) Expected and observed density in the combined treatments, for the different antibiotic addition times. Expected density extrapolated from single treatments, assuming additive action of antibiotic and phage.

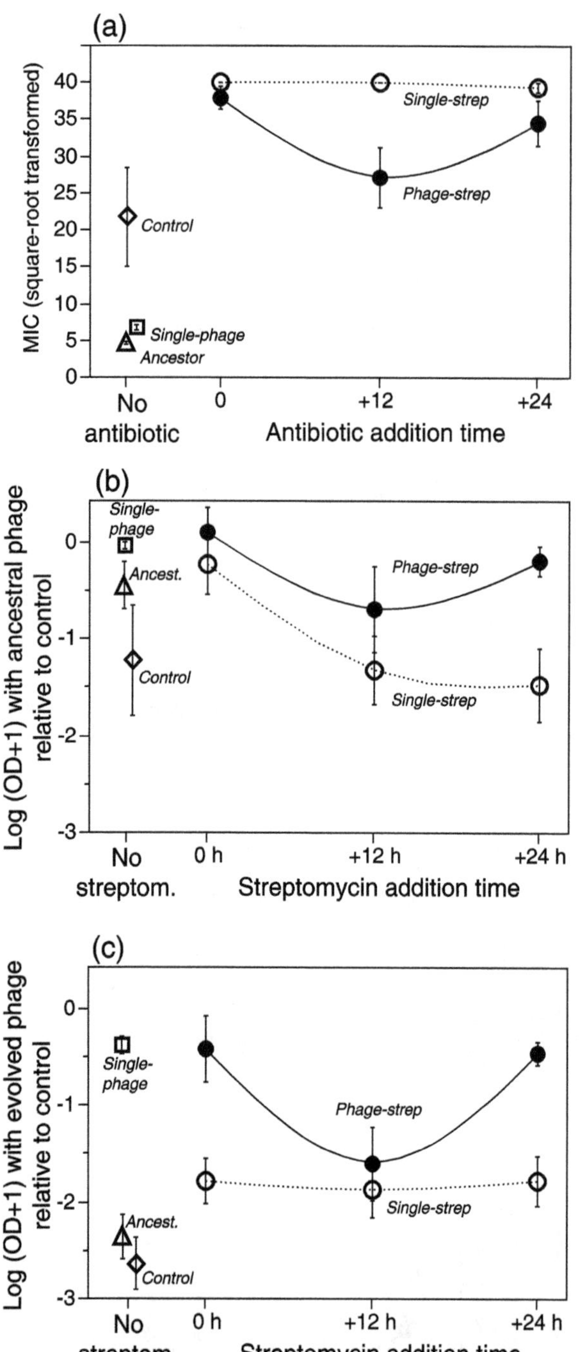

Figure 4. Mean (± SE) resistance of final bacterial populations (70 h) from single (strep or phage) and combined (phage-strep) treatments. Control lines were untreated, and ancestral bacteria regrown from frozen stocks for the assay (a) Streptomycin resistance, assessed as the Minimal Inhibitory Concentration (MIC, in µg/mL). (b) Resistance to ancestral phage, calculated as the difference in optical density (OD, log-transformed) (+1) of bacteria in the presence and absence of phage, measured in a 24-h growth assay. (c) Resistance to evolved phage, as measured in (b). Note that the lines connect final densities for independently tested addition times, and do not represent time series of bacterial density.

Finally, we found no evidence for a trade-off between antibiotic resistance and phage resistance in the combined treatment. In fact, the across-population correlation between the two traits was positive rather than negative ($r = 0.41$, $n = 34$, $p = 0.0154$), indicating that higher levels of phage resistance were associated with higher antibiotic resistance.

Discussion

There is increasing attention on alternative treatments against bacterial pathogens, due to the inevitability of antibiotic resistance and difficulties in developing new antibiotics [25]. The combination of antibiotics and phages for clinical or environmental applications is a tantalizing possibility, but it is not known whether phage therapy alone or in combination with antibiotics will improve on antibiotics alone in the short term, and reduce or prevent resistance in the long term. In this work, we study the combined effect of an antibiotic and phages on *Pseudomonas aeruginosa* bacterial population density and levels of evolved resistance, testing different application sequences and antibiotic doses. We show that combined treatments result in synergistic suppression of bacterial density and less resistance than either treatment alone, but also that the application sequence of both antimicrobials, and not antibiotic dose, is key to minimize the levels of resistance.

We found that combining phage and antibiotic results in lower bacterial density than expected from the addition of the respective single treatment effects (Figure 3). This positive synergism is consistent with previous observations in *P. fluorescens* [9] [26], and probably due to resistance mutation limitation as a result of lowered bacterial population size [27]. A possible explanation for this synergistic effect is a demographic feedback produced by the addition of phage [18], limiting the capacity of the bacteria to resist antibiotic exposure, as suggested by the detailed bacterial density dynamics (Figure 2). Bacteria were most affected when the antibiotic was applied when the phages themselves had their strongest impact on bacterial population density (+12 h addition time), suggesting an optimal window of opportunity in the implementation of combined therapies to restrain pathogens. For the other two application times, the synergistic effect was reduced. When the two agents are applied simultaneously, streptomycin is likely to constrain the efficacy of phages due to intensive host damage by means of protein synthesis inhibition [28]. If inhibition reduces the per-host cell output of phage, then overall phage titer may not be sufficient to cause massive reductions in bacterial cell density. Conversely, when the antibiotic is applied 24 hours after the phage, we argue that bacterial populations recover before being submitted to the antibiotic. This demographic feedback mechanism is consistent with 'evolutionary rescue', which links the demographic dynamics of population decline with the genetic dynamics of adaptation under rapid environmental deterioration [29]. Our results suggest that understanding the population dynamics and evolutionary biology of multiple interactive agents is important for the success of new therapies [18].

A possible evolutionary risk of antimicrobial compounds producing a synergistic effect in a combined treatment is that resistance mutations have a larger selective advantage compared to the single treatments, and as such the rate of adaptation will be higher [30] [31]. In our combined treatments, resistance to both phage and antibiotic increased relative to the ancestral bacteria, indicating positive responses to selection by both agents (as already shown separately for phages and antibiotics in *P. aeruginosa*; [32] [33], respectively). However, resistance levels did not exceed those of single treatments: they were equal or even lower (Figure 4). This suggests that our combined treatments did not lead to faster adaptation of the bacteria. Consistent with previous work [34], we observed an increase in antibiotic resistance in non-treated control populations relative to ancestral bacteria, possibly associated with biofilm formation, adaptation to the nutrient media or the emergence of low frequency antibiotic resistant mutants in the large populations occurring in our microcosms ($>10^9$ cells/mL).

Interestingly, we show that an intermediate time delay between application of phage followed by an antibiotic leads to lower levels of bacterial resistance to either agent, as compared to shorter or longer delays, or to the application of either agent separately. Indeed, the sequential application of combined therapies has been suggested to generate lower levels of resistance compared to simultaneous addition, especially when the antimicrobial agents have different bacterial targets, as in a recent study employing phages that use different host receptors [35]. We also provide experimental support that in synergistic combinations it is possible to reduce antibiotic doses and still reduce bacterial populations significantly whilst limiting resistance [11]. Given the prediction that demographic and genetic changes interact, it is not unexpected that the effects of addition time on resistance mirrored those on bacterial density. The capacity of treated bacterial populations to recover and attain high densities is directly related to the increasing frequency of resistant mutants, and thus the higher mean population resistance levels observed at the end of our experiment. Nevertheless, in another study using distinct combinations of phages against bacteria, Hall and colleagues [35] argue that the effectiveness of multiphage therapy depends on the order and type of phages combined, indicating a more mechanistic constraint rather than the demographic one suggested here.

More generally, combined phage-antibiotic therapy may be expected to have an advantage over antibiotic cocktails of less cross-resistance because phage and antibiotics are fundamentally different regarding cellular mechanisms affected and the genetic changes resulting in resistance [18]. In particular, streptomycin resistance mutations typically involve the ribosomal protein S12 and 16S rRNA [36], whereas resistance mutations to a *Podovirus* such as LUZ7 usually require the alteration of lipopolysaccharide (LPS) components, the phage receptor on the bacterial outer membrane [37]. Here, we observed a general positive association between mean levels of phage and antibiotic resistance for the combined treatments, suggesting potentially unconstrained multiple resistance evolution, or in other words, that trade-offs between phage and antibiotic resistance do not appear to play a role [38]. It should be noted, however, that we measured resistance at the population level; a more precise analysis of the relationship would require measurements of individual bacterial genotypes to establish genetic correlations.

As an antimicrobial agent, bacteriophages are different from antibiotics in that the former can evolve or even coevolve with the bacteria, and therefore potentially limit resistance evolution during treatment [9] [18] [26]. How this (co)evolutionary component influences the efficiency and predictability of treatment outcomes is still largely unclear [23] [35] [39]. Here, we find evidence for the evolution of bacterial resistance to phage, but also evolutionary change in phage infectivity, in agreement with recent study [23]. Overall, bacterial resistance to phage from the end of the experiment was lower than that to the ancestral phage, clearly suggesting evolution of the phage towards increased infectivity. Levels of resistance to ancestral and evolved phage were highly correlated, indicating considerable coherence in treatment effects regarding resistance evolution. Future study needs to evaluate to what extent these patterns involve coevolutionary specificity, and what the implications are for longer-term pathogen control.

We acknowledge that *in vitro* studies such as ours will be limited in predicting outcomes in a clinical context, in which other important drivers of selection for pathogenic microbes include the host immune system and its spatial structure [40], and bacterial densities might be significantly lower [41]. In addition, we should expect that resistance to either phages and/or antibiotics will entail fitness costs for bacteria that could be accentuated *in vivo* [26]. Interestingly, phage resistance may lead to selection for less virulent bacterial variants, for example, through the loss of surface phage receptors that are also virulence determinants, as in the case of *Yersinia pestis* [42]. This possibility should be explored both *in vitro* and in hospital settings to evaluate if combined approaches at disinfection are also able to reduce the pathogenicity of bacteria surviving such treatments.

The use of combined antimicrobial therapies for the treatment of highly resistant pathogens has been applied in the clinic, for example, to combat *Mycobacterium tuberculosis*, HIV, and the malaria pathogen *Plasmodium falciparium* [43]. Even though it has achieved considerable success, the potential to reduce the rate of evolution of resistance to combination therapies need to reach more problematically resistant infectious diseases in the future [44]. A better understanding of the pharmacodynamics of combining phages and antibiotics will be vital to the eventual implementation of new therapeutic strategies targeting multiresistant nosocomial infections [19]. Our work shows that at an intermediate application time there is a window of opportunity, where mortality due to the antibiotic results in lower absolute populations and reduced resistance levels because bacterial populations are at low density, and both antibiotic and phage resistance mutations are less likely to be present [29].

Acknowledgments

We thank P.J. Ceyssens for supplying the phage and M.A. Brockhurst for providing the PAO1 strain of *P. aeruginosa*. All data are available at figshare: http://dx.doi.org/10.6084/m9.figshare.1120505. This is ISEM 2014-116 article.

Author Contributions

Conceived and designed the experiments: CTB OK MEH. Performed the experiments: FIAS MV JR. Analyzed the data: OK CTB. Contributed reagents/materials/analysis tools: MEH. Wrote the paper: CTB OK MEH.

References

1. Rodriguez-Rojas A, Rodriguez-Beltran J, Couce A, Blazquez J (2013) Antibiotics and antibiotic resistance: A bitter fight against evolution. Int. J. Med. Microbiol. 303, 293–297.
2. Garcia-Quintanilla M, Pulido MR, Lopez-Rojas R, Pachon J, McConnell MJ (2013) Emerging therapies for multidrug resistant *Acinetobacter baumannii*. Trends Microbiol. 21, 157–163.
3. Allen HK, Levine UY, Looft T, Bandrick M, Casey TA (2013) Treatment, promotion, commotion: antibiotic alternatives in food-producing animals. Trends Microbiol. 21, 114–119.
4. Pirnay JP, Verbeken G, Rose T, Jennes S, Zizi M, et al. (2012) Introducing yesterday's phage therapy in today's medicine. Future Virol. 7, 379–390.
5. Wright A, Hawkins CH, Anggard EE, Harper DRA (2009) Controlled clinical trial of a therapeutic bacteriophage preparation in chronic otitis due to

antibiotic-resistant *Pseudomonas aeruginosa*; a preliminary report of efficacy. Clin. Otolaryngol. 34, 349–35.
6. Chan BK, Abedon ST, Loc-Carrillo C (2013) Phage cocktails and the future of phage therapy. Future Microbiol. 8, 769–783.
7. Bourguet D (2013) Heterogeneity of selection and the evolution of resistance. Trends Ecol. Evol. 28, 110–118.
8. Lu TK, Collins JJ (2009) Engineered bacteriophage targeting gene networks as adjuvants for antibiotic therapy. Proc. Natl. Acad. Sci. USA 106, 4629–4634.
9. Escobar-Paramo P, Gougat-Barbera C, Hochberg ME (2012) Evolutionary dynamics of separate and combined exposure of *Pseudomonas fluorescens* SBW25 to antibiotics and bacteriophage. Evol. Appl. 5, 583–592.
10. Ward H, Perron GG, Maclean RC (2009) The cost of multiple drug resistance in *Pseudomonas aeruginosa*. J. Evol. Biol. 22, 997–1003.
11. Yeh PJ, Hegreness MJ, Aiden AP, Kishony R (2009) Drug interactions and the evolution of antibiotic resistance. Nat. Rev. Microbiol. 7, 460–466.
12. Michel J-B, Yeh PJ, Chait R, Moellering RC Jr, Kishony R (2008) Drug interactions modulate the potential for evolution of resistance. P. Natl. Acad. Sci. USA 105, 14918–14923 .
13. Gumbo T, Louie A, Deziel MR, Liu W, Parsons LM, et al. (2007) Concentration-dependent *Mycobacterium tuberculosis* killing and prevention of resistance by rifampin. Antimicrob. Agents Chem. 51, 3781–3788.
14. Zinner SH, Gilbert D, Greer K, Portnoy YK, Firsov AA (2013) Concentration-resistance relationships with *Pseudomonas aeruginosa* exposed to doripenem and ciprofloxacin in an in vitro model. J. Antimicrob. Chem. 68, 881–887.
15. Canton R, Morosini MI (2011) Emergence and spread of antibiotic resistance following exposure to antibiotics. Fems Microbiol. Rev. 35, 977–991.
16. Read AF, Day T, Huijben S (2011) The evolution of drug resistance and the curious orthodoxy of aggressive chemotherapy. Proc. Natl. Acad. Sci. USA 108, 10871–10877.
17. Ryan EM, Gorman SP, Donnelly RF, Gilmore BF (2011) Recent advances in bacteriophage therapy: how delivery routes, formulation, concentration and timing influence the success of phage therapy. J. Pharm. Pharmacol. 63, 1253–1264.
18. Levin BR, Bull JJ (1996) Phage therapy revisited: The population biology of a bacterial infection and its treatment with bacteriophage and antibiotics. Am. Nat. 147, 881–898.
19. Breidenstein EBM, de la Fuente-Nunez C, Hancock REW (2011) *Pseudomonas aeruginosa*: all roads lead to resistance. Trends Microbiol. 19, 419–426.
20. Zembower R, Noskin GA, Postelnick MJ, Nguyen N, Peterson LR (1998) The utility of aminoglycosides in an era of emerging drug resistance. Int. J. Antimicrob. Agents. 10 (2): 95–105.
21. Poole K (2005) Aminoglycoside resistance in *Pseudomonas aeruginosa*. Antimicrobial agents and Chemotherapy 49 (2): 479–487.
22. Ceyssens PJ, Brabban A, Rogge L, Lewis MS, Pickard D, et al. (2010) Molecular and physiological analysis of three *Pseudomonas aeruginosa* phages belonging to the "N4-like viruses". Virology 405, 26–30.
23. Betts A, Vasse M, Kaltz O, Hochberg ME (2013) Back to the future: evolving bacteriophages to increase their effectiveness against the pathogen *Pseudomonas aeruginosa* PAO1. Evol. Appl. 6, 1054–1063.
24. SAS. JMP statistics and graphics guide (version 10) (2012) In. SAS Institute Cary, N.C.
25. Arias CA, Murray BE (2009) Antibiotic-Resistant Bugs in the 21st Century – A Clinical Super-Challenge. New Engl. J. Med. 360, 439–443.
26. Zhang Q-G, Buckling A (2012) Phages limit the evolution of bacterial antibiotic resistance in experimental microcosms. Evol. Appl. 5, 575–582.
27. Martin M, Aguilée R, Ramsayer J, Kaltz O, Ronce O (2013) The probability of evolutionary rescue: towards a quantitative comparison between theory and evolution experiments. Phil. Trans. R. Soc. B. 368, 20120088.
28. Kohanski MA, Dwyer DJ, Collins JJ (2010) How antibiotics kill bacteria: from targets to networks. Nat. Rev. Microbiol. 8, 423–435.
29. Ramsayer J, Kaltz O, Hochberg ME (2013) Evolutionary rescue in populations of *Pseudomonas fluorescens* across an antibiotic gradient. Evol. Appl. 6, 608–616.
30. Chait R, Craney A, Kishony R (2007) Antibiotic interactions that select against resistance. Nature 446, 668–671.
31. Hegreness M, Shoresh N, Damian D, Hartl D, Kishony R (2008) Accelerated evolution of resistance in multidrug environments. Proc. Natl. Acad. Sci. USA 105, 13977–13981.
32. Brockhurst MA, Buckling A, Rainey PB (2005) The effect of a bacteriophage on diversification of the opportunistic bacterial pathogen, *Pseudomonas aeruginosa*. Proc. Roy. Soc. B-biol. Sci. 272, 1385–1391.
33. Hall AR, Iles JC, MacLean RC (2011) The fitness cost of rifampicin resistance in *Pseudomonas aeruginosa* depends on demand for RNA polymerase. Genetics 187, 817–822.
34. Drenkard E, Ausubel FM (2002) Pseudomonas biofilm formation and antibiotic resistance are linked to phenotypic variation". Nature 416 (6882): 740–743.
35. Hall AR, De Vos D, Friman V-P, Pirnay J-P, Buckling A (2012) Effects of sequential and simultaneous applications of bacteriophages on populations of *Pseudomonas aeruginosa* in vitro and in wax moth larvae. Appl. Environ. Microb. 78, 5646–5652.
36. Springer B, Kidan YG, Prammananan T, Ellrott K, Böttger EC, et al. (2001) Mechanisms of streptomycin resistance: Selection of mutations in the 16S rRNA gene conferring resistance. Antimicrob. Agents Chem. 45, 2877–2884.
37. Samson JE, Magadan AH, Sabri M, Moineau S (2013) Revenge of the phages: defeating bacterial defences. Nat. Rev. Microbiol. 11, 675–687.
38. De Paepe M, Taddei F (2006) Viruses' life history: Towards a mechanistic basis of a trade-off between survival and reproduction among phages. Plos Biol. 4, e193.
39. Cairns BJ, Timms AR, Jansen VAA, Connerton IF, Payne RJH (2009) Quantitative models of in vitro bacteriophage-host dynamics and their application to phage therapy. Plos Pathog. 5, e1000253.
40. Brown EM, Sadarangani M, Finlay BB (2013) The role of the immune system in governing host-microbe interactions in the intestine. Nat. Immunol. 14, 660–667.
41. Leggett HC, Cornwallis CK, West SA (2012) Mechanisms of pathogenesis, infective dose and virulence in human parasites. PLoS Pathog 8(2): e1002512.
42. Filippov AA, Sergueev KV, He Y, Huang XZ, Gnade BT, et al. (2011) Bacteriophage-resistant mutants in *Yersinia pestis*: identification of phage receptors and attenuation for mice. Plos One 6, e25486.
43. Fischbach MA (2011) Combination therapies for combating antimicrobial resistance. Curr. Opin. Microbiol. 14, 519–523.
44. Cottarel G, Wierzbowski J (2007) Combination drugs, an emerging option for antibacterial therapy. Trends Biotechnol. 25, 547–555.

15

Effective Removal of Staphylococcal Biofilms by the Endolysin LysH5

Diana Gutiérrez, Patricia Ruas-Madiedo, Beatriz Martínez, Ana Rodríguez, Pilar García*

Instituto de Productos Lácteos de Asturias (IPLA-CSIC), Villaviciosa, Asturias, Spain

Abstract

Staphylococcal biofilms are a major concern in both clinical and food settings because they are an important source of contamination. The efficacy of established cleaning procedures is often hindered due to the ability of some antimicrobial compounds to induce biofilm formation, and to the presence of persister cells, a small bacterial subpopulation that exhibits multidrug tolerance. Phage lytic enzymes have demonstrated antimicrobial activity against planktonic and sessile bacteria. However, their ability to lyse and/or select persister cells remains largely unexplored so far. In this work, the lytic activity of the endolysin LysH5 against *Staphylococcus aureus* and *Staphylococcus epidermidis* biofilms was confirmed. LysH5 reduced staphylococcal sessile cell counts by 1–3 log units, compared with the untreated control, and sub-inhibitory concentrations of this protein did not induce biofilm formation. LysH5-surviving cells were not resistant to the lytic activity of this protein, suggesting that no persister cells were selected. Moreover, to prove the lytic ability of LysH5 against this subpopulation, both *S. aureus* exponential cultures and persister cells obtained after treatment with rifampicin and ciprofloxacin were subsequently treated with LysH5. The results demonstrated that besides the notable activity of endolysin LysH5 against staphylococcal biofilms, persister cells were also inhibited, which raises new opportunities as an adjuvant for some antibiotics.

Editor: Holger Rohde, Universitätsklinikum Hamburg-Eppendorf, Germany

Funding: This research study was supported by grants AGL2012-40194-C02-01 (Ministry of Science and Innovation, Spain) and PIE200970I090 (CSIC, Spain). DG is a fellow of the Ministry of Science and Innovation, Spain. The funders had no role in study design, data collection and analysis, decision to publish, or preparation of the manuscript.

Competing Interests: The authors have declared that no competing interests exist.

* Email: pgarcia@ipla.csic.es

Introduction

Two staphylococcal species, *Staphylococcus aureus* and *Staphylococcus epidermidis*, are the main cause of hospital-associated infections [1,2]. *S. epidermidis* is a common etiological agent of nosocomial infections, mostly occurring in immune compromised patients with implanted medical devices [3]. Methicillin-resistant *S. aureus* (MRSA) strains are also responsible for both health care and community-associated infections, mostly involving skin and wound infections, pneumonia, severe sepsis and endocarditis [4]. In addition, *S. aureus* is one of the major bacterial agents causing foodborne diseases in humans due to the production of enterotoxins [5].

Pathogenicity of both species is clearly associated with their ability to form biofilms on biotic and abiotic surfaces, providing high resistance to host defenses and antibiotics, and also to cleaning and disinfection processes [2]. The ability to form biofilms allows *S. aureus* to survive in hostile environments, such as food industry surfaces [6], and this enhances the recurrence of food contamination. The ability to form biofilms in both species is due to the production of an extracellular material that contributes to intercellular aggregation and attachment to surfaces. In most cases, this matrix is composed of exopolysaccharides such as poly-N-acetyl-β-(1,6)-glucosamine (PIA/PNAG), teichoic acids, DNA and specific proteins [2]. *S. aureus* strains that are not reliant on polysaccharides for biofilm formation have also been identified. Specific proteins such as Bap [7], Spa [8], FnBPA, FnBPB [9], and SasG [10], are surface adhesins involved in biofilm development. In *S. epidermidis* an extracellular matrix-binding protein (Embp) is necessary and sufficient to promote protein-dependent biofilm formation [11]. In addition, the accumulation-associated protein (Aap) also mediates intercellular and surface adhesion and is considered the most important factor contributing to protein-dependent biofilm formation in *S. epidermidis* [12].

Bacteria embedded in biofilms are considerably less susceptible to antibiotics than their planktonic counterparts, mainly due to the limited access which the antibiotic has into the biofilm [13]. Moreover, several studies have shown that *S. aureus* biofilm formation can be stimulated by sub-inhibitory concentrations (sub-MICs) of some antibiotics [14]. Some bacteria within the biofilm also show a reduced susceptibility to antibiotics due to their dormant phenotype. These bacteria, named persister cells, are genetically identical to susceptible bacteria and give rise to a new sensitive population after the removal of antibiotic pressure [15]. Persistance is a main cause of concern as it has been observed in most bacterial species, in relation to different classes of antibiotics, with persister cells being involved in recurrent infections [16]. In this regard, several strategies for the clearance of staphylococcal biofilms, other than disinfectants and antibiotics, have been assayed. These include PIA-degrading enzymes like dispersin B [17], the peptidoglycan-degrading enzyme lysostaphin [18], bacteriocins [19] and bacteriophages [20–23].

Recently, phage lytic proteins (endolysins and virion-associated peptidoglycan hydrolases) have shown a potent antimicrobial activity against planktonic bacteria [24,25]. They have been investigated as therapeutic agents and biopreservatives against a range of pathogens, such as *Bacillus anthracis*, *Streptococcus pneumoniae*, *S. aureus*, and *Streptococcus suis*. Moreover, *S. aureus*, *S. suis*, *S. pneumonia* and *Streptococcus pyogenes* biofilms have been successfully removed by endolysins [26–29]. We have previously identified and characterized the endolysin LysH5 encoded by the *S. aureus* phage vB_SauS-phiIPLA88, which showed a high degree of similarity (97%) with the endolysin from phage phi11, for which anti-biofilm activity has previously been reported [30]. LysH5 has two catalytic domains (CHAP domain and amidase-2 domain) and a cell wall binding domain (SH3b). This protein is able to lyse a wide range of staphylococci, including bovine and human *S. aureus* and *S. epidermidis* [31]. A synergistic antimicrobial activity was previously observed with both the bacteriocin nisin [32] and the virion-associated peptidoglycan hydrolase HydH5 and its derivative fusion proteins [33].

The aim of this study was to determine the efficacy of the endolysin LysH5 to remove the biofilms formed by *S. aureus* and *S. epidermidis* strains. We have studied the ability of LysH5 to induce biofilms, to select persister cells, and finally, to see if LysH5 was also able to kill persister cells previously selected by antibiotics.

Material and Methods

Bacterial strains, growth media and proteins

Staphylococcal strains (Table 1) were routinely cultured in TSB broth (Tryptic Soy Broth, Scharlau, Barcelona, Spain) at 37°C with shaking, or in TSB plates containing 2% (w/v) bacteriological agar (TSA).

LysH5 purification was performed as previously described [32]. The LysH5 specific activity against the different strains was calculated as the ΔOD per mg of protein per min [31]. Lysostaphin was obtained from Sigma (Sigma, Missouri, USA).

Biofilm assays

For biofilm formation the protocol of Herrera et al., (2007), was used with some modifications [34]. Briefly, overnight (o/n) staphylococcal cultures were diluted in fresh TSBg (TSB supplemented with 0.25% w/v D-(+)-glucose) up to 10^6CFU/ml, and 200 µl were poured into TC Microwell 96U w/lid nunclon D SI plates (Thermo Scientific, NUNC, Madrid, Spain), and incubated at 37°C for 24 h. Wells were washed twice with sterile phosphate-buffered saline (PBS buffer) (137 mM NaCl, 2.7 mM KCl, 10 mM Na_2HPO_4 and 2 mM KH_2PO_4; pH 7.4).

To determine the counts of bacteria attached to each well once the biofilm was developed, the well was scratched twice with a sterile swab and then immersed into 9 ml of PBS buffer. A vigorous shaking for 1 min allowed the disaggregation of the biofilm [34]. Finally, several decimal dilutions were plated onto TSA and incubated at 37°C.

Likewise, the biofilm that had adhered to the surfaces of the wells was observed by staining with crystal violet (0.1% w/v) for 15 min, followed by a gentle wash with water and de-staining in acetic acid (33% v/v). Absorbance was measured at a wavelength of 595 nm [9]. All the assays were performed using two biological replicates.

Treatment of staphylococcal biofilms with LysH5

24 h-old biofilms were developed as described previously. After the washing step with PBS buffer, 0.15 µM of LysH5, 0.2 µM of lysostaphin, or 200 µl of sodium phosphate buffer (50 mM pH 7) for control purposes, were added to each well. Plates were incubated for 6 h at 37°C and cell counts were measured as described above. Surviving cells were recovered after treatment by scratching the well with a sterile swab and transferred to 2 ml of TSB medium. For two-step treatment assay, 24 h-old biofilms were first treated with LysH5 (0.15 µM) for 6 h, then washed and treated again with LysH5 (0.15 µM) for 12 h. Cell counts were measured as described above.

Determination of minimal inhibitory concentration (MIC) of LysH5 and biofilm induction

The MIC of LysH5 was determined in duplicate by a conventional broth microdilution technique in TSB. Two-fold dilution of LysH5 (0.4 µM) was made in microtiter plates and each well was inoculated with 10^6 CFU. The MIC was defined as the lowest protein concentration that inhibited visible bacterial growth after 24 h of incubation at 37°C.

The ability of LysH5 to induce the formation of biofilms was tested using sub-inhibitory concentrations of the protein, ranging from 0 to 6 µg/ml (0–0.1 µM) for *S. aureus* 15981, *S. aureus* 132, *S. aureus* IPLA1, *S. aureus* IPLA6 and *S. epidermidis* YLIC17;

Table 1. Strains used in this work, origin and relevant properties.

	Strain	Origin	Relevantproperties	Reference
S. aureus	15981	Clinical isolate	Produces PNAG	[42]
	ISP479r	Clinical isolate		
	V329	Bovine subclinical mastitis	Expresses*bap*protein	[7]
	132	Clinical isolate	Produces PNAG	[43]
	IPLA1	Dairy industry surface	Genes *icaA* and *icaD*	[6]
	IPLA16	Meat industry surface		
	SA113	-	NCTC8325 derivative, agr–, 11-bp deletion in *rsbU*	[44]
	Newman	-	Sub-inhibitory concentrations of meticillin induces biofilm formation.	[45]
S. epidermidis	B	Breast milk of women suffering infectious mastitis	Genes *icaA* and *icaD*	[46]
	YLIC17			
	DG2n			

and from 0 to 3 μg/ml (0–0.05 μM) for *S. aureus* ISP479r, *S. aureus* V329, *S. epidermidis* B and *S. epidermidis* DG2n. *S. aureus* Newman treated with 0–10 μg/ml of methicillin was used as a positive control of biofilm formation under sub-inhibitory concentrations of this antibiotic [14]. The biofilm induction value for each strain (expressed as relative absorbance) was defined as the absorbance value measured after crystal violet staining at each concentration of LysH5, or methicillin, divided by the absorbance value in the absence of the antimicrobials. For CFU determination of non-adhered cells, planktonic cells of each well were diluted and plated onto TSA plates. All experiments were conducted using two biological replicates.

Biochemical characterization of the biofilm matrix

24 h-old staphylococcal biofilms were washed with PBS and then treated for 1 h 37°C either with a solution of 10 mM sodium metaperiodate in 50 mM sodium acetate buffer (pH 4.5) (sodium metaperiodate treatment is used to disrupt the extracellular polysaccharides), with 100 μg/ml of proteinase K (Sigma, Madrid, Spain) in 20 mM Tris (pH 7.5) and 100 mM NaCl, or with 100 μg/ml of DNAseI (Sigma, Madrid, Spain) in 150 mM of NaCl and 1 mM $CaCl_2$. After treatments, the biofilms were washed with water, stained with crystal violet, and the absorbance measured as described above.

LysH5 activity against persister cells

To determine if treatment of planktonic cells with LysH5 yielded a subpopulation of persister cells, overnight cultures of *S. aureus* 15981 were used to inoculate TSB fresh medium and grown at 37°C up to $OD_{600} = 0.5$. Then, LysH5 was added at 1 to 5-fold MIC (0.1–0.5 μM, respectively) and cultures were incubated at 37°C for 3 h.

Lytic activity of LysH5 against previously selected persister cells was also tested as follows. Persister cells of *S. aureus* SA113 were selected as described previously [35] after the treatment of exponential cultures ($OD_{600} = 0.5$) with 2 μg/ml of rifampin or 3 μg/ml of ciprofloxacin (100 and 10-fold MIC, respectively) for 4 h. LysH5 was then added at 0.5 μM and incubated at 37°C for 4 additional hours.

For CFU determinations, samples were taken before and during the antimicrobial challenge and appropriate dilutions plated onto TSA plates. All experiments were conducted using two biological replicates.

Low-temperature scanning electron microscopy (LTSEM)

LTSEM was used to visualize the structure of the 24 h-old biofilms before and after treatment with the endolysin LysH5. The biofilms were grown on a glass cover lid and visualized at −135°C with a DSM 960 Zeiss scanning electron microscope as previously described [36].

Statistical analysis

Statistical analysis was performed in order to establish any significant differences between the total bacteria number in the control and treated biofilms, and between the control and treated planktonic cultures. The differences were expressed as the mean ± standard error and were determined by one-way analysis of variance (ANOVA) and the LSD test was used for a comparison of means at a level of significance $P<0.05$.

Results

LysH5 is effective against staphylococcal biofilms

The ability of LysH5 to remove staphylococcal biofilms and kill sessile cells was determined on six *S. aureus* and three *S. epidermidis* strains. Three *S. aureus* clinical strains producing a polysaccharide matrix and one protein-dependent biofilm strain from bovine origin were tested. Additionally, two *S. aureus* strains isolated from a food environment and three *S. epidermidis* of clinical origin, all of them carrying biofilm related genes, were included in the assay (Table 1). 24 h-old biofilms developed on 96-well polystyrene plates were treated with LysH5 (0.15 μM) and lysostaphin (0.2 μM) and biofilm removal was determined after 6 h of incubation. LysH5 showed a notable disrupting activity against biofilms of both *S. aureus* and *S. epidermidis* strains, as judged by the reduction of the attached bacterial counts by 1–3 log units per well (Fig. 1). The activity of this protein turned out to be higher than that of lysostaphin, with the exception of the biofilms developed by *S. aureus* 15981 and *S. aureus* ISP479r, against which lysostaphin was more effective. As expected, lysostaphin was quite ineffective against *S. epidermidis* biofilms, while LysH5 showed similar viable count reduction percentages against strains of both species.

The variability observed in the activity of LysH5 against the different biofilms might be determined by the susceptibility of the planktonic cells to LysH5 but also by the composition of the extracellular material which might impair or limit the access of LysH5 to the bacterial cells. However, we could not establish a correlation between the activity of LysH5 against each biofilm and the intrinsic susceptibility of each strain or the composition of the biofilm matrix. As shown in Table 2, the specific activity of LysH5 on exponentially growing *S. epidermidis* cells was rather low (1.5–3.8 ΔOD/mg min) in contrast to that shown by *S. aureus* cells (2.1–16.5 ΔOD/mg min). However, LysH5 removed staphylococcal biofilms regardless of the chemical composition of the extracellular matrix (polysaccharidic, protein or DNA based biofilms) (Table 3). Moreover, attached viable cells were recovered from the wells after a single exposure to LysH5 and the biofilm formation ability of these isolates was determined by crystal violet staining. Absorbance values identical to those of untreated strains were observed (Table S1). Thus, the surviving cells do not seem to have an increased biofilm-forming ability. Of note, a second LysH5 treatment resulted in the decrease of the staphylococcal populations to undetectable levels (<10 CFU/well) (Fig. 1). This suggests that the viable cells remaining attached to the well surface after the first LysH5 treatment were not persister cells as they remained susceptible to the endolysin.

LTSEM supports biofilm disruption by LysH5

Electron micrographs reinforced our results as they showed unstructured biofilms as a consequence of the treatment with LysH5 (Fig. 2). In the case of *S. aureus* 15981, a polysaccharide-dependent biofilm producer, cells were arranged in layers and covered by a thin matrix that kept them attached to surface (Fig. 2A); after LysH5 treatment, no cells or extracellular material were observed (Fig. 2B). *S. epidermidis* YLIC17 biofilm structure showed cell aggregates producing a compact biofilm covered by a polysaccharidic layer (Fig. 2C) which was not removed by the endolysin treatment but whole cells were no longer present (Fig. 2D). This observation suggests that LysH5 was able to penetrate through the matrix structure.

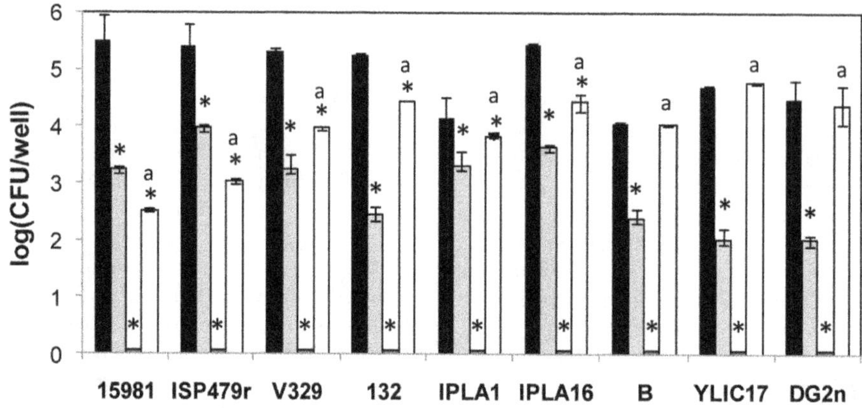

Figure 1. Removal of 24 h-old biofilms of *S. aureus* and *S. epidermidis*. Biofilms of *S. aureus* 15981, ISP479r, V329, 132, IPLA1 and IPLA16 and *S. epidermidis* B, YLIC17and DG2n were treated with 0.15 μM of LysH5 (light grey) and 0.2 μM of lysostaphin (white) during 6 hours. Alternatively, biofilms were treated with 0.15 μM of LysH5 during 6 hours, followed by another treatment with the endolysin for 12 hours (gross line in X axis). Control biofilms are represented in black. Adhered cell counts were expressed as log CFU/well. Bacteria detection threshold (<10 CFU/ml). Means and standard deviations were calculated from two biological replicates. Bars having an asterisk are significantly different from the control (ANOVA; $P<0.05$) and bars with a lower case 'a' indicates a significantly difference between the lysostaphin treatment and the treatment with LysH5 (ANOVA; $P<0.05$).

Sub-inhibitory concentrations of LysH5 do not induce staphylococcal biofilm and prevent its formation in some strains

To evaluate a putative inducing effect of LysH5 on biofilm formation, *S. aureus* strains were grown in the presence of LysH5 at concentrations lower than the MIC (Table 2) and checked for biofilm formation. As a control for antibiotic-induced biofilm, the strain *S. aureus* Newman was used [14]. As expected, a relative absorbance increase of 18-fold was observed for this strain treated with sub-inhibitory concentrations of methicillin and further stained with crystal violet. Significantly, induction in biofilm formation was also observed for strains *S. aureus* 132 and *S. aureus* IPLA1 (1.6 and 1.3 times, respectively) (Fig. S1). In the presence of sub-MIC concentrations of LysH5, bacterial growth was not inhibited and biofilm induction was not observed in any of the tested strains (Fig. S2). Interestingly, the staphylococcal strains behaved differently in the presence of sub-inhibitory LysH5 concentrations (Fig. 3). In fact, biofilm formation by *S. aureus* 15981, *S. aureus* 132 and *S. epidermidis* B was completely inhibited (Fig. 3A), while biofilm growth of *S. aureus* ISP479r, *S. aureus* V329, *S. aureus* IPLA1 and *S. epidermidis* YLIC17 was reduced, but not completely inhibited when growing at 0.25× to 0.5×MIC of LysH5 (Fig. 3B). For the remaining strains, *S. aureus* IPLA16, *S. aureus* Newman and *S. epidermidis* DG2n, no inhibition of biofilm formation was observed under sub-inhibitory concentrations (Fig. 3C).

LysH5 is active against *S. aureus* planktonic persister cells

To assess if the treatment of staphylococcal planktonic cultures with LysH5 could select persister cells, *S. aureus* 15981 exponential cultures were treated with increasing LysH5 concentrations. Bacterial counts decreased as LysH5 concentration increased, and no surviving bacteria were detected at 0.5 μM endolysin (Fig. 4A). Similar results were obtained when *S. aureus* SA113 was treated with increasing concentrations of LysH5 (data not shown). To further confirm the effectiveness of LysH5,

Table 2. Sensitivity of staphylococcal strains to LysH5.

	Strain	Specific activity (ΔOD/mg min)	MIC (μM)
S. aureus	15981	13.4±0.8	0.1
	ISP479r	16.5±0.3	0.05
	V329	8.7±0.1	0.05
	132	14.3±0.4	0.1
	IPLA1	2.1±0.6	0.1
	IPLA16	3.0±0.1	0.05
	Newman	11.6±0.3	0.1
	SA113	12.1±0.2	0.1
S. epidermidis	B	3.8±0.1	0.05
	YLIC17	1.9±0.2	0.1
	DG2n	1.5±0.3	0.05

Specific activity of LysH5 against staphylococcal exponential cultures (ΔOD per mg of protein per min) and MIC values of LysH5 determined in staphylococcal broth cultures. The values are means ± standard deviations two biological experiments.

Table 3. Ability of specific treatments to remove biofilm formed by staphylococcal strains.

	Strain	Removed biofilm		
		$NaIO_4$	Proteinase K	DNAse I
S. aureus	15981	+++	–	–
	ISP479r	++	++	+
	V329	–	+++	++
	132	+	+	+
	IPLA1	++	+	–
	IPLA16	++	++	–
S. epidermidis	B	+++	–	–
	YLIC17	+++	–	+
	DG2n	+++	–	+

Extracellular components of the biofilm matrix were estimated from the percentage of removed biofilm with specific treatments; sodium metaperiodate ($NaIO_4$), proteinase K and DNAseI (+ = 30%; ++ = 30–70% and +++>70%).

planktonic persister cells were isolated from *S. aureus* SA113 by treatment with either rifampicin or ciprofloxacin [35]. Fig. 4B shows the selection of persister cells by rifampicin and ciprofloxacin with the typical biphasic killing curve, and their further behavior after the addition of 0.5 μM LysH5. Indeed, no viable bacteria were detected after 2 h of incubation at 37°C, supporting the LysH5 lytic potential against this subpopulation of bacteria that can survive the antibiotic challenge.

Discussion

The endolysin LysH5 has a remarkable activity to lyse *S. aureus* and *S. epidermidis* planktonic cells [31], which has been extended against staphylococcal biofilms in the current work. Low concentrations of LysH5 (0.15 μM) were effective to decrease viable bacteria inside biofilms formed by either *S. aureus* or *S. epidermidis* strains. No correlation was observed between the

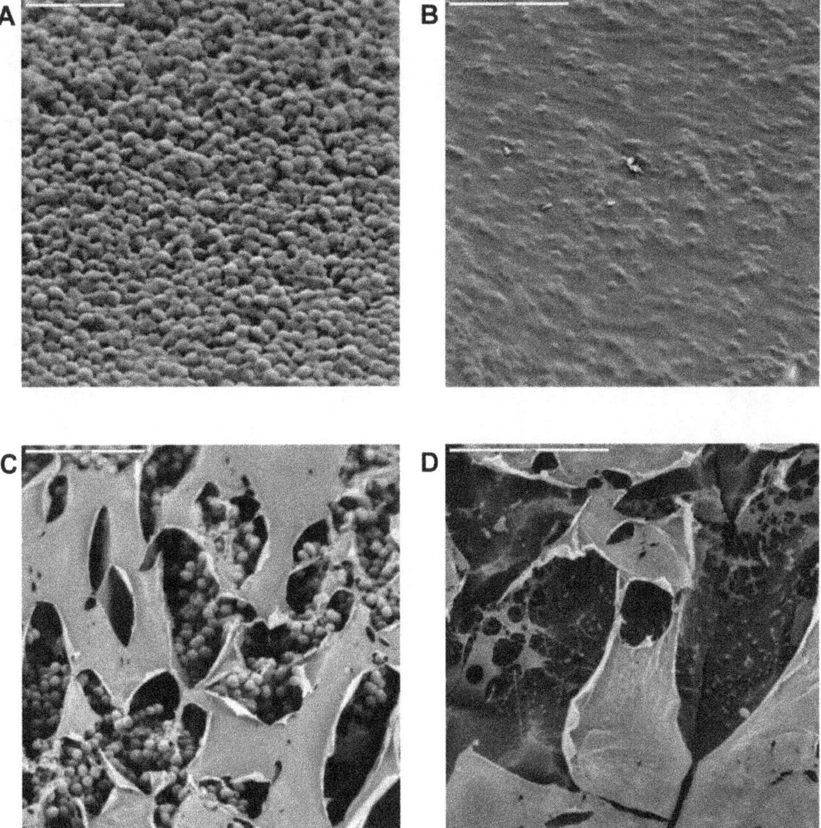

Figure 2. LTSEM micrographs of 24 h-old biofilms formed by *S. aureus* and *S. epidermidis*. Not treated biofilms of *S. aureus* 15981 and *S. epidermidis* YLIC17 are represented in A and C, respectively; biofilms after treatment with LysH5 are represented in B and D.

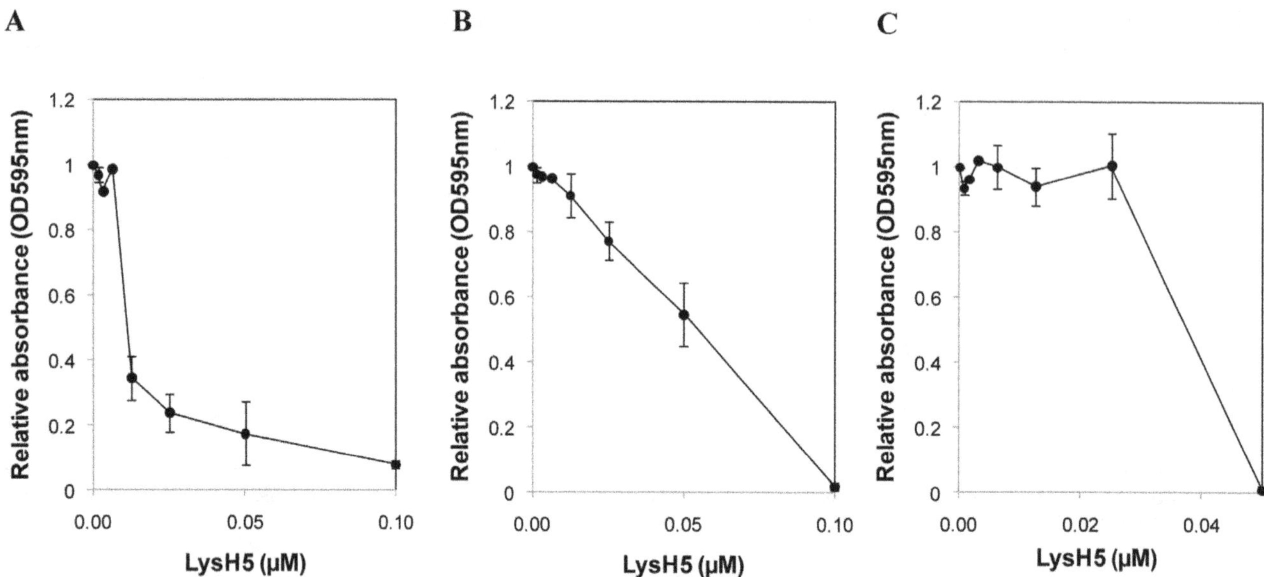

Figure 3. Behavior of 24 h-old biofilms formed by *S. aureus* and *S. epidermidis* strains, grown in the presence of sub-inhibitory concentrations of LysH5. Three strains were selected as representative of biofilm behavior (A) Prevention of the biofilm formation (*S. aureus* 15981); (B) Biofilm reduction (*S. aureus* ISP479r) and (C) no effect in biofilm formation (*S. aureus* IPLA16). Biofilm formation was expressed as relative absorbance (595 nm) of crystal violet stained cultures (treated/untreated cultures)(●). LysH5 concentration is expressed in μM. Each value is the mean ± standard deviation of two biological replicates.

ability of LysH5 to remove biofilms and the origin of the strains (clinical, food), the extracellular matrix (polysaccharide, protein/DNA) or the peptidoglycan composition, although the latter could have an impact on LysH5 activity, as shown by the higher specific activity of LysH5 against *S. aureus* versus *S. epidermidis* planktonic cells [31] (Table 2). However, low diffusion of LysH5 inside biofilms might represent the main limitation to protein activity, as judged by the similar activity of LysH5 against *S. aureus* and *S. epidermidis* biofilms (Fig. 1).

Beyond the ability of endolysins to remove biofilms, the aim of this work was to study the effects of LysH5 on both biofilm formation and putative surviving cells after the LysH5 treatment. In this regard, we have determined the impact of the expected low diffusion of LysH5 through the extracellular material and the consequent sub-inhibitory concentration inside the biofilms. Our results showed no increase in biofilm formation at low concentrations of LysH5, neither in *S. aureus* nor in *S. epidermidis* strains. On the contrary, sub-inhibitory concentrations of many antibiotics have been reported to enhance the production of PIA in *S. epidermidis* biofilms [37] or extracellular DNA in *S. aureus* biofilms [14]. It is not totally clear which mechanism mediates biofilm induction by some antibiotics but it is thought to be a

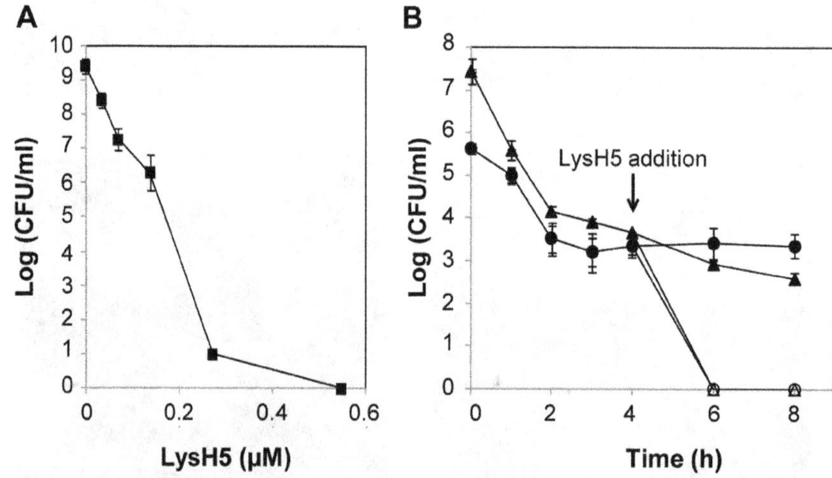

Figure 4. *S. aureus* persister cells selection and elimination by LysH5. (A) Activity of different concentrations of LysH5 against exponential cultures of *S. aureus* 15981. Means and standard deviations were calculated from two independent assays. Bacteria detection threshold (<10 CFU/ml). (B) Exponential cultures of *S. aureus* 113 treated with 2 μg/ml of rifampicin (●), and 3 μg/ml of ciprofloxacin (▲) for 4 h to select persister cells and subsequent treatment with 0.5 μM LysH5 (○ and Δ, respectively). Bacteria detection threshold (<10 CFU/ml). Each value represents the mean ± standard deviation of two biological replicates.

consequence of a global stress response [37]. Furthermore, we found that endolysin LysH5 was able to inhibit biofilm formation in some strains, even at 0.125× MIC. A similar effect was previously described in gallidermin, which represses transcription of genes involved in primary adhesion and exopolysaccharide production [19].

One of the main issues in biofilm control is the presence of persister cells, a small subpopulation of cells that spontaneously enter a dormant state and, consequently, survive bactericidal antibiotic treatment. Persisters have been found for almost every type of antibiotic tested so far [16]. They can reestablish the population once the stress is removed [38], and are behind recalcitrant chronic infections [39]. With this in mind, it is remarkable that no persister bacteria were detected in staphylococcal biofilms after LysH5-treatment. Surviving bacteria were still sensitive to the endolysin, confirming previous results of the lack of resistance after the exposure of *S. aureus* to LysH5 [33]. These results are also in accordance with the absence of reports on bacteria resistant to phage endolysins, despite several attempts to select them [40].

Keeping in mind that approximately 0.001–0.1% of cells of an isogenic bacterial population display tolerance to antibiotics [14], we also approached the selection of persister cells from exponential cultures. In a typical assay for selection of persister cells by antibiotics, the number of cells in the population is initially decreased by the antibiotic action, but in a second phase, despite increasing the time or the antibiotic concentration, a subpopulation of persister cells remain (biphasic curve) [41]. In a similar assay carried out with LysH5, a lack of persister bacteria after treatment of *S. aureus* exponential cultures with LysH5 was confirmed. The survival rates of bacteria treated with increasing concentrations of LysH5 failed to show a biphasic curve and no surviving bacteria were detected at high LysH5 concentration. In this regard, the fact that LysH5 is able to eliminate persister bacteria previously selected by two antibiotics, rifampicin and ciprofloxacin, is particularly interesting. As previously reported [35], treatment of *S. aureus* SA113 exponential cultures with 10×MIC ciprofloxacin and 100×MIC rifampicin resulted in a drug-tolerant population of about 10^3 CFU/ml, which were further eliminated by LysH5. Our findings suggest that the endolysin target is accessible in a dormant state in contrast to antibiotics that need an active target only available in growing cells. Therefore, delivering endolysin LysH5 as adjuvant to some antibiotics would be beneficial in the treatment of chronic bacterial infections. Regarding the disinfection process of industrial food facilities, a two-step treatment could guarantee the removal of any persistent contamination and open the way to the use of LysH5 as part of the routine process exploiting putative synergistic action with other disinfectants/antimicrobials. Despite these promising results, the total duration of the two-step treatment (18 h) should be shortened, increasing the concentration of endolysin in order to achieve the desire efficacy, avoiding the interruption of the production in food industries. Moreover, our data should be confirmed in dynamic biofilm models, we provide additional proof of concept supporting the role of phage endolysins, not only as antibacterial agents against staphylococcal biofilms, but also as anti-persister agents, and therefore being an effective weapon to combat bacterial infections.

Supporting Information

Figure S1 Biofilms formed by *S. aureus* and *S. epidermidis* strains grown in the presence of sub-inhibitory concentrations of meticillin. (A) Strains with antibiotic-induction of the biofilm; (B) strains with no effect in biofilm formation. Biofilm formation was expressed as relative absorbance (595 nm) of crystal violet stained cultures (treated/untreated cultures) (•). Each value is the mean ± standard deviation of two biological replicates.

Figure S2 Biofilms formed by *S. aureus* and *S. epidermidis* strains grown in the presence of sub-inhibitory concentrations of LysH5. (A) Strains showing prevention of the biofilm formation; (B) strains showing a biofilm reduction and (C) strains with no effect in biofilm formation. Biofilm formation was expressed as relative absorbance (595 nm) of crystal violet stained cultures (treated/untreated cultures) (•). Each value is the mean ± standard deviation of two biological replicates.

Table S1 Biofilm formation ability of the *S. aureus and S. epidermidis* strains. Comparison of 24 h biofilm growth using o/n cultures of the staphylococcal strains and o/n cultures of surviving cells recovered after biofilms LysH5 treatment. Values expressed as absorbance units (595 nm) of crystal violet stained cultures are the means ± standard deviations of two independent experiments.

Acknowledgments

We thank Dr. A. Toledo-Arana (Instituto de Agrobiotecnología, CSIC-Universidad Pública de Navarra, Spain) for providing the *S. aureus* strains.

Author Contributions

Conceived and designed the experiments: PG AR BM PR-M DG. Performed the experiments: DG PG. Analyzed the data: PG AR BM PR-M. Wrote the paper: PG AR BM PR-M DG. Supervised the LTSEM analysis: PR-M.

References

1. Lowy FD (1998) *Staphylococcus aureus* infections. N Engl J Med 339: 520–532.
2. Otto M (2008) Staphylococcal biofilms. Curr Top Microbiol Immunol 322: 207–228.
3. Jabbouri S, Sadovskaya I (2010) Characteristics of the biofilm matrix and its role as a possible target for the detection and eradication of *Staphylococcus epidermidis* associated with medical implant infections. FEMS Immunol Med Microbiol 59: 280–291.
4. Dulon M, Haamann F, Peters C, Schablon A, Nienhaus A (2011) MRSA prevalence in European healthcare settings: a review. BMC Infect Dis 11: 138.
5. Le Loir Y, Baron F, Gautier M (2003) *Staphylococcus aureus* and food poisoning. Genet Mol Res 2: 63–76.
6. Gutierrez D, Delgado S, Vazquez-Sanchez D, Martinez B, Cabo ML, et al. (2012) Incidence of *Staphylococcus aureus* and analysis of associated bacterial communities on food industry surfaces. Appl Environ Microbiol 78: 8547–8554.
7. Cucarella C, Solano C, Valle J, Amorena B, Lasa I, et al. (2001) Bap, a *Staphylococcus aureus* surface protein involved in biofilm formation. J Bacteriol 183: 2888–2896.
8. Merino N, Toledo-Arana A, Vergara-Irigaray M, Valle J, Solano C, et al. (2009) Protein A-mediated multicellular behavior in *Staphylococcus aureus*. J Bacteriol 191: 832–843.
9. O'Neill E, Pozzi C, Houston P, Humphreys H, Robinson DA, et al. (2008) A novel *Staphylococcus aureus* biofilm phenotype mediated by the fibronectin-binding proteins, FnBPA and FnBPB. J Bacteriol 190: 3835–3850.
10. Corrigan RM, Rigby D, Handley P, Foster TJ (2007) The role of *Staphylococcus aureus* surface protein SasG in adherence and biofilm formation. Microbiology 153: 2435–2446.
11. Christner M, Franke GC, Schommer NN, Wendt U, Wegert K, et al. (2010) The giant extracellular matrix-binding protein of *Staphylococcus epidermidis*

mediates biofilm accumulation and attachment to fibronectin. Mol Microbiol 75: 187–207.
12. Hussain M, Herrmann M, von Eiff C, Perdreau-Remington F, Peters G (1997) A 140-kilodalton extracellular protein is essential for the accumulation of *Staphylococcus epidermidis* strains on surfaces. Infect Immun 65: 519–524.
13. Davies D (2003) Understanding biofilm resistance to antibacterial agents. Nat Rev Drug Discov 2: 114–122.
14. Kaplan JB, Izano EA, Gopal P, Karwacki MT, Kim S, et al. (2012) Low levels of beta-lactam antibiotics induce extracellular DNA release and biofilm formation in *Staphylococcus aureus*. MBio 3: e00198–00112.
15. Kint CI, Verstraeten N, Fauvart M, Michiels J (2012) New-found fundamentals of bacterial persistence. Trends Microbiol 20: 577–585.
16. Lewis K (2010) Persister cells. Annu Rev Microbiol 64: 357–372.
17. Kaplan JB, Ragunath C, Velliyagounder K, Fine DH, Ramasubbu N (2004) Enzymatic detachment of *Staphylococcus epidermidis* biofilms. Antimicrob Agents Chemother 48: 2633–2636.
18. Wu JA, Kusuma C, Mond JJ, Kokai-Kun JF (2003) Lysostaphin disrupts *Staphylococcus aureus* and *Staphylococcus epidermidis* biofilms on artificial surfaces. Antimicrob Agents Chemother 47: 3407–3414.
19. Saising J, Dube L, Ziebandt AK, Voravuthikunchai SP, Nega M, et al. (2012) Activity of gallidermin on *Staphylococcus aureus* and *Staphylococcus epidermidis* biofilms. Antimicrob Agents Chemother 56: 5804–5810.
20. Gutierrez D, Martinez B, Rodriguez A, Garcia P (2012) Genomic characterization of two *Staphylococcus epidermidis* bacteriophages with anti-biofilm potential. BMC Genomics 13: 228.
21. Kelly D, McAuliffe O, Ross RP, Coffey A (2012) Prevention of *Staphylococcus aureus* biofilm formation and reduction in established biofilm density using a combination of phage K and modified derivatives. Lett Appl Microbiol 54: 286–291.
22. Cerca N, Oliveira R, Azeredo J (2007) Susceptibility of *Staphylococcus epidermidis* planktonic cells and biofilms to the lytic action of *Staphylococcus* bacteriophage K. Lett Appl Microbiol 45: 313–317.
23. Curtin JJ, Donlan RM (2006) Using bacteriophages to reduce formation of catheter-associated biofilms by *Staphylococcus epidermidis*. Antimicrob Agents Chemother 50: 1268–1275.
24. Nelson DC, Schmelcher M, Rodriguez-Rubio L, Klumpp J, Pritchard DG, et al. (2012) Endolysins as antimicrobials. Adv Virus Res 83: 299–365.
25. Rodriguez-Rubio L, Martinez B, Donovan DM, Rodriguez A, Garcia P (2013) Bacteriophage virion-associated peptidoglycan hydrolases: potential new enzybiotics. Crit Rev Microbiol 39: 427–434.
26. Fenton M, Keary R, McAuliffe O, Ross RP, O'Mahony J, et al. (2013) Bacteriophage-Derived Peptidase CHAP(K) Eliminates and Prevents Staphylococcal Biofilms. Int J Microbiol 2013: 625341.
27. Meng X, Shi Y, Ji W, Zhang J, Wang H, et al. (2011) Application of a bacteriophage lysin to disrupt biofilms formed by the animal pathogen *Streptococcus suis*. Appl Environ Microbiol 77: 8272–8279.
28. Domenech M, Garcia E, Moscoso M (2011) In vitro destruction of *Streptococcus pneumoniae* biofilms with bacterial and phage peptidoglycan hydrolases. Antimicrob Agents Chemother 55: 4144–4148.
29. Shen Y, Koller T, Kreikemeyer B, Nelson DC (2013) Rapid degradation of *Streptococcus pyogenes* biofilms by PlyC, a bacteriophage-encoded endolysin. J Antimicrob Chemother 68: 1818–1824.
30. Sass P, Bierbaum G (2007) Lytic activity of recombinant bacteriophage phi11 and phi12 endolysins on whole cells and biofilms of *Staphylococcus aureus*. Appl Environ Microbiol 73: 347–352.
31. Obeso JM, Martinez B, Rodriguez A, Garcia P (2008) Lytic activity of the recombinant staphylococcal bacteriophage PhiH5 endolysin active against *Staphylococcus aureus* in milk. Int J Food Microbiol 128: 212–218.
32. Garcia P, Martinez B, Rodriguez L, Rodriguez A (2010) Synergy between the phage endolysin LysH5 and nisin to kill *Staphylococcus aureus* in pasteurized milk. Int J Food Microbiol 141: 151–155.
33. Rodriguez-Rubio L, Martinez B, Rodriguez A, Donovan DM, Gotz F, et al. (2013) The phage lytic proteins from the *Staphylococcus aureus* bacteriophage vB_SauS-phiIPLA88 display multiple active catalytic domains and do not trigger staphylococcal resistance. PLoS One 8: e64671.
34. Herrera JJ, Cabo ML, Gonzalez A, Pazos I, Pastoriza L (2007) Adhesion and detachment kinetics of several strains of *Staphylococcus aureus* subsp. aureus under three different experimental conditions. Food Microbiol 24: 585–591.
35. Lechner S, Lewis K, Bertram R (2012) Staphylococcus aureus persisters tolerant to bactericidal antibiotics. J Mol Microbiol Biotechnol 22: 235–244.
36. Moscoso M, Garcia E, Lopez R (2006) Biofilm formation by *Streptococcus pneumoniae*: role of choline, extracellular DNA, and capsular polysaccharide in microbial accretion. J Bacteriol 188: 7785–7795.
37. Kaplan JB (2011) Antibiotic-induced biofilm formation. Int J Artif Organs 34: 737–751.
38. Keren I, Kaldalu N, Spoering A, Wang Y, Lewis K (2004) Persister cells and tolerance to antimicrobials. FEMS Microbiol Lett 230: 13–18.
39. Fauvart M, De Groote VN, Michiels J (2011) Role of persister cells in chronic infections: clinical relevance and perspectives on anti-persister therapies. J Med Microbiol 60: 699–709.
40. Fischetti VA (2008) Bacteriophage lysins as effective antibacterials. Curr Opin Microbiol 11: 393–400.
41. Keren I, Mulcahy LR, Lewis K (2012) Persister eradication: lessons from the world of natural products. Methods Enzymol 517: 387–406.
42. Valle J, Toledo-Arana A, Berasain C, Ghigo JM, Amorena B, et al. (2003) SarA and not sigmaB is essential for biofilm development by *Staphylococcus aureus*. Mol Microbiol 48: 1075–1087.
43. Vergara-Irigaray M, Valle J, Merino N, Latasa C, Garcia B, et al. (2009) Relevant role of fibronectin-binding proteins in *Staphylococcus aureus* biofilm-associated foreign-body infections. Infect Immun 77: 3978–3991.
44. Iordanescu S, Surdeanu M (1976) Two restriction and modification systems in *Staphylococcus aureus* NCTC8325. J Gen Microbiol 96: 277–281.
45. Duthie ES, Lorenz LL (1952) Staphylococcal coagulase; mode of action and antigenicity. J Gen Microbiol 6: 95–107.
46. Delgado S, Arroyo R, Jimenez E, Marin ML, del Campo R, et al. (2009) *Staphylococcus epidermidis* strains isolated from breast milk of women suffering infectious mastitis: potential virulence traits and resistance to antibiotics. BMC Microbiol 9: 82.

16

Benchmarking Undedicated Cloud Computing Providers for Analysis of Genomic Datasets

Seyhan Yazar[1], George E. C. Gooden[1], David A. Mackey[1,2], Alex W. Hewitt[1,2,3]*

1 Centre for Ophthalmology and Visual Science, University of Western Australia, Lions Eye Institute, Perth, Western Australia, Australia, 2 School of Medicine, Menzies Research Institute Tasmania, University of Tasmania, Hobart, Tasmania, Australia, 3 Centre for Eye Research Australia, University of Melbourne, Department of Ophthalmology, Royal Victorian Eye and Ear Hospital, Melbourne, Victoria, Australia

Abstract

A major bottleneck in biological discovery is now emerging at the computational level. Cloud computing offers a dynamic means whereby small and medium-sized laboratories can rapidly adjust their computational capacity. We benchmarked two established cloud computing services, Amazon Web Services Elastic MapReduce (EMR) on Amazon EC2 instances and Google Compute Engine (GCE), using publicly available genomic datasets (*E.coli* CC102 strain and a Han Chinese male genome) and a standard bioinformatic pipeline on a Hadoop-based platform. Wall-clock time for complete assembly differed by 52.9% (95% CI: 27.5–78.2) for *E.coli* and 53.5% (95% CI: 34.4–72.6) for human genome, with GCE being more efficient than EMR. The cost of running this experiment on EMR and GCE differed significantly, with the costs on EMR being 257.3% (95% CI: 211.5–303.1) and 173.9% (95% CI: 134.6–213.1) more expensive for *E.coli* and human assemblies respectively. Thus, GCE was found to outperform EMR both in terms of cost and wall-clock time. Our findings confirm that cloud computing is an efficient and potentially cost-effective alternative for analysis of large genomic datasets. In addition to releasing our cost-effectiveness comparison, we present available ready-to-use scripts for establishing Hadoop instances with Ganglia monitoring on EC2 or GCE.

Editor: Maureen J. Donlin, Saint Louis University, United States of America

Funding: This work was supported by funding from the BrightFocus Foundation, the Ophthalmic Research Institute of Australia and a Ramaciotti Establishment Grant. CERA receives operational infrastructure support from the Victorian government. The funders had no role in study design, data collection and analysis, decision to publish, or preparation of the manuscript.

Competing Interests: The authors have declared that no competing interests exist.

* Email: hewitt.alex@gmail.com

Introduction

Through the application of high-throughput sequencing, there has been a dramatic increase in the availability of large-scale genomic datasets [1]. With reducing sequencing costs, small and medium-sized laboratories can now easily amass many gigabytes of data. Given this dramatic increase in the volume of data generated, researchers are being forced to seek efficient and cost-effective measures for computational analysis [2]. Cloud computing offers a dynamic means whereby small and medium-sized laboratories can rapidly adjust their computational capacity, without concern about its physical structure or ongoing maintenance [3–6]. However, transitioning to a cloud environment presents with unique strategic decisions [7], and although a number of general benchmarking results are available (http://serverbear.com/benchmarks/cloud; https://cloudharmony.com/; Accessed 2014 Aug 7), there has been a paucity of comparisons of cloud computing services specifically for genomic research.

We undertook a performance comparison on two established cloud computing services: Amazon Web Services EMR on Amazon EC2 instances and GCE. Paired-end sequence reads of publicly available genomic datasets (*Escherichia coli* CC102 strain and a Han Chinese male genome) were analysed using Crossbow, a genetic annotation tool, on Hadoop-based platforms with equivalent system specifications [8–10]. A standard analytical pipeline was run simultaneously on both platforms multiple times (Figure 1 and 2). The performance metrics of both platforms were recorded using Ganglia, an open-source high performance computing monitoring system [11].

Results

Wall-clock time for complete mapping and SNP calling differed by 52.9% (95% CI: 27.5–78.2) and 53.5% (95% CI: 34.4–72.6) for *E.coli* and human genome alignment and variant calling, respectively, with GCE being more efficient than EMR. Table 1 displays the key metrics for data analysis using both services. The proportion of central processing unit (CPU) usage by Crossbow differed between platforms when aligning and SNP calling each genome, with GCE having better utilisation as the genome size increased. There was considerably more free memory on GCE for the smaller *E.coli* dataset and on EMR for larger human genome runs. The CPU idle percentage, the percentage of time where the CPU was idle without waiting for disk input/output (I/O), was greater on EMR for the human genome while CPU waiting for I/O (WIO) was considerably lower on the same platform. The CPU idle and CPU WIO percentages were both significantly higher on EMR for the *E.coli* genome. The cost of running this Crossbow

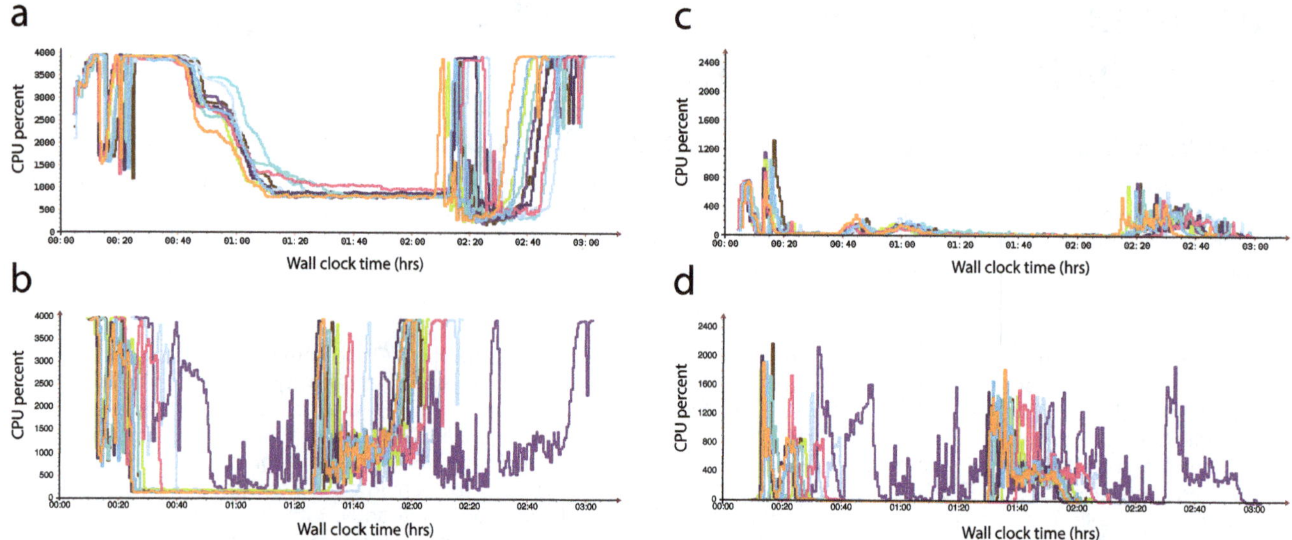

Figure 1. Comparison of undedicated cloud computing performances. The panel includes results of Amazon Web Services Elastic MapReduce (EMR) on Amazon EC2 instances (panels a & c) versus Google Compute Engine (GCE) (panels b & d) for human genome alignment and variant calling. In this 40 node cluster the total CPU percent for CPU idle (a and b) and waiting for disk input/output (c and d) is displayed. Note the greater consistency in performance of Crossbow, though generally longer wall clock times for complete analysis, on EMR compared to GCE.

pipeline on EMR and GCE also differed significantly ($p<0.001$), with the costs on EMR being 257.3% (95% CI: 211.5–303.1) and 173.9% (95% CI: 134.6–213.1) more expensive than GCE for *E.coli* and human assemblies, respectively. For ~36x coverage of a human genome, at a current sequencing cost of ~US$1000, the median cost for computation on GCE was US$29.81 (range: US$28.86 to US$45.99), whilst on EMR with a fixed hourly rate it was US$69.60 (range: US$69.60 to US$92.80).

Although runtime variability was inevitable and present in both platforms when assembling each genome, GCE had a considerably greater variability with the larger human genome compared to EMR (coefficient of variation $(COV)_{EMR} = 4.48\%$ vs - $COV_{GCE} = 16.72\%$). We identified a single outlier in run time on GCE during the human genome analysis. This occurred due to the virtual cluster having a slower average network connection (1.55 MB/s compared to the average of the other GCE clusters of 2.02 MB/s) and a higher CPU WIO percentage than the average for the other GCE runs (9.56% versus 3.52%). The variation in cluster performance likely reflects an increase in network congestion amongst GCE servers.

Runtime predictably is an important issue in undedicated cloud computing. The existing workload of the cloud at the time of service usage is one of the main determinants of variability in runtime of undedicated services [12]. In our benchmarking, EMR was more consistent, though slower, in overall wall-clock time compared to GCE. This may suggest that GCE is more susceptible

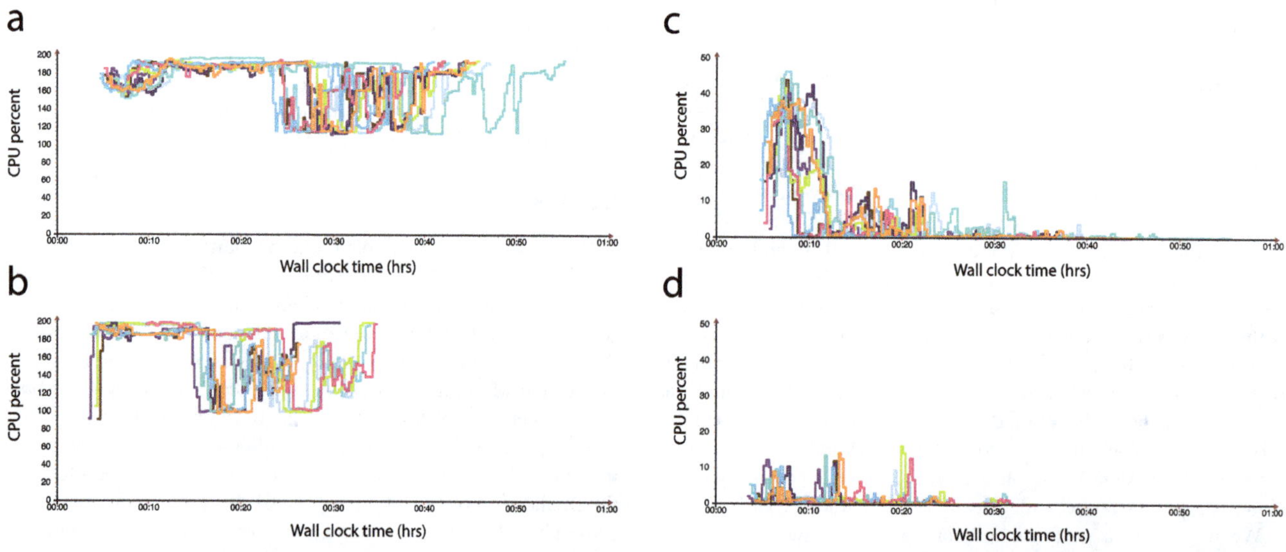

Figure 2. Comparison of undedicated cloud performance of Amazon Web Services Elastic MapReduce (EMR) on Amazon EC2 instances (panels a & c) versus Google Compute Engine (GCE) (panels b & d) for *E.coli* genome alignment and variant calling. In this two node cluster the total CPU percent for CPU idle (a and b) and waiting for disk input/output (c and d) is displayed. Note the shorter wall clock times for complete analysis on GCE compared to EMR.

Table 1. Comparison of performance metrics for genomic alignment and SNP calling.

	E.coli Genome			Human Genome		
Metric	**EMR (n = 10)**	**GCE (n = 10)**	**p-value***	**EMR (n = 10)**	**GCE (n = 10)**	**p-value***
Wall clock time (mean)	0:46:30	0:31:50	<0.001	2:58:24	2:14:12	<0.001
Pre-processing short reads time (mean)	0:14:37	0:12:46	0.109	0:07:29	0:06:23	0.116
Alignment with Bowtie time (mean)	0:07:04	0:05:03	<0.001	1:51:06	1:15:07	0.003
Calling SNPs with SOAPsnp time (mean)	0:05:05	0:02:51	<0.001	0:35:31	0:29:31	0.033
Post-processing time (mean)	0:04:51	0:00:57	<0.001	0:01:23	0:01:03	<0.001
CPU user (mean %)	17.44±1.30	22.31±3.14	<0.001	43.80±1.87	58.05±6.20	<0.001
CPU idle (mean %)	72.75±1.23	65.76±4.63	<0.001	47.48±2.30	22.17±3.14	<0.001
CPU wio (mean %)	3.88±1.06	0.70±0.16	<0.001	1.86±0.19	4.54±1.82	0.001
Bytes in (MB/sec)	1.15±0.09	2.12±0.42	<0.001	1.58±0.07	2.00±0.19	<0.001
Memory free (GB)	2.19±0.13	6.17±0.42	<0.001	0.91±0.07	0.70±0.03	<0.001

All times are presented as hr:min:sec and remaining metrics are shown as mean ± standard deviation.
*Calculated by paired *t*-test.

to server congestion than EMR; though service usage data is difficult to obtain.

Discussion

Our findings confirm that cloud computing is an efficient and potentially cost-effective alternative for analysis of large genomic datasets. Cloud computing offers a dynamic, economical and versatile solution for large-scale computational analysis. There have been a number of recent advances in bioinformatic methods utilising cloud resources [4,9,13], and our results suggest that a standard genomic alignment is generally faster in GCE compared to EMR. The time differences identified could be attributed to the hardware used by the Google and Amazon for their cloud services. Amazon offers a 2.0 GHz Intel Xeon Sandy Bridge CPU, whilst Google uses a 2.6 GHz Intel Xeon Sandy Bridge CPU. This clock speed variability is considered the main contributing factor to the difference between the two undedicated platforms. It must also be noted that the resource requirements of Ganglia may have had a small impact on completion times [11].

There are a number of technical differences between GCE and EMR, which are important to consider when running standard bioinformatic pipelines. Running Crossbow on Amazon Web Services was simplified by an established support service, which provides an interface for establishing and running Hadoop clusters (Text S1). In contrast, there is currently no built-in support for GCE in Crossbow (Text S2). The current process to run a Crossbow job on GCE requires users to complete various steps such as installing and configuring the required software on each node in the cluster, transferring input data onto the Hadoop Distributed File System (HDFS), downloading results from the HDFS and terminating the cluster on completion. All of these steps are automatically performed by Crossbow on EMR. Python scripts offering similar functionality for GCE that Crossbow provides for EMR were created and are available (https://github.com/hewittlab/Crossbow-GCE-Hadoop).

While our findings confirm that cloud computing is an attractive alternative to the limitations imposed by the local environment, it is noteworthy that better performance metrics and lower cost were found with GCE compared to its established counterpart, Amazon's EMR. Currently, a major limitation of these services remains at the initial transfer of large datasets onto the hosted cloud platform [14]. To circumvent this in the future, sequencing service providers are likely to directly deposit data to a designated cloud service provider, thereby eliminating the need for the user to double handle the data transfer [15]. Once this issue is resolved, it is foreseen that demand for these services is likely to increase considerably, given the low cost, broad flexibility and good customer support for cloud services [15]. The development of

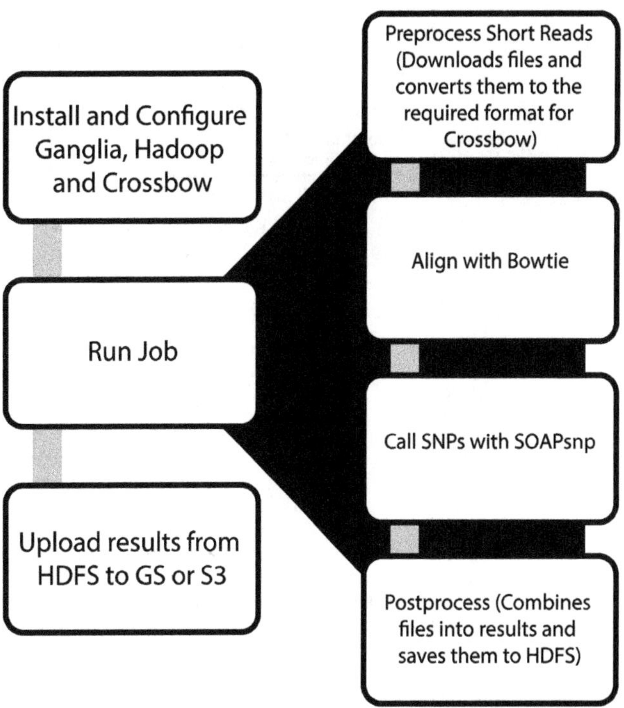

Figure 3. Analytical pipeline demarcating each step required to complete the Crossbow job in the cloud.

Table 2. Specification of used computational nodes for each system.

	Virtual Cores	Memory (GB)	Included Storage (GB)	Price (USD/Hour)^
Amazon Elastic Compute Cloud (EC2) + Elastic MapReduce (EMR) [c1.xlarge]	8	7	4×420	$0.640
Google Compute Engine [n1-highcpu-8]	8	7.2	0#	$0.352

^Date accessed: April to June 2014; prior to this period, pricing was $0.700 and $0.520 in Amazon and Google respectively.
#for each instance we added the minimum storage quota of 128 GB.

additional tools specific to genomic analysis in the cloud, which offer flexibility in choice of providers, is clearly required.

Methods

Datasets and Analytical Pipeline

We benchmarked two platforms by a single job that completed read alignment and variant calling stages of next generation sequencing analysis simultaneously on two independent cloud platforms. To investigate the impact of data size on undedicated cluster performance, one small (*Escherichia coli* CC102 strain (3 GB SRA file; Accession: SRX003267) and one large (a Han Chinese male genome (142 GB Fastq files; Accession: ERA000005) publicly available genomic dataset was selected for analysis [8,10]. For each job in this experiment, a parallel workflow was designed using Crossbow. This workflow included the following four steps: (1) Download and conversion of files; (2) Short read alignment with Bowtie; (3) SNP call with SOAPsnp; and (4) Combination of the results (Figure 3). Crossbow was the preferred genetic annotation tool in this experiment, as it has built in support for running via Amazon's EMR and Hadoop clusters [16].

Cluster construction and architecture

Instances were simultaneously established on Amazon's EMR (http://aws.amazon.com/ec2/; Accessed 2014 Aug 7) and GCE (http://cloud.google.com/products/compute-engine.html; Accessed 2014 Aug 7). Undedicated clusters were optimized by selection of computational nodes as suggested for Crossbow [9]. Nodes with equivalent specifications were selected for each system (Table 2), these being c1.xlarge node in EMR and the closest specification node n1-highcpu-8 in GCE. For the *E.coli* genome, two nodes (one master and one slave) were used on each platform. On the other hand, for the human genome, the cluster was built with 40 nodes (one master and 39 slaves). As GCE did not provide any included storage for each instance, a 128 GB drive (the default storage quota provided by GCE) was added for each node. This was at the additional cost of $0.04/GB/Month or $0.000056/GB/Hour (Jan to June 2014).

Each cluster was run using Apache Hadoop, an open-source implementation of the MapReduce algorithm [17]. MapReduce was used to organise distributed servers, manage the communication between servers and provide fault tolerance allowing tasks to be performed in parallel [18].

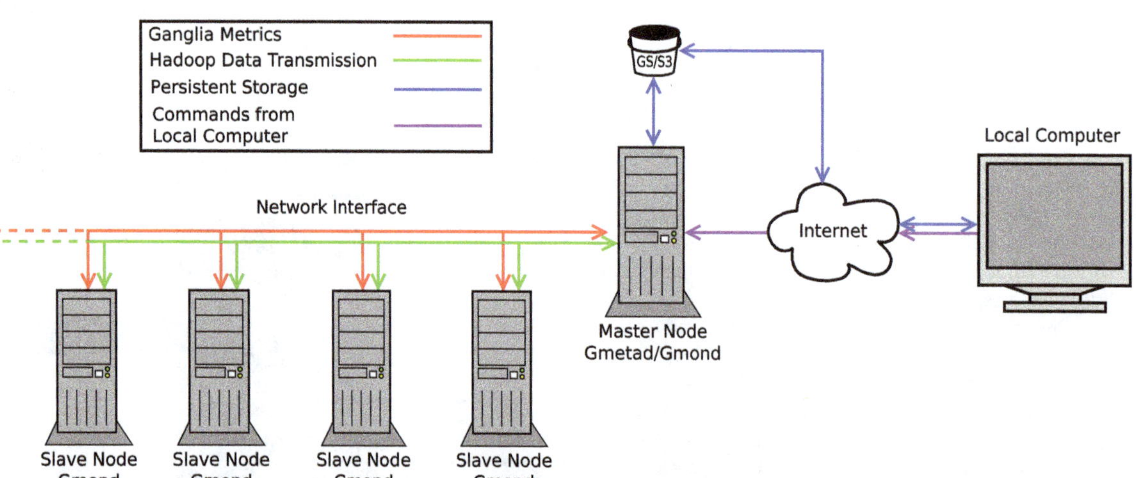

Figure 4. Directions and types of network transfers in our cloud-computing model. There are a variety of different network transfers between the nodes for each of the services in use in our model. Hadoop requires a bidirectional transmission of data between the master node and the slave nodes. This is required to coordinate the parallel processing of the cluster, and to allow for data transfer between nodes. Ganglia uses a unidirectional connection from the slave nodes to the master node to transfer the recorded metrics for storage and visualization. The persistent storage (provided by Amazon S3 (Simple Storage Service) or Google Storage, or an alternative method such as an FTP server) is accessed via the master node. The master node uses it to download input files for Crossbow, such as the manifest file and the reference Jar, and to use for persistent storage of the results of the Crossbow job as the instances destroy their storage on termination. Our local computer can also access the persistent storage via the Internet to allow access to upload the input files, or to download the results. The local computer needs to access the master node to initiate Crossbow. In EMR, this is replaced by a web interface and a JavaScript Object Notation Application Programming Interface (JSON API). In GCE, the user is required to remotely log in via Secure Shell (SSH) to commence the job.

To explore the effect of network activity differences between the platforms, each job was run simultaneously; same day (including weekdays and weekends) and same time. Detailed description of the set up and scripts to run the jobs can be found in Text S1 and Text S2.

Cluster Monitoring

In both EMR and GCE, multiple components of cloud infrastructure including CPU utilisation, memory usage and network speeds were monitored and recorded for each node using Ganglia. The default setting of Ganglia for distributing incoming requests is multicast mode; however, since EMR and GCE environments do not currently support multicast Ganglia, it was configured in unicast mode (Figure 4). The metric output files constructed in.rrd format were converted into.csv format with a Perl script (Text S3). For comparison between performance and costs between platforms, the Student t-test was undertaken using the statistical software R (R Foundation for Statistical Computing version 3.0.2; http://www.r-project.org/). In the analysis, cost of each run was calculated using current pricing (June 10th 2014); however, all *E.coli* runs and one human genome run were performed prior to a recent decrease in price on both platforms. The COV for runtime variability was calculated as the ratio of the standard deviation to the mean time (mins) for each system.

Supporting Information

Text S1 Uploading data and setting up an Amazon Web Services Elastic MapReduce (EMR) cluster.

Text S2 Scripts for configuration and running jobs on Google Compute Engine (GCE).

Text S3 Transformation of metric outputs from. RRD to. CSV format.

Author Contributions

Conceived and designed the experiments: SY AWH. Performed the experiments: SY GECG. Analyzed the data: SY GECG. Contributed reagents/materials/analysis tools: DAM AWH. Contributed to the writing of the manuscript: SY GECG DAM AWH.

References

1. Marx V (2013) Biology: The big challenges of big data. Nature 498: 255–260.
2. Patro R, Mount SM, Kingsford C (2014) Sailfish enables alignment-free isoform quantification from RNA-seq reads using lightweight algorithms. Nat Biotechnol 32: 462–464.
3. Schatz MC, Langmead B, Salzberg SL (2010) Cloud computing and the DNA data race. Nat Biotechnol 28: 691–693.
4. Angiuoli SV, White JR, Matalka M, White O, Fricke WF (2011) Resources and Costs for Microbial Sequence Analysis Evaluated Using Virtual Machines and Cloud Computing. PLoS ONE 6: e26624.
5. Fusaro VA, Patil P, Gafni E, Wall DP, Tonellato PJ (2011) Biomedical Cloud Computing With Amazon Web Services. PLoS Comput Biol 7: e1002147.
6. Drake N (2014) Cloud computing beckons scientists. Nature 509: 543–544.
7. Marx V (2013) Genomics in the clouds. Nat Meth 10: 941–945.
8. Parkhomchuk D, Amstislavskiy V, Soldatov A, Ogryzko V (2009) Use of high throughput sequencing to observe genome dynamics at a single cell level. Proc Natl Acad Sci USA 106: 20830–20835.
9. Langmead B, Schatz MC, Lin J, Pop M, Salzberg SL (2009) Searching for SNPs with cloud computing. Genome Biol 10: R134.
10. Wang J, Wang W, Li R, Li Y, Tian G, et al. (2008) The diploid genome sequence of an Asian individual. Nature 456: 60–65.
11. Massie ML, Chun BN, Culler DE (2004) The ganglia distributed monitoring system: design, implementation, and experience. Parallel Comput 30: 817–840.
12. Schad J, Dittrich J, Quiané-Ruiz J-A (2010) Runtime measurements in the cloud: observing, analyzing, and reducing variance. Proceedings VLDB Endowment 3: 460–471.
13. Onsongo G, Erdmann J, Spears MD, Chilton J, Beckman KB, et al. (2014) Implementation of Cloud based Next Generation Sequencing data analysis in a clinical laboratory. BMC Res Notes 7: 314.
14. Schadt EE, Linderman MD, Sorenson J, Lee L, Nolan GP (2010) Computational solutions to large-scale data management and analysis. Nat Rev Genet 11: 647–657.
15. Stein LD (2010) The case for cloud computing in genome informatics. Genome Biol 11: 207.
16. Crossbow project homepage. Available: http://bowtie-bio.sourceforge.net/crossbow/index.shtml. Accessed 2014 Aug 7.
17. Hadoop - Apache Software Foundation project homepage. Available: http://hadoop.apache.org/. Accessed 2014 Aug 7.
18. Dean J, Ghemawat S (2008) MapReduce. Commun ACM 51: 107–113.

Selective Microbial Genomic DNA Isolation Using Restriction Endonucleases

Helen E. Barnes, Guohong Liu, Christopher Q. Weston, Paula King, Long K. Pham, Shannon Waltz, Kimberly T. Helzer, Laura Day, Dan Sphar, Robert T. Yamamoto, R. Allyn Forsyth*

FLIR Systems, Inc., La Jolla, California, United States of America

Abstract

To improve the metagenomic analysis of complex microbiomes, we have repurposed restriction endonucleases as methyl specific DNA binding proteins. As an example, we use DpnI immobilized on magnetic beads. The ten minute extraction technique allows specific binding of genomes containing the DpnI G^{m6}ATC motif common in the genomic DNA of many bacteria including γ-proteobacteria. Using synthetic genome mixtures, we demonstrate 80% recovery of *Escherichia coli* genomic DNA even when only femtogram quantities are spiked into 10 µg of human DNA background. Binding is very specific with less than 0.5% of human DNA bound. Next Generation Sequencing of input and enriched synthetic mixtures results in over 100-fold enrichment of target genomes relative to human and plant DNA. We also show comparable enrichment when sequencing complex microbiomes such as those from creek water and human saliva. The technique can be broadened to other restriction enzymes allowing for the selective enrichment of trace and unculturable organisms from complex microbiomes and the stratification of organisms according to restriction enzyme enrichment.

Editor: Gunnar F Kaufmann, The Scripps Research Institute and Sorrento Therapeutics, Inc., United States of America

Funding: Funding for this research was provided in full through contract HSHQDC-10-C-00019 to FLIR Systems Inc. by the Department of Homeland Security, Science and Technology Directorate, http://www.dhs.gov/. Dr. Hultgren was the DHS technical representative who reviewed and approved the manuscript for publication. The funders had no additional roles in study design, data collection, analysis or preparation of the manuscript.

Competing Interests: The authors have read the journal's policy and have the following competing interests: HEB, GL, CQW, PK, LKP, SW, KTH, LD, DS, RTY and AF are all affiliated with the commercial company FLIR Systems Inc. which executed the research.

* Email: rforsyth@mail.sdsu.edu

Introduction

Next Generation Sequencing (NGS) has reinvigorated the understanding of the role that bacteria play as symbionts and pathogens of plants [1], insects [2], vertebrates [3] and in the environment [4,5]. NGS has broadened the study of the prokaryotic world beyond the small fraction of bacteria (less than 1%) thought to be culturable [6,7,8]. Using NGS for metagenomic studies, in which an entire sample of mixed organismal DNA is sequenced, has the advantage of querying the entire population of isolated DNA and overcomes many biases of other metagenomic methods such as microarray analysis or multiplex PCR. However, there are some drawbacks to using NGS metagenomic strategies. First, sensitivity to microbes may be decreased in the presence of large amounts of non-informative DNA (e.g. eukaryotic DNA). Second, typical metagenomic samples can contain hundreds of bacterial species making it difficult to parse and assemble genomes [9].

Recently developed methods to selectively enrich prokaryotic DNA exploit the 5-methylcytosine (5mC) in CpG sites of eukaryotes (mCpG), a modification largely absent in the bacterial world. One method uses a methyl-binding protein/Fc fusion protein to bind eukaryotic mCpG containing DNA and remove it from the mixture [10]. In an alternate approach, a truncated version of the human cytidylate-phosphate-deoxyguanylate protein has been used to bind non-methylated CpG sequences in bacterial DNA [11]. Bacteria have other stable epigenetic modifications in addition to 5mC including 6-methyladenine (6 mA) and 4-methylcytosine (4mC). The 6 mA modification was shown to occur at 94.1% of the 41,791 GATC sites in the *Escherichia coli* genome [12] and is widespread in prokaryotes but is otherwise reported only in ciliates and lower eukaryotes [13]. The DNA adenine methyltransferase (DamMT) directs adenine methylation within the context of GATC sequences and is found in at least one clade of bacteria consisting of the orders Enterobacteriales, Vibrionales, Aeromonadales, Pasteurellales and Alteromonadales [14]. In *E. coli*, GATC methylation influences chromosome replication, gene expression and mismatch repair. In *Vibrio cholerae* it is required for viability and in *Salmonella enterica* and *Haemophilus influenzae* it may act as a virulence factor [14]. 6 mA is also generated by some methyltransferases (MTases) as part of restriction modification systems [15]. Restriction endonucleases rely on methylation patterns to combat invasive genomes, particularly phage, while avoiding digestion of host DNA. Evolution has thus selected for enzymes with exquisite methylation sensitivity.

Here we present a restriction endonuclease-mediated DNA enrichment approach. DpnI is a methyl-directed restriction endonuclease that restricts DNA only when it is methylated on adenine residues within the GATC sequence [16,17]. We therefore anticipated that DpnI could distinguish bacterial genomes containing the G^{m6}ATC DNA modification from other bacterial and eukaryotic DNA. By manipulating the reaction conditions, we can use it to bind DNA without cutting. Since DpnI binds to DNA only when it is adenine methylated within GATC sites we predicted little or no binding to eukaryotic DNA and highly specific binding to DNA from DamMT+ bacteria. We demonstrate that DpnI can selectively enrich microbial DNA from synthetic and real-world samples. We extend our approach to a second restriction enzyme, DpnII that specifically enriches non-methylated GATC DNA (*e.g.* human genome). DNA enriched by this method can be used for PCR, qPCR and NGS analysis. The technique can enable the targeted enrichment of genomes from various microbiomes or the specific identification of pathogens from complex samples. We envision the use of restriction endonuclease binders to stratify complex metagenomic samples into groupings based on methylome signatures. This could link DNA fragments in otherwise poorly assembled contigs, aiding the reconstruction of genomes from unculturable organisms.

Materials and Methods

Genomic DNA was obtained from the ATCC with the exception of the following: *E. coli* K12 (Affymetrix, Santa Clara, CA); *Yersinia pestis*, *Franscisella tularensis*, *Burkholderia mallei*, *Burkholderia cepacia*, *Brucella abortus*, *Bacillus anthracis* (BEI Resources, Manassas, VA); and Human, Arabidopsis and Rice (Zyagen, San Diego, CA). Commercially available DpnI and pUC19 were purchased from NEB (Ipswich, MA).

DpnI purification and biotinylation

DpnI was purified essentially as described [18] with some modifications. BL21(DE3)A cells transformed with pLS252 were obtained from ATCC. Following a 5 hour expression, cells were harvested, resuspended in 20 mM Tris (pH 7.6), 0.5 M NaCl, 0.1 mM EDTA, 1 mM BME and lysed. Following centrifugation, nucleic acids were removed by polyethyleneimine (PEI) treatment. The PEI supernatant was treated with 75% ammonium sulfate and subjected to centrifugation. The pellet was resuspended in 20 mM Tris pH 7.6, 100 mM NaCl, 0.1 mM EDTA, 1 mM BME and dialyzed against Buffer A (20 mM Tris pH 7.6, 150 mM NaCl, 0.1 mM EDTA, 5 mM BME). The dialysate was loaded onto a phosphocellulose column and eluted with buffer B (20 mM Tris pH 7.6, 1 M NaCl, 0.1 mM EDTA, 5 mM BME). Fractions containing DpnI were pooled, dialyzed against buffer A and loaded onto an EMD sulfate column. Fractions containing DpnI were again pooled, dialyzed against buffer A and loaded onto an EMD sulfate column to remove any remaining contaminates.

DpnI was biotin labeled with the EZ-Link Sulf-NHS-biotin kit (Pierce, Rockford, IL) following the manufacturer's protocol. The extent of biotinylation was evaluated using the HABA assay (Pierce). Each mole of DpnI was found to contain 4-5 mole of biotin.

Restriction activity assay

1 µg of pUC19 was digested in the presence of 100 ng of purified DpnI, DpnI-biotin or with 20 U of commercial DpnI in 20 mM Tris-HCl (pH 7.6), 50 nM NaCl, 10 mM $CaCl_2$, with or without 20 mM $MgCl_2$ for 1 hour at 37°C. Reactions were stopped by the addition of loading buffer containing SYBR green (Life Technologies, Carlsbad, CA). DNA was separated on a 1.5% TBE agarose gel.

Generation of template DNA

DNA was PCR amplified from pUC19 using primers (IDT, San Diego, CA) that resulted in a 477 nt fragment (Forward- TC-TGCGCTCTGCTGAAGCCAGTTAC; reverse- GCTGATAA-ATCTGGAGCCGGTGAGC) or a 651 nt fragment (forward-GGCAGCAGCCACTGGTAACAGGATT; reverse- GATG-GAGGCGGATAAAGTTGCAGGA). The 477 nt fragment was treated with *dam* methyltransferase (NEB) resulting in DNA containing the G^{m6}ATC modification. All fragments were gel-purified using agarose gel electrophoresis and the MinElute Gel Extraction kit (Qiagen, Venlo, Limburg).

Electrophoretic mobility shift assay

EMSA was carried out as previously described [19] with some modifications. FAM-labeled duplex oligonucleotide containing one G^{m6}ATC site with the top strand sequence FAM-GCAGG-^{m6}ATCAACAGTCACACT (TriLink, San Diego, CA) was incubated with DpnI (or DpnI-biotin) in the presence of 20 mM Tris-HCl, 50 mM NaCl, 10 mM $CaCl_2$, 1 mg/ml BSA and 10 µg/ml salmon sperm DNA for 30 minutes at room temperature. Glycerol was added to a final concentration of 10% and the samples loaded onto a 20% TBE acrylamide gel (Life Technologies) that had been pre-run for 2 hours at 4°C with TBE. Samples were subjected to separation at 200 V for 1.75 hours. FAM-labeled DNA was imaged using an AlphaImager (Protein Simple, Santa Clara, CA).

DpnI pull-down assay

Preparation of DpnI-coated magnetic beads. 20 µl streptavidin magnetic beads (NEB) were washed twice with Binding Buffer (10 mM Tris pH 7.9, 50 mM NaCl, 10 mM $CaCl_2$, 0.01% Tween 20). Biotinylated DpnI was added to the beads at 10 ng DpnI/µl beads. After mixing by pipetting, the beads were washed twice with Binding Buffer and used for binding reactions.

DNA pull-down. DNA samples were prepared in Binding Buffer. The assay was performed either in 1.7 ml microcentrifuge tubes or in a 96-well plate. 50 µl DNA samples were added to the DpnI coated beads. The beads were mixed by end-over-end rotation or on a plate shaker for 5 minutes to 1 hour. Magnetic beads were separated using either a tube magnetic stand (Life Technologies) or a plate magnet (Millipore, Billerica, MA). The beads were washed once with Wash Buffer (10 mM Tris pH 7.9, 500 mM NaCl, 10 mM $CaCl_2$, 0.1% Tween 20) followed by one Binding Buffer wash. Beads were resuspended in 50 µl of Binding Buffer for qPCR analysis.

For gel analysis and next-generation library preparation, the DNA was eluted from beads by incubation with 50 µl 5 M guanidinium thiocyanate at room temperature for 5 minutes. The eluent was transferred to a 3500 MWCO dialysis tube (Thermo Scientific, Waltham, MA) and dialyzed against distilled water for 1 hour at room temperature.

Genomic DNA qPCR analysis

Primers were synthesized by IDT and probes were made by Life Technologies. Reactions were prepared using the QuantiProbe FAST PCR Kit (Qiagen) except for the DYZ assay which was prepared with TaqMan Universal Master Mix (Life Technologies). Reactions were cycled once at 95°C for 3 minutes followed by 40 cycles of 95°C for 3 seconds and 60°C for 30 seconds on an ABI 7300. The universal bacterial 16S assay has been described previously [20]. Assays specific for Human RNaseP, human TERT and Arabidopsis ACT2 gene were obtained from Life Tech. *E. coli* 16S assay: forward -CCAGGGCTACACACGT-GCTA; reverse - TCTCGCGAGGTCGCTTCT; probe - AAT-GGCGCATACAAA. Human DYZ assay: forward - TCGAGTG-CATTCCATTCCG; reverse - ATGGAATGGCATCAAACG-GAA; probe - TGGCTGTCCATTCCA. Relative abundance was calculated using either a standard curve or the delta Ct method. For the universal 16S assay, standard curves were generated using the genomic DNA of the organism being tested to correct for the varied copy number of the 16S gene.

Preparation of synthetic mixture

Bacterial genomes were obtained through the ATCC or BEI as listed and concentrations determined using the Qubit dsDNA HS assay (Life Technologies). Bacterial genomes were diluted with water to obtain the desired concentrations (Table 1) and validated again using Qubit dsDNA HS assay before assembly of the final synthetic mix.

DNA isolation from saliva

The PowerSoil DNA isolation kit (MO BIO Laboratories, Carlsbad, CA) was used to extract DNA from 1 ml of pooled human saliva (BioReclamation, Farmingdale, NY). The DNA was eluted in DpnI Binding Buffer and 400 ng of the DNA was subjected to the DpnI pull-down assay. The input, unbound, and bound/eluted fractions were used to prepare sequencing libraries.

DNA isolation from creek water

A 1000 ml water sample was collected from a creek 25 meters downstream from a sedimentation pond used for primary passive treatment of ground water run-off. A 100 ml aliquot was filtered over a 0.2 µm Nalgene sterile analytical filter unit (Thermo Scientific) prior to DNA extraction with the PowerWater DNA Isolation Kit (MO BIO Laboratories). A 150 ng aliquot of the DNA was subjected to the DpnI pull-down assay. The input, unbound, and bound/eluted fractions were used to prepare sequencing libraries.

Library preparation and sequencing

The Nextera DNA Sample Preparation Kit (Illumina, San Diego, CA) was used to prepare libraries from input, unbound, and bound/eluted fractions from DpnI pull-down assays. Manufacturer's instructions were followed for the library preparation except for recommended number of PCR cycles, which were varied according to the amount of DNA. For the synthetic mixture, they were as follows: Input – 7 cycles, DpnI bound – 10 cycles, DpnI unbound – 7 cycles. Libraries were sequenced following the manufacturer's instructions for the HiSeq 2500 Rapid Run mode to obtain 50 nucleotide read lengths. The files

Table 1. DpnI pulls down genomic DNA from different organisms with varying efficiency.

Family	Organism	Strain	Gram	DamMT	DpnI Pull Down Efficiency*
Aeromonadaceae	*Aeromonas hydrophila*	ATCC 7966	-	+	++
Enterobacteriaceae	*Enterobacter cloacae*	ATCC 13047	-	+	++
	Escherichia coli	K12	-	+	++
	Klebsiella pneumoniae	ATCC 700721	-	+	++
	Proteus mirabilis	ATCC 12453	-	+	++
	Salmonella typhimurium	SU453	-	+	++
	Serratia marcescens subsp. marcescens	ATCC 13880	-	+	++
	Yersinia pestis	China CDC	-	+	++
	Yersinia pseudotuberculosis	ATCC 13979	-	+	++
Pasterellaceae	*Haemophilus influenzae*	ATCC 51907	-	+	++
	Haemophilus parahaemolyticus	ATCC 10014	-	+	++
	Haemophilus parainfluenzae	ATCC 33392	-	+	++
Legionellaceae	*Legionella pneumophila*	ATCC 33152	-	(+)	++
Campylobacteraceae	*Campylobacter jejuni subsp. jejuni*	ATCC 700819	-	(+)	+
Helicobacteraceae	*Helicobacter pylori*	ATCC 700824, J99	-	(-)	+
Burkholderiaceae	*Burkholderia mallei*	CRP 23344	-	(-)	+
	Burkholderia cepacia	CRP BRUK102	-	-	+
Brucellaceae	*Brucella abortus*	CRP 2308	-	-	+
Pseudomonadaceae	*Pseudomonas aeruginosa*	ATCC 47085	-	-	+
Bacillaceae	*Bacillus anthracis*	Sterne	+	-	-
Enterococcaceae	*Enterococcus faecium*	ATCC 51559	+	-	-
Eukaryota - Fungi	*Aspergillus fumigatus*	MYA-4609	N/A	-	+/-
Eukaryota - Brassicaceae	*Arabidopsis thaliana*		N/A	-	-
Eukaryota - Hominidae	*Homo sapiens*, male		N/A	-	-

*Recovery as compared to input by qPCR.
-Less than 2%, +/-2–10%, +10–50%, ++50–100%.

corresponding to all the raw reads generated in this study are publicly available at the NCBI Short Read Archive (SRP044748).

Sequence analysis

For microbial taxa identification, Illumina data sets were analyzed by an automated pipeline (ZovaSeq from Zova Systems, LLC, San Diego CA) in which identifying sequence reads are assigned to specific microbial taxa when a given read length is found to occur uniquely within the taxa as defined by the NCBI taxonomy database [21,22]. Relative abundance was calculated using two methods which gave equivalent results: tallying the number of ZovaSeq identifying reads for each bacterial taxa or by using Bowtie 1.0.0 to map reads to all identified organisms in the sample by perfect match. For known higher eukaryotes in the sample (*Homo sapiens*, *Oryza sativa*) reads were mapped using Bowtie 1.0.0 with parameters allowing 2 mismatches in a 28 bp seed region.

Relative enrichment of the DpnI bound versus input samples were determined by the following equation:

$$Enrichment = \frac{BacterialREADCount_{BOUND} \div TotalREADCount_{BOUND}}{BacterialREADCount_{INPUT} \div TotalREADCount_{INPUT}}$$

Relative enrichment as compared with human DNA was determined by dividing DpnI enrichment for the organism of interest by DpnI enrichment for human.

Results

6mA is a frequent prokaryotic DNA modification that has only rarely been reported in eukaryotic genomes [13]. Since DpnI is one of a limited number of methyl-directed Type II restriction endonucleases that depend on the presence of 6 mA to bind and cut its target DNA sequence [16,17], we surmised that it could effectively bind $G^{m6}ATC$ containing genomes for enrichment, allowing segregation away from non-methylated GATC DNA. To test this, we covalently bound biotin to DpnI to facilitate immobilization of the enzyme onto streptavidin coated particles. This necessitated purification of DpnI since commercial sources for the enzyme are dilute and contain other proteins that prevent us from selectively biotinylating the restriction enzyme. The activity of purified DpnI both before and after biotinylation was analyzed by restriction digestion of pUC19 isolated from DamMT+ *E. coli*. DpnI and DpnI-biotin were both found to be active when compared to commercially available enzyme, with a slight reduction in activity observed when the protein was biotinylated (Figure 1A).

To effectively bind and separate $G^{m6}ATC$ DNA fragments from a mixture, the cleavage activity of DpnI must be prevented. We tested DpnI digestion of pUC19 in the absence of magnesium ions and did not observe cleavage activity, as previously reported [16]. Since the absence of magnesium might also affect the binding of DpnI to its target, we tested both DpnI and DpnI-biotin in an electrophoretic mobility shift assay. A FAM-labeled oligonucleotide duplex containing a single $G^{m6}ATC$ sequence was incubated with increasing amounts of DpnI and DpnI-biotin. Both DpnI and DpnI-biotin are able to bind and shift $G^{m6}ATC$ containing DNA in the absence of magnesium and no noticeable decrease in the binding affinity is observed when DpnI is biotinylated (Figure 1B).

To test our hypothesis that DpnI could be used to separate $G^{m6}ATC$ containing DNA from fragments without $G^{m6}ATC$ sites, we used a mixture of a 477 bp Dam-methylated DNA fragment and a 651 bp non-methylated fragment. The two fragments both contained seven GATC sites and were derived from overlapping regions in pUC19 to minimize bias caused by sequence differences. DpnI-biotin was immobilized onto streptavidin-magnetic particles and titrated into a mixture of the two fragments. DNA that bound to the DpnI-coated particles was eluted and desalted. All fractions were separated by electrophoresis on an agarose gel. An increase in the amount of DpnI-beads resulted in further depletion of the 477 bp fragment. The eluted fractions contained only the 477 bp fragment (Figure 1C, lanes 6–9) leaving the non-methylated 651 bp fragment in the supernatant (Figure 1C, lanes 2–5). Thus immobilized DpnI specifically bound $G^{m6}ATC$ containing DNA (477 bp) which could be purified away from other fragments.

After observing efficient segregation of specific $G^{m6}ATC$ DNA fragments, we investigated whether DpnI-biotin was suitable for isolating a $G^{m6}ATC$-containing genome when mixed with GATC-containing genomes. A synthetic mix containing 1 ng *E. coli* and 500 ng of Human genomic DNA was prepared and incubated with immobilized DpnI. After separation, fractions were analyzed using qPCR. We found that DpnI-coated particles isolated *E. coli* genomic DNA with high efficiency (Figure 2A), binding nearly 80% in 5 minutes. Enrichment was also specific, with 99.6% of Human DNA remaining unbound. Comparable isolation efficiency was observed for the DNA mixtures prepared in buffers ranging from pH 4 to 10 (Figure 2B). Additionally when fragment sizes were at least 3 kb, DpnI binding was not significantly affected, but did decrease with smaller fragments (Figure S3).

The relative genomic composition of complex samples varies widely. We therefore tested the limits of DpnI separation by incubating various amounts of *E. coli* and human DNA with immobilized DpnI. To test the sensitivity of DpnI separation, the level of human DNA was held constant at 1 μg and *E. coli* DNA was titrated from 1 ng to 10 fg. We observed approximately 80% recovery of *E. coli* DNA and rejection of 99.5% of human DNA. Sensitivity was observed to 10 fg *E. coli* DNA, the detection limit of the qPCR assay used (Figure 2C). This demonstrates efficient separation by DpnI of $G^{m6}ATC$ containing DNA when present at as low as 10^{-8} of the level of eukaryotic DNA.

We next tested the ability of DpnI to exclude human DNA present at high concentrations. When the concentration of *E. coli* DNA was held constant at 1 ng while increasing the concentration of human DNA, we observed *E. coli* DNA recovery as high as 82% and exceeding 60% even in the presence of 10 μg of human DNA, a 10,000-fold difference (Figure 2D). These results demonstrate that DpnI DNA segregation is effective and efficient with differing ratios of target versus non-target DNA.

We next examined how efficiently DpnI binds genomes from a variety of organisms including some that are clinically relevant [23]. For each organism of interest, 1 ng of bacterial genomic DNA was combined with 1 μg of human DNA. DNA mixtures were incubated with immobilized DpnI. Following segregation, DNA in the DpnI bound and unbound fractions were analyzed by qPCR. DpnI successfully bound and separated genomic DNA from gram-negative organisms known to express DamMT (Table 1). The range of recovery was between 50% and 100% of the measured input. For gram-negative bacteria not known to have a DamMT gene, recovery was lower, from 10% to 45% of the measured input, but still significantly higher than binding to human DNA. Binding of gram-positive bacterial DNA was less than 3% and binding to eukaryotic DNA was below 0.5%. We conclude that DpnI can be used to efficiently bind and segregate

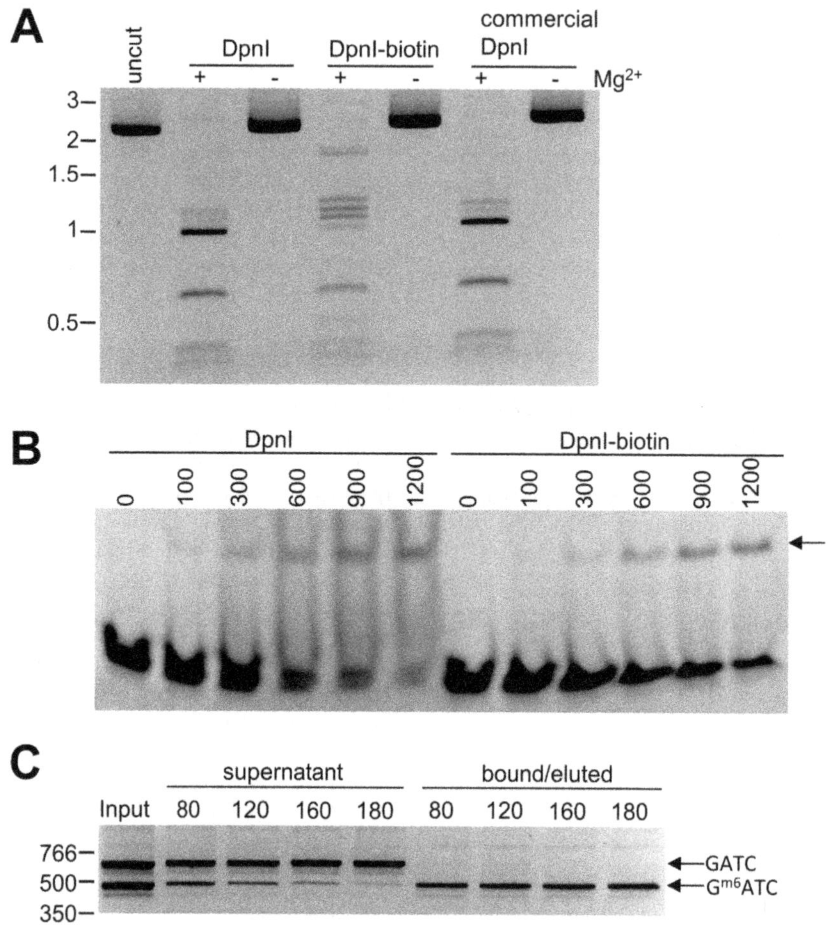

Figure 1. Analysis of biotinylated DpnI. (A) pUC19 was incubated with DpnI, DpnI-biotin or commercially sourced DpnI in the presence or absence of 10 mM magnesium chloride. The digested fragments were separated on a 1.5% agarose gel. (B) A FAM-labeled DNA duplex containing one $G^{m6}ATC$ site was incubated with increasing amounts of DpnI or DpnI-biotin (0 to 1200 ng). The reactions were separated on a 20% TBE gel and analyzed with fluorescence imaging. (C) An unmethylated 651 bp DNA fragment and a Dam-methylated 477 bp DNA fragment were combined and incubated with increasing amounts of immobilized DpnI-biotin (80–180 µl). DNA was eluted using GTC and desalted. All fractions were separated on a 3% agarose gel.

genomes from a wide variety of organisms with very little binding to eukaryotic DNA.

To test how well DpnI enrichment can improve the coverage and read depth of prokaryotic DNA in a mixture, we designed a synthetic mixture of genomic DNA that included both eukaryotic and prokaryotic DNA (Table 2). Human DNA made up the bulk of the mixture at over 97% by weight. DNA from rice (1%) and *Aspergillus* (1%) was added to represent plant and fungal genomes, respectively. Microbe genomes were added in a pair-wise fashion. Each pair consisted of an equal amount of DNA from a DamMT+ and a DamMT- organism, and subsequent pairs were diluted ten-fold to test the limit of DpnI enrichment. The DNA mixture was subjected to DpnI segregation. The DNA from the bulk mixture, the unbound fraction and the bound/eluted fraction were used to prepare sequencing libraries. We found that the number of reads from eukaryotes was dramatically reduced in the DpnI-bound fraction (Figure S5). Reads mapping to the human genome made up 59% of the mapped reads in the synthetic mix input but only 5% in the DpnI-bound fraction. The reads mapping to *Oryza* (rice) were also greatly reduced, from 31% of the mapped reads in the input sample to 2.5% of the mapped reads in the bound fraction (Figure 3A).

Surprisingly, we observed that DNAs from all microbial organisms, not just from DamMT+ bacteria, were enriched compared to human and rice (Figure 3B, 3C and Figure S5). DNA from DamMT+ bacteria was most effectively enriched, up to 70-fold compared to input levels and up to 800-fold when directly compared to human (Figure 3B). The *E. coli* DNA in the mixture was enriched from comprising less than 1% of the reads in the sample input to over 50% of the reads in the bound fraction. This resulted in improved sequencing coverage of the *E. coli* genome. Only 67% of the *E. coli* genome sequence was covered by reads in the input sample. Following DpnI enrichment, >99% of the *E. coli* genome sequence was covered, with a depth of coverage averaging 40 reads. Furthermore, there was no discernable coverage bias in the enriched genomes (Figure 4B), indicating that DpnI enrichment can be used to greatly improve whole genome sequencing. A similar pattern of enrichment was observed for the remaining DamMT+ organisms.

As an exemplar clinical sample, DNA in saliva is overwhelmingly derived from human cells [24], with prokaryotic DNA making up less than 4%. We isolated DNA from saliva and performed a DpnI separation. The input, bound/eluted and unbound fractions were sequenced. Whereas human reads made

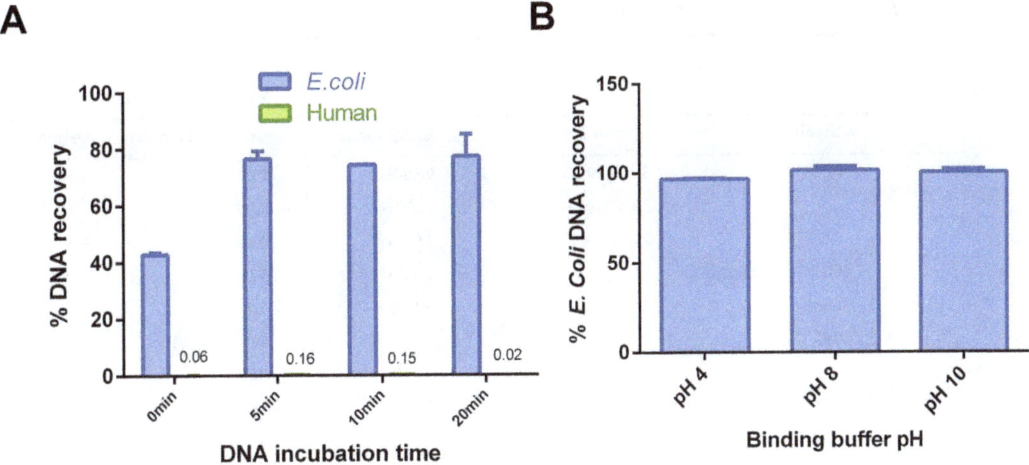

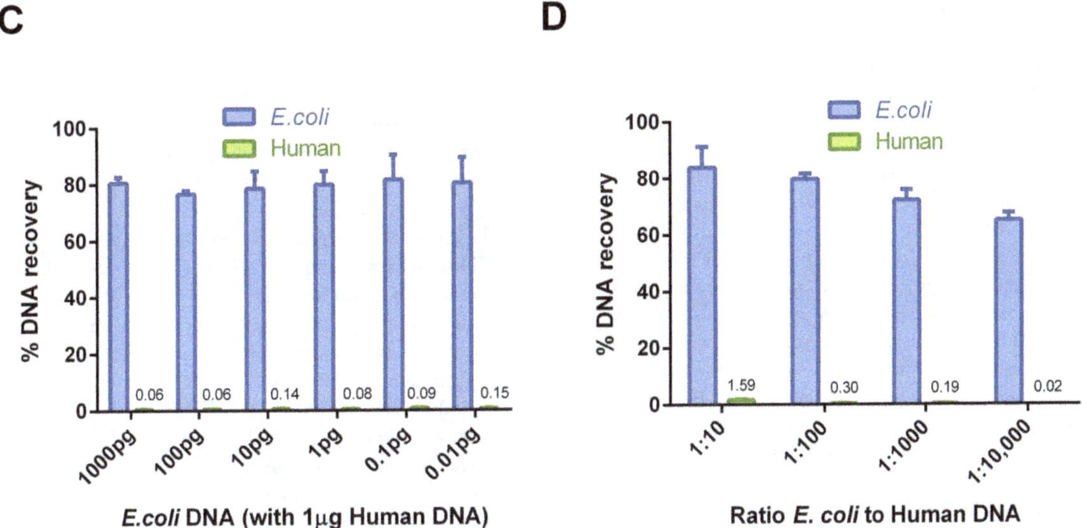

Figure 2. Efficiency and range of DpnI pull-down. (A) Immobilized DpnI was incubated with a mixture of *E. coli* and human DNA for varying amounts of time. 40% of *E. coli* DNA binding occurs on less than one minute. Less than 0.2% of human DNA binds to DpnI. (B) Immobilized DpnI was incubated with *E. coli* DNA in buffers with pH of 4, 8 or 10. Almost all *E.coli* DNA was recovered in the range of pH tested. (C) A fixed amount of human DNA (1 μg) was mixed with decreasing levels of *E. coli* DNA and then incubated with immobilized DpnI. Approximately 80% of *E. coli* DNA is recovered down to levels of 10 fg. All data shown is the average of three experiments. (D) A fixed amount of *E. coli* DNA (1 ng) was mixed with increasing amounts of human DNA and then incubated with immobilized DpnI. There is a slight decrease in the recovery of *E. coli* DNA with increasing amounts of human DNA. However, even when human DNA is present at 10,000x, DpnI recovers over 70% of *E. coli* DNA.

up over 75% of the total reads in the input sample, following DpnI enrichment less than 5% of the total reads were human (Figure 5A). Prokaryotic reads increased from less than 5% of the total reads to over 50% in the DpnI-enriched fraction. There are a significant number of reads that could not be assigned to any organism. This is likely due to the high number of unsequenced organisms in the sample. The most abundant genera in the sample were *Haemophilus, Neisseria, Veillonella, Prevotella* and *Streptococcus*. Together these five genera comprised 87% of reads mapped to prokaryotes. As expected, a subset of the organisms was highly enriched in the bound fraction while some organisms were not enriched and yet another set were depleted (Figure 5B). *Haemophilus, Aggregatibacter, Actinobacillus, Vibrio* and *Treponema* were all enriched ten-fold in the bound fraction compared to input (Figure 5B). *Haemophilus parainfluenzae* was a major component of both the input and bound fractions and was enriched 36-fold compared to input. Though not enriched, *Prevotella*, an organism closely associated with dental carries [25], is still a major component of the bound fraction. Other organisms were undetectable in the input fraction but had mapped reads in the bound fraction (Figure 5B and Table S1).

Table 2. Genome mix used for sequencing and relative enrichment results.

Organism in Input Mixture				Relative Enrichment	
Species	**Strain**	**Genome Size**	**% by mass**	**Bound vs. Input**	**Organism vs. Human**
Homo sapiens		3,209,290,000	96.80%	0.09	N/A
Oryza sativa		382,780,000	1.00%	0.08	0.9
Aspergillus fumigatus	Af293	29,390,000	1.00%	0.8	9.1
Escherichia coli	O157:H7 str. EDL933	5,620,000	1.00%	57	654
Bacillus anthracis	Sterne	5,228,663	0.10%	1.2	13.5
Salmonella enterica	Ty2	4,790,000	0.10%	58	666.1
Streptococcus pneumoniae	R6	2,038,615	0.01%	0.7	7.8
Shigella flexneri	2457T	4,600,000	0.01%	58.9	676.2
Staphylococcus aureus	Mu50	2,903,147	0.001%	0.9	9.8
Yersinia pestis	A1122	4,660,000	0.001%	72.1	827.2
Enterococcus faecalis	V583	3,360,000	0.0001%	ND*	ND*
Vibrio cholera	N16961	3,745,000	0.0001%	75.4	865.2
Pantoea ananatis		N/A**	N/A**	55.9	641.7

ND: Not determined. N/A: Not applicable.
**E. faecalis* was not detectable in the Input fraction.
***P. ananatis* was not knowingly added in the sample mix but is a probable contaminant of the rice genome (*O. sativa*).

We next isolated DNA from a water sample collected from a creek after a heavy rain and subjected it to segregation by DpnI. The identified genera segregated into three distinct groups in the bound fraction: highly enriched, slightly enriched and non-enriched (Figure 6A). Eleven genera were enriched over 20-fold compared to input. Of these, *Aeromonas, Shewanella, Pantoea, Enterobacter* and *Rahnella* were the most abundant in the bound fraction. For example, we found a high number of identifying reads in the bound fraction that mapped to the fish pathogen *Aeromonas salmonicida* (over 18% of mapped reads and 0.48% of the total reads). The same organism represented less than 6% of mapped reads and 0.014% of the total reads in the input (Figure S1). The coverage we observed suggests that the sequenced organism is a close relative of *Aeromonas salmonicida*. DpnI segregation resulted in nearly 35-fold enrichment of this organism's DNA.

Having succeeded in efficiently segregating DNA genomes with DpnI, we investigated whether this approach might be applicable to other restriction enzymes. DpnII is known to have the opposite activity of DpnI in that it recognizes and cuts only non-methylated GATC sequences and DpnII activity is blocked by 6 mA. We therefore expected DpnII to bind to human, but not *E. coli* genomic DNA. Similar to our experiments with DpnI, we immobilized DpnII to test its ability to separate a mixture of 1 ng of human DNA and 500 ng of *E. coli* DNA. DpnII was able to enrich the human DNA with minimal binding to *E. coli* DNA (Figure S2). Therefore restriction endonuclease-mediated DNA separation is not limited to DpnI.

Discussion

Type II restriction endonucleases have been selected during evolution to ensure they do not cut their own DNA, a suicidal event, while quickly binding to and digesting any foreign DNA that lacks the correct methylation pattern [26]. We demonstrate that manipulation of *in vitro* conditions enables DpnI to bind but not cut DNA containing its target sequence. While the binding affinity of DpnI has not been determined, several restriction enzymes have been measured in the picomolar [27,28] to nanomolar range [29,30] and our results support the use of restriction enzymes as strong and specific DNA binding proteins.

DpnI binding to target DNA was rapid, with 75% of *E. coli* DNA bound after only 5 minutes (Figure 2A). We also observed highly specific binding with over 99.5% of human DNA excluded and over 80% of targeted *E. coli* DNA binding (Figure 2A, C and D). This rapid and exquisite target discrimination by DpnI *in vitro* is a reflection of the natural ability of restriction endonucleases to quickly scan and locate target sequences in large amounts of DNA *in vivo* [31]. Immobilized DpnI can be used to differentially bind and segregate prokaryotic DNA present at 1/10,000 the level of eukaryotic DNA (Figure 2D). Efficient removal of background human genetic material enables pathogen DNA to be concentrated to achieve sensitive detection which could be particularly useful for un-culturable pathogenic bacteria. This feature could be exploited for the diagnostic detection of trace amount of pathogens in clinical samples such as blood from patients with septicemia, a serious infection that lacks an early detection method [32].

One critique of using a methyl-directed binding protein to enrich DNA is that the process may introduce coverage bias with more reads observed in close proximity to the protein binding site. However, when samples were separated by DpnI and then analyzed by NGS, DpnI enrichment resulted in very low sequence coverage biases (Figure 4). The even coverage is likely due to the frequency and distribution of DpnI binding sites in target DNA. For example, in *E. coli* O157:H7, there are approximately 42,000 GATC sites, 94% of which have been shown by SMRT sequencing to be adenine methylated with an average gap between GATC sites of about 250 bp [12]. Additionally, DpnI segregation generated low biases when input DNA fragments were above 3 kb (Figure S3). Thus typical DNA isolation procedures are sufficient to achieve efficient DpnI segregation. Biases could arise however if smaller bacterial fragments, from degraded DNA for instance, are present.

We predicted little or no binding to eukaryotic DNA and highly specific binding to DNA from DamMT+ bacteria. We did not

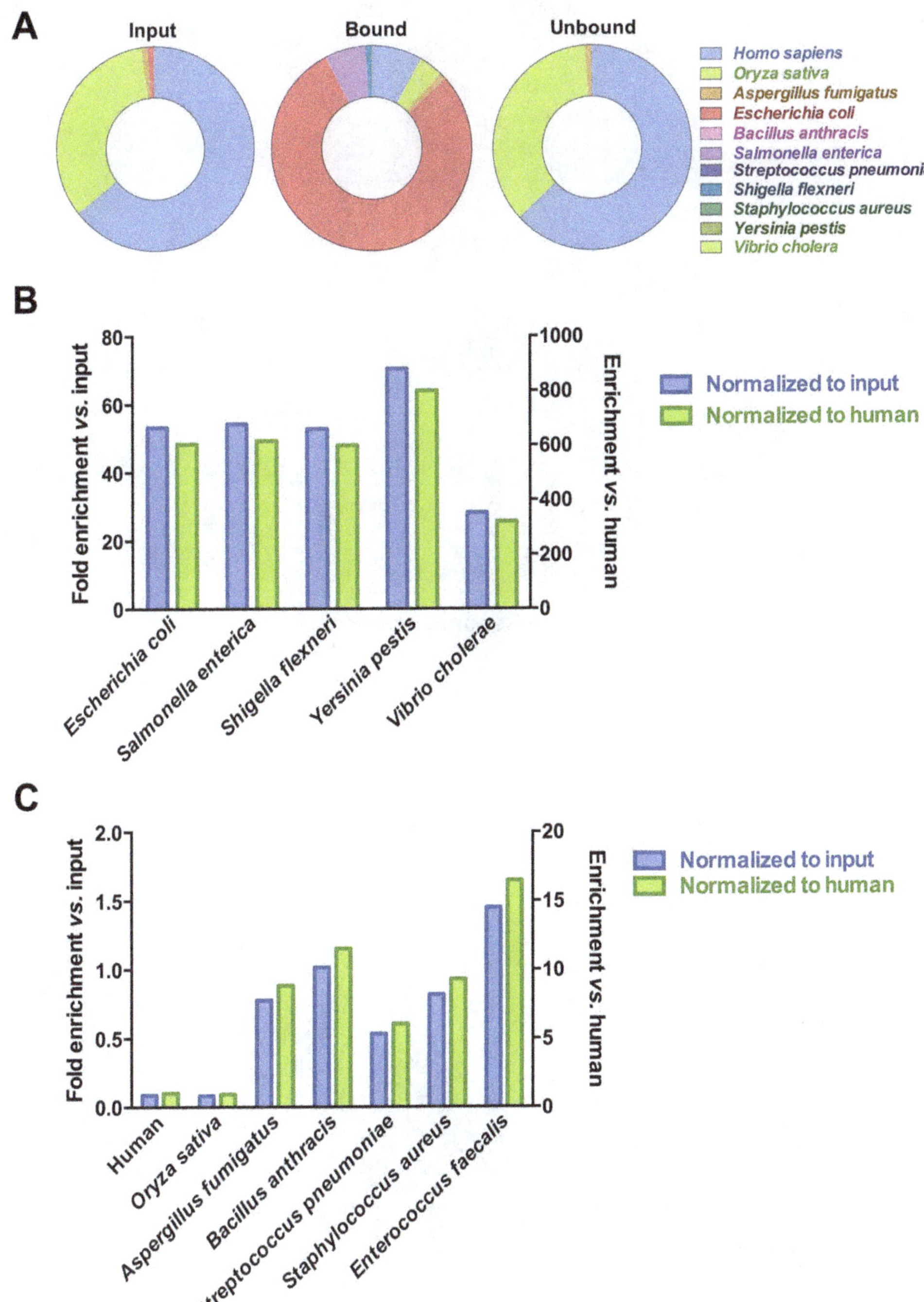

Figure 3. DpnI enriches prokaryotic DNA as determined by NGS. (A) NGS reads from the input, bound and unbound fractions of the synthetic mix. Reads from the input map overwhelmingly to human and rice, with less than 10% mapping to prokaryotes in the synthetic mixture. Less than 10% of the reads from the bound fraction map to human with the majority mapping to *E. coli*. (B) DamMT+ genomes are enriched 30 to 70-fold versus their input levels, and 300 to 800-fold versus human DNA. (C) Genomes that lack DamMT are enriched when compared to human and rice.

anticipate the low level binding of DpnI to micro-organisms not known to contain $G^{m6}ATC$ sites (Figure 3B). This *in vitro* non-canonical binding may simply reflect a difference in DNA binding affinity compared to the more rigorously studied specificity of restriction activity. Published factors known to affect restriction specificity of DpnI include the presence of non-GATC sequences that contain a methylated adenine residue [33] and DNA topology effects [34]. Alternatively, DNA modifications other than 6mA may be affecting DpnI binding specificity. Although DpnI needs a $G^{m6}ATC$ site to cut, it appears that at least some amount of binding occurs when that pattern is absent and that binding decreases in the presence of CpG methylation. We observed that when the *Aspergillus fumigatus* genome which is not known to contain $G^{m6}ATC$ is treated with a CpG methyltransferase, binding drops significantly (Figure S4). It is unknown whether this is a differential feature of binding versus digestion or an artifact of biotinylation. A more in-depth study of DpnI binding

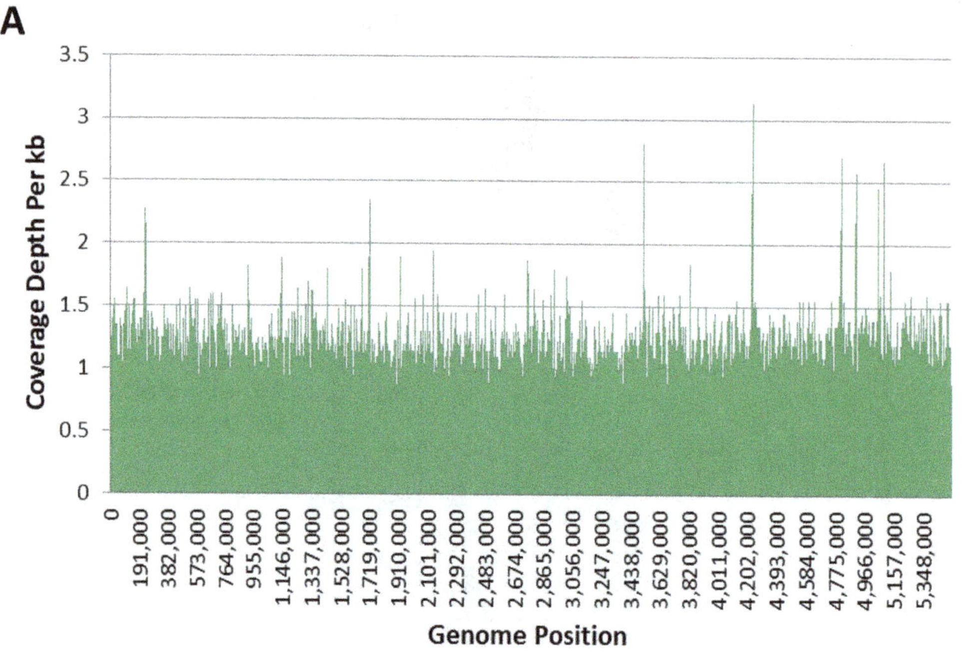

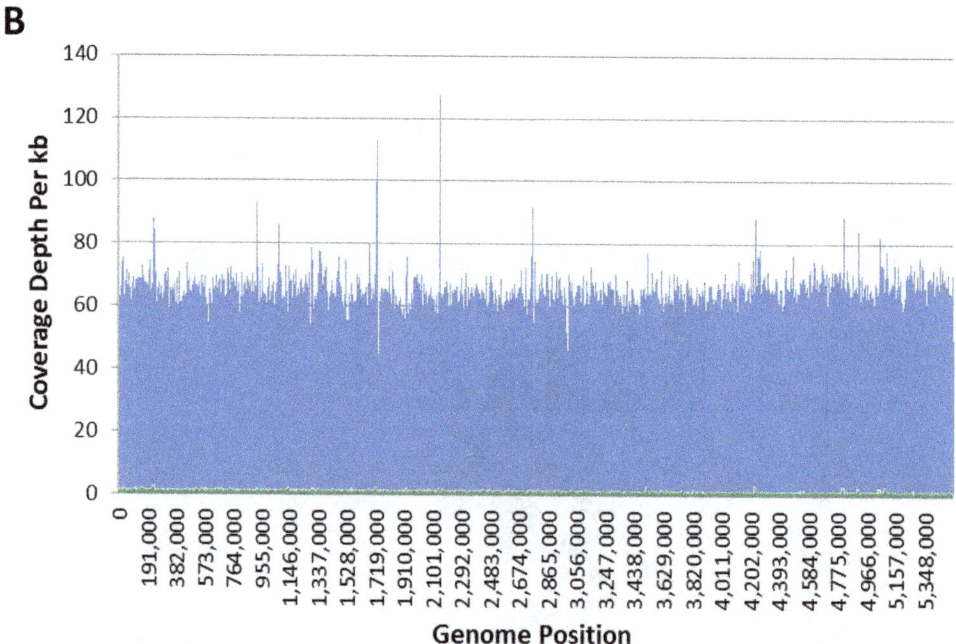

Figure 4. NGS coverage maps for *E. coli* from input (A) and bound (B) fractions of the synthetic mixture. Reads were mapped to *E. coli* O157:H7 EDL933 and binned into 1000 bp bins. (A) The average depth of coverage is 0.5 for *E. coli* in the input fraction (green), with 62% of the genome covered. (B) For the bound fraction (blue), the average depth of coverage increases to 60 and 99.5% of the genome is covered. The input fraction (green) is also plotted here for comparison to the bound fraction at the same scale.

patterns is needed to better understand the binding to DNA from DamMT- organisms.

Observations to date suggest that methyl signatures created by restriction modification systems are only sporadically distributed amongst microbial taxa [26,35]. In contrast, orphan MTases, such as DamMT, are often conserved across extensive groups of bacteria which rely on these methylation patterns to control crucial cellular processes like chromosome replication [14,36]. We consider DamMT+ bacteria to be part of a more expansive methylome which would include organisms which methylate at GATC sites in other contexts (*e.g. B. amyloliquefaciens*, BamHI $GG^{6m}ATCC$). The broad and deep genomic coverage consistently observed when sequencing DpnI enriched DamMT+ bacterial DNA (Figure 4) suggests that the binding kinetics are equivalent across these organisms. We hypothesize that with regard to $G^{6m}ATC$, organisms may divide into genomes that (A) have a

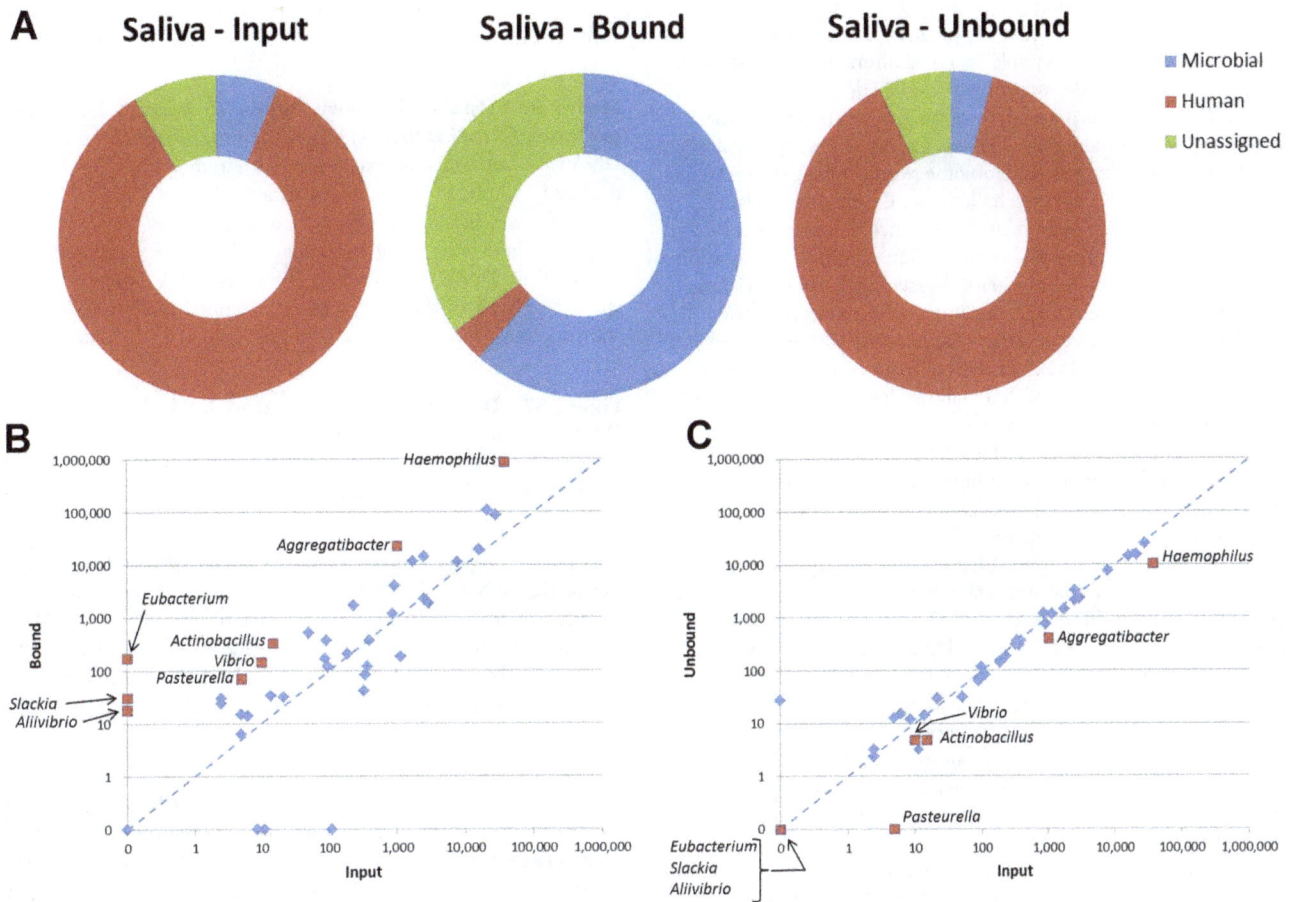

Figure 5. NGS data of saliva samples. (A) Donut plots depicting relative abundance of identifying reads for microbial and human genomes in input, bound and unbound samples. (B, C) Pairwise plots of sample fractions versus input. (B) Bound vs. input. (C) Unbound vs. input. Plotted points are identifying reads of genera. To facilitate direct visual comparison between samples reads were normalized to 10 M total.

DamMT-like density of G^{6m}ATC sites and are highly enriched, (B) lower site density that are only slightly enriched and (C) those genomes with no G^{6m}ATC sites. This last category may be greatly discriminated against if it possess mCpG sites, as does human DNA, or may result in an equal in abundance in the bound fraction and the input sample when CpG sites are absent (Figure 6A), as is the case for most bacteria.

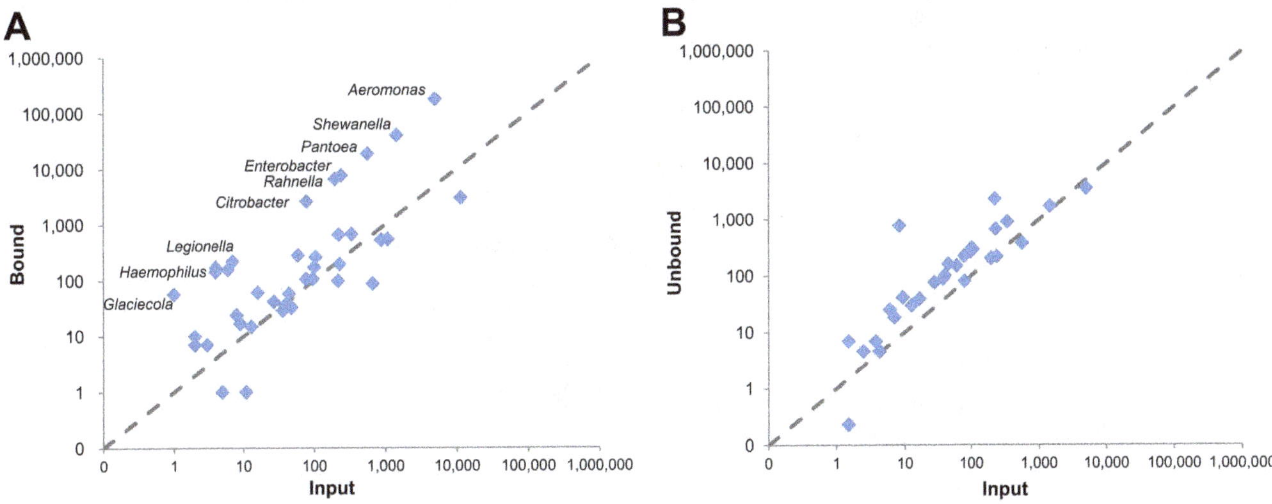

Figure 6. NGS data of creek samples. Pairwise plots of sample fractions versus input. (A) Bound vs. input. (B) Unbound vs. input. Plotted points are identifying reads of genera arbitrarily normalized to 10 M reads total.

We demonstrated that by purifying DNA by methylome, enrichment exceeding 50-fold of specific genomes is possible. In the case of the water sample, an organism closely related to *Aeromonas salmonicida* was highly enriched, with hundreds of thousands of non-normalized reads in the DpnI bound fraction compared to approximately 5000 in the input library. Typically, the high complexity of a microbiome would make reassembling genomes of unknown species challenging. Existing methods rely on bioinformatics, using alignment to reference genomes, nucleotide composition [37], differential coverage binning [38], or variations in gene count [9] to achieve partial assemblies. Our enrichment approach increases coverage, facilitates informatics processes and provides opportunities to characterize previously unsequenced and unculturable microbial taxa in diverse microbial communities.

Enrichment upstream of NGS allows for better coverage and increased certainty of the presence of organisms. This may be useful for samples with a very high load of eukaryotic DNA, such as those from the throat, buccal mucosa, or saliva [24]. The DpnI enrichment of pathogen DNA from saliva has several potential applications. Bacterial populations in saliva change in response to many disease conditions [39]. Identification and quantification of bacterial profiles may be important for detection of oral and/or systemic disease. With only about 100 cultivable strains out of the over 700 oral microbiota taxa [39], DpnI enrichment may provide a reliable way to identify novel bacterial species present in saliva using NGS. For example, *Aggregatibacter actinomycetemcomitans*, a strain known to be involved in periodontitis [40] was enriched 27-fold over input (Table S1). *Treponema denticola*, another strain implicated in periodontitis [40], was undetectable in the input fraction but had over 300 associated reads in the bound fraction (Table S1).

DpnI is unique in that it is a methyl-directed type II enzyme that can be used as a tool to bind DNA of a broad clade of widely studied bacteria with impacts on human health. Our demonstration that DpnII, a methyl *inhibited* type II endonuclease can also be used for differential selection of DNA opens the door to using alternative enzymes for DNA segregation. Over 300 restriction endonucleases with methyl-specific recognition specificities have been catalogued [41] potentially offering many more opportunities to discriminate genomes based on methylation patterns. By choosing restriction endonucleases with different methylation specificities, we envision the ability to stratify complex genomic mixtures into various methylomes, thus simplifying the experimental characterization of any microbiome.

The discovery of restriction endonucleases enabled the biotech revolution. These enzymes now offer a new technical utility, expanding on their natural role as discriminators of their own genomes to allow isolation of genomes from unculturable bacterial genomes present at low levels from diverse hosts and environments. Careful consideration of 6 mA, 4mC and 5mC directed or blocked endonucleases has led us to use these molecular biological tools in new ways and to develop new methodologies that promise additional insights into the natural and pathogenic microbiomes of our world.

Supporting Information

Figure S1 NGS coverage maps for *Aeromonas salmonicida*. NGS reads from the creek input (orange) and bound (blue) fractions were mapped to *Aeromonas salmonicida*, grouped into 1000 nt bins and plotted.

Figure S2 DpnII efficiently binds to human DNA and excludes *E. coli* DNA. (A) Equal amounts of human and *E. coli* DNA were combined and separated with immobilized DpnII. The majority of human DNA bound to immobilized DpnII while *E. coli* DNA was left behind. (B) An excess of *E. coli* DNA (500 ng) was combined with 1 ng of human DNA then separated with immobilized DpnII. Almost 80% of human DNA bound to DpnII while nearly all of the *E. coli* DNA remained unbound. All data shown is the average of three experiments.

Figure S3 DpnI binding to sheared DNA. *E. coli* genomic DNA was sheared by sonication and the size confirmed by gel electrophoresis. Sheared and intact (genomic) DNA was subjected to DpnI separation and binding was assessed by qPCR.

Figure S4 DpnI segregation of *Aspergillus fumigatus* genomic DNA. *Aspergillus* DNA was subjected to DpnI segregation and the fractions analyzed by qPCR. When DNA was treated with M.SssI, a CpG specific methyltransferase, the amount of *Aspergillus* DNA recovered by DpnI decreased from 2% to 0.7% compared to input levels. Data shown is the average of 4 experiments.

Figure S5 Pairwise plots showing reads mapped to synthetic genomic mix input DNAs and normalized to total reads for DpnI Bound versus Input fractions. There are three methylomes represented: those with $G^{m6}ATC$ are highly enriched (above line); those that are present at the same levels of input (on line) and those that are excluded (below line).

Table S1 DpnI enriches saliva organisms that are known to be involved in oral diseases. DNA isolated from saliva was segregated with DpnI and the DNA used to prepare NGS libraries. The fold of enrichment was calculated based on NGS mapped reads normalized to total reads in the DpnI bound versus input fractions. In the cases where there were no mapped reads in the input fraction (*T. denticola*) the normalized mapped read counts are listed.

Acknowledgments

We are grateful to Kurt Klimpel for his critical review of this manuscript. We also thank the staff of GHC Technologies for early technical support.

Author Contributions

Conceived and designed the experiments: HEB GL CQW LKP KTH RTY RAF. Performed the experiments: HEB GL CQW PK LKP SW LD. Analyzed the data: HEB GL DS RTY RAF. Contributed reagents/materials/analysis tools: PK KTH DS RTY RAF. Wrote the paper: HEB GL CQW PK SW LKP DS RTY RAF.

References

1. Philippot L, Raaijmakers JM, Lemanceau P, van der Putten WH (2013) Going back to the roots: the microbial ecology of the rhizosphere. Nat Rev Microbiol 11: 789–799.
2. Engel P, Moran NA (2013) The gut microbiota of insects - diversity in structure and function. FEMS Microbiol Rev 37: 699–735.
3. Kostic AD, Howitt MR, Garrett WS (2013) Exploring host-microbiota interactions in animal models and humans. Genes Dev 27: 701–718.

4. Ntougias S, Bourtzis K, Tsiamis G (2013) The microbiology of olive mill wastes. Biomed Res Int 2013: 784591.
5. Alvarez B, Lopez MM, Biosca EG (2007) Influence of native microbiota on survival of Ralstonia solanacearum phylotype II in river water microcosms. Appl Environ Microbiol 73: 7210–7217.
6. Vartoukian SR, Palmer RM, Wade WG (2010) Strategies for culture of 'unculturable' bacteria. FEMS Microbiol Lett 309: 1–7.
7. Stewart EJ (2012) Growing unculturable bacteria. J Bacteriol 194: 4151–4160.
8. Amann RI, Ludwig W, Schleifer KH (1995) Phylogenetic identification and in situ detection of individual microbial cells without cultivation. Microbiological Reviews 59: 143–169.
9. Carr R, Shen-Orr SS, Borenstein E (2013) Reconstructing the genomic content of microbiome taxa through shotgun metagenomic deconvolution. PLoS Comput Biol 9: e1003292.
10. Feehery GR, Yigit E, Oyola SO, Langhorst BW, Schmidt VT, et al. (2013) A Method for Selectively Enriching Microbial DNA from Contaminating Vertebrate Host DNA. PLoS One 8: e76096.
11. Sachse S, Straube E, Lehmann M, Bauer M, Russwurm S, et al. (2009) Truncated Human Cytidylate-Phosphate-Deoxyguanylate-Binding Protein for Improved Nucleic Acid Amplification Technique-Based Detection of Bacterial Species in Human Samples. Journal of Clinical Microbiology 47: 1050–1057.
12. Fang G, Munera D, Friedman DI, Mandlik A, Chao MC, et al. (2012) Genome-wide mapping of methylated adenine residues in pathogenic Escherichia coli using single-molecule real-time sequencing. Nat Biotechnol 30: 1232–1239.
13. Ratel D, Ravanat JL, Berger F, Wion D (2006) N6-methyladenine: the other methylated base of DNA. Bioessays 28: 309–315.
14. Lobner-Olesen A, Skovgaard O, Marinus MG (2005) Dam methylation: coordinating cellular processes. Curr Opin Microbiol 8: 154–160.
15. Loenen WA, Raleigh EA (2013) The other face of restriction: modification-dependent enzymes. Nucleic Acids Res 42: 56–69.
16. Lacks S, Greenberg B (1975) A deoxyribonuclease of Diplococcus pneumoniae specific for methylated DNA. Journal of Biological Chemistry 250: 4060–4066.
17. Vovis GF, Lacks S (1977) Complementary action of restriction enzymes endo R Â·DpnI and endo R Â· DpnII on bacteriophage f1 DNA. Journal of Molecular Biology 115: 525–538.
18. de la Campa AG, Springhorn SS, Kale P, Lacks SA (1988) Proteins encoded by the DpnI restriction gene cassette. Hyperproduction and characterization of the DpnI endonuclease. J Biol Chem 263: 14696–14702.
19. Xu SY, Schildkraut I (1991) Isolation of BamHI variants with reduced cleavage activities. J Biol Chem 266: 4425–4429.
20. Bispo PJ, de Melo GB, Hofling-Lima AL, Pignatari AC (2010) Detection and gram discrimination of bacterial pathogens from aqueous and vitreous humor using real-time PCR assays. Invest Ophthalmol Vis Sci 52: 873–881.
21. Sayers EW, Barrett T, Benson DA, Bryant SH, Canese K, et al. (2009) Database resources of the National Center for Biotechnology Information. Nucleic Acids Res 37: D5–15.
22. Benson DA, Karsch-Mizrachi I, Lipman DJ, Ostell J, Sayers EW (2009) GenBank. Nucleic Acids Res 37: D26–31.
23. Low DA, Weyand NJ, Mahan MJ (2001) Roles of DNA Adenine Methylation in Regulating Bacterial Gene Expression and Virulence. Infection and Immunity 69: 7197–7204.
24. Methé BA NK, Pop M, Creasy HH, Giglio MG, Huttenhower C, et al. (2012) A framework for human microbiome research. Nature 486: 215–221.
25. Yang F, Zeng X, Ning K, Liu KL, Lo CC, et al. (2012) Saliva microbiomes distinguish caries-active from healthy human populations. ISME J 6: 1–10.
26. Vasu K, Nagaraja V (2013) Diverse Functions of Restriction-Modification Systems in Addition to Cellular Defense. Microbiology and Molecular Biology Reviews 77: 53–72.
27. Lynch TW, Kosztin D, McLean MA, Schulten K, Sligar SG (2002) Dissecting the molecular origins of specific protein-nucleic acid recognition: hydrostatic pressure and molecular dynamics. Biophys J 82: 93–98.
28. Wong DL, Pavlovich JG, Reich NO (1998) Electrospray ionization mass spectrometric characterization of photocrosslinked DNA-EcoRI DNA methyltransferase complexes. Nucleic Acids Res 26: 645–649.
29. Taylor JD, Badcoe IG, Clarke AR, Halford SE (1991) EcoRV restriction endonuclease binds all DNA sequences with equal affinity. Biochemistry 30: 8743–8753.
30. Sud'ina AE, Zatsepin TS, Pingoud V, Pingoud A, Oretskaya TS, et al. (2005) Affinity modification of the restriction endonuclease SsoII by 2'-aldehyde-containing double stranded DNAs. Biochemistry (Mosc) 70: 941–947.
31. Bonnet I, Biebricher A, Porte PL, Loverdo C, Benichou O, et al. (2008) Sliding and jumping of single EcoRV restriction enzymes on non-cognate DNA. Nucleic Acids Res 36: 4118–4127.
32. Yagupsky P, Nolte FS (1990) Quantitative aspects of septicemia. Clin Microbiol Rev 3: 269–279.
33. Siwek W, Czapinska H, Bochtler M, Bujnicki JM, Skowronek K (2012) Crystal structure and mechanism of action of the N6-methyladenine-dependent type IIM restriction endonuclease R.DpnI. Nucleic Acids Res 40: 7563–7572.
34. Kingston IJ, Gormley NA, Halford SE (2003) DNA supercoiling enables the type IIS restriction enzyme BspMI to recognise the relative orientation of two DNA sequences. Nucleic Acids Res 31: 5221–5228.
35. Seshasayee AS, Singh P, Krishna S (2012) Context-dependent conservation of DNA methyltransferases in bacteria. Nucleic Acids Res 40: 7066–7073.
36. Marinus MG, Casadesus J (2009) Roles of DNA adenine methylation in host–pathogen interactions: mismatch repair, transcriptional regulation, and more. FEMS Microbiology Reviews 33: 488–503.
37. Herlemann DP, Lundin D, Labrenz M, Jurgens K, Zheng Z, et al. (2013) Metagenomic de novo assembly of an aquatic representative of the verrucomicrobial class Spartobacteria. MBio 4: e00569–00512.
38. Albertsen M, Hugenholtz P, Skarshewski A, Nielsen KL, Tyson GW, et al. (2013) Genome sequences of rare, uncultured bacteria obtained by differential coverage binning of multiple metagenomes. Nat Biotechnol 31: 533–538.
39. Malamud D (2011) Saliva as a diagnostic fluid. Dent Clin North Am 55: 159–178.
40. Zhang L, Henson BS, Camargo PM, Wong DT (2009) The clinical value of salivary biomarkers for periodontal disease. Periodontol 2000 51: 25–37.
41. Roberts RJ, Vincze T, Posfai J, Macelis D (2010) REBASE—a database for DNA restriction and modification: enzymes, genes and genomes. Nucleic Acids Res 38: D234–236.

18

Effects of Clonal Integration on Microbial Community Composition and Processes in the Rhizosphere of the Stoloniferous Herb *Glechoma longituba* (Nakai) Kuprian

Ningfei Lei[1], Jun Li[1], Shijun Ni[1]*, Jinsong Chen[2]*

1 Chengdu University of Technology, Chengdu, China, 2 Sichuan Normal University, Chengdu, China

Abstract

The effects of rhizodeposition on soil C and N availabilities lead to substantial changes of microbial community composition and processes in the rhizosphere of plants. Under heterogeneous light, photosynthates can be translocated or shared between exposed and shaded ramets by clonal integration. Clonal integration may enhance the rhizodeposition of the shaded ramets, which further influences nutrient recycling in their rhizosphere. To test the hypothesis, we conducted a pot experiment by the stoloniferous herb *Glechoma longituba* subjected to heterogeneous light. Microbial biomass and community composition in the rhizosphere of shaded offspring ramets, assessed by phospholipid fatty acids (PLFAs) analysis, were markedly altered by clonal integration. Clonal integration positively affected C, N availabilities, invertase and urease activities, N mineralization (N_{min}) and nitrification rates (N_{nitri}) in the rhizosphere of shaded offspring ramets. However, an opposite pattern was also observed in phenoloxidase (POXase) and peroxidase (PODase) activities. Our results demonstrated that clonal integration facilitated N assimilation and uptake in the rhizosphere of shaded offspring ramets. The experiment provides insights into the mechanism of nutrient recycling mediated by clonal integration.

Editor: Fei-Hai Yu, Beijing Forestry University, China

Funding: This project was supported by the National Natural Science Foundation of China (Grant No. 31170502) and the Study On the Breed of Trichosanthes kirilowii New Varieties and Supporting Technology (Grant No. 2011NZ0098-12-07). The funders had no role in study design, data collection and analysis, decision to publish, or preparation of the manuscript.

Competing Interests: The authors have declared that no competing interests exist.

* Email: cjs74@163.com (JSC); nsj@cdut.edu.cn (SJN)

Introduction

Rhizosphere, a zone of usually high microbial turnover and activity, has been coined to describe the soil adjacent to and influenced by plant roots [1]. Plant-derived root exudates are primary sources of labile C inputting to soil [2,3]. These labile C sources rapidly metabolized by microorganisms may generally stimulate their growth or succession in the rhizosphere [4,5]. So, plant roots exert strong influences on the rhizosphere through 'rhizodeposition' (root exudation such as sugars, amino acids, organic acids and hormones, as well as mucilage, enzymes, sloughed root cells and C allocated to root-associated symbionts) [6].

In the form of rhizodeposition, photosynthates released into soil by plant roots are a major source of carbon, energy or structural material for soil microorganisms and affect the microbial community composition in the rhizosphere [7–11]. Fungi, especially ectomycorrhizal (ECM) fungi involved in nitrogen turnover (e.g. mineralization and nitrification), prefer the substrates with larger C/N ratios [12]. Microbial processes, such as extracellular enzyme activities [1,13,14], N mineralization and nitrification [15,16], are mediated by specific groups of microorganisms in the rhizosphere. Two experiments to disrupt root exudation into the soil demonstrated that decreased resource availability negatively affected nitrogen mineralization and nitrification in the rhizosphere via rhizodeposition from plant root [1,15]. So, microbial processes are highly sensitive to the availabilities of labile C and N in the rhizosphere [5].

Clonal plants can translocate or share resources, such as carbohydrates, water and nutrients among interconnected ramets through clonal integration [17]. Shading may have negative effects on photosynthetic capacity and growth performance of plants [18,19]. Clonal integration may alter resource levels of ramets under heterogeneous habitats [20,21]. So, enhanced photosynthates availability caused by clonal integration may have a significant influence on microbial community composition and processes in the rhizosphere of the ramets subjected to low light availability stress. Further, microorganisms present in the rhizosphere may mediate nutrient availability for plants by carrying out a wide spectrum of decomposition processes. As mentioned-above, rhizosphere processes may play a vital role in community or ecosystem nutrient cycling [1]. However, studies on the mechanism of nutrient recycling mediated by clonal integration are rare in the rhizosphere. A pot experiment was conducted by the stoloniferous herb *Glechoma longituba* subjected to heterogeneous light (mother ramets suffering from full sun versus offspring ones suffering from 80% shade). Comparing with severed offspring ramets, we predicted that connected offspring ramets displayed (1) higher C and N availabilities in the rhizosphere. Based on effects of C and N availabilities on microbial community composition and processes, we expected that connected offspring ramets exhibited

(2) higher microbial biomass and different microbial community composition in the rhizosphere; (3) higher extracellular enzymes activities in the rhizosphere; (4) greater N mineralization and nitrification rates in the rhizosphere.

Material and Methods

Plant species and experimental design

G. longituba (Lamiaceae) is a stoloniferous perennial herb. Its monopodial stolons are able to creep on the ground. Ramets can develop on all stolon nodes. A genet or fragment consists of a number of ramets connected by stolons for a certain period of time. Each ramet has two zygomorphic single leaves originating from a stolon node. Every leaf axil bears one bud that may grow into a secondary stolon. The plant is generally found in forests, on roadsides or by creeks and distributed all over China except for the Northwest [22].

In May 2012, ten original clonal fragments of *G. longituba* were collected from a forest understory in Suining City (30°10′~31°10′N; 105°03′~106°59′E), Sichuan Province, China. The sampling site did not belong to the part of any farms or national parks. *G. longituba* is widespread in China and not an endangered or protected species, so we did not need any relevant permissions/permits for plant sample collection. These original plants were at least 100 m apart from one another. They were propagated in a greenhouse with a mean temperature of 22±8°C. The plants were watered and fertilized as needed.

In June 2013, each clonal fragment consisting of a mother and an offspring ramet with similar size was selected. The two ramets were planted separately in two adjacent plastic pots (10 cm in diameter, 8.5 cm in height) filled with a 3:1 mixture of humus soil and sand. Plants were watered regularly with distilled water to prevent water stress. After two weeks of growth, offspring ramets were subjected to a 80% shading treatment and the other mother ramets were grown in full sun, whereas stolons between the mother and the offspring ramets were either severed or remained intact (Fig. 1). No new offspring ramets were produced during the two weeks of recovery. Shading was imposed by placing small shade cages covered with black cloth above the pots. The mesh was covered on the top of each pot to avoid potential effects of litters. Only original ramets were allowed to root during the experiment. Each treatment was replicated 10 times and all treatments included clonal fragments from the 10 original plants. All replicates of each treatment were randomly located on benches in a greenhouse. Because soil closely adhering to the roots (up to 2 mm around the root) was considered as rhizosphere soil, the experimental procedure was repeated 10 times to collect enough soil sample. The experiment lasted for 10 weeks.

Soil sampling and assays

At the end of experiment, rhizosphere soil was sampled according to the shaking root method [23]. The rhizosphere soil of shaded offspring ramets was separated from roots by hand, sieved (<2 mm mesh) and stored at −20°C.

Soil microbial biomass carbon (C_{mic}) and microbial biomass nitrogen (N_{mic}) were analyzed using chloroform fumigation-extraction (CFE) method [24,25]. Briefly, 20 g of fresh, sieved soil was used for the fumigation and non-fumigation treatments, both extracted using 0.5 M K_2SO_4 at a ratio of 1:4 (w/v), shaken for 30 min and filtrated through a Whatman no. 42 filter paper. The K_2SO_4-extract of both fumigated and non-fumigated samples were analyzed immediately for dissolved organic carbon (DOC) and dissolved organic nitrogen (DON) using a TOC/TN analyzer (elementar vario TOC SELECT, Germany). C_{mic} and N_{mic} were calculated using the following equations: C_{mic} (or N_{mic}) = $2.22 \times E_B$ [26], where E_B was the difference of carbon (or nitrogen) extracted from fumigated soil between non-fumigated soil. Soil moisture was detected gravimetrically, i.e. a sample of 20 g was oven-dried at 105°Cfor 48 h until a constant weight. Total soil organic carbon (TOC) and total nitrogen (TN) were determined using an elemental analyzer (elementar vario MACRO CUBE, Germany). Soil pH was measured in a ratio of 1: 2.5 (soil: water, w/v).

N mineralization and nitrification

N mineralization and nitrification were assessed by the modified anaerobic incubation method [27]. Briefly, fresh soil samples (5 g) were placed into a 200 mL plastic bottles, 10 mL deionized water was added to the bottles to thoroughly submerge the soil. The plastic bottles were sealed with stopper to avoid water evaporation during incubation, and placed in a constant temperature (40°C) incubator for 7 days. At the beginning of the incubation experiment, pre-incubation soil was sampled to measure the initial concentrations of NH_4^+-N and NO_3^--N. After a week of incubation, the post-incubation soil samples were mixed with 40 mL of 2 M KCl using a 1: 8 soil: extractant (w/v) ratio, shaken for 30 min on a reciprocal shaker; then the extracts were filtered through prewashed Whatman no. 42 filter papers and supernatants were stored at −20°C until analysis of NH_4^+-N and NO_3^--N concentrations. The NH_4^+-N and NO_3^--N concentrations were separately measured by spectrophotometry using the ammonium indophenol blue method and the cadmium reduction method [28]. All concentrations of NH_4^+-N and NO_3^--N were based on dry soil weight and expressed on a mg·g^{-1} (*DW*). The N mineralization rate (N_{min}) was calculated as the changes in the inorganic N (NH_4^+-N, NO_3^--N) content from time zero to 7 days. A similar formula was used to calculate N nitrification rate (N_{nitri}) [29,30]:

$$N_{min} = \frac{\textit{inorganic N content after incubation} - \textit{inorganic N content before incubation}}{\textit{incubation days}}$$

$$N_{nitri} = \frac{NO_3^- - N \textit{ content after incubation} - NO_3^- - N \textit{ content before incubation}}{\textit{incubation days}}$$

Extracellular enzymes activities assays

Invertase activity was measured by the modified method [31]. Briefly, fresh soil (1.0 g) was added to 5 mL 0.1 M saccharose in 0.1 M Na-acetate buffer (pH 4.65) in a 50 mL reaction flask and incubated for 1 h at 30°C. After of incubation, the mixture was transferred immediately to a freezer for 10 min to stop the enzymatic reaction, centrifuged at 3500 *g* for 10 min and the reaction products were determined in the supernatants. The concentrations of glucose produced by saccharose hydrolysis were determined by the Nelson-Somoji reagent at 660 nm [32]. Invertase activity was expressed in μg glucose g^{-1}*DW* h^{-1}.

For the determination of urease activity, the modified procedure was adopted according to the description [33]. Briefly, 5 g fresh soil was incubated with 2.5 mL 0.08 M urea solution and 20 mL borate buffer (pH 10.0) for 2 h at 37°C. Released ammonium was extracted using 50 mL 2 M KCl solution, and determined colorimetrically at 690 nm.

POXase activity was measured according to an improved procedure [34]. Briefly, 1.0 g of fresh soil was added to 3 mL of

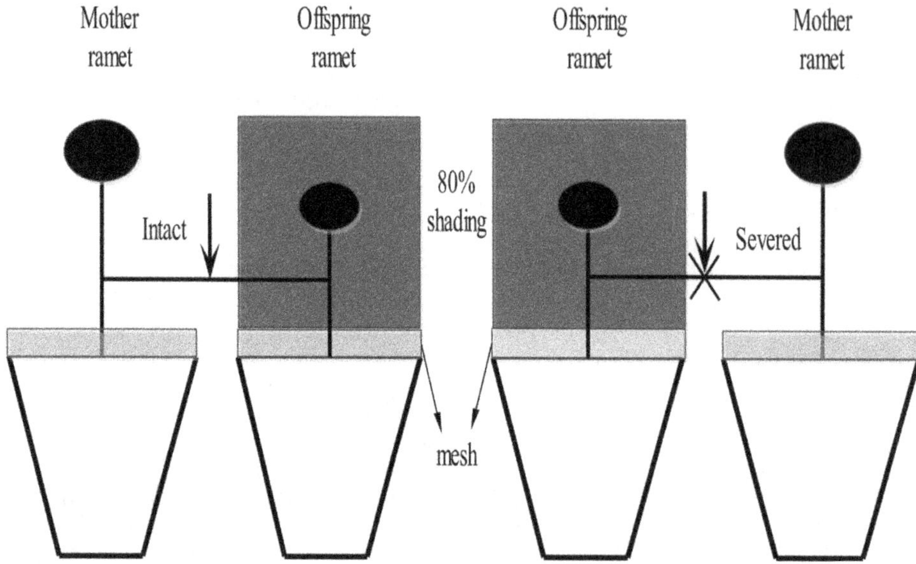

Figure 1. Schematic diagram of the experimental design. The clonal fragment consists of mother ramet and its successional offspring ramet. The offspring ramets were subjected to the 80% shading treatment, and the mother ramets were exposed to full sun. The stolon between mother and offspring ramets remained intact or severed. The mesh was covered on the top of each pot to avoid potential effects of litters.

reagent solution (obtained by mixing 1.5 mL of catechol solution with 1.5 mL of proline solution) and 2 mL of phosphate buffer (0.1 M, pH 6.5). The suspension was swirled and incubated at 37°C for 1 h, then reaction was stopped by cooling and adding 5 mL of ethanol. The mixture was centrifuged at 4000 *g* at 4°C for 5 min. The absorbance of the supernatant fraction was determined at 525 nm. Assays without soil and catechol were carried out simultaneously as controls. POXase activity was expressed as μmol oxidized catechol (*o*-catechol) $g^{-1}DW\ h^{-1}$.

PODase activity was determined with 3,3′,5,5′-tetramethylbenzidine (TMB) as the substrates [35]. Samples of fresh soil (4 g) were mixed with 200 mL cold acetate buffer (5°C, 50 mM, pH 5.0) on a vortex mixer for 30 s at high speed. The soil suspension was then diluted 20-fold in acetate buffer. Aliquots of 0.25 mL were transferred to 2 mL centrifuge tubes. 0.5 mL preheated TMB reagent (25°C) was added to the centrifuge tubes and the tubes were incubated in a constant temperature (25°C) incubator. After 2 h, peroxidase reaction was terminated by adding 1.2 mL sulfuric acid (0.2 M). The tubes were then centrifuged under dark and absorbance of the supernatants were read at 450 nm. Controls were performed using acetate buffer substituted TMB to confirm that there was no photo-oxidation of TMB.

Microbial community composition

Microbial community composition was assessed using phospholipid fatty acids (PLFAs) [1]. (1) Extraction: A sample of 8.0 g fresh soil was extracted with a mixture of citrate buffer (0.15 M, pH 4.0 with NaOH), chloroform and methanol at a ratio of 0.8: 1: 2 (v/v/v). Suspension was shaken darkly at 25°C for 30 min, centrifuged at 10000 *g* for 0.5 h and the supernatant was transferred to new vials. Chloroform and citrate buffer were added to supernatant for separation of the phases. After 18 h, the organic phases were removed and dried under a stream of dry N_2. (2) Chromatography: lipids were redissolved in chloroform and neutral lipids were separated from phospholipids on silica columns by elution with chloroform (5 mL), acetone (10 mL) and methanol (5 mL) gradually. Methanol-phase was collected and dried under N_2 stream. (3) Methyl esterification: phospholipids were subsequently converted to fatty acids methyl esters (FAMEs) by alkaline methanolysis. Phospholipids were dissolved in 1 mL of methanolic 0.2 M KOH and 1 mL of methanol-toluene (1:1, v/v) and incubated for 15 min at 37°C, then mixed with 2 mL of deionized water, 0.3 mL of acetic acids (0.2 M) and 2 mL of hexane, swirled and centrifuged for 10 min. The hexane-phase was removed and dried under N_2 stream. After adding 100 μL of a solution of methyl-nonadecanoate (C19:0, 25 ng μL^{-1}) as an internal standard, FAMEs were dissolved in C19:0 and analyzed by capillary gas chromatography (Agilent Technologies, 6850N-GC System, USA). Concentration of single FAMEs was calculated using the internal standard (C19:0) peak as a reference according to the following formula:

$$PLFAs(ng\ g^{-1}DW) = \frac{A * 25\ ng\ \mu L^{-1} * 100\mu L}{B * W}$$

Where A and B were the peak areas of each fatty acid methyl ester and internal standard, respectively; W was the oven-dry soil weight (*DW*).

PLFAs used as biomarkers for specific groups of soil microorganisms and were designated according to an standard nomenclature: ((*a,i,cy*)*X: YωZ(c,t)*), where the *X* referred to the number of C atoms, the *Y* indicated the number of double bonds followed by the position (*ω*) and distance (*Z*) of the double bonds from the methyl end. The prefixes *a*, *i* indicated *anteiso*- and *iso*-branching; the suffixes *c*, *t* referred to *cis*- and *trans*-double bonds; *cy* represented cyclopropyl-group. The fatty acids used as biomarkers for specific groups of soil organisms were listed in Table 1.

Statistical analysis

Soil pH, moisture, C and N availabilities, microbial biomass, extracellular enzymes activities, N mineralization and nitrification rates in the rhizosphere of shaded offspring ramets were investigated by one-way ANOVA. Microbial community composition of shaded offspring ramets was analyzed by a principal

Table 1. PLFAs biomarkers of different microbial groups were used in the study.

Soil microbial groups	PLFAs biomarkers	References
Bacteria	i13:0, i14:0, 14:0, 15:0, 16:0, 17:0, i15:0, i16:0, i17:0, a15:0, 16:1ω7c, a17:0, 18:0, cy17:0, cy19:0, 16:1ω7c, cy19:0ω9c, 10Me16:0	[13,43,63–65]
Actinomycete	10Me18:0	[65]
Fungi	18:1ω9, 18:2ω6,9	[13]
Ectomycorrhizal fungi	18:2ω6,9	[12]
Gram-positive bacteria	i13:0, i14:0, 14:0, i15:0, i16:0, i17:0, a15:0, a17:0, 10Me16:0	[13,64]
Gram-negative bacteria	18:0, cy17:0, cy19:0, 16:1ω7c, cy19:0ω9c	[43,64]
Bacteria/Fungi	(i13:0, i14:0, 14:0, 15:0, 16:0, 17:0, i15:0, i16:0, i17:0, a15:0, 16:1ω7c, a17:0, 18:0, cy17:0, cy19:0, 16:1ω7c, cy19:0ω9c, 10Me16:0)/(18:1ω9, 18:2ω6,9)	

component analysis (PCA) using specific PLFAs biomarkers. For PCA analysis, scores of different soil microbial groups were expressed as percentage of the total PLFAs in the sample. Pearson correlations were used for relating PLFAs concentrations of different microbial groups to extracellular enzymes activities, N mineralization and nitrification, C and N availabilities. Significance was set at $p = 0.05$ level. If needed, data were natural logarithm-transformed or arcsine-transformed in order to achieve normality and homogeneity of variance. All statistical analyses were performed using SPSS 20.0 software (SPSS, Chicago, IL, USA).

Results

Changes in soil properties

Clonal integration significantly increased TOC, DOC and DON concentrations in the rhizosphere of shaded offspring ramets as well as C/N, whereas no effects of clonal integration on soil moisture, pH and TN were observed in the rhizosphere of shaded offspring ramets (Table 2). C_{mic} and N_{mic} were significantly higher in the rhizosphere of shaded, connected offspring ramets compared to shaded, severed offspring ramets (Table 2). Concentrations of inorganic nitrogen (NH_4^+-N and NO_3^--N) in the rhizosphere of shaded offspring ramets were also markedly increased by clonal integration. Meanwhile, NH_4^+-N concentration was evidently higher than NO_3^--N concentration, regardless of stolon connection or severing (Table 2).

Changes in microbial community composition

A principal component analysis (PCA) based on PLFAs biomarkers of different microbial groups revealed that microbial community composition was clearly distinct in the rhizosphere of shaded, connected offspring ramets compared to shaded, severed offspring ramets (Fig. 2). The results of PCA were further supported by the absolute PLFAs concentrations for different microbial groups (Fig. 3). The PLFAs concentrations of bacteria, fungi, Gram-positive bacteria and Gram-negative bacteria were significantly increased by clonal integration as well as the total PLFAs concentration (Fig. 3a,b,c,e,f). The biomarker (18:2ω6,9) accounted for 86% of total fungal PLFAs concentrations in the rhizosphere of shaded, connected offspring ramets, the proportion was slightly decreased (75%) in shaded, severed offspring ones. Specially, clonal integration distinctly increased the PLFAs concentration of ECM fungi (18:2ω6,9) by 72% (Fig. 3g). However, effects of clonal integration on the PLFAs concentration of actinomycete and Ba/Fu were not observed (Fig. 3d, h).

Changes in microbial processes

Compared to shaded, severed offspring ramets, invertase and urease activities in the rhizosphere of shaded, connected offspring ramets were increased by 59.8% ($p<0.001$) and 38.9% ($p = 0.005$) respectively (Fig. 4a,b). On the contrary, POXase and PODase activities in the rhizosphere of shaded, connected offspring ramets were decreased by 47.8% ($p<0.001$) and 12.7% ($p<0.001$) respectively (Fig. 4c,d). N_{nitri} accounted for only 17% of N_{min} in the rhizosphere of shaded, connected offspring ramets, compared to 28% in the rhizosphere of shaded, severed offspring ones ($p = 0.034$). N_{min} and N_{nitri} in the rhizosphere of shaded offspring ramets were markedly increased by clonal integration (Fig. 5a,b).

Correlations between the PLFAs concentrations of different microbial groups and soil properties or microbial processes

Soil properties (such as DOC, DON, TOC, NH_4^+, NO_3^-) were positively correlated to the PLFAs concentrations of most microbial groups except for actinomycete (Table 3). Similarly, microbial processes (such as N_{nitri}, N_{min}, invertase and urease activities) were positively correlated to the PLFAs concentrations of most microbial groups (Table 3). On the contrary, POXase and PODase activities were negatively correlated to the PLFAs concentrations of most microbial groups (Table 3). In addition, Ba/Fu and G^-/G^+ ratios were not significantly correlated with any of the measured soil properties and microbial processes (Table 3).

Discussion

Changes in C and N availabilities

A major source of labile C inputting to soil is the root exudates [3,36,37]. Because girdling blocked the flow of newly formed photosynthates to the roots and to mycorrhizal fungi, it significantly decreased concentration of DOC and C_{mic} in the rhizosphere or bulk soils [1,13,38]. Clonal plants can translocate or share photosynthates from exposed ramets to shaded ramets by clonal integration [39]. So, effects of clonal integration on labile C may be similar with those reported in previous girdling experiments. Root exudation is associated with increased N availability in the rhizosphere [14]. DON and N_{mic} were decreased in the girdled plots [38]. The decrease in DOC and DON in girdled plots was probably caused by the decrease in root exudation because a substantial portion of DOC and DON originated from photosynthates [14]. In our study, the increase of belowground

Table 2. Effects of clonal integration on soil properties in the rhizosphere of shaded offspring ramets.

Soil properties	Treatments	
	Connected	Severed
TOC ($g \cdot kg^{-1}$)	18.42±0.19	17.87±0.06 **
TN ($g \cdot kg^{-1}$)	1.96±0.03	2.04±0.08^{ns}
TOC/TN (C/N)	9.42±0.22	8.75±0.37 *
DOC ($mg \cdot g^{-1}$)	1.06±0.038	0.96±0.04 **
DON ($mg \cdot g^{-1}$)	0.58±0.037	0.52±0.006 **
C_{mic} (μg $C \cdot g^{-1}$)	403.55±23.23	292.17±40.28 ***
N_{mic} (μg $N \cdot g^{-1}$)	351.57±20.21	74.60±20.19 ***
soil moisture (%)	16.49±2.35	17.91±1.37^{ns}
pH (soil: water = 1: 5)	7.75±0.07	7.47±0.15^{ns}
NH_4^+-N ($mg \cdot kg^{-1}$)	35.67±5.14	22.54±2.46 *
NO_3^--N ($mg \cdot kg^{-1}$)	13.40±0.16	6.61±0.27 ***

Values are means ± SE (standard errors). The significant differences between connected offspring and severed offspring ramets were indicated by *** ($p<0.001$), ** ($p<0.01$), * ($p<0.05$) and ns (not significant); n = 10. Abbreviations: TOC, total organic carbon; TN, total nitrogen; DOC, dissolved organic carbon; DON, dissolved organic nitrogen; C_{mic}, microbial biomass carbon; N_{mic}, microbial biomass nitrogen.

carbon allocation caused by clonal integration may improve C and N availability in the rhizosphere of shaded offspring ramets.

Increased dissolved inorganic N concentrations in response to girdling have been found in other studies [14,40]. On the contrary, we suspect that clonal integration, by increasing the supply of available C to microorganisms, may stimulate N mineralization and nitrificantion in the rhizosphere of shaded offspring ramets. Alternatively, effects of available C supply on soil dissolved inorganic N concentrations may depend on the species-specific. Compared with NH_4^+-N, NO_3^--N concentration was lower in the rhizosphere of shaded offspring ramets, regardless of stolon connection or severing (Table 2). This is most likely due to the high mobility of NO_3^- in soil [40].

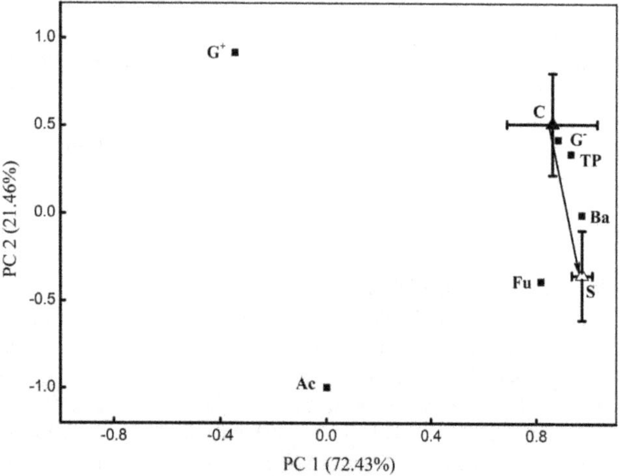

Figure 2. Microbial community composition described by a principal component analysis of the concentrations of PLFAs. Error bars showed standard error of the mean of PCA weighted loading values for connected (C) and severed (S) shaded offspring ramets. Black squares represented PCA weighted loading values of microorganisms. Microbial groups abbreviations: TP, total PLFAs; Ba, bacteria; Fu, fungi; Ac, actinomycete; G^+, gram-positive bacteria; G^-, gram-negative bacteria, n = 10.

Changes in microbial biomass and community composition

DOC is an available C source for microbes and affects their abundance, composition and activity [41]. A larger portion of easily assimilable C was derived from photosynthates produced by plants and shaped a specific microbial community composition [9,11,42,43]. Tree girdling did not affect the total PLFAs concentration and increased the concentration of bacterial PLFAs [38]. In addition, the concentrations of PLFAs for bacterial groups were also related to soil pH [44]. Although no effect of clonal integration on soil pH was observed, the total, bacteria, Gram-positive bacteria and Gram-negative bacteria PLFAs concentrations in the rhizosphere of shaded offspring ramets were significant increased by clonal integration (Table 2; Fig. 3). The similar patterns were observed in another girdling experiment [1].

Specially, clonal integration distinctly increased the PLFAs concentration of ECM fungi (18:2ω6,9) by 72% (Fig. 3g). This was consistent with a previous girdling experiment [13]. Our results further confirm that the dramatic increase in fungi caused by clonal integration is associated with a increase in C supply and that fungi depend to a much higher degree on belowground C allocation [12].

The increase of total fungal biomass caused by clonal integration was mainly related to the increase of ECM fungal biomass [38,41]. Compared to Gram-positive bacterial PLFAs, the concentrations of Gram-negative bacterial PLFAs were higher in the rhizosphere of shaded offspring ramets, regardless of stolon connection or severing (Fig. 3e,f). This is consistent with the suggestion that Gram-negative bacteria are generally favored by the labile C substrates released by rhizosdeposition and more frequent in the rhizosphere [1]. The effects of clonal integration on actinomycetes were not observed (Fig. 3d). A possible explanation is that actinomycetes are less stimulated in the rhizosphere [45]. Association relationships between different microbial groups (bacteria, Gram-negative bacteria, Gram-positive bacteria, fungi and ECM) and DOC or TOC concentrations may imply their

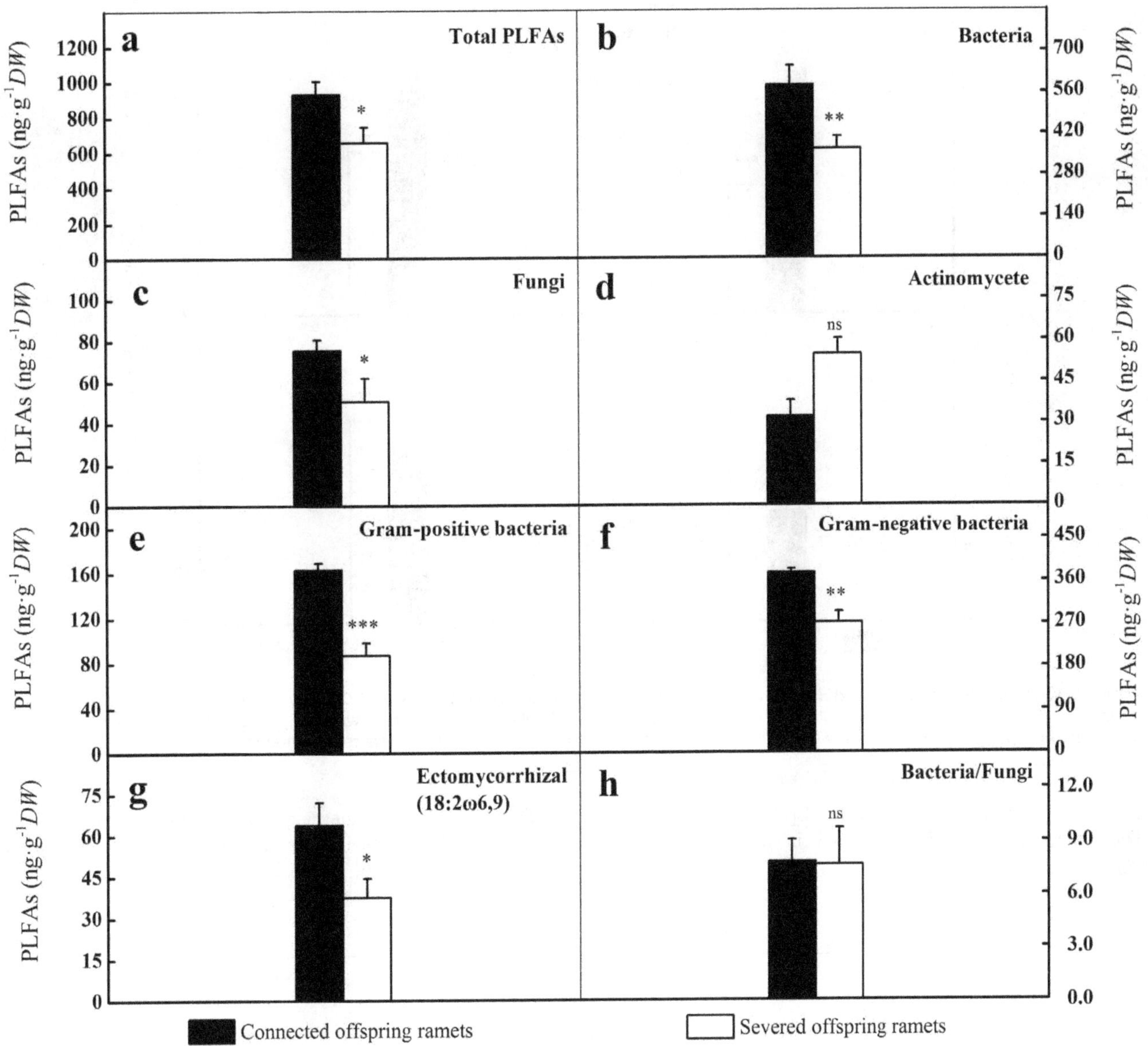

Figure 3. Concentrations or ratios of soil microbial groups PLFAs in the rhizosphere of shaded offspring ramets. Significant differences between connected offspring ramets (black bars) and severed offspring ones (open bars) were indicated by *** ($p<0.001$), ** ($p<0.01$), * ($p<0.05$) and ns (not significant). Error bars represented standard errors, n = 10.

strong dependence (Table 3). We tentatively conclude that the effects of clonal integration on soil C and N availability lead to substantial changes in microbial biomass and community composition in the rhizosphere of shaded offspring ramets.

Changes in microbial processes

Soil enzyme activities can be used as potential indicators of nutrient cycling processes. Invertase catalyzes hydrolytic processes of SOM [44]. Urease is generally produced by bacteria, filamentous fungi and yeasts, thus enhancing N mineralization [46,47]. Plant root exudates may provide a constant energy supply, thereby creating optimal conditions for SOM degraders [48]. The increased microbial biomass caused by enhancing root exudation could increase extracellular enzymes activities and the release of N from SOM [49,50]. The suggestions were further supported by the positive correlations between invertase or urease activities and most soil microbial groups in the rhizosphere of shaded offspring ramets (Table 3).

Phenoloxidase (POXase) and peroxidase (PODase) are the lignolytic enzyme involved in the degradation of recalcitrant SOM (e.g. lignin) [51]. POXase and PODase are generally produced by slow-growing specialist decomposers (e.g. saprotrophic fungi) [52,53]. Competition between microbial groups could also have been responsible for the shift of enzyme activities. Mycorrhizal fungi are known to dominate the rooted soil layers as a result of a competitive advantage gained through access to root C, whereas saprotrophic fungi are thought to be more competitive in the litter layer [53,54]. Clonal integration greatly increased the abundance of mycorrhizal fungi, thereby possibly also giving saprotrophic fungi a competitive disadvantage in the rooting zone. The suggestions were further confirmed by the negative correlations between POXase or PODase activities and most soil microbial

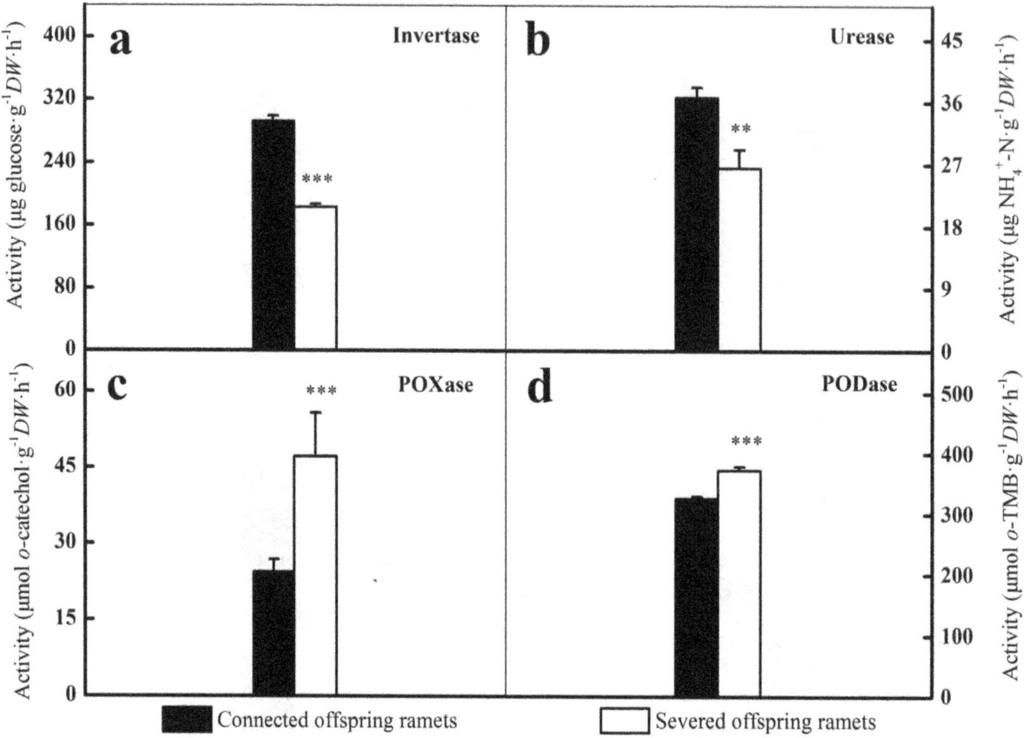

Figure 4. Extracellular enzymes activities involved in the depolymerization of C, N from SOM. Invertase (a), Urease (b), POXase (c) and PODase (d) were measured in the rhizosphere of shaded offspring ramets. Significant differences between connected offspring ramets (black bars) and severed offspring ones (open bars) were indicated by *** ($p<0.001$) and ** ($p<0.01$). Error bars represented standard errors, n = 10.

groups in the rhizosphere of shaded offspring ramets (Table 3). The similar patterns were observed in a previous girdling experiment [1].

Dissolved organic matter (e.g. DOC, DON) was considered to influence soil microbial processes, such as soil respiration/C mineralization [55,56] and N mineralization [57,58]. N mineralization and nitrification were regulated by a variety of

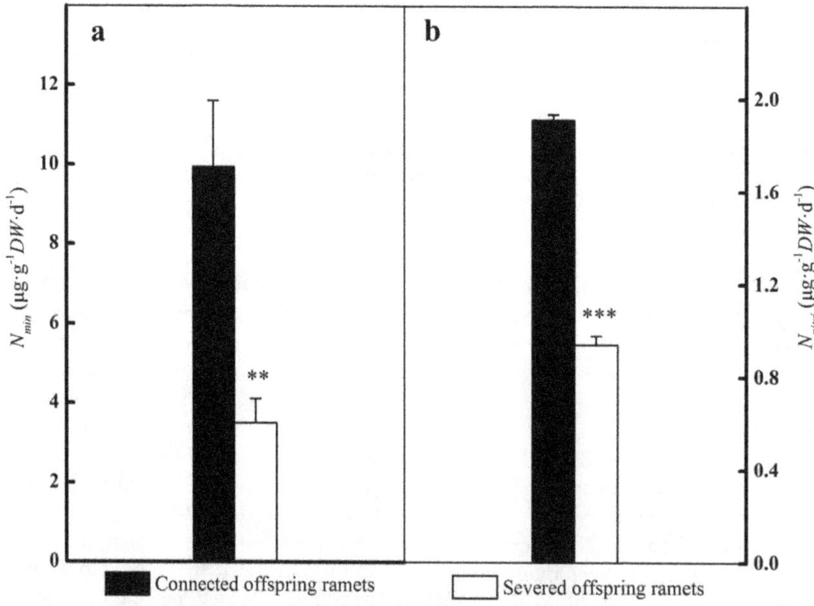

Figure 5. N mineralization rate (a) and nitrification rate (b) in the rhizosphere of shaded offspring ramets were measured by anaerobic incubation methods. Significant differences between connected offspring ramets (black bars) and severed offspring ones (open bars) were indicated by *** ($p<0.001$), ** ($p<0.01$). Error bars represented standard errors, n = 10.

Table 3. Correlations between concentrations or ratios of soil microbial groups PLFAs and soil properties or microbial processes in the rhizosphere of shaded offspring ramets.

Soil Microbialgroups	Soil properties						Microbial processes							
	DOC	DON	TOC	TN	pH	moisture	NH_4^+	NO_3^-	N_{min}	N_{nitri}	Invertase	Urease	POXase	PODase
TP	0.87*	0.89*	0.83*	−0.63	0.56	−0.48	0.86*	0.88*	0.95**	0.88*	0.89*	0.90*	−0.65	−0.83*
Ba	0.90*	0.88*	0.82*	−0.38	0.70	−0.39	0.82*	0.92**	0.94**	0.92**	0.94**	0.94**	−0.82*	−0.92**
Fu	0.86*	0.89*	0.92**	−0.75$^+$	0.97**	−0.59	0.77$^+$	0.87*	0.82*	0.87*	0.85*	0.72	−0.92**	−0.84*
ECM	0.88*	0.92**	0.98***	−0.64	0.96**	−0.73	0.74$^+$	0.92*	0.88*	0.92**	0.89**	0.85*	−0.94**	−0.90*
Ac	0.69	0.64	0.54	−0.16	0.42	0.01	0.69	0.70	0.76$^+$	0.70	0.75$^+$	0.74$^+$	−0.57	−0.69
G^+	0.97**	0.95**	0.88*	−0.49	0.79$^+$	−0.57	0.88*	0.98***	0.96**	0.98***	0.98**	0.97**	−0.90*	−0.98**
G^-	0.84*	0.83*	0.74$^+$	−0.48	0.44	−0.59	0.67	0.85*	0.88*	0.85*	0.84*	0.91*	−0.60	−0.83*
Ba/Fu	0.08	−0.03	−0.14	0.55	−0.34	0.18	0.02	0.05	0.17	0.05	0.09	0.30	0.13	−0.10
G^-/G^+	−0.43	−0.40	−0.45	0.11	−0.80$^+$	0.12	−0.56	−0.45	−0.34	−0.46	−0.47	−0.32	0.74$^+$	0.49

Correlation coefficients and significant levels were shown. Significant levels were indicated by *** ($p<0.001$), ** ($p<0.01$), * ($p<0.05$) and $^+$ ($p<0.1$); n = 10. Abbreviations: TP, total PLFAs; Ba, bacteria; Fu, fungi; ECM, ectomycorrhizal fungi; Ac, actinomycete; G^+, gram-positive bacteria; G^-, gram-negative bacteria; Ba/Fu, bacteria/fungi; G^-/G^+, gram-negative bacteria/gram-positive bacteria; N_{min}, nitrogen mineralization rate; N_{nitri}, nitrogen nitrification rate; POXase, phenol oxidase; PODase, peroxidas.

heterotrophic bacteria and fungi via using labile C source or SOM [40]. N mineralization and nitrification rates were strongly increased by clonal integration in the rhizosphere of shaded offspring ramets (Fig. 5). Positive correlations between N mineralization or nitrification rates and most microbial groups were observed in the rhizosphere of shaded offspring ramets (Table 3). The similar patterns were found in a previous girdling experiment [1]. Notably, N_{nitri} tended to be much lower than N_{min} in the rhizosphere of shaded offspring ramets, regardless of stolon connection or severing (Fig. 5). This could be explained by the fact that NO_3^- was derived from the oxidizing of NH_4^+ (i.e. nitrification) by chemoautotrophic bacteria or heterotrophic microorganisms [59,60]. In addition, NO_3^- was water soluble and seldom present in detectable amounts [61].

Because of a high availability of easily assimilable carbon and nutrients, microbial community composition were modified by clonal integration in the rhizosphere of shaded offspring ramets. Invertase and urease activitties, N mineralization and nitrification rates were enhanced by clonal integration in the rhizosphere of shaded offspring ramets. So, clonal integration may facilitate N assimilation and uptake in the rhizosphere of shaded offspring ramets. A field study investigated the effects of clonal integration on nutrient recycling of the Serengeti grassland communities [62]. Our experiment provides insights into the mechanism of nutrient recycling mediated by clonal integration. To allow a robust generalization, however, more experimental studies, especially those conducted in the field, are required.

Author Contributions

Conceived and designed the experiments: NFL JL JSC. Performed the experiments: NFL JL. Analyzed the data: SJN JSC. Contributed reagents/materials/analysis tools: NFL JL. Wrote the paper: NFL JL. JL SJN JSC.

References

1. Koranda M, Schnecker J, Kaiser C, Fuchslueger L, Kitzler B, et al. (2011) Microbial processes and community composition in the rhizosphere of European beech - The influence of plant C exudates. Soil Biology and Biochemistry 43: 551–558.
2. Hütsch BW, Augustin J, Merbach W (2002) Plant rhizodeposition—an important source for carbon turnover in soils. Journal of Plant Nutrition and Soil Science 165: 397–407.
3. Kuzyakov Y (2002) Review: Factors affecting rhizosphere priming effects. Journal of Plant Nutrition and Soil Science 165: 382–396.
4. Butler JL, Bottomley PJ, Griffith SM, Myrold DD (2004) Distribution and turnover of recently fixed photosynthate in ryegrass rhizospheres. Soil Biology and Biochemistry 36: 371–382.
5. Schimel JP, Weintraub MN (2003) The implications of exoenzyme activity on microbial carbon and nitrogen limitation in soil: a theoretical model. Soil Biology and Biochemistry 35: 549–563.
6. Bais HP, Weir TL, Perry LG, Gilroy S, Vivanco JM (2006) The role of root exudates in rhizosphere interactions with plants and other organisms. Annual Review of Plant Biology 57: 233–266.
7. Barea J-M, Pozo MJ, Azcon R, Azcon-Aguilar C (2005) Microbial cooperation in the rhizosphere. Journal of Experimental Botany 56: 1761–1778.
8. Fu S, Cheng W (2004) Defoliation affects rhizosphere respiration and rhizosphere priming effect on decomposition of soil organic matter under a sunflower species: *Helianthus annuus*. Plant and Soil 263: 345–352.
9. Grayston S, Vaughan D, Jones D (1997) Rhizosphere carbon flow in trees, in comparison with annual plants: the importance of root exudation and its impact on microbial activity and nutrient availability. Applied Soil Ecology 5: 29–56.
10. Kuzyakov Y, Cheng W (2001) Photosynthesis controls of rhizosphere respiration and organic matter decomposition. Soil Biology and Biochemistry 33: 1915–1925.
11. Rajaniemi T, Allison V (2009) Abiotic conditions and plant cover differentially affect microbial biomass and community composition on dune gradients. Soil Biology and Biochemistry 41: 102–109.
12. Olsson PA (1999) Signature fatty acids provide tools for determination of the distribution and interactions of mycorrhizal fungi in soil. FEMS Microbiology Ecology 29: 303–310.
13. Kaiser C, Koranda M, Kitzler B, Fuchslueger L, Schnecker J, et al. (2010) Belowground carbon allocation by trees drives seasonal patterns of extracellular

enzyme activities by altering microbial community composition in a beech forest soil. New Phytologist 187: 843–858.
14. Weintraub MN, Scott-Denton LE, Schmidt SK, Monson RK (2007) The effects of tree rhizodeposition on soil exoenzyme activity, dissolved organic carbon, and nutrient availability in a subalpine forest ecosystem. Oecologia 154: 327–338.
15. Norton JM, Firestone MK (1996) N dynamics in the rhizosphere of *Pinus ponderosa* seedlings. Soil Biology and Biochemistry 28: 351–362.
16. Priha O, Grayston SJ, Pennanen T, Smolander A (1999) Microbial activities related to C and N cycling and microbial community structure in the rhizospheres of *Pinus sylvestris*, *Picea abies* and *Betula pendula* seedlings in an organic and mineral soil. FEMS Microbiology Ecology 30: 187–199.
17. Du J, Wang N, Alpert P, Yu MJ, Yu FH, et al. (2010) Clonal integration increases performance of ramets of the fern Diplopterygium glaucum in an evergreen forest in southeastern China. Flora 205: 399–403.
18. Roiloa SR, Retuerto R (2007) Responses of the clonal *Fragaria vesca* to microtopographic heterogeneity under different water and light conditions. Environmental and Experimental Botany 61: 1–9.
19. He WM, Alpert P, Yu FH, Zhang LL, Dong M (2011) Reciprocal and coincident patchiness of multiple resources differentially affect benefits of clonal integration in two perennial plants. Journal of Ecology 99: 1202–1210.
20. Alpert P, Mooney H (1986) Resource sharing among ramets in the clonal herb, *Fragaria chiloensis*. Oecologia 70: 227–233.
21. Evans JP (1991) The effect of resource integration on fitness related traits in a clonal dune perennial, *Hydrocotyle bonariensis*. Oecologia 86: 268–275.
22. Liao M, Yu F, Song M, Zhang S, Zhang J, et al. (2003) Plasticity in R/S ratio, morphology and fitness-related traits in response to reciprocal patchiness of light and nutrients in the stoloniferous herb, *Glechoma longituba* L. Acta Oecologica 24: 231–239.
23. Riley D, Barber S (1970) Salt accumulation at the soybean (*Glycine max* (L.) Merr.) root-soil interface. Soil Science Society of America Journal 34: 154–155.
24. Vance E, Brookes P, Jenkinson D (1987) An extraction method for measuring soil microbial biomass C. Soil biology and Biochemistry 19: 703–707.
25. Brookes P, Landman A, Pruden G, Jenkinson D (1985) Chloroform fumigation and the release of soil nitrogen: a rapid direct extraction method to measure microbial biomass nitrogen in soil. Soil Biology and Biochemistry 17: 837–842.
26. Wu J, Joergensen R, Pommerening B, Chaussod R, Brookes P (1990) Measurement of soil microbial biomass C by fumigation extractionan - automated procedure. Soil Biology and Biochemistry 22: 1167–1169.
27. Carter MR, Gregorich EG (1993) Soil sampling and methods of analysis. Canadian: Canadian Society of Soil Science. 599–605 p.
28. Carter MR, Gregorich EG (1993) Soil sampling and methods of analysis. Canadian: Canadian Society of Soil Science. 71–80 p.
29. Zhou W, Chen H, Zhou L, Lewis BJ, Ye Y, et al. (2011) Effect of freezing-thawing on nitrogen mineralization in vegetation soils of four landscape zones of Changbai Mountain. Annals of Forest Science 68: 943–951.
30. Ren W, Chen F-s, Hu X-f, Yu M-q, Feng X (2011) Soil nitrogen transformations varied with plant community under Nanchang urban forests in mid-subtropical zone of China. Journal of Forestry Research 22: 569–576.
31. Gianfreda L, Sannino F, Violante A (1995) Pesticide effects on the activity of free, immobilized and soil invertase. Soil Biology and Biochemistry 27: 1201–1208.
32. Nelson N (1944) A photometric adaptation of the Somogyi method for the determination of glucose. Journal of Biological Chemistry 153: 375–379.
33. Kandeler E, Gerber H (1988) Short-term assay of soil urease activity using colorimetric determination of ammonium. Biology and Fertility of Soils 6: 68–72.
34. Perucci P, Casucci C, Dumontet S (2000) An improved method to evaluate the *o*-diphenol oxidase activity of soil. Soil Biology and Biochemistry 32: 1927–1933.
35. Johnsen AR, Jacobsen OS (2008) A quick and sensitive method for the quantification of peroxidase activity of organic surface soil from forests. Soil Biology and Biochemistry 40: 814–821.
36. Jones DL, Hodge A, Kuzyakov Y (2004) Plant and mycorrhizal regulation of rhizodeposition. New Phytologist 163: 459–480.
37. Jones D, Nguyen C, Finlay R (2009) Carbon flow in the rhizosphere: carbon trading at the soil–root interface. Plant and Soil 321: 5–33.
38. Chen D, Zhou L, Wu J, Hsu J, Lin Y, et al. (2012) Tree girdling affects the soil microbial community by modifying resource availability in two subtropical plantations. Applied Soil Ecology 53: 108–115.
39. Alpert P (1999) Effects of clonal integration on plant plasticity in *Fragaria chiloensis*. Plant Ecology 141: 99–106
40. Zeller B, Liu J, Buchmann N, Richter A (2008) Tree girdling increases soil N mineralisation in two spruce stands. Soil Biology and Biochemistry 40: 1155–1166.
41. Högberg MN, Högberg P, Myrold DD (2007) Is microbial community composition in boreal forest soils determined by pH, C-to-N ratio, the trees, or all three? Oecologia 150: 590–601.
42. Brant JB, Myrold DD, Sulzman EW (2006) Root controls on soil microbial community structure in forest soils. Oecologia 148: 650–659.
43. Yarwood SA, Myrold DD, Högberg MN (2009) Termination of belowground C allocation by trees alters soil fungal and bacterial communities in a boreal forest. FEMS Microbiology Ecology 70: 151–162.
44. Gu Y, Wang P, Kong CH (2009) Urease, invertase, dehydrogenase and polyphenoloxidase activities in paddy soil influenced by allelopathic rice variety. European Journal of Soil Biology 45: 436–441.
45. Zhang LZ, Niu W, Niu Y, Niu XW (2009) Impact of *Caragana* Fabr. plantation on plant community and soil properities of saline-alkali wasteland. Acta Ecologica Sinica 29(9):4693–4699.
46. Mobley H, Hausinger R (1989) Microbial ureases: significance, regulation, and molecular characterization. Microbiological Reviews 53: 85–108.
47. Gianfreda L, Antonietta Rao M, Piotrowska A, Palumbo G, Colombo C (2005) Soil enzyme activities as affected by anthropogenic alterations: intensive agricultural practices and organic pollution. Science of the Total Environment 341: 265–279.
48. Fontaine S, Mariotti A, Abbadie L (2003) The priming effect of organic matter: a question of microbial competition? Soil Biology and Biochemistry 35: 837–843.
49. Cheng W, Kuzyakov Y (2005) Root effects on soil organic matter decomposition. American society of Agronomy 48: 119–143.
50. Blagodatskaya E, Blagodatsky S, Dorodnikov M, Kuzyakov Y (2010) Elevated atmospheric CO_2 increases microbial growth rates in soil: results of three CO_2 enrichment experiments. Global Change Biology 16: 836–848.
51. Sinsabaugh RL (2010) Phenol oxidase, peroxidase and organic matter dynamics of soil. Soil Biology and Biochemistry 42: 391–404.
52. Baldrian P (2009) Ectomycorrhizal fungi and their enzymes in soils: is there enough evidence for their role as facultative soil saprotrophs? Oecologia 161: 657–660.
53. Hobbie EA, Horton TR (2007) Evidence that saprotrophic fungi mobilise carbon and mycorrhizal fungi mobilise nitrogen during litter decomposition. New Phytologist 173: 447–449.
54. Lindahl BD, Ihrmark K, Boberg J, Trumbore SE, Högberg P, et al. (2007) Spatial separation of litter decomposition and mycorrhizal nitrogen uptake in a boreal forest. New Phytologist 173: 611–620.
55. Chantigny MH, Angers DA, Prévost D, Simard RR, Chalifour F-P (1999) Dynamics of soluble organic C and C mineralization in cultivated soils with varying N fertilization. Soil Biology and Biochemistry 31: 543–550.
56. Gregorich E, Rochette P, McGuire S, Liang B, Lessard R (1998) Soluble organic carbon and carbon dioxide fluxes in maize fields receiving spring-applied manure. Journal of Environmental Quality 27: 209–214.
57. Appel T, Mengel K (1993) Nitrogen fractions in sandy soils in relation to plant nitrogen uptake and organic matter incorporation. Soil Biology and Biochemistry 25: 685–691.
58. Murphy D, Macdonald A, Stockdale Ea, Goulding K, Fortune S, et al. (2000) Soluble organic nitrogen in agricultural soils. Biology and Fertility of Soils 30: 374–387.
59. De Boer W, Kowalchuk G (2001) Nitrification in acid soils: micro-organisms and mechanisms. Soil Biology and Biochemistry 33: 853–866.
60. Brierley E, Wood M, Shaw P (2001) Influence of tree species and ground vegetation on nitrification in an acid forest soil. Plant and Soil 229: 97–104.
61. Keeney DR, Nelson D (1982) Methods of soil analysis. Part 2. Chemical and microbiological properties, Nitrogen—inorganic forms. Ameirican: American Society of Agronomy. 643–698.
62. Wilsey B (2002) Clonal plants in a spatially heterogeneous environment: effects of integration on Serengeti grassland response to defoliation and urine-hits from grazing mammals. Plant Ecology 159: 15–22.
63. Rinnan R, Michelsen A, Bååth E, Jonasson S (2007) Fifteen years of climate change manipulations alter soil microbial communities in a subarctic heath ecosystem. Global Change Biology 13: 28–39.
64. Green CT, Scow KM (2000) Analysis of phospholipid fatty acids (PLFA) to characterize microbial communities in aquifers. Hydrogeology Journal 8: 126–141.
65. Vestal JR, White DC (1989) Lipid analysis in microbial ecology. Bioscience 39: 535–541.

19

A Suitable Streptomycin-Resistant Mutant for Constructing Unmarked In-Frame Gene Deletions Using *rpsL* as a Counter-Selection Marker

Yu-Kuo Tsai[1], Ci-Hong Liou[2], Jung-Chung Lin[2], Ling Ma[1], Chang-Phone Fung[3], Feng-Yee Chang[2,4]*, L. Kristopher Siu[1,2,5]*

1 Institute of Infectious Diseases and Vaccinology, National Health Research Institutes, Miaoli, Taiwan, **2** Division of Infectious Diseases and Tropical Medicine, Department of Internal Medicine, Tri-Service General Hospital, National Defense Medical Center, Taipei, Taiwan, **3** Section of Infectious Diseases, Department of Medicine, Taipei Veterans General Hospital and National Yang-Ming University, Taipei, Taiwan, **4** Taiwan Centres for Disease Control, Taipei, Taiwan, **5** Graduate Institute of Basic Medical Science, China Medical University, Taichung, Taiwan

Abstract

The streptomycin counter-selection system is a useful tool for constructing unmarked in-frame gene deletions, which is a fundamental approach to study bacteria and their pathogenicity at the molecular level. A prerequisite for this system is acquiring a streptomycin-resistant strain due to *rpsL* mutations, which encodes the ribosomal protein S12. However, in this study no streptomycin resistance was found to be caused by *rpsL* mutations in all 127 clinical strains of *Klebsiella pneumoniae* isolated from liver abscess patients. By screening 107 spontaneous mutants of streptomycin resistance from a clinical strain of *K. pneumoniae*, nucleotide substitution or insertion located within the *rpsL* was detected in each of these strains. Thirteen different mutants with varied S12 proteins were obtained, including nine streptomycin-dependent mutants. The virulence of all four streptomycin-resistant mutants was further evaluated. Compared with the parental strain, the K42N, K42T and K87R mutants showed a reduction in growth rate, and the K42N and K42T mutants became susceptible to normal human serum. In the mice LD_{50} (the bacterial dose that caused 50% death) assay, the K42N and K42T mutants were ~1,000-fold less lethal (~2×10^5 CFU) and the K87R mutant was ~50-fold less lethal (~1×10^4 CFU) than the parental strain (~2×10^2 CFU). A K42R mutant showed non-observable effects on the above assays, while this mutant exhibited a small cost ($P<0.01$) in an *in vitro* growth competition experiment. In summary, most of the *K. pneumoniae* strains with streptomycin resistance caused by *rpsL* mutations are less virulent than their parental strain in the absence of streptomycin. The K42R mutant showed similar pathogenicity to its parental strain and should be one of the best choices when using *rpsL* as a counter-selection marker.

Editor: Christophe Herman, Baylor College of Medicine, United States of America

Funding: This work was supported by grants from the National Health Research Institutes and the National Science Council of Taiwan (NSC 102-2314-B-016-053-MY3). The funders had no role in study design, data collection and analysis, decision to publish, or preparation of the manuscript.

Competing Interests: The authors have declared that no competing interests exist.

* Email: fychang@ndmctsgh.edu.tw (FYC); lksiu@nhri.org.tw (LKS)

Introduction

Streptomycin, the first aminoglycoside antibiotic, was reported in 1944 [1] and can inhibit protein synthesis by binding to the 30S ribosomal subunit. The most common mechanisms contributing to streptomycin resistance are aminoglycoside-modifying enzymes and target gene mutations [2], while 16S rRNA methylases, which have recently emerged, confer high-level resistance to all clinically available aminoglycosides except streptomycin [3–5]. Adenylyltransferase encoded by *aadA* and phosphoryltransferase encoded by *strA* or *strB* have been demonstrated to be responsible for streptomycin resistance and are widely disseminated among *Enterobacteriaceae* [2,6,7]. On the other hand, streptomycin resistance caused by its target genes is mediated by mutations in 16S rRNA encoded by *rrs* or ribosomal protein S12 encoded by *rpsL* [8]. However, a single mutation in an *rrs* gene has only a limited effect on bacteria with multicopy *rrs* gene [9], such as that in *Escherichia coli* and *Klebsiella pneumoniae*. The mutations of the S12 protein have often been found to confer a high-level streptomycin resistance by preventing streptomycin binding and/or conferring ribosomal hyperaccuracy [8,10,11]. Some mutations could cause a strongly hyperaccurate phenotype and lead to streptomycin dependence [11]. The streptomycin-dependent mutants need streptomycin to keep their survival because their ribosomal proteins require the binding of streptomycin for properly function during protein synthesis [11].

K. pneumoniae is a common cause of infections. Most community-acquired *K. pneumoniae* infections cause pneumonia or urinary tract infections, while a distinct invasive syndrome that causes liver abscesses has been increasingly reported in the past two decades [12]. The construction of unmarked in-frame gene deletions in bacteria is a fundamental approach to study bacteria at a molecular level, including the study of their pathogenicity.

Counter-selection markers have been used for the positive selection of the deletion mutant, and the streptomycin counter-selection system is one of the most frequently used [13]. A prerequisite for this system is to acquire a streptomycin-resistant strain due to *rpsL* mutations. When both wild-type and mutant alleles of *rpsL* are expressed in the same strain, it confers a streptomycin-sensitive phenotype in streptomycin-resistant strains, which enables positive selection with streptomycin to detect the loss of the vector that contains the wild-type allele of *rpsL* [13,14]. Previous studies have demonstrated that *rpsL* can be used as a counter-selection marker in various bacteria, such as *K. pneumoniae* [15–20], *Vibrio cholera* [14], *Borrelia burgdorferi* [21], *Corynebacterium glutamicum* [22], *Streptococcus pneumoniae* [23] or mycobacteria [24,25], but little is known about which streptomycin resistance-inducing *rpsL* mutations should be used. Previous studies have found that some *rpsL* mutations could reduce the virulence of bacteria [26,27], and an *rpsL* mutant of *K. pneumoniae* showing lower virulence than its parental liver abscess isolate has been selected for use with this technology [17–20]. However, our results demonstrated that *rpsL* mutations were rare in *K. pneumoniae* that was isolated from liver abscess patients, and previous studies have showed that different protein expression levels could be caused by *rpsL* mutations [26,28]. These results suggest that a low or no-cost *rpsL* mutant should be chosen for the use of *rpsL* as a counter-selection marker. In this study, thirteen different streptomycin resistance-inducing mutations of the S12 protein were obtained, and the costs of these mutations were evaluated.

Materials and Methods

Ethics Statement

All studies with human blood samples were conducted within the National Defense Medical Center (Taiwan), and blood samples from healthy volunteers were obtained after written informed consent was provided. *K. pneumoniae* strains isolated from patients with liver abscesses were received from already-existing collections [29,30]. The use of these samples was approved by the Institutional Review Board of the National Defense Medical Center (Taiwan), and the identification number is B-102-13. All samples were anonymized and identified only by study subject number.

All animal studies were conducted in strict accordance with the recommendations in the Guide for the Care and Use of Laboratory Animals of the National Research Council. The protocol was approved by the Institutional Animal Care and Use Committee of the National Defense Medical Center (Taiwan), and the identification number is IACUC-13-171.

Strains, Growth Rate, Streptomycin Susceptibility and Resistance Genes

The 127 strains of *K. pneumoniae* investigated were isolated from liver abscess patients from 2002 to 2009 [29,30], and 48, 55 and 24 isolates were collected from Taiwan, Singapore and Hong Kong, respectively. Unless otherwise noted, *K. pneumoniae* and its derivatives were cultured at 37°C in Luria-Bertani (LB) broth or brain heart infusion (BHI) broth with appropriate antibiotics. The growth rate was evaluated using a Bioscreen C MBR, an automated microbiology growth analysis system (Oy Growth Curves Ab, Helsinki, Finland). Cells from overnight cultures were transferred to fresh LB broth without or with streptomycin (50 or 500 μg/ml) to give an initial OD_{600} of 0.005. The cultures were incubated in the Bioscreen C system at 37°C with continuous shaking between measurements, and the growth was quantified every 20 min based on the OD_{600}. Each experimental run was conducted with an associated negative control sample containing blank medium and a positive control sample with wild-type cells. The maximum doubling time was determined from the logarithms of the values measured in the mid-exponential phase. Minimal inhibitory concentrations (MICs) of streptomycin were determined using the E-test (Biodisk AB, Sweden). To evaluate the reasons for streptomycin resistance, *aadA*, *strA*, *strB* and *rpsL* were detected via PCR and then sequenced using primer pairs as shown in Table S1. The specific primers for *aadA*, *strA* and *strB* were designed based on the conserved regions of their open reading frames in previous studies [7,31], while the specific primers for *rpsL* were designed according to the conserved regions of its franking region in this study.

Selection and Identification of Spontaneous Mutations

K. pneumoniae NVT1001, capsular serotype 1, was one of the isolates from the liver abscess patients in Taiwan. This strain was grown in LB broth at 37°C to the late-exponential growth phase; the culture was then diluted 10^{-8}-fold in 4 ml fresh LB broth. After being incubated at 37°C overnight until the stationary growth phase, at which point the viable cell number was estimated to be $\sim 1.6 \times 10^9$ cells/ml, the cells of 1 ml culture were harvested via centrifugation at 12,000 *g* for 5 min, washed prior to being resuspended in 0.1 ml sterile physiological saline (0.9% sodium chloride), and then spread on LB agar plates supplemented with 50 or 500 μg/ml streptomycin. The cells were incubated at 37°C, and spontaneous mutants were obtained from colonies that arose within 3 days. Independent cultures were grown and plated as described above. Only one mutant colony was picked from each culture every day. The DNA fragment containing the complete *rpsL* gene was amplified via PCR using the primers rpsL.for and rpsL.rev (Table S1) for the selected mutants. The mutations in the *rpsL* gene were identified via sequencing and comparison with the nucleotide sequence of the wild-type strain.

Construction of Revertants

Plasmid pUT-kmy, which consists of an R6K origin of replication, an mobRP4 origin of transfer, and a kanamycin resistance cassette [32], was ligated with a *sacB* gene to generate plasmid pUT-KB for constructing revertants. Plasmid pUT-KB is a suicide vector containing a counter-selection marker, *sacB*, which originates from *Bacillus subtilis* [13]. When this gene is expressed on the integrated pUT-KB, it confers a sucrose-sensitivity phenotype, which enables positive selection with sucrose to detect the loss of the vector.

The allelic exchange method was used to restore the wild-type *rpsL* gene in the *K. pneumoniae* K42R, K42N, K42T, and K87R mutants (Figure S1). Briefly, DNA fragments of the entire *rpsL* with their flanking regions were amplified from *K. pneumoniae* NVT1001 using PCR with the primers ApaL.for and ApaL.rev (Table S1). The 2.4-kb PCR fragment generated was digested with ApaLI and then cloned into pUT-KB that was similarly digested, resulting in plasmid pRpsL-WT. For homologous recombination, plasmid pRpsL-WT was then transformed into *E. coli* S17-1 *λpir* [14] using the heat shock method and mobilized into the *K. pneumoniae* K42R, K42N, K42T and K87R mutants via conjugation. Single-crossover strains were selected from brilliant green containing inositol-nitrate-deoxycholate (BIND) plates supplemented with kanamycin (50 μg/ml), while the growth of the non-*K. pneumoniae* strains was effectively suppressed on the BIND plates [33]. The kanamycin-resistant transconjugant was selected, and the insertion of pRpsL-WT was verified via PCR. After being incubated in 20 ml BHI for 6 hours in the absence of kanamycin at 37°C, the fully grown cultures were spread onto LB plates supplemented with 10% sucrose. After double crossover occurred, the sucrose-resistant and kanamycin-sensitive colonies were selected, and the restorations of wild-type *rpsL* gene were confirmed via DNA sequencing.

In Vitro Growth Competition Assay

Competitive growth *in vitro* was performed as previously described [34,35] with minor modifications. Briefly, cells were grown overnight in LB broth at 37°C to the stationary phase, and equal densities of *K. pneumoniae* NVT1001 and one of its streptomycin-resistant mutants were mixed and then diluted 2^{17}-fold into 4 ml of fresh LB broth each to a final concentration of $\sim 6 \times 10^3$ CFU/ml. This culture was incubated at 37°C with 200 rpm shaking for 24 hours to complete a growth cycle ($\sim 1.6 \times 10^9$ cells/ml, ~17 generations). Each successive growth cycle was initiated by diluting the mixture 2^{17}-fold into 4 ml of fresh LB broth. After the initial mixing and each growth cycle, appropriate dilutions of the mixture were plated on LB agar plates for colony counts. The number of streptomycin-resistant bacteria was determined by plating the mixture on LB agar plates supplemented with 500 µg/ml streptomycin. The number of parental streptomycin-susceptible bacteria was calculated as the total number of bacteria minus the number of streptomycin-resistant bacteria. Each competition assay was performed in triplicates of four competition cycles, and the serial dilutions were plated in triplicate The difference in fitness between two competing strains was calculated using the following function as previously described [34,35]:

$$S_t = \ln\left[\left(\frac{r_t/s_t}{r_{t-1}/S_{t-1}}\right)^{1/17}\right]$$

where r_t and s_t denote the number of drug-resistant and drug-susceptible cells at a given time *t*, respectively, and r_{t-1} and s_{t-1} denote the number of drug-resistant and drug-susceptible cells at the preceding time point, respectively. The quotient of the ratios of the cell numbers was standardized with the exponent 1/17 because cell numbers were determined approximately every 17 generations. S_t is called the selection coefficient at time *t*. S_t is equal to 0 if there is no difference in fitness between the competing strains. If antibiotic resistance reduces bacterial fitness, S_t is negative, while S_t is positive if resistance increases bacterial fitness. The data are presented as relative bacterial fitness (fit_t), which was defined by Sander *et al.* as $\text{fit}_t = 1+S_t$ [34].

Neutrophil Phagocytosis Assay

Phagocytosis was measured using a standard assay [36]. A FACScan (Becton Dickinson Immunocytometry Systems, San Jose, CA, USA) was used to measure the phagocytic rate. The labeling of bacteria with fluorescein isothiocyanate was performed as described by Heinzelmann *et al.* [36], and the isolation of neutrophils from three healthy volunteers was performed as previously described [37]. A mixture of labeled bacteria, the neutrophil suspension, the pooled normal human serum and PBS was incubated in a shaking water bath at 37°C. The percentage of neutrophils that had phagocytosed bacteria was counted at 15 and 30 min. An unincubated tube served as the 0 min time point. The experimental procedures and FACS settings have been described previously [38].

Serum Bactericidal Assay

Normal human serum (NHS) pooled from healthy volunteers was divided into equal volumes and stored at −70°C prior to use. The serum bactericidal activity was measured using the method described by Podschun *et al.* [39], which was modified using a 10-fold reaction volume and 2-fold bacteria concentration [15]. Briefly, bacteria were grown in BHI broth until an OD_{600} of 0.35 ($\sim 10^8$ cells/ml) was reached. The cultures were washed and then diluted 25-fold using phosphate buffered saline (PBS). Two hundred fifty microliters of the cell suspension and 750 µl of pooled human serum were placed into 1.5 ml Eppendorf tubes, mixed, and incubated at 37°C. To determine the number of viable bacteria, an aliquot of each bacterial suspension was removed immediately and after 3 hours of incubation. The number of viable bacteria was determined via dilution and plating on LB agar for colony counts. The results were expressed as a percentage of the inoculum, and strains with survival rates >100% in NHS after 3 hours of incubation were considered resistant; those with survival rates <100% were considered susceptible. NHS was decomplemented via heating at 56°C for 30 min. The classical and lectin complement pathways were selectively blocked via chelation with 10 mM ethylene glycol-bis(β-aminoethyl ether)-*N,N,N'N'*-tetraacetic acid (EGTA) plus 10 mM $MgCl_2$ (NHSMgEGTA) [40–42]. To selectively inhibit the alternative complement pathway, factor B was inactivated via the incubation of NHS at 50°C for 20 min (NHS50°C) [42,43].

Mouse Lethality Assay

Pathogen-free, 6- to 8-week old, male BALB/c mice were obtained from the National Laboratory Animal Center (Taiwan) and maintained in the pathogen-free vivarium of the Laboratory Animal Center of National Defense Medical Center (Taiwan). The tested bacteria were cultured overnight at 37°C in BHI broth and were then diluted (1:100) in fresh BHI broth. The culture was incubated until the mid-exponential growth phase, and the cells were then washed once, resuspended in PBS, and adjusted to the desired concentrations according to OD_{600}. The actual concentrations were verified by plating the cells to determine viable counts. Six mice for each group were injected intraperitoneally with 0.1 ml of the cell suspension, and the mice were monitored daily for 14 days to measure survival and the severity of illness.

Statistical Analysis

A two-tailed *t*-test or a log-rank test (for survival analysis) was used for statistical analysis. A *P* value of <0.05 was considered statistically significant.

Results

Streptomycin Susceptibility and Resistance Genes

The 127 *K. pneumoniae* strains obtained from the liver abscess patients in different geographic localities were investigated as described in the Materials and Methods. The minimal inhibitory concentrations (MICs) of streptomycin values ranged from 1 to 64 µg/ml, while the MIC_{50} and MIC_{90} of streptomycin were 3 and 6 µg/ml, respectively. Eighteen *K. pneumoniae* strains (MICs≥6 µg/ml) were used to study the mechanisms of streptomycin resistance (Table 1). The *aadA* gene or *strA-strB* genes were detected in the 6 strains that had MICs≥12 µg/ml. No streptomycin resistance was found to be caused by the *rpsL* mutations. *K. pneumoniae* NVT1001, with a 2 µg/ml streptomycin MIC value, was used to select spontaneous mutants with streptomycin resistance (Table 2).

Table 1. Streptomycin resistance and target genes in 18 *K. pneumoniae* strains with MIC values ranging from 6 to 64 μg/ml.

Gene	Strain number			
	6[b]	12	16	64
aadA	0	2	2	0
strA+strB	0	0	0	2
None[a]	12	0	0	0
Wild-type *rpsL*	12	2	2	2

[a]The *aadA, strA* and *strB* genes were not detected.
[b]The MIC (μg/ml) of streptomycin was determined using the E-test.

Isolation and Characteristics of the S12 Mutants

Spontaneous mutants of *K. pneumoniae* that were resistant to streptomycin were selected on LB agar plates supplemented with 50 or 500 μg/ml streptomycin, and 107 mutants were isolated. The *rpsL* genes of all of these mutants were sequenced, and fourteen different mutations on the *rpsL* were obtained, which cause thirteen varied S12 protein (Table 2). The examination of streptomycin resistance for these different mutants using the E-test led to the identification of ten streptomycin-dependent mutants and four streptomycin-resistant mutants (Table 2). Different concentrations (6–96 μg/ml) of streptomycin were needed for the survival of the ten streptomycin-dependent mutants (Table 2).

Growth Rate and *In Vitro* Growth Competition Assay

Compared to the parental strain cultured in antibiotic-free LB broth, the ten streptomycin-dependent mutants showed ~1.5- to 3.7-fold slower growth rates ($P<0.01$) when cultured in LB broth supplemented with 500 μg/ml streptomycin. Four of these ten mutants were also cultured in LB broth supplemented with 50 μg/ml streptomycin and the four streptomycin-dependent mutants showed ~2.0- to 2.7-fold slower growth rates ($P<0.01$) compared to the parental strain cultured in antibiotic-free LB broth. When comparing the four streptomycin-resistant mutants to the parental strain, which were all cultured in antibiotic-free LB broth, the K42R mutant showed a similar growth rate, while the K42N, K42T and K87R mutants showed ~1.1- to 1.3-fold slower growth rates ($P<0.05$). Unlike other streptomycin-resistant mutants, the K87R mutant showed a ~1.3-fold slower growth rate ($P<0.05$) when cultured in the presence of 500 μg/ml streptomycin than that without streptomycin. The growth phenotype of the streptomycin-resistant mutants can be restored by replacing the wild-type *rpsL* to these mutants using an allelic exchange method (Table 2 and Figure S1).

A highly sensitive *in vitro* growth competition experiment was used to further evaluate the fitness cost of the four streptomycin-resistant mutants (Table 3). Compared with the parental strain (with a fitness of 1), the K42N, K42T and K87R mutations reduced fitness to ~0.82, ~0.85 and ~0.94 respectively ($P<0.0001$), while the K42R mutation also slightly reduced fitness to ~0.97 ($P<0.01$). Competition assays were also performed for the four streptomycin-resistant mutants using their revertants as references, and the results demonstrated that the decreased fitness could be restored to original levels following complementation (Table 3).

Phagocytosis and Serum Bactericidal Assays

No significant difference was found between strain NVT1001 and its four streptomycin-resistant mutants in the neutrophil phagocytosis assay (Table 4). In the serum bactericidal assay, the K42R and K87R mutants both showed resistance to normal human serum (NHS) as their parental strain, while the K42N and K42T mutants showed susceptibility to NHS and a ~1.8-fold smaller survival rate ($P<0.05$) in decomplemented 75% NHS compared with the parental strain (Table 4). Although NHS resistance was observed in the K87R mutant, this mutant demonstrated a ~1.3-fold smaller survival rates ($P<0.05$) in 75% NHS compared with the parental strain. Similar result was observed when comparing the survival rates of these two strains cultured in decomplemented 75% NHS. This result indicates that the smaller survival rate of K87R mutant in 75% NHS was mainly caused by its slower growth. The results of serum bactericidal assay have been further validated by testing the revertant strains (Table 4).

The two serum-susceptible strains, the K42N and K42T mutants, were used to further investigate the contribution of each complement pathway to complement-mediated bacterial killing. In comparison with culturing in decomplemented 75% NHS, the two mutants cultured in 75% NHS50°C showed a ~1.2-fold smaller survival rate ($P<0.05$), while their survival rates were strongly reduced ($P<0.01$) in 75% NHSMgEGTA (as in 75% NHS) (Table 5). These results suggest that the alternative pathway has a crucial role in complement activation.

Virulence in Mice

To further assess the effects of the *rpsL* mutations on virulence, a mouse peritonitis model was used. A non-significant difference (log-rank test, $P>0.05$) was found between NVT1001 strain and its K42R mutant regarding the survival of 12 mice per strain, and their LD_{50} (the bacterial dose that caused 50% death) were both ~2×10^2 CFU (Figure 1A). When the experiment was performed at a higher dose (~2×10^3 CFU) in 6 mice per strain, no mice survived within 14 days after being inoculated with the NVT1001 strain or its K42R mutant (data not shown). Compared with the NVT1001-inoculated mice, the mice (n = 6 mice per strain) inoculated with the K42N or K42T mutants showed a ~1,000-fold increased LD_{50} (~2×10^5 CFU), while that inoculated with the K87R mutant showed a ~50-fold increased LD_{50} (~1×10^4 CFU) (Figure 1A). When the experiment was performed at a lower dose in 6 mice per strain, all mice survived within 14 days after being inoculated with K42N (~2×10^4 CFU), K42T (~2×10^4 CFU) or K87R mutant (~2×10^2 CFU) (data not shown). These results have been further confirmed via the intraperitoneal injection of 2×10^2 CFU of each revertant (K42R, K42N, K42T or K87R revertant) in 6 mice per strain (Figure 1B). A non-significant difference (log-rank test, $P>0.05$) was found between these revertants and NVT1001 strain regarding mouse lethality.

Table 2. Characteristics of *K. pneumoniae* NVT1001 and its derived strains.

Strain	Description[a]	Codon change[b]	Isolation frequency	Streptomycin MIC (μg/ml)	Phenotype[c]	Doubling time (min)[d]		
						Sm 0	Sm 50	Sm 500
NVT1001	Clinical isolate (WT)			2	Sm^S	18.8±0.4	ND	ND
K42R mutant	*rpsL* K42R	128A→G	8/107	>1,024	Sm^R	19.1±0.4	19.6±0.3	20.2±0.4
K42N mutant	*rpsL* K42N	129G→T	9/107	>1,024	Sm^R	**24.7±1.1**	24.4±0.7	23.8±0.9
K42T mutant	*rpsL* K42T	128A→C	5/107	>1,024	Sm^R	**21.7±0.5**	22.1±0.6	22.2±0.4
K87R mutant	*rpsL* K87R	263A→G	3/107	>1,024	Sm^R	**20.4±0.6**	21.5±0.9	26.4±1.1
P41L mutant	*rpsL* P41L	125C→T	1/107	>1,024; 96[e]	Sm^D	ND	ND	61.6±3.1
K42Q mutant	*rpsL* K42Q	127A→C	3/107	>1,024; 96[e]	Sm^D	ND	ND	58.0±2.9
K43E mutant	*rpsL* K43E	130A→G	2/107	>1,024; 96[e]	Sm^D	ND	ND	68.7±3.5
+K87 mutant	*rpsL* +K87	262-264::AAA	1/107	>1,024; 96[e]	Sm^D	ND	ND	61.1±2.1
D88E.1 mutant	*rpsL* D88E	267C→A	1/107	>1,024; 96[e]	Sm^D	ND	ND	61.7±2.6
D88E.2 mutant	*rpsL* D88E	267C→G	5/107	>1,024; 96[e]	Sm^D	ND	ND	61.5±2.9
P90Q mutant	*rpsL* P90Q	272C→A	2/107	>1,024; 8[e]	Sm^D	ND	38.2±1.2	28.9±1.2
P90R mutant	*rpsL* P90R	272C→G	1/107	>1,024; 6[e]	Sm^D	ND	45.5±2.4	36.1±2.5
P90L mutant	*rpsL* P90L	272C→T	36/107	>1,024; 8[e]	Sm^D	ND	50.2±3.6	53.3±2.5
G91D mutant	*rpsL* G91D	275G→A	30/107	>1,024; 8[e]	Sm^D	ND	47.7±1.5	53.0±1.9
K42R revertant	WT-*rpsL* replacement strain of K42R mutant			2	Sm^S	18.9±0.5	ND	ND
K42N revertant	WT-*rpsL* replacement strain of K42N mutant			2	Sm^S	19.1±0.8	ND	ND
K42T revertant	WT-*rpsL* replacement strain of K42T mutant			2	Sm^S	18.6±0.5	ND	ND
K87R revertant	WT-*rpsL* replacement strain of K87R mutant			2	Sm^S	19.0±0.6	ND	ND

[a]WT, wild type; Amino acid replacements are listed; +, an insertion.
[b]Numbering begins with the start codon (ATG) of the ORF.
[c]Sm^S, streptomycin-sensitive; Sm^R, streptomycin-resistant; Sm^D, streptomycin-dependent.
[d]Doubling times were measured in Luria-Bertani broth without or with streptomycin (50 or 500 μg/ml) and are the average of three independent experiments. The doubling times of growth in the absence of streptomycin were further bolded when a significant ($P<0.05$) difference was found between NVT1001 and its derived strains. ND, not determined. These strains would not grow in either the presence or absence of streptomycin according to the results of the E-test.
[e]The minimal streptomycin concentration that enables growth was determined using the E-test, which was also used to determine the streptomycin MICs.

Table 3. Determination of relative fitness via *in vitro* growth competition assay.

Strain	Fitness (fit_r ± SD)[a]	*P* value[b]	N[c]
Relative to NVT1001			
K42R mutant	0.97±0.027	0.0091	12
K42N mutant	0.82±0.011	<0.0001	12
K42T mutant	0.85±0.015	<0.0001	12
K87R mutant	0.94±0.010	<0.0001	12
Relative to its revertant			
K42R mutant	0.98±0.016	0.0005	12
K42N mutant	0.83±0.010	<0.0001	12
K42T mutant	0.86±0.013	<0.0001	12
K87R mutant	0.95±0.016	<0.0001	12

[a]Fitness relative to *K. pneumoniae* NVT1001 or its revertant (with a fitness of 1); SD, standard deviation.
[b]*P* value, statistical significance of difference in fitness relative to *K. pneumoniae* NVT1001 or its revertant.
[c]N, number of fit_r on which the average fit_r value is based.

Discussion

In the 18 *K. pneumoniae* strains that had MICs≥6 μg/ml, streptomycin resistance genes could only be found in the 6 strains that had MICs≥12 μg/ml. This result was similar to the previous study in which the streptomycin epidemiological cut-off values were recommended as WT≤16 μg/ml for *Salmonella* and WT≤ 8 μg/ml for *E. coli* [2]. The *aadA* gene was found in the four *Klebsiella* strains exhibiting streptomycin MIC 12 or 16 μg/ml. The low-level resistance to streptomycin conferred by *aadA* can also be found in *Salmonella* and *E. coli* [2,7,44]. No streptomycin resistance was found to be caused by *rpsL* mutations in any of the 127 clinical strains of *K. pneumoniae*. This result suggests that the mutations of S12 protein were uncommon in *K. pneumoniae* that was isolated from liver abscess patients.

Compared with the parental strain in this study, the K42N and K42T mutants showed reduced virulence, while the K42R mutant showed similar virulence. These results are consistent with those reported in *Salmonella typhimurium* using the competition experiments in mice [27] and those reported in *Erwinia carotovora* using the potato tuber tests [26]. A reduced virulence of the streptomycin-resistant K87R mutant was further found in this study. Compared with the parental strain, the K87R mutant showed a 50-fold increase in the LD_{50} in a mouse peritonitis model, while the K42N and K42T mutants showed a 1,000-fold increase. The reduced virulence of these mutants should be partially caused by metabolic fitness, which was indicated by their slower growth than the parental strain when cultured in LB broth and decomplemented 75% NHS. The slower growth rate may cause them to be more easily eliminated by the immune system *in vivo*.

In the serum bactericidal assay, the K42N and K42T mutants were found to become susceptible to serum, while the alternative pathway should be the crucial role in complement activation. Whether this result is due to different levels of protein expression caused by the *rpsL* mutations requires further study, while this difference has been found between the K42T mutant and its parental strain in *E. carotovora* using a quantitative proteomic analysis [26]. Becoming serum-susceptible should explain the lowest virulence found for the K42N and K42T mutants

Table 4. Effect of mutations on phagocytosis and susceptibility to normal human serum (NHS)[a].

Strain	Ingested bacterial (%)[b]		Survival rate (%)[c]	
	15 min	30 min	75% NHS	Decomplemented
NVT1001	25.6±4.1	46.9±6.0	199±27	204±17
K42R mutant	20.4±3.7	47.7±9.7	198±12	193±16
K42N mutant	25.0±3.9	43.9±7.1	**14±7**	**112±4**
K42T mutant	20.6±5.9	44.3±9.0	**17±8**	**118±9**
K87R mutant	28.2±5.0	45.3±10.9	**151±13**	**142±11**
K42R revertant	ND	ND	194±16	183±11
K42N revertant	ND	ND	203±7	198±13
K42T revertant	ND	ND	201±20	183±5
K87R revertant	ND	ND	206±29	203±10

[a]Each value represents the means of three independent experiments ± the standard deviation. Boldface numbers indicate a significant (*P*<0.05) difference between NVT1001 and its derived strains. ND, not determined.
[b]*K. pneumoniae* strains were incubated with neutrophils for 15 or 30 min.
[c]Percent survival of the cells after 3 h of serum contact. Decomplemented, NHS was decomplemented via heating at 56°C for 30 min.

Table 5. Comparison of the bactericidal activity of normal human serum (NHS) with different treatments against serum-susceptible *K. pneumoniae* mutants.

NHS[a]	Survival rate (%)[b]	
	K42N mutant	K42T mutant
75% NHS	13±3	17±9
75% NHSMgEGTA	**32±6**	**43±10**
75% NHS50°C	**95±10**	**98±14**
Decomplemented 75% NHS	**120±9**	**115±10**

[a]NHSMgEGTA, NHS plus inhibition of the classical and lectin complement pathways; NHS50°C, NHS with inhibition of the alternative complement pathway. NHS was decomplemented via heating at 56°C for 30 min.

[b]Percent survival of cells after 3 h of serum contact. Each value represents the means of three independent experiments ± the standard deviation. Boldface numbers indicate a significant ($P<0.05$) difference in survival rate compared with the same strain incubated in 75% NHS.

compared with the parental strain, as well as the K42R and K87R mutants, which all showed serum resistance.

Based on the *in vitro* and animal experiments, the K42R mutant only showed a small fitness cost in a highly sensitive *in vitro* growth competition experiment. This mutant is also found at the highest frequency in streptomycin-resistant clinical isolates of *Mycobacterium tuberculosis*, while the K87R mutant was also prevalent [9,45]. Because the properties of Arginine are very similar to that of Lysine, such as they both are positively-charged amino acids, this substitution could cause the subtle effect on protein structure and function [46].

Compared with the parental strain, all four streptomycin-resistant mutants exhibited reduced fitness, and three of them showed decreased virulence, while the ten streptomycin-dependent mutants needed streptomycin to maintain their growth. Only the K42R mutant showed similar pathogenicity to its parental strain as previously described in other *Enterobacteriaceae* [26,27,47]. This similar pathogenicity can also been found in a highly virulent *K. pneumoniae* strain; its K42R mutant has an $LD_{50}<10$ CFU in a mouse peritonitis model as its parental strain and was used to construct unmarked in-frame gene deletions in our previous study [15]. In conclusion, the streptomycin resistance caused by *rpsL* mutations usually made the bacteria less competitive than its wild-type strain in the absence of streptomycin. The K42R mutant should be one of the best choices when using *rpsL* as a counter-selection marker.

Supporting Information

Figure S1 Creation of revertant strains for *rpsL* mutants.

Table S1 Oligonucleotide primers used in this study.

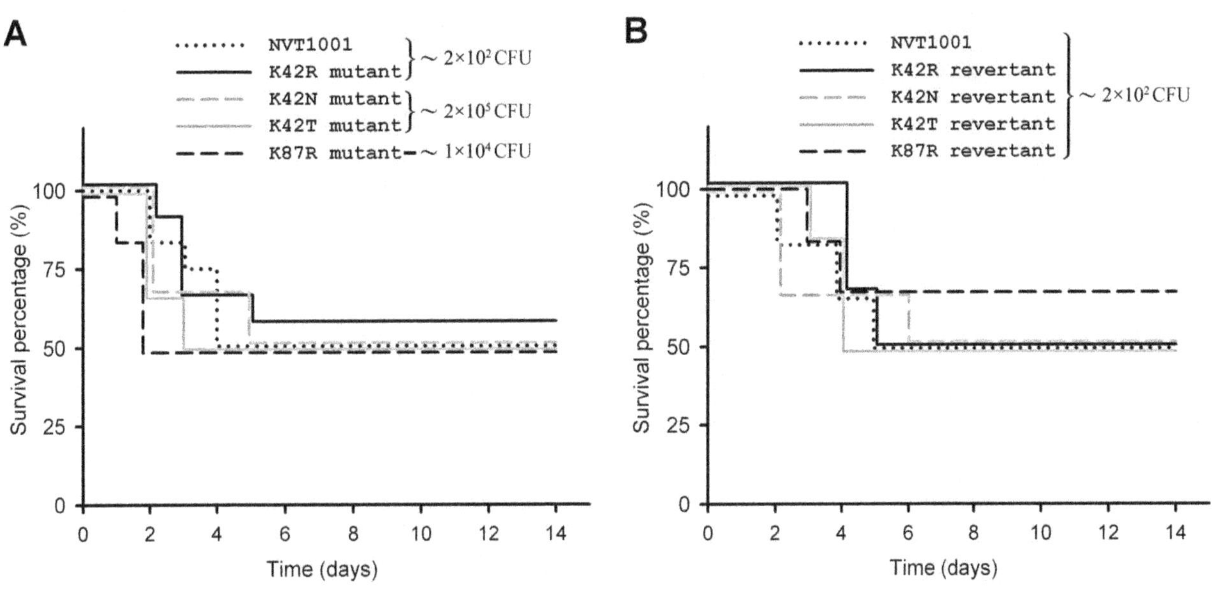

Figure 1. The effect of *rpsL* mutations on mouse lethality. The survival after intraperitoneal injection with 2×10^2 to 2×10^5 CFU of *K. pneumoniae* NVT1001, streptomycin-resistant mutant (A) or the revertant (B) was documented over 14 days. The data points represent the percentage of mice surviving in each experimental group over time (n = 6 mice per strain), while the data illustrated for *K. pneumoniae* NVT1001 and the K42R mutant in Figure 1A are pooled from two independent experiments. The LD_{50} of *K. pneumoniae* NVT1001 and the K42R mutant were similar, both approximately 2×10^2 CFU. The K42N, K42T and K87R mutants were all less virulent than NVT1001, while the LD_{50} of the K42N and K42T mutants were approximately 2×10^5 CFU and the LD_{50} of the K87R mutant was approximately 1×10^4 CFU.

Author Contributions

Conceived and designed the experiments: YKT FYC LKS. Performed the experiments: YKT CHL. Analyzed the data: YKT CHL JCL FYC LKS. Contributed reagents/materials/analysis tools: JCL LM CPF. Wrote the paper: YKT FYC LKS.

References

1. Schatz A, Bugle E, Waksman SA (1944) Streptomycin, a substance exhibiting antibiotic activity against gram-positive and gram-negative bacteria. Proc Soc Exp Biol Med 55: 66–69.
2. Garcia-Migura L, Sunde M, Karlsmose S, Veldman K, Schroeter A, et al. (2012) Establishing streptomycin epidemiological cut-off values for *Salmonella* and *Escherichia coli*. Microb Drug Resist 18: 88–93.
3. Galimand M, Sabtcheva S, Courvalin P, Lambert T (2005) Worldwide disseminated *armA* aminoglycoside resistance methylase gene is borne by composite transposon Tn1548. Antimicrob Agents Chemother 49: 2949–2953.
4. Doi Y, Arakawa Y (2007) 16S ribosomal RNA methylation: emerging resistance mechanism against aminoglycosides. Clin Infect Dis 45: 88–94.
5. Durante-Mangoni E, Grammatikos A, Utili R, Falagas ME (2009) Do we still need the aminoglycosides? Int J Antimicrob Agents 33: 201–205.
6. Madsen L, Aarestrup FM, Olsen JE (2000) Characterisation of streptomycin resistance determinants in Danish isolates of *Salmonella* Typhimurium. Vet Microbiol 75: 73–82.
7. Sunde M, Norström M (2005) The genetic background for streptomycin resistance in *Escherichia coli* influences the distribution of MICs. J Antimicrob Chemother 56: 87–90.
8. Finken M, Kirschner P, Meier A, Wrede A, Böttger EC (1993) Molecular basis of streptomycin resistance in *Mycobacterium tuberculosis*: alterations of the ribosomal protein S12 gene and point mutations within a functional 16S ribosomal RNA pseudoknot. Mol Microbiol 9: 1239–1246.
9. Hughes D, Andersson DI (2001) Target alterations mediating antibiotic resistance. *In* Antibiotic Development and Resistance, Hughes D. and Andersson D.I. (eds.), Taylor and Francis, London and New Tork, pp 23–36.
10. Chiou CS, Jones AL (1995) Molecular analysis of high-level streptomycin resistance in *Erwinia amylovora*. Phytopathology 85: 324–328.
11. Carter AP, Clemons WM, Brodersen DE, Morgan-Warren RJ, Wimberly BT, et al. (2000) Functional insights from the structure of the 30S ribosomal subunit and its interactions with antibiotics. Nature 407: 340–348.
12. Siu LK, Yeh K-M, Lin J-C, Fung C-P, Chang F-Y (2012) *Klebsiella pneumoniae* liver abscess: a new invasive syndrome. Lancet Infect Dis 12: 881–887.
13. Reyrat J-M, Pelicic V, Gicquel B, Rappuoli R (1998) Counterselectable markers: untapped tools for bacterial genetics and pathogenesis. Infect Immun 66: 4011–4017.
14. Skorupski K, Taylor RK (1996) Positive selection vectors for allelic exchange. Gene 169: 47–52.
15. Tsai YK, Fung CP, Lin JC, Chen JH, Chang FY, et al. (2011) *Klebsiella pneumoniae* outer membrane porins OmpK35 and OmpK36 play roles in both antimicrobial resistance and virulence. Antimicrob Agents Chemother 55: 1485–1493.
16. Wu Q, Stewart V (1998) NasFED proteins mediate assimilatory nitrate and nitrite transport in *Klebsiella oxytoca (pneumoniae)* M5al. J Bacteriol 180: 1311–1322.
17. Huang CJ, Wang ZC, Huang HY, Huang HD, Peng HL (2013) YjcC, a c-di-GMP phosphodiesterase protein, regulates the oxidative stress response and virulence of *Klebsiella pneumoniae* CG43. PLoS One 8: e66740.
18. Chiang MK, Lu MC, Liu LC, Lin CT, Lai YC (2011) Impact of Hfq on global gene expression and virulence in *Klebsiella pneumoniae*. PLoS One 6: e22248.
19. Cheng HY, Chen YS, Wu CY, Chang HY, Lai YC, et al. (2010) RmpA regulation of capsular polysaccharide biosynthesis in *Klebsiella pneumoniae* CG43. J Bacteriol 192: 3144–3158.
20. Lai YC, Peng HL, Chang HY (2003) RmpA2, an activator of capsule biosynthesis in *Klebsiella pneumoniae* CG43, regulates K2 *cps* gene expression at the transcriptional level. J Bacteriol 185: 788–800.
21. Drecktrah D, Douglas JM, Samuels DS (2010) Use of *rpsL* as a counterselectable marker in *Borrelia burgdorferi*. Appl Environ Microbiol 76: 985–987.
22. Kim IK, Jeong WK, Lim SH, Hwang IK, Kim YH (2011) The small ribosomal protein S12P gene *rpsL* as an efficient positive selection marker in allelic exchange mutation systems for *Corynebacterium glutamicum*. J Microbiol Methods 84: 128–130.
23. Sung CK, Li H, Claverys JP, Morrison DA (2001) An *rpsL* cassette, janus, for gene replacement through negative selection in *Streptococcus pneumoniae*. Appl Environ Microbiol 67: 5190–5196.
24. Sander P, Springer B, Bottger EC (2001) Gene replacement in *Mycobacterium tuberculosis* and *Mycobacterium bovis* BCG using *rpsL* as a dominant negative selectable marker. Methods Mol Med 54: 93–104.
25. Sander P, Meier A, Bottger EC (1995) *rpsL+*: a dominant selectable marker for gene replacement in mycobacteria. Mol Microbiol 16: 991–1000.
26. Barnard AM, Simpson NJ, Lilley KS, Salmond GP (2010) Mutations in *rpsL* that confer streptomycin resistance show pleiotropic effects on virulence and the production of a carbapenem antibiotic in *Erwinia carotovora*. Microbiology 156: 1030–1039.
27. Björkman J, Hughes D, Andersson DI (1998) Virulence of antibiotic-resistant *Salmonella typhimurium*. Proc Natl Acad Sci USA 95: 3949–3953.
28. Chumpolkulwong N, Hori-Takemoto C, Hosaka T, Inaoka T, Kigawa T, et al. (2004) Effects of *Escherichia coli* ribosomal protein S12 mutations on cell-free protein synthesis. Eur J Biochem 271: 1127–1134.
29. Siu LK, Fung CP, Chang FY, Lee N, Yeh KM, et al. (2011) Molecular typing and virulence analysis of serotype K1 *Klebsiella pneumoniae* strains isolated from liver abscess patients and stool samples from noninfectious subjects in Hong Kong, Singapore, and Taiwan. J Clin Microbiol 49: 3761–3765.
30. Yeh KM, Kurup A, Siu LK, Koh YL, Fung CP, et al. (2007) Capsular serotype K1 or K2, rather than *magA* and *rmpA*, is a major virulence determinant for *Klebsiella pneumoniae* liver abscess in Singapore and Taiwan. J Clin Microbiol 45: 466–471.
31. Gebreyes WA, Altier C (2002) Molecular characterization of multidrug-resistant *Salmonella enterica* subsp. *enterica* Serovar Typhimurium isolates from swine. J Clin Microbiol 40: 2813–2822.
32. Yeh KM, Lin JC, Yin FY, Fung CP, Hung HC, et al. (2010) Revisiting the importance of virulence determinant *magA* and its surrounding genes in *Klebsiella pneumoniae* causing pyogenic liver abscesses: exact role in serotype K1 capsule formation. J Infect Dis 201: 1259–1267.
33. Ohtomo R, Saito M (2003) A new selective medium for detection of *Klebsiella* from dairy environments. Microbes Environ 18: 138–144.
34. Sander P, Springer B, Prammananan T, Sturmfels A, Kappler M, et al. (2002) Fitness cost of chromosomal drug resistance-conferring mutations. Antimicrob Agents Chemother 46: 1204–1211.
35. Nielsen KL, Pedersen TM, Udekwu KI, Petersen A, Skov RL, et al. (2012) Fitness cost: a bacteriological explanation for the demise of the first international methicillin-resistant *Staphylococcus aureus* epidemic. J Antimicrob Chemother 67: 1325–1332.
36. Heinzelmann M, Gardner SA, Mercer-Jones M, Roll AJ, Polk HC Jr (1999) Quantification of phagocytosis in human neutrophils by flow cytometry. Microbiol Immunol 43: 505–512.
37. Campbell PA, Canono BP, Dervets DA, editors (1996) Isolation and functional analysis of neutrophils. New York: Wiley.
38. Lin JC, Chang FY, Fung CP, Xu JZ, Cheng HP, et al. (2004) High prevalence of phagocytic-resistant capsular serotypes of *Klebsiella pneumoniae* in liver abscess. Microbes Infect 6: 1191–1198.
39. Podschm R, Sievers D, Foscher A, Ullmann U (1993) Serotypes, hemagglutinins, siderophore synthesis, and serum resistance of *Klebsiella* isolates causing human urinary tract infections. J Infect Dis 168: 1415–1421.
40. Fine DP (1977) Comparison of ethyleneglycoltetraacetic acid and its magnesium salt as reagents for studying alternative complement pathway function. Infect Immun 16: 124–128.
41. Fine DP, Marney SR Jr, Colley DG, Sergent JS, Des Prez RM (1972) C3 shunt activation in human serum chelated with EGTA. J Immunol 109: 807–809.
42. Bugla-Ploskonska G, Korzeniowska-Kowal A, Guz-Regner K (2011) Reptiles as a source of *Salmonella* O48–clinically important bacteria for children: the relationship between resistance to normal cord serum and outer membrane protein patterns. Microb Ecol 61: 41–51.
43. Eidinger D, Bello E, Mates A (1977) The heterocytotoxicity of human serum: I. Activation of the alternative complement pathway by heterologous target cells. Cell Immunol 29: 174–186.
44. Sunde M (2005) Prevalence and characterization of class 1 and class 2 integrons in *Escherichia coli* isolated from meat and meat products of Norwegian origin. J Antimicrob Chemother 56: 1019–1024.
45. Nhu NT, Lan NT, Phuong NT, Chau N, Farrar J, et al. (2012) Association of streptomycin resistance mutations with level of drug resistance and *Mycobacterium tuberculosis* genotypes. Int J Tuberc Lung Dis 16: 527–531.
46. Betts MJ, Russell RB (2003) Amino acid properties and consequences of substitutions. Bioinformatics for Geneticists: John Wiley & Sons, Ltd. pp. 289–316.
47. Enne VI, Delsol AA, Davis GR, Hayward SL, Roe JM, et al. (2005) Assessment of the fitness impacts on *Escherichia coli* of acquisition of antibiotic resistance genes encoded by different types of genetic element. J Antimicrob Chemother 56: 544–551.

20

Evidence of Carbon Fixation Pathway in a Bacterium from Candidate Phylum SBR1093 Revealed with Genomic Analysis

Zhiping Wang[1,2], Feng Guo[1], Lili Liu[1,3], Tong Zhang[1]*

1 Environmental Biotechnology Laboratory, The University of Hong Kong, Hong Kong, **2** School of Environmental Science and Engineering, Shanghai Jiao Tong University, Shanghai, China, **3** State Environmental Protection Key Laboratory of Environmental Risk Assessment and Control on Chemical Process, East China University of Science and Technology, Shanghai, China

Abstract

Autotrophic CO_2 fixation is the most important biotransformation process in the biosphere. Research focusing on the diversity and distribution of relevant autotrophs is significant to our comprehension of the biosphere. In this study, a draft genome of a bacterium from candidate phylum SBR1093 was reconstructed with the metagenome of an industrial activated sludge. Based on comparative genomics, this autotrophy may occur via a newly discovered carbon fixation path, the hydroxypropionate-hydroxybutyrate (HPHB) cycle, which was demonstrated in a previous work to be uniquely possessed by some genera from *Archaea*. This bacterium possesses all of the thirteen enzymes required for the HPHB cycle; these enzymes share 30~50% identity with those in the autotrophic species of *Archaea* that undergo the HPHB cycle and 30~80% identity with the corresponding enzymes of the mixotrophic species within *Bradyrhizobiaceae*. Thus, this bacterium might have an autotrophic growth mode in certain conditions. A phylogenetic analysis based on the 16S rRNA gene reveals that the phylotypes within candidate phylum SBR1093 are primarily clustered into 5 clades with a shallow branching pattern. This bacterium is clustered with phylotypes from organically contaminated environments, implying a demand for organics in heterotrophic metabolism. Considering the types of regulators, such as FnR, Fur, and ArsR, this bacterium might be a facultative aerobic mixotroph with potential multi-antibiotic and heavy metal resistances. This is the first report on *Bacteria* that may perform potential carbon fixation via the HPHB cycle, thus may expand our knowledge of the distribution and importance of the HPHB cycle in the biosphere.

Editor: Gabriel Moreno-Hagelsieb, Wilfrid Laurier University, Canada

Funding: This research work is financed by the Research Grants Council of Hong Kong (GRF7190/12E) and National Natural Science Foundation of China (51108262). Dr. Wang would thank Hong Kong Scholar Program for financial support (XJ2012030), and Dr. Guo would thank the University of Hong Kong (HKU) for the postdoctoral fellowship. The funders had no role in study design, data collection and analysis, decision to publish, or preparation of the manuscript.

Competing Interests: The authors have declared that no competing interests exist.

* Email: zhangt@hku.hk

Introduction

As the most diverse and abundant cellular life forms in the biosphere, microorganisms play key roles in nearly all biogeochemical processes. However, most microorganisms are not available in pure cultures and can only be detected with culture-independent molecular surveys, which greatly inhibits our comprehension of their roles in ecological and biogeochemical processes. The genomic sequencing of these microorganisms is significant in the construction of blueprints for evolutionary and metabolic diversity [1]. With advances in next generation sequencing (NGS) and bioinformatics, draft genomes of uncultured bacteria can be reconstructed from various complex environmental samples via single-cell genome sequencing [2] or genome binning [3]. Therefore, metabolic deductions and evolutionary analyses can be performed based on the reconstructed genomes and comparative genomics [4], which may greatly expand our understanding of microbial metabolism and its potential role in ecology and biogeochemistry.

SBR 1093 was established as a candidate phylum using several 16S rRNA gene clones in phosphate-removing activated sludge from a sequencing batch reactor [5] that was supplied with sodium acetate for phosphate removal. Thereafter, they were continuously detected in an industrial wastewater treatment system receiving low-molecular-weight organic acids and short-chain alcohols [6], activated sludge from coking wastewater treatment, chlorinated hydrocarbon-contaminated soil and hydrocarbon-contaminated soil [7]. All of these environments were associated with short-chain fatty acids, which implied that the bacteria within this candidate phylum may proliferate effectively with short-chain fatty acids. In addition to the contaminated environment, 16S rRNA clones within candidate phylum SBR1093 were also detected in samples from ocean environments, such as ocean crust from the East Pacific Rise [8], polymetallic nodules and the surrounding sediments, oceanic surface sediment [9], sponges [10], etc. Considering these specific niches, deficiency of light, O_2 and organics, the most probable metabolism for these bacteria may be chemoautotrophy rather than heterotrophy. This is consistent with a report on a stalactite microbial community found in a desert cave [11] in which SBR1093-like 16S rRNA gene sequences comprised up to 10% of the total bacterial 16S rRNA gene sequences. Thus far, the metabolism of bacteria within candidate phylum SBR1093 remains elusive because there are no available pure cultures or enrichments from experiments or genomes. Because their abundance in the known microbial community is very low (less than 1% [11]), the metabolism of SBR1093 in these artificial and biogeochemical processes is difficult to deduce. Therefore, genome binning using the metagenome of a microbial community enriched with a member from this phylum could shed light on its metabolic properties and ecological functions.

As opposed to microbial communities in municipal wastewater treatment plants, which are fed with a mixture of natural organics and dominated by bacteria within *Proteobacteria*, *Bacteroidetes*, *Actinobacteria*, etc. [12], those in industrial wastewater treatment plants show unique populations in each plant [13]. Shaped by the specific substrates and physical-chemical conditions, microbial communities in industrial wastewater treatment plants are often enriched with uncultured microorganisms with specific metabolisms [14], and their metabolisms are associated with the biotransformation and biodegradation of specific substrates. Considering their relatively high abundance in these systems, draft genomes of the dominant populations could be reconstructed via the genome binning of the metagenome [15,16] in an attempt to elucidate their physiological and ecological functions in the microbial community (as well as their taxonomy) [17]. Based on a survey of 454 pyrosequencing for the microbial community pyrosequencings in industrial-activated sludge (data not shown here), a bacterium of candidate phylum SBR1093 was enriched in a full industrial wastewater treatment plant (WWTP), which fed with morpholine distilling-wastewater and performed an alternating anoxic/aerobic process. The objective of this study is to reconstruct the draft genome of a bacterium from candidate phylum SBR1093 with the metagenome of activated sludge from this WWTP. This may shed light on its taxonomic identity, metabolic properties and ecological role, thus be helpful in determining potential conditions for its cultivation and isolation.

Materials and Methods

Sample collection and DNA extraction

Activated sludge samples were collected from a local industrial wastewater treatment plant: two samples from anoxic and aerobic tanks, respectively, which were fed with morpholine distilling-wastewater and that operated in alternate anoxic/aerobic processes (There is no specific permission required for the collection of sludge samples. This sampling site is located at the Shanghai Industrial Park, N31.01, E121.41, and the field studies did not involve endangered or protected species.). The collected samples were fixed onsite with absolute ethanol at a volume ratio of 1:1, and then transported in an icebox and stored at −20°C prior to DNA extraction. For the DNA extraction, the microbial cells in the samples were collected after centrifugation and washed twice with phosphate-buffered solution (PBS). The DNA extraction was performed according to the protocol of the FastDNA SPIN Kit for soil (Qbiogene Inc., CA, USA), which was verified as the most suitable method to extract DNA from the activated sludge samples [18].

Metagenomic sequencing

With the extracted DNA, libraries with insert sizes of 200 bp and 800 bp were constructed for each sample according to the manufacturer's instructions (Illumina, San Diego, USA). Then, the metagenomic sequencing was performed with an Illumina HiSeq 2000 Platform (Illumina, San Diego, USA) using the 101 bp paired-end (PE101) strategy (BGI, Shenzhen, China). With a 2/5 Illumina sequencing run for the 200-bp library and a 1/5 sequencing run for the 800-bp library, approximately 101 and 35 million sequencing reads (100 bp) were obtained, respectively. Raw reads containing any ambiguous bases or those with an average quality score lower than 20 were removed prior to the following analysis.

De novo assembly

The filtered reads were imported into the CLC genomic workbench (version 4.9), and the paired-end sequences were used for the following *de novo* assembly in the CLC genomic workbench. The K parameter (k-mer size) was set to 51 (half of the PE sequencing length) during the assembly. Only contigs longer than 500 bp were output as well as the corresponding mapping reads for further analysis. More than 50% of the reads were assembled into contigs >500 bp (98,505 contigs), with a maximum length of 349,894 bp. As a test to examine the potential errors in the assembly, the coverage consistence of the assembled contigs was checked according to the previous report [15].

Genome binning

Genome binning was performed according to the previous work [19], based on a plot of coverage and GC ratio of contigs, including PE-tracking and reassembly, which was further refined with Metacluster 4.0 [20]. Then, the integrity and redundancy of the binning draft genome were assessed via the comparison of essential single copy genes (ESCGs) of most organisms in the domain *Bacteria* [15,19].

Gene annotation and comparison

Open reading frames (ORFs) were predicted online with MetaGeneMark [21], and the deduced amino acid (AA) sequences were obtained for BLASTp against the NCBI nr database (released on July 18, 2013) with an E-value of 10^{-5} and minimum alignment of 50 AA, respectively [22]. The results were taxonomically assigned with the lowest common ancestor (LCA) algorithm and functionally annotated based on KEGG using MEGAN 4.0 [23]. A Pfam search with an E-value of 10^{-5} was performed based on the Hidden Markol Model and against PfamA database version 26.0 [24], which could also be used for the comparison of gene clusters and verification of MEGAN annotation.

Results and Discussion

Genome binning and completeness assessment

A draft genome containing 94 scaffolds with a total size of 3,099,643 bp with GC contents of 56.4% was reconstructed (Figure 1). According to the prediction of MetaGeneMark, 3,228 ORFs were presented and 3,037 were in full length with the sole initiator and terminator, implying that the *de novo* assembly was accurately performed and that only the ORFs at the ends of the contigs were incomplete. Considering the functional assignment, 2,532 ORFs shared a mean similarity of 51.2% with the known enzymes in the nr-database (released on July 18, 2013), which is nearly at full align length (with a mean cover ratio of 0.87). All of the 40 universally occurring clusters of orthologous groups (COGs) [25] and tRNAs for all 20 amino acids are presented in this draft genome (Table S1 in File S1), which implies that it is near completeness. Additionally, based on the Hidden Markov Model (HMM) search, 102 unique ESCGs were found in this draft genome (Table S2 in File S1), indicating a completeness

>96% [26]. Among the four suspected repetitive ESCGs, TIGR00436 and PF00750 are not always a single copy gene [19], and TIGR02350 is hit with the duplicate genes located within contig_838, implying no contamination from other bacteria. The only suspected duplicate ESCGs (PF00162) are distributed in different contigs and in best hit on the NCBI web to sequences in heterotrophic *Anoxybacillus* sp. (YP002316858) and autotrophic *Nitrososphaera* sp. (YP006862459), respectively. These two sequences are assigned as glyceraldehyde 3-phosphate dehydrogenase and triosephosphate isomerase, which are responsible for glycolysis and gluconeogenesis, respectively, and thus may coexist in an autotrophic or mixotrophic bacterium (consistent with the following metabolism analysis). In summary, this draft genome has no verifiable contamination from other bacteria.

Phylogenetic and biogeographic characterization

This draft genome contains a complete rRNA operon (16S, 5S and 23S, 3,379–8,941 bp) on contig_439, and the 16S rRNA gene (1,567 bp, 6,992–8,558 bp) is used for the taxonomic identification with BLASTn against NCBI and the Greengenes 16S rRNA gene database (released at May, 2013). The genome shares only 85.9% similarity with the 16S rRNA gene of pure culture *Vibrio* sp. Gp-3-5.1 (HF912444) but approximately 94.9% similarity with the first nominated SBR1093 sequence (AF269002) and 99.9% with the uncultured bacterium (HE646343). Therefore, this draft genome should represent a bacterium from candidate phylum SBR1093, named as SBR1093 HKSP. The neighbor-joining and maximum-likelihood phylogenetic tree of the 16S rRNA gene in this bacterium and strains from relative phyla revealed that the candidate phylum SBR1093 represented by this bacterium is close to *Proteobacteria* (Figure S1 in File S1). Additionally, as shown in the phylogenetic tree of this 16S rRNA gene and the reference sequences (Figure 2), phylotypes from candidate phylum SBR1093 are primarily divided into two subdivisions, terrestrial and marine. They are further clustered into 5 clades within which SBR1093 HKSP belongs to clade I. The shallow branching pattern within this phylum (the largest distance between these phylotypes is less than 0.12) implies that bacteria within SBR1093 are recently radiated [27]. It should be noted that some sequences named SBR1093 (GQ348518, 350258, 350948, AY907765, EF573230) are clustered with *Vibrio* sp. (HF912444) and should be excluded from this candidate phylum. Clades within this candidate phylum are distinguished clearly according to their biogeographic distributions, which are differed from dissolve oxygen, salinity, pH, as well as organic nutrients. Bacteria in clades I and II are primarily found in terrestrial environments associated with fertile organics, whereas those in clade IV and V are primarily found in barren marine environments. The 16S rRNA gene of this draft genome SBR1093 HKSP is clustered with phylotypes found in the activated sludge or contaminated environment (with similarity > 99.9%), conditions with plenty of organics, thus implying the potential of organic metabolism (see details below).

In addition to the taxonomy of the 16S rRNA gene, proteins encoding in the genome may be another important resource in determining the phylogenetic position of an unknown bacterium. With the results of BLASTp checked against those in the nr-database, MEGAN may be used to classify these proteins using an LCA algorithm. Therefore, the ancestry of this bacterium can be speculated according to the phylogenetic relationship of proteins encoding in this draft genome. Of the 2,532 proteins in SBR1093 HKSP that have homologs with proteins in the nr-database, the largest section is clustered within the phylum level of *Proteobacteria* (n = 549), which is followed by *Firmicutes* (n = 70), *Cyanobacteria* (n = 53) and *Bacteroidetes* (n = 42) (Figure S2 in File S1). On the genus level, the most hits (n = 104) belong to *Geobacter* in *Deltaproteobacteria*, and the rests are evenly distributed among more than 50 genera. Therefore, candidate phylum SBR1093 represented by this bacterium may be close to *Proteobacteria* and *Firmicutes* but nonetheless separate from them, which is consistent with the taxonomy analysis of the 16S rRNA gene.

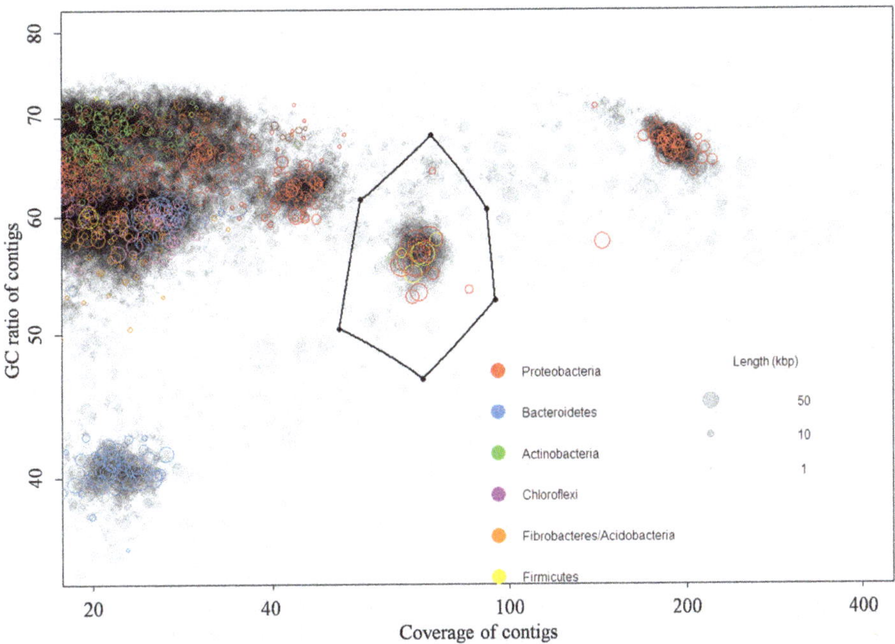

Figure 1. Genome binning of the dominant population with a plot of assembly contigs (based on coverage versus GC ratio). The circles represent the contigs with the size of the square root of their length. Clusters of contigs with similar color present potential genome bins, and contigs cluster with a coverage of approximately 80 (enclosed with black line) were collected for genome binning in this study.

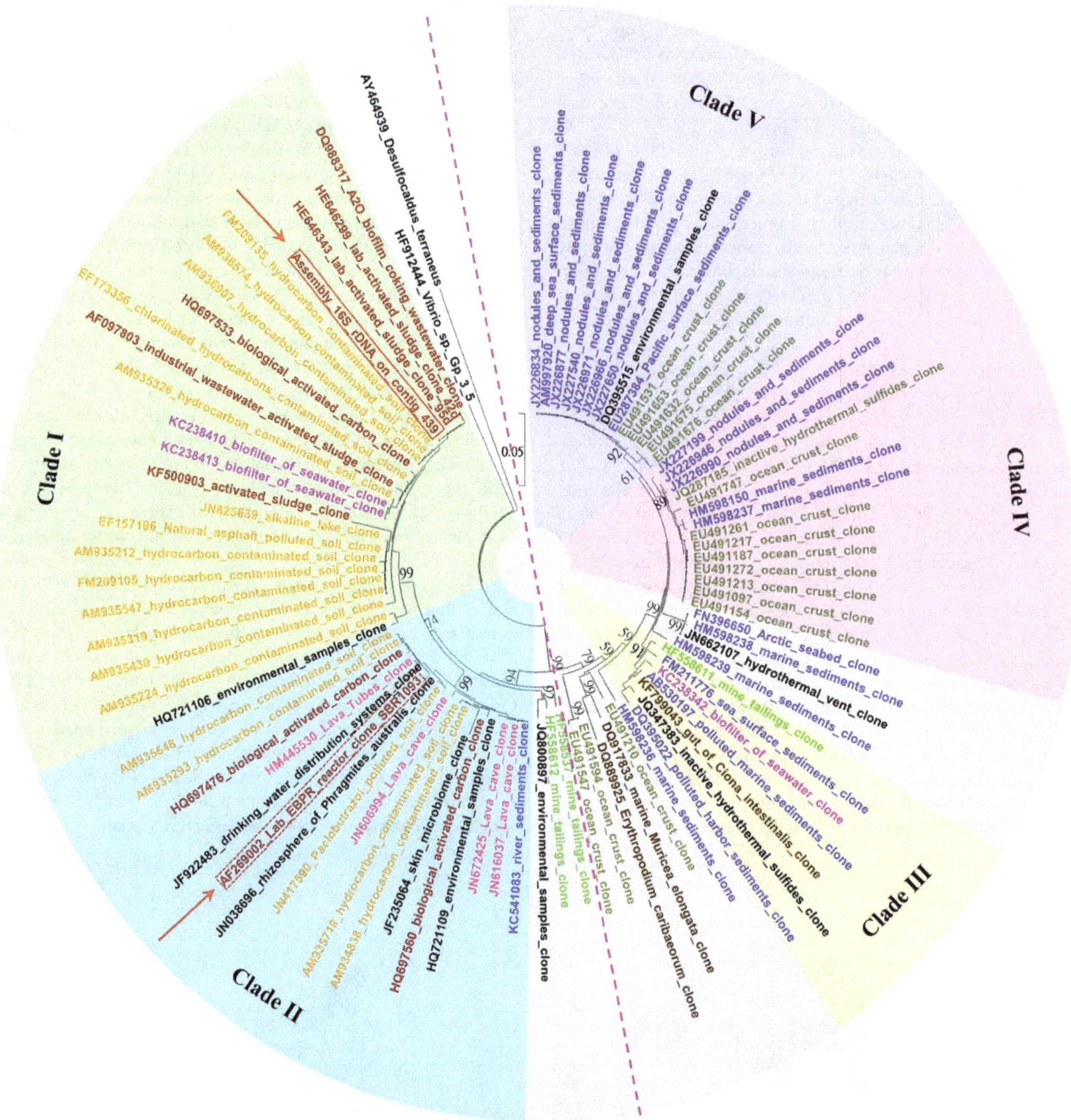

Figure 2. Phylogeny of phylotypes affiliated with the candidate phylum SBR1093. This phylogenetic tree is constructed with 16S rRNA gene sequences based on the neighbor-joining method with Jukes and Cantor distances. The main clades with nodes supported by a bootstrap value of >50% are labeled and marked with different background colors (Clade I green, Clade II blue, Clade III yellow, Clade IV pink and Clade V purple). The phylotypes derived from different sources are labeled with the following: dark red, activated sludge; orange, soil; blue, sediments; dark green, ocean crust; pink, lava; purple, seawater; green, mine tailings; brown, marine organisms; black, others. The phylotype SRB1093 HKSP obtained in this study is enclosed with a solid red line, whereas the first reported phylotype is enclosed with a red dashed line. The scale bar represents 0.05 nucleotide substitutions per site.

Bacterial morphology and cell wall type

This SBR1093 bacterium might be rod-shape, for there is a complete set of genes for rod shape proteins identified in this draft genome (gene_1559–1563). Considering with the cell wall type, a complete set of genes responsible for peptidoglycan biosynthesis (gene_1066–1072) and outer membrane lipoprotein encoding (gene_41, 44, 411 and 2121) are identified, which are only present in Gram-negative bacteria, suggesting this bacterium might be Gram negative. During the life cycle, the bacterium may form spores in adverse conditions and germinate in more suitable conditions, because there is a complete set of genes for spore formation, maturation and germination (gene_1062, 1522 and 265).

Primary metabolism

According to the KEGG annotation in MEGAN, approximately half of the predicted proteins (1,153 of the total 2,329) belong to metabolism, including 268 for carbohydrate metabolism, 223 for amino acid metabolism and 76 for lipid metabolism. The genes responsible for glycolysis/gluconeogenesis and the citrate cycle are complete in this draft genome, as well as a set for oxidative phosphorylation (Table S3 in File S1), indicating that this bacterium should be an aerobic heterotroph. However, two genes encoding the CRP/FNR family regulator and five genes for the Fur family regulators were also identified in this draft genome, indicating the potential anoxic metabolism of this bacterium. Therefore, it appears to be a facultative aerobic bacterium. This is consistent with the conditions of the anoxic/aerobic process from which these samples were collected, as well as the biogeographic distribution of the phylotypes from candidate phylum SBR1093. Additionally, full genes responsible for the upstream metabolism of glycolysis, such as the metabolism of glycogen/starch and cellulose, are also identified in this draft genome, but there were no genes related to the uptake of glucose or metabolism of sucrose, maltose or xylose. Therefore, this bacterium may use glycolysis only for the catabolism of self-producing sugar rather than external sugar.

Based on the genetic analysis, this bacterium may reduce nitrate via an assimilation path and nitrite via an assimilation or dissimilation path (Figure S3 in File S1), which indirectly verifies the facultative aerobic metabolism. Although most of the amino acid can be synthetized with the assimilated ammonia, no genes for asparagine synthesis were detected, implying the necessity for asparagine supplementation in the enrichment of this bacterium. This is consistent with the branched-chain amino acid ABC transporters. Additionally, only genes encoding sulfite reductase exist in this draft genome (none for sulfate reductase), indicating that this bacterium may survive in niches with relatively high redox potential rather than strictly anaerobic conditions. This is also consistent with the conditions in which the microbial community enriched this bacterium (i.e., cycling between the anoxic and aerobic tanks).

Regarding the utilization of the substrates, microbes enriched in a specific condition may adapt to this environment and interact closely with their habitat through metabolic reactions [28]. Because this bacterium is enriched in an anoxic/aerobic process and is fed with morpholine distilling wastewater, it may be involved in the metabolism of the main organics in this wastewater. However, the organics in this influent are primarily in the form of morpholine (or its derivatives), which is reported to decompose into glyoxylate and glycolate only by some species of *Mycobacterium* containing a complex of cytochrome p450, ferredoxin and ferredoxin reductase [29]. Based on the genetic analysis, there is no p450 gene in this draft genome. Therefore, this bacterium may not be active in the ring-opening of morpholine and is more likely involved in the metabolism of the intermediates of morpholine decomposition such as glycolate and glyoxylate. This is consistent with multiple genes responsible for the conversion of glycolate (gene_1890, 1891, 2884, 2885, 2887) and glyoxylate (gene_55), as well as the fusion protein responsible for the glyoxylate shunt (Table S3 in File S1, gene_720), which combines the glyoxylate with acetyl-CoA for the synthesis of malate.

Identification of the carbon fixation pathway

Interestingly, this draft genome possesses genes encoding phosphoenolpyruvate (PEP), acetyl-CoA and propionyl-CoA carboxylase, which catalyze carbon fixation, indicating that this bacterium might have autotrophic metabolism potential. Further comparative genetics reveal that this draft genome contains all of the 13 enzymes required for the hydroxypropionate–hydroxybutyrate (HPHB) cycle (Figure 3), a carbon fixation pathway possessed only by *Archaea* [30]. The enzymes in this draft genome share 30~50% amino acid identity with those involved in the HPHB cycle in *Metallosphaera sedula*, a typical strain of *Archaea* that undergoes the HPHB cycle (Table 1). In this carbon fixation pathway, two molecules CO_2 are assimilated in form of bicarbonate with the carboxylation of a bifunctional biotin-dependent acetyl-CoA/propionyl-CoA carboxylase, and the produced acetyl-CoA is transferred to the citrate cycle for the synthesis of the amino acids and fatty acids or alternatively glycogen via gluconeogenesis. For other carbon fixation pathways, some key enzymes are absent from this draft genome (Figure S4 in File S1), such as pyruvate synthase (EC:1.2.7.1) for the 3-dicarboxylate-hydroxybutyrate (DCHP) pathway, ATP-citric lyase (EC:2.3.3.8) and 2-oxoglutarate synthase (EC:1.2.7.3) for the reductive citric acid cycle pathway, NADP+ dependent formate dehydrogenase (EC:1.2.1.43) and formyltetrahydrofolate synthetase (EC:6.3.4.3) for the reductive acetyl-CoA pathway, malyl-CoA lyase (EC:4.1.3.24) for the 3-hydroxypropionate cycle pathway, and ribulose-bisphosphate carboxylase (EC:4.1.1.39) for the reductive pentose phosphate cycle pathway [31]. Therefore, SBR1093 HKSP may have autotrophic metabolism via the HPHB carbon fixation pathway, whereas the phosphoenolpyruvate carboxylase may function to maintain the balance of intermediates within the citrate cycle [32]. Additionally, enzymes employed for this carbon fixation pathway in *Archaea* are oxygen-tolerant, which is also consistent with the living conditions of this bacterium.

However, microbes possessing the HPHB pathway may evolve as facultative autotrophs rather than obligate autotrophs because this carbon fixation pathway is the most energy-consuming pathway, requiring nine ATP equivalents for one pyruvate [33]. Organisms harboring this carbon fixation pathway may proliferate via a heterotrophic rather than autotrophic pathway as long as there are available organic substrates, which is the nutritional strategy of microbes [34]. For example, although many species within *Sulfolobales* are described as autotrophs or mixotrophs via this carbon fixation pathway, some strains deposited in the public culture collection could lose their autotrophic ability after continuous laboratory cultivation in nutrient-rich media [33]. According to the above discussion, SBR1093 HKSP may possess potential of both autotrophic and heterotrophic metabolisms, thus should be a mixotroph. This is consistent with the enriching conditions, as well as the habitats of strains in the same taxonomic clade. Mixotrophy has been demonstrated to be a widespread and important phenomenon in the biosphere [35], which combines the traits of autotrophs and heterotrophs to utilize both inorganic and organic resources to enable the host survive in oligotrophic or transient environments. In this study, SBR1093 HKSP is enriched in an anoxic/aerobic process, which cycled microbes continuously from feast to fast according to anoxic and aerobic conditions. The microorganisms thus adapt to the transient environment and strive for the dominant position in this microbial community.

Additionally, in this draft genome, a key enzyme for this HPHB pathway, 4-hydroxybutyryl-CoA dehydratase (gene_213), is diverged from those in *Archaea* possessed the HPHB pathway, but clustered with facultative autotrophic bacteria such as *Bradyrhizobium* and *Afipia* within *Bradyrhizobiaceae* (Figure 4). This suggests that this gene is unlikely to be horizontally obtained across a recent domain for SBR1093. A further comparison analysis reveals that in addition to the identified autotrophic

Table 1. Genes responsible for the HPHB cycle in the identified genomes and their identity with those in SBR1093.

SBR1093 HKSP	Functional enzymes	*Ac. hospitalis*		*M. sedula*		*S. tokodaii*		*N. gargensis*		*Af. felis*		*B. sp.* STM3843		*O.carboxidovorans*	
		Acc. No.	Identity	Acc. No.	Identity	Acc. No.	Identity	Acc. No.	Identity	Acc. No.	Identity	Acc. No.	Identity	Acc. No.	Identity
gene_764	ACR	58889	32%	91508	31%	78610	32%	63583	37%	19219	37%	39404	35%	31729	36%
gene_2584	**ACC**	**59288**	**44%**	**90248**	**45%**	**76481**	**45%**	**60794**	**51%**	**19087**	**53%**	**33415**	**51%**	**32993**	**52%**
gene_2098	HPCD	58277	45%	92065	47%	77478	44%	62785	46%	19777	41%	36654	42%	32328	35%
gene_2045	MCE	59386	40%	90738	38%	76441	44%	63084	43%	18664	42%	33379	38%	33030	42%
gene_2581	MSR	59445	41%	91505	41%	78088	43%	61994	39%	21469	44%	35006	54%	34115	43%
gene_5	MCM	59385	47%	90737	47%	76440	48%	63082	48%	//	//	36638	63%	//	//
gene_1024	HBCS	58189	29%	91107	31%	77036	30%	62874	25%	20234	36%	37419	37%	31206	37%
gene_451	HPCS	58189	48%	91435	53%	78009	52%	62874	54%	20930	57%	37438	59%	31167	55%
gene_2450	ACK	59345	41%	90755	43%	76400	41%	62339	28%	21237	59%	33768	59%	31369	58%
gene_1582	**MCR**	**58545**	**28%**	**91779**	**26%**	**77174**	**29%**	**60718**	**31%**	**20991**	**55%**	**38304**	**55%**	**31211**	**55%**
gene_595	**CCH**	**59445**	**24%**	**90500**	**29%**	**75917**	**30%**	**61994**	**23%**	**21942**	**27%**	**35636**	**29%**	**33680**	**27%**
gene_213	HBCD	59205	30%	91403	31%	77631	30%	64012	34%	19793	82%	33076	80%	32741	82%
gene_2933	SSR	59446	31%	91506	30%	78341	29%	63668	37%	19498	32%	38465	35%	32182	30%

The identified genomes are downloaded from NCBI FTP with the accession names *Acidianus hospitalis* W1 uid66875 (*Ac. hospitalis*), *Metallosphaera sedula* DSM 5348 uid58717 (*M. sedula*), *Sulfolobus tokodaii* 7 uid57807 (*S. tokodaii*), Candidatus *Nitrososphaera gargensis* Ga9 2 uid176707 (*N. gargensis*), *Afipia felis* ATCC 53690 uid179396 (*Af. felis*), Bradyrhizobium STM 3843 uid80711 (*B. sp.* STM3843), *Oligotropha carboxidovorans* OM5 uid72795 (*O. carboxidovorans*). The enzymes in bold are bifunctional enzymes in the HPHB pathway, and columns filled with//indicate that there were no hit enzymes in the genome. *M. sedula*, *S. tokodaii* and *Ac. hospitalis* are members of *Sulfolobales* and have the potential for autotrophic metabolism via the reported HPHB pathway, whereas *B. sp.* STM3843, *Af. felis* and *O. carboxidovorans* are members from *Bradyrhizobiaceae* with facultative autotrophic metabolism via an unknown carbon fixation pathway. The accession number for *Acidianus hospitalis* should have 'YP0044' before the presented numbers, and *Metallosphaera* should have 'YP0011', *Sulfolobus* 'NP3', *Nitrososphaera* 'YP0068', *Afipia* 'ZP114', *Bradyrhizobium* 'ZP0943', and *Oligotropha* 'YP0046'.

Abbreviations for functional enzymes: Acryloyl-CoA reductase (ACR); Acetyl/propionyl-CoA carboxylase (ACC); 3-Hydroxypropionyl-CoA dehydratase (HPCD); Methylmalonyl-CoA epimerase (MCE); Malonic semialdehyde reductase (MSR); Methylmalonyl-CoA mutase (MCM); 4-Hydroxybutyrate-CoA ligase (HBCS); 3-Hydroxypropionate-CoA ligase (HPCS); Acetoacetyl-CoA β-ketothiolase (ACK); Succinyl/Malonyl-CoA reductase (MCR); Crotonyl-CoA hydratase (CCH); 4-Hydroxybutyryl-CoA dehydratase (HBCD); Succinic semialdehyde reductase (SSR).

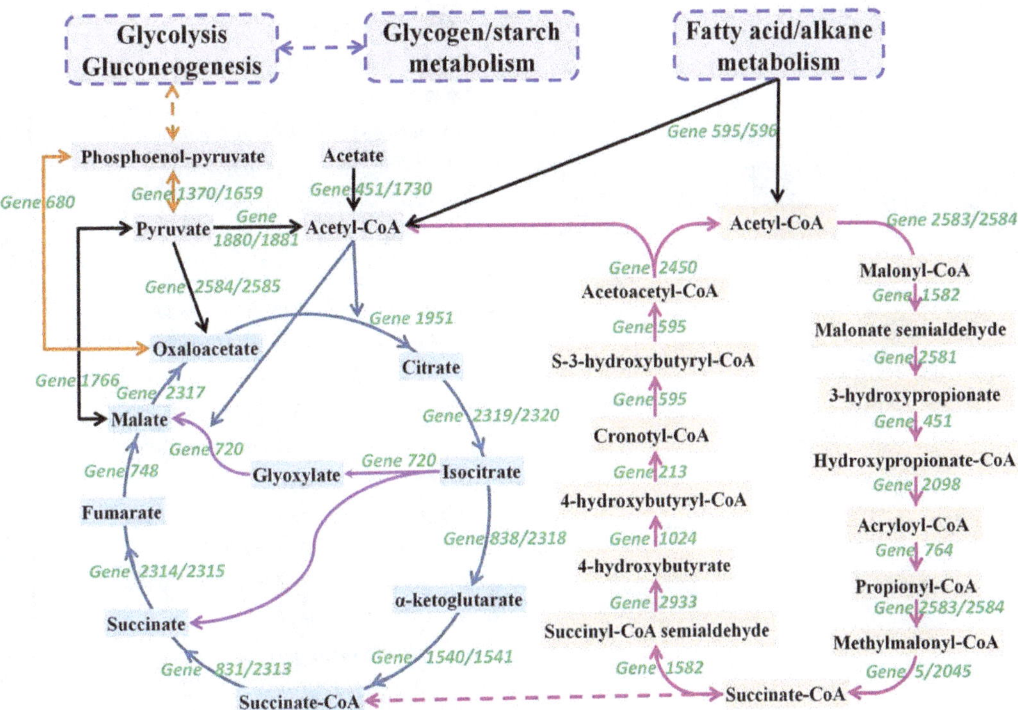

Figure 3. Putative metabolic pathway of SBR1093 HKSP (based on the genetic analysis). Carbon fixation with the HPHB cycle is used for the biomass synthesis via the transfer of acetyl-CoA to the citrate cycle or gluconeogenesis, and the genes responsible for each step are marked with green words. Intermediates connected with colored lines represent different metabolic pathways.

Archaea, all of the 13 enzymes involved in the HPHB pathway are also identified in some species within *Bradyrhizobiaceae* (Table 1), implying the wide distribution of the HPHB pathway in the biosphere and the potential importance of carbon fixation in soils and oceans. A phylogenetic analysis based on concatenated amino acid sequences, that are responsible for the transfer of succinyl-CoA to acetoacetyl-CoA and shared by DCHP & HPHB pathway, also reveals that this bacterium is closer to bacteria within *Bradyrhizobiaceae* than *Archaea* (those responsible for the transfer of crotonyl-CoA to acetoacetyl-CoA are not included because they are fusion proteins of enoyl-CoA hydratase and hydroxybutyryl-CoA dehydrogenase in bacteria but in reverse order in *Archaea*). Therefore, this bacterium may have a nutrition pattern similar to that of bacteria within *Bradyrhizobiaceae*.

Compared with the oligotrophic niches in which *Archaea* reside, *Bradyrhizobiaceae* bacteria are often found in fertile habitats rich in nutrients. Bacteria within *Bradyrhizobiaceae* often appear as heterotrophs rather than autotrophs, although most can grow chemolithoautotrophically using hydrogen, thiosulfate or sulfide as an electron donor [36]. Similarly, the bacterium SBR1093 HKSP is enriched in an industrial WWTP with an influent chemical oxygen demand (COD) concentration of up to 7,000 mg L^{-1} and ammonia concentration up to 100 mg L^{-1}; thus, it may also be a facultative heterotroph or obligate mixotroph.

Antibiotics and heavy metal resistance

It is also interesting to note that this bacterium may have multidrug resistance, such as penicillin, tetracycline, methylenomycin A, chloramphenicol and some macrolide-specific drugs, because abundance of genes associated with drug resistance and efflux transporters are identified in the draft genome. For example, 13 genes encoding a putative drug exporter in the RND superfamily may catalyze antibiotics efflux via an H+ antiport mechanism (Table S4 in File S1), and 19 genes encoding MFS transporter may have antibiotic resistance potential.

Additionally, a complete set gene associated with arsenate reduction and transport has also been identified in this draft genome, as well as 12 genes encoding the ArsR family transcriptional regulator, which may repress the expression of operons linked to stress of di- and multivalent heavy metal ions [37]. Therefore, a variety of heavy metal resistances and transporter genes can be expected in the genome, including the resistance genes for copper and mercury and transport genes for Cu^{2+} and chromate (Table S4 in File S1). The variety resistance to a variety of heavy metals is consistent with the relevant geographical distribution, ocean crust, marine sediment, contaminated soil and activated sludge.

Conclusions

In summary, a draft genome of uncultivated bacteria from candidate phylum SBR1093 was reconstructed with the metagenome of a microbial community from a full-scale industrial wastewater treatment plant. According to the phylogenetic analysis, this bacterium belongs to clade I of candidate phylum SBR1093, which is associated with a contaminated environment and indicates the demand of organics for metabolism. Genome analysis indicates that the bacterium SBR1093 in this phylum may grow autotrophically via the HPHB cycle, a carbon fixation pathway recently found only in some genera from *Archaea*. Enzymes in this draft genome involved in carbon fixation are diverged from those in *Archaea* but share obvious homology to those found in *Bradyrhizobiaceae*. Therefore, this indicates that this bacterium may be a mixotroph. So far, all of the metabolic properties of this SBR1093 HKSP are deduced only based on the

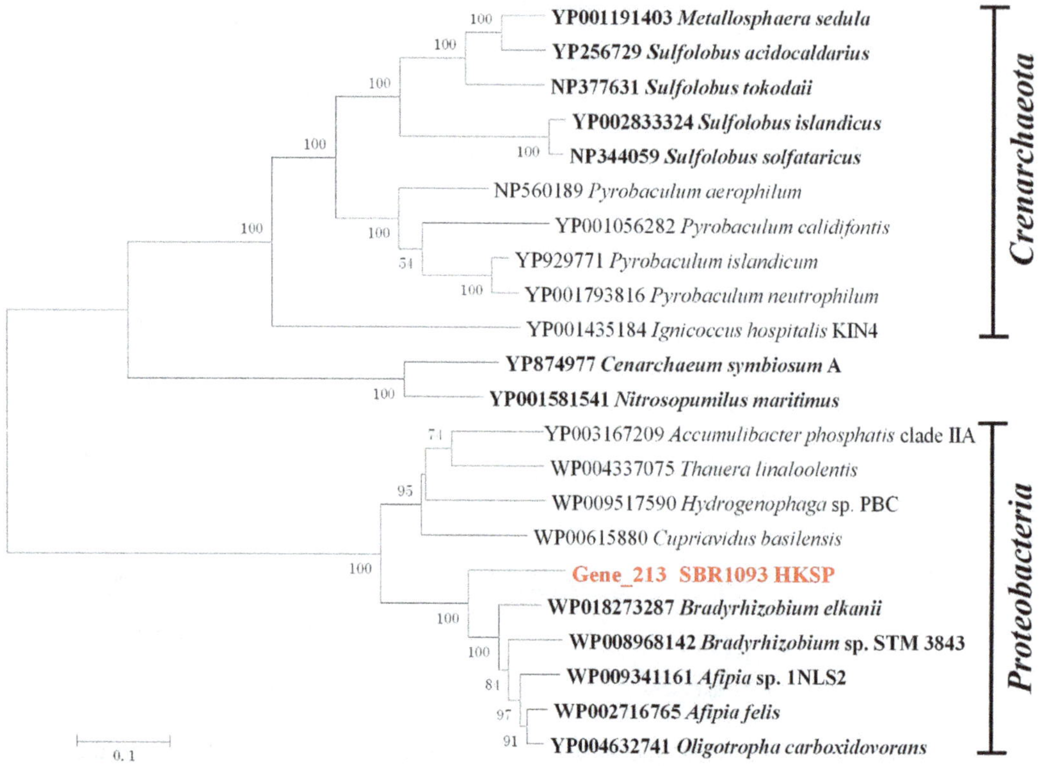

Figure 4. Phylogenetic tree of 4-hydroxybutyryl-CoA dehydratase proteins. The 4-hydroxybutyryl-CoA dehydratases in this draft genome and others in identified autotrophic *Achaea* were retrieved from NCBI to build the phylogenetic tree. The hosts of these marked (in bold) are suspected autotrophic microbes that undergo the HPHB cycle. The number in front of the taxonomy presents the accession number in NCBI. The tree topography and evolutionary distances are given via the neighbor-joining method with a Poisson correction. The numbers at the nodes indicate the percentage bootstrap values for the clade of this group in 1,000 replications. The scale bar represents a difference of 0.1 substitutions per site.

genomic analysis and comparative genomics. Further understanding of the ecological role of this candidate phylum will be obtained through its effective enrichment in the laboratory and the investigation on pure culture.

Supporting Information

File S1 Supporting Figures and Tables. Figure S1. Phylogenetic tree of sequences of SBR1093 and the reference sequences from representative phyla. This phylogenetic tree is constructed with 16S rRNA gene sequences based on the maximum-likelihood method with Jukes and Cantor distances. Only bootstrap value of >50% is labeled and the phylotypes of SRB1093 are marked with red in bold. The scale bar represents 0.1 nucleotide substitutions per site. **Figure S2. Taxonomy of proteins in this SBR1093 HKSP.** Proteins are converted with the predicted gene, and performed BLASTp against NCBI nr database (released at July 18, 2013). Therefore, it is imported into Megan for taxonomic classification. **Figure S3. Putative nitrogen metabolic pathway of this SBR1093 HKSP (Adapted from KEGG). Figure S4. Suspected carbon fixation pathway possessed by this SBR1093 HKSP (Adapted from KEGG).** Columns with solid fill indicate these enzymes are identified in this draft genome, which are connected with colorful lines and arrows and each color represents a type of carbon fixation pathway. Only the red one are full filled, which represents the HPHB cycle. **Table S1. Universally occurring clusters of orthologous groups and tRNAs for amino acids identified in this draft genome. Table S2. List of ESCGs identified in this draft genome and their taxonomy. Table S3. Suspected genes in this draft genome that involving in the primary metabolism. Table S4. Suspected genes in this draft genome that involving in the antibiotics and heavy metal resistance or export.**

Author Contributions

Conceived and designed the experiments: ZW TZ. Performed the experiments: ZW LL. Analyzed the data: ZW FG. Contributed reagents/materials/analysis tools: ZW LL TZ. Contributed to the writing of the manuscript: ZW FG LL TZ.

References

1. Rinke C, Schwientek P, Sczyrba A, Ivanova NN, Anderson IJ, et al. (2013) Insights into the phylogeny and coding potential of microbial dark matter. Nature 499: 431–437.
2. Lasken RS (2012) Genomic sequencing of uncultured microorganisms from single cells. Nat Rev Microbiol 10(9): 631–640.
3. Tyson GW, Chapman J, Hugenholtz P, Allen EE, Ram RJ, et al. (2004) Community structure and metabolism through reconstruction of microbial genomes from the environment. Nature 428: 37–43.
4. Flowers JJ, He SM, Malfatti S, del Rio TG, Tringe SG, et al. (2013) Comparative genomics of two *Candidatus Accumulibacter*' clades performing biological phosphorus removal. ISME J 7: 2301–2314.

5. Bond PL, Hugenholtz P, Keller J, Blackall LL (1995) Bacterial Community Structures of phosphate-removing and non-phosphate-removing activated sludges from sequencing batch reactors. Appl Environ Microbiol 61(5): 1910–1916.
6. Layton AC, Karanth PN, Lajoie CA, Meyers AJ, Gregory IR, et al. (2000) Quantification of *Hyphomicrobium* populations in activated sludge from an Industrial wastewater treatment system as determined by 16S rRNA analysis. Appl Environ Microbiol 66(3): 1167–1174.
7. Militon C, Boucher D, Vachelard C, Perchet G, Barra V, et al. (2010) Bacterial community changes during bioremediation of aliphatic hydrocarbon- contaminated soil. FEMS Microbiol Ecol 74: 669–681.
8. Santelli CM, Orcutt BN, Banning E, Bach W, Moyer CL, et al. (2008) Abundance and diversity of microbial life in ocean crust. Nature 453: 653–656.
9. Li H, Yu Y, Luo W, Zeng Y, Chen B (2009) Bacterial diversity in surface sediments from the Pacific Arctic Ocean. Extremophiles 13(2): 233–246.
10. Schmitt S, Tsai P, Bell J, Fromont J, Ilan M, et al. (2012) Assessing the complex sponge microbiota: core, variable and species-specific bacterial communities in marine sponges. ISME J 6: 564–576.
11. Ortiz-Ortiz M (2012) Kartchner caverns: habitat scale community diversity and function in a carbonate cave. Thesis. The University of Arizona. 65: 99.
12. Zhang T, Shao MF, Ye L (2012) 454 Pyrosequencing reveals bacterial diversity of activated sludge from 14 sewage treatment plants. ISME J 6: 1137–1147.
13. Ju F, Xia Y, Guo F, Wang ZP, Zhang T (2014) Taxonomic relatedness shapes bacterial assembly in activated sludge of globally distributed wastewater treatment plants. Environmental Microbiology DOI: 10.1111/1462-2920.12355.
14. Werner JJ, Knights D, Garcia ML, Scalfone NB, Smith S, et al. (2010) Bacterial community structures are unique and resilient in full-scale bioenergy systems. PNAS 108(10): 4158–4163.
15. Hess M, Sczyrba A, Egan R, Kim TW, Chokhawala H, et al. (2011) Metagenomic discovery of biomass-degrading genes and genomes from cow rumen. Science 331: 463–467.
16. Sharon I, Banfield JF (2013) Genomes from metagenomics. Science 342: 1057–1058.
17. Strous M, Pelletier E, Mangenot S, Rattei T, Lehner A, et al. (2006) Deciphering the evolution and metabolism of an anammox bacterium from a community genome. Nature 440: 790–794.
18. Guo F, Zhang T (2013) Biases during DNA extraction of activated sludge samples revealed by high throughput sequencing. Appl Microbiol Biotechnol 97: 4607–4616.
19. Albertsen M, Hugenholtz P, Skarshewski A, Nielsen KL, Tyson GW, et al. (2013) Genome sequences of rare, uncultured bacteria obtained by differential coverage binning of multiple metagenomes. Nat Biotechnol 31: 533–538.
20. Wang Y, Leung HCM, Yiu S, Chin FYL (2012) MetaCluster 4.0: a novel binning algorithm for NGS reads and huge number of species. J Comput Biol 19: 241–249.
21. Zhu W, Lomsadze A, Borodovsky M (2010) Ab initio gene identification in metagenomic sequences. Nucleic Acids Res 38(12): e132.
22. Wong MT, Zhang D, Li J, Hui RK, Tun HM, et al. (2013) Towards a metagenomic understanding on enhanced biomethane production from waste activated sludge after pH 10 pretreatment. Biotechnol Biofuels 6(1): 38.
23. Yu K, Zhang T (2013) Construction of customized sub-databases from NCBI-nr database for rapid annotation of huge metagenomic datasets using a combined BLAST and MEGAN Approach. PLoS ONE 8(4): e59831.
24. Sonnhammer EL, Eddy SR, Birney E, Bateman A, Durbin R (1998) Pfam: multiple sequence alignments and HMM-profiles of protein domains. Nucleic Acids Res 26(1): 320–322.
25. Stark M, Berger SA, Stamatakis A, von Mering C (2010) MLTreeMap-accurate Maximum Likelihood placement of environmental DNA sequences into taxonomic and functional reference phylogenies. BMC Genomics 11: 461.
26. Dupont CL, Rusch DB, Yooseph S, Lombardo MJ, Richter RA, et al. (2012) Genomic insights to SAR86, an abundant and uncultivated marine bacterial lineage. ISME J 6: 1186–1199.
27. Hunt DE, David LA, Gevers D, Preheim SP, Alm EJ, et al. (2008) Resource partitioning and sympatric differentiation among closely related *Bacterioplankton*. 320: 1080–1085.
28. Finlay B, Clark K (1999) Ubiquitous dispersal of microbial species. Nature 400: 828–828.
29. Trigui MPS, Truffaut N, Thomas D, Poupin P (2004) Molecular cloning, nucleotide sequencing and expression of genes encoding a cytochrome P450 system involved in secondary amine utilization in *Mycobacterium* sp. strain RP1. Res Microbiol 155: 1–9.
30. Ramos-Vera WH, Berg IA, Fuchs G (2009) Autotrophic carbon dioxide assimilation in *Thermoproteales* revisited. J Bacteriol 191: 4286–4297.
31. Berg IA (2011) Ecological aspects of the distribution of different autotrophic CO2 fixation pathways. Appl Environ Microbiol 77(6): 1925–1936.
32. Dunn MF (2011) Anaplerotic function of phosphoenolpyruvate carboxylase in *Bradyrhizobium japonicum* USDA110. Curr Microbiol 62: 1782–1788.
33. Berg IA, Ramos-Vera WH, Petri A, Huber H, Fuchs G (2010) Study of the distribution of autotrophic CO_2 fixation cycles in *Crenarchaeota*. Microbiology 156: 256–269.
34. Tittel J, Bissinger V, Zippel B, Gaedke U, Bell E, et al. (2003) Mixotrophs combine resource use to outcompete specialists: Implications for aquatic food webs. PNAS 100(22): 12776–12781.
35. Zubkov MV, Tarran GA (2008) High bacterivory by the smallest phytoplankton in the North Atlantic Ocean. Nature 455: 224–227.
36. Fuhrmann S, Ferner M, Jeffke T, Henne A, Gottschalk G, et al. (2003) Complete nucleotide sequence of the circular megaplasmid pHCG3 of *Oligotropha carboxidovorans*: function in the chemolithoautotrophic utilization of CO, H_2 and CO_2. Gene 322: 67–75.
37. Busenlehner LS, Pennella MA, Giedroc DP (2003) The SmtB/ArsR family of metalloregulatory transcriptional repressors: structural insights into prokaryotic metal resistance. FEMS Microbiol Rev 17: 131–143.

21

Space-Dependent Formation of Central Pair Microtubules and Their Interactions with Radial Spokes

Yuki Nakazawa[1¤a], Tetsuro Ariyoshi[1¤b], Akira Noga[1], Ritsu Kamiya[1,2], Masafumi Hirono[1]*

1 Department of Biological Sciences, Graduate School of Science, University of Tokyo, Bunkyo-ku, Tokyo, Japan, 2 Department of Life Science, Faculty of Science, Gakushuin University, Toshima-ku, Tokyo, Japan

Abstract

Cilia and flagella contain nine outer doublet microtubules and a pair of central microtubules. The central pair of microtubules (CP) is important for cilia/flagella beating, as clearly shown by primary ciliary dyskinesia resulting from the loss of the CP. The CP is thought to regulate axonemal dyneins through interaction with radial spokes (RSs). However, the nature of the CP-RS interaction is poorly understood. Here we examine the appearance of CPs in the axonemes of a *Chlamydomonas* mutant, *bld12,* which produces axonemes with 8 to 11 outer-doublets. Most of its 8-doublet axonemes lack CPs. However, in the double mutant of *bld12* and *pf14*, a mutant lacking the RS, most 8-doublet axonemes contain the CP. Thus formation of the CP apparently depends on the internal space limited by the outer doublets and RSs. In 10- or 11-doublet axonemes, only 3–5 RSs are attached to the CP and the doublet arrangement is distorted most likely because the RSs attached to the CP pull the outer doublets toward the axonemal center. The CP orientation in the axonemes varies in double mutants formed between *bld12* and mutants lacking particular CP projections. The mutant *bld12* thus provides the first direct and visual information about the CP-RS interaction, as well as about the mechanism of CP formation.

Editor: Takashi Toda, Cancer Research UK London Research Institute, United Kingdom

Funding: This work was supported by JSPS KAKENHI Grant Numbers 23657046, 24370079, and 25113503. The funders had no role in study design, data collection and analysis, decision to publish, or preparation of the manuscript.

Competing Interests: The authors have declared that no competing interests exist.

* Email: hirono@bs.s.u-tokyo.ac.jp

¤a Current address: Center for Neuroscience, University of California Davis, Davis, California, United States of America
¤b Current address: Department of Neurobiology, Graduate School of Medicine, University of Tokyo, Bunkyo-ku, Tokyo, Japan

Introduction

Motile cilia and flagella are ancient organelles present in various eukaryotic organisms including humans. Defects in structure or motility of cilia and flagella result in a category of diseases called primary ciliary dyskinesia [1,2]. The axoneme of motile cilia and flagella has a strikingly conserved structure consisting of nine outer doublet microtubules and two central-pair (CP) microtubules. These microtubules have various types of projections, such as outer and inner dynein arms, radial spokes (RSs), CP projections and nexin links, which directly or indirectly interact with each other [3–8]. The interactions between these projections must play fundamental roles for maintaining the cylindrical arrangement of the outer doublets, for producing the motive force, and for converting the force to ciliary/flagellar bending waves [9–11]. However, the nature of interaction between these structures is not well understood.

The ninefold symmetrical arrangement of the outer doublets is determined by the centriole (basal body), which has a cylindrical shape consisting of nine short triplet microtubules. Axonemal doublet microtubules assemble onto the A- and B-tubules of each triplet microtubule through a junction called the transition zone [11]. In contrast to the outer doublets assembling on clearly defined template structures, the assembly process of the CP is largely unknown. In *Chlamydomonas* flagella, the distal end of the CP is capped with a plate while the proximal end does not attach to any recognizable structure [12,13]. In *Tetrahymena* cilia, the distal end is also capped with a plate, while the proximal end of one CP microtubule is covered with the axosome, an amorphous structure observed in ciliate axonemes, and the other CP microtubule is apparently associated with no distinct structures [14,15]. Neither of the ends appears to function as the nucleation site for CP assembly [16–20]. However, what nucleates the CP assembly and what determines the CP number in the axoneme are not known.

The CP and RS appear to function as a regulatory system for ciliary and flagellar motility [21–25]. For generation of axonemal beating, a subset of the axonemal dyneins must be activated at a specific phase of beating, and the region of the activated dynein must move along the axoneme as the bending propagates toward the tip. Although the mechanism of this dynein regulation is yet to be elucidated, the CP/RS system clearly plays a crucial role in coordinating dynein activities [9,10,23]. Several lines of evidence suggest that, in some protists including *Chlamydomonas* and *Paramecium*, the CP assumes a twisted conformation and rotates within the axoneme once per beating cycle. The signal of CP orientation is most likely transmitted to the dynein arms through the RS [9,10,26–28]. The CP-RS interaction probably involves mechanical force since the RS is the structure that keeps the CP at the center of the axoneme [24]. In accordance with the probable mechanical nature of interaction, a recent study has suggested that the RS pushes the CP in beating axonemes [29]. However, the evidence is rather indirect and thus whether the RS pushes or pulls the CP appears to need further studies.

We previously reported a *Chlamydomonas* mutant, *bld12*, that has severe defects in a subcentriolar structure termed the cartwheel [30]. The cartwheel, consisting of a central hub and nine spokes, is located at the proximal end of the centriole as a stack of several layers [31,32]. The mutant *bld12* lacks the central part of the cartwheel due to a null mutation in the gene coding for SAS-6, a component of the cartwheel [30]. X-ray crystallography and biochemical analyses showed that SAS-6 forms a dimer having two globular heads and a coiled coil tail, and the dimers assemble into a ring through their hydrophobic interaction between the heads [33,34].

Lack of the cartwheel in *bld12* causes severe defects in the centriole assembly: ~80% of the centrioles observed are split into fragments and only ~20% are assembled in the cylindrical structure. Interestingly, the number of the triplets varies from seven to eleven in the cylindrical centrioles. As a consequence, flagellar axonemes produced in a small fraction of *bld12* cells contain variable numbers of outer doublet microtubules ranging from eight to eleven [30]. In this study, we investigated the effects of the variation in the outer doublet number on the appearance of the CP within the axoneme. The results revealed the importance of the spatial factor for the formation of the CP, and furthermore, provided evidence for the presence of attractive force between the CP and RS.

Material and Methods

Strains

Chlamydomonas strains CC124 (wild-type), *pf6*, *pf14*, and *pf16* were obtained from the *Chlamydomonas* Resource center, and *cpc1* from Dr. Mitchell of State University of New York Upstate Medical University [35]. The mutant *bld12* (*bld12-1*) was previously isolated in our laboratory (Nakazawa et al., 2007). The double and triple mutants, *bld12pf6*, *bld12pf14*, *bld12pf16*, *bld12cpc1*, *bld12cpc1pf6*, *pf14pf6*, *pf14pf16*, and *pf14cpc1*, were produced by genetic crosses [36]. Cells were grown in Tris-acetate-phosphate (TAP) media [37] with aeration on a 12 h/12 h light/dark cycle, or under constant illumination with agitation.

Preparation of flagellar axonemes

Flagella were isolated from *bld12-1* or *bld12-1cw92* cells by the dibucaine method of Witman [38]. Detached flagella were collected by centrifugation at 10,000×g, overlaid on a sucrose cushion (25% sucrose, 20 mM HEPES pH 7.4) and centrifuged at 1,000×g for 10 min at 4°C. Flagella at the boundary between the upper phase and the sucrose solution was collected, and demembranated by treatment with 0.5% Nonidet P-40 in HMDE solution (30 mM HEPES, 5 mM $MgCl_2$, 1 mM dithiothreitol (DTT), 1 mM EGTA, pH 7.4) [38]. The axonemes collected by centrifugation were resuspended with HMDE solution. For analysis of flagellar length, isolated flagella were collected and their lengths were measured using ImageJ software.

Electron microscopy

Axonemes were prefixed with 2% glutaraldehyde and 1% tannic acids in 50 mM phosphate buffer pH 7.2 for 1–2 h at 0°C, and postfixed with 1% OsO_4 in the phosphate buffer for 1 h. The samples were dehydrated by passing through graded concentrations of alcohol solutions, and embedded in EPON 812. Thin sections (50–70 nm) were stained with aqueous uranyl acetate and Reynold's lead citrate, and observed with a JEM1010 electron microscope. Images were obtained using a film-base camera or MegaView III digital camera (JEOL, Tokhyo). For analysis of axonemal cross sections, images were chosen on the basis of clear appearance of all outer doublets, to ensure that the axoneme under examination was cut almost normal to the axoneme axis. The diameter and inner-doublet spacing were measured using ImageJ software. Correlation between the diameter and the doublet number was analyzed using a linear regression model. Group difference in inter-doublet spacing was analyzed using ANOVA. Statistical significance for the test was set at $P<0.05$.

Analysis of CP-RS association and helical properties of the CP

For analyses of the CP-RS association, cross section images of 10- or 11-doublet axonemes of *bld12*, *bld12pf6*, *bld12cpc1*, *bld12cpc1pf6*, and *bld12pf16* were chosen and collected based on the clarity of the C1 and C2 microtubules. In the case of *bld12pf16*, images of the 9-doublet axonemes were used for the analysis in addition to the 10- or 11-doublet axonemes. Each image was oriented with the dynein arms projecting clockwise. The CP surface in the image, which was approximated to a circle, was equally divided into 12 sectors. The spokeheads attached to or detached from the CP were identified by visual inspection. The center of the consecutive spokeheads attached to the CP was defined as the spokehead-interaction site. Distribution of the interaction sites on the CP surface was represented by polar histograms.

In the CP orientation analysis in cross section images of *pf14*, *pf14pf6*, *pf14cpc1*, *pf14pf16* axonemes, distribution of the sector closest to the outer doublet wall was represented by polar histograms. Statistical significance of the difference in distribution was evaluated by the χ^2 test. Statistical significance was set at $P<0.05$.

Results

Abnormal axonemes in *bld12*

As we previously reported, ~10% of *bld12* cells produced one or two flagella when cell walls were removed by treatment with autolysin [30]. Under this condition, about 8% of cells were uniflagellated and 2% biflagellated. Of the biflagellated cells, about 50% had flagella of unequal lengths. The flagellar length was variable but always shorter than that of wild type (Figure 1A). The *bld12* flagella displayed a variety of motility phenotypes, ranging from complete paralysis to sporadic twitching to almost normal beating.

Electron microscopy showed that the percentage of 8-, 9-, 10-, and 11-doublet axonemes was, respectively, ~5%, ~90%, ~5%, and ~0.1% (Figure 1B) [30]. While the diameter of the axoneme increased with the doublet number, the space between the adjacent doublets was constant (Figure 1C), suggesting that the inter-doublet structures such as the inner- and outer-dynein arms and the nexin links were not distorted in the abnormal axonemes. In the images of these axonemes, we noticed two remarkable features that are not seen in normal axonemes: ~95% of the 8-doublet axonemes had no CP microtubules; and, in 10- or 11-doublet axonemes, the circular arrangement of the doublets was distorted because only three to five RSs were attached to the CP. Similarly distorted axonemes were also observed in intact flagella of this mutant (Yuki Nakazawa, unpublished observation).

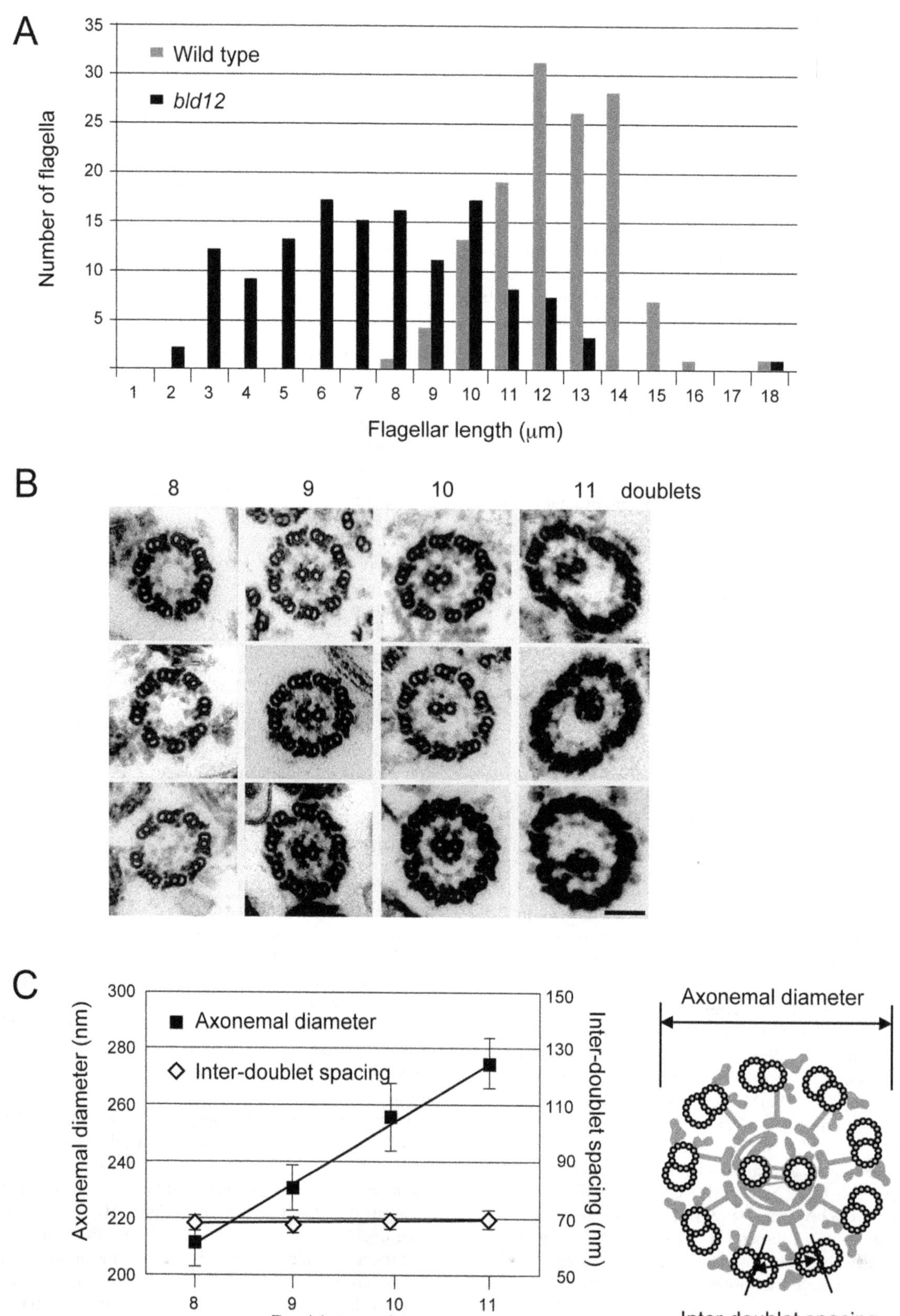

Figure 1. Abnormal features of *bld12* axonemes. (A) Length distributions of wild type (gray) and *bld12* (black) flagella. (B) Axonemes with 8, 9, 10, and 11 outer doublet microtubules. Bar, 100 nm. (C) Diameter and inter-doublet spacing in axonemes with various doublet numbers. Diameter increases with the doublet numbers, while the inter-doublet spacing is constant. The linear regression slope of the diameter is 22.14±1.69 nm ($R^2 = 0.900$). The difference between the inter-doublet spacing is not significant (ANOVA, $P = 0.59$).

Effects of RS removal on CP formation

Absence of the CP in most of the 8-doublet axonemes led us to assume that these flagella do not have enough room to accommodate a CP in the central area. To test this hypothesis, we produced the double mutant of *bld12* and *pf14,* a mutant that lacks the RSs and has a larger internal space [39]. This mutant, *bld12pf14,* also produced axonemes with a variable number of doublets, ranging from 7 to 11 (Figure 2). As expected, all of the 8-doublet axonemes and even 7-doublet axonemes had the CP (Figure 2, Table 1). In addition, ~5% of the 9-doublet axonemes or ~74% of the 10-doublet axonemes contained three or four central microtubules (Figure 2, Table 1). In *pf14* also, axonemes with three central microtubules were observed although the occurrence was rare (~0.7% in 2000 cross-section images). These observations indicate that the formation of the CP depends on the size of the space limited by the RSs and the outer doublet microtubules.

Spokeheads preferentially bind to distinct sites on the CP surface

Cross section images of the 10- or 11-doublet axonemes of *bld12* showed that five or six doublets were tethered by RSs to the CP to form a semicircle of a normal diameter. The rest of the doublets, not tethered to the CP due to the detachment of RSs, are arranged in another semicircle bulging outward (Figure 1B). Whether an RS was attached or detached could be easily judged from the position of the bulky spokehead. The doublet arrangement was distorted at the junctions of the two semicircles. In contrast, no such distortion was observed in the RS-lacking axonemes of *bld12pf14* (Figure 2). These observations suggest that the RS binds to the CP to help the doublets align in a circular arrangement of a constant diameter, and that the CP-RS binding is strong enough to distort the arrangement in 10- or 11-doublet axonemes.

To investigate whether this RS binding occurs on particular regions on the CP surface, we examined cross-section images of 10- or 11-doublet axonemes for the possible location on the CP where the spokeheads preferentially attach. We divided the CP image in cross section into 12 sectors, and scored the frequency of each sector to locate at the center of the group of CP-associated RSs (Figure 3, A–C). An analysis of 56 cross-section images revealed that the CP surface had two preferred sites for association with the RS: one near the C1a and the other near the C1b projection. These two preferred sites must bind to the spokeheads more strongly than the other regions of the CP.

Previous studies showed that the *Chlamydomonas* CP, when released from the axoneme, forms a helical complex; when contained in the axoneme, it must be forced to assume a straight form with a 360 degree twist per the length of the flagellum [40,41]. This tendency of the CP to assume a helix might bias the distribution of the apparent spoke-interaction sites in the 10- or 11-doublet axonemes (Figure 3, B and C). To address this possibility, we examined the helical tendency of the CP in the spoke-less *pf14* axonemes, in which the CP should assume a small-amplitude helical form facing its outer surface to the outer doublet wall (Figure 3, D and E). An analysis of 80 cross sections of *pf14* axonemes indicated that the C1 microtubule is located outer side of the CP helix, i.e. closest to the doublet wall. This distribution pattern is clearly different from the pattern in the axonemes of *bld12*, which showed two preferred orientation regions (Figure 3, B–E). We therefore concluded that the helical tendency of the CP did not mask the CP orientation resulting from its interaction with RSs.

Removal of CP projections identifies multiple weak CP-spoke association sites

We next examined CP-RS interactions in double mutants between *bld12* and each one of four mutants that lack specific CP projections (Figure 4). The CP mutants used were *pf6* lacking the prominent projection C1a [42]; *cpc1* lacking C1b and C2b [35]; *pf6cpc1* double mutant lacking these three projections; and *pf16* lacking the C1 microtubule [42]. Interestingly, the spokeheads preferentially bound to the C1b–C2b region when the C1a projection was absent, whereas they bound to the C1a–C2a region when C1b and C2b were absent. In *pf6cpc1* or *pf16*, the spokeheads tended to bind to the C1d and C2c regions, or to the C2b region, where binding was only infrequently observed in *bld12* axonemes (Figure 4, A and B). As controls, we also analyzed the axonemal images of double mutants *pf14pf6*, *pf14cpc1*, and *pf14pf16* because we were particularly concerned that the mutant CPs might have varied tendencies to assume helical forms, which might affect the results of our analysis. The histograms obtained in the control experiments showed that the helical tendency was not largely affected by the *pf6*, *cpc1*, or *pf16* mutations (Figure 4C). For example, the *pf14pf16* CP, which lacks the C1 microtubule, displayed the same tendency as the *pf14* CP (Figure 3D); that is, the side of the C2 that would be positioned closest to the C1 in the wild type CP was still positioned on the outside of the helix. These results suggest that, although different regions of the CP surface differ in the spokehead binding affinity, almost the whole area of the CP surface can bind to the spokeheads.

Table 1. The number of axonemes with particular numbers of outer doublets and central microtubules observed in *bld12pf14*.

	# of central microtubules per axoneme		
# of outer doublets per axoneme	0 or 1	2	3 or 4
7	1	2	0
8	0	465	0
9	15	801	42
10	3	5	23
11	0	1	1

(Total number of axonemes counted, 1,369).

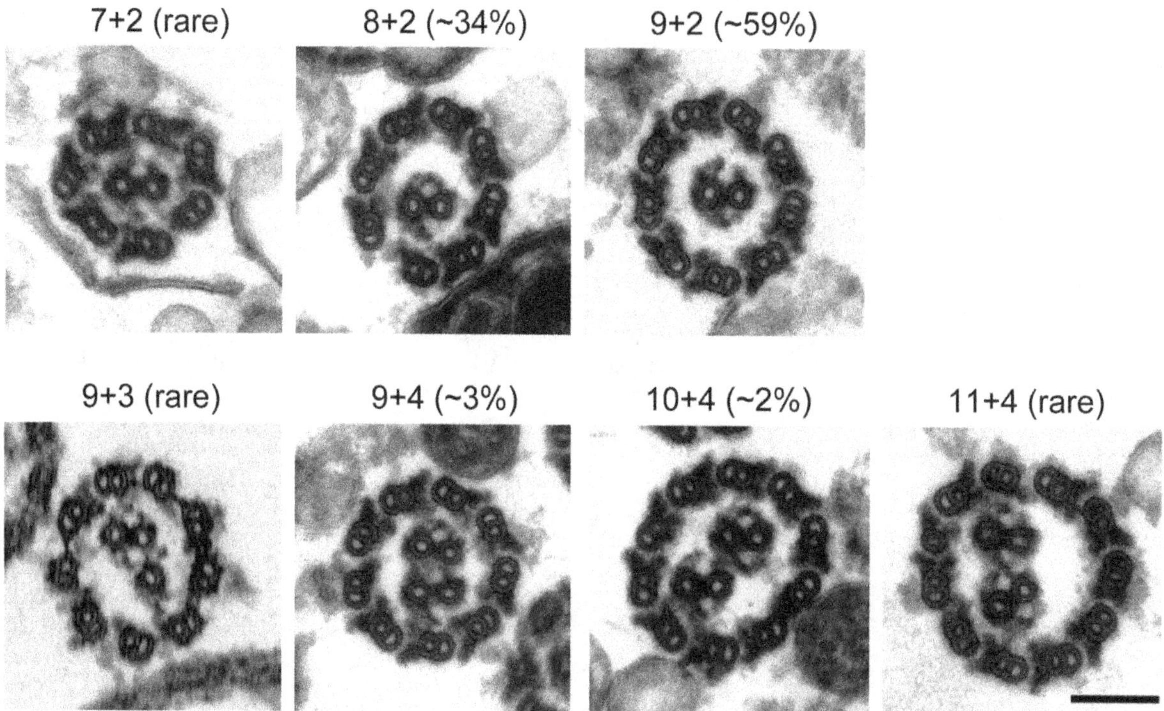

Figure 2. CPs in RS-lacking axonemes. Cross section images of *bld12pf14* axonemes with 7 to 10 outer doublets are shown. The frequency of each pattern in the observed images is indicated in the parenthesis (n = 1,369, Table 1). All CPs have the normal polarity in the axoneme as judged by their CP projections. Bar, 100 nm.

Discussion

Formation of the CP

We showed that the CP did not assemble in 8-doublet axonemes of *bld12*, but assembled in those of *bld12pf14* lacking the RS. Furthermore, while 9-doublet axonemes of *bld12pf14* only rarely contained two CPs, its 10-doublet axonemes frequently contained two CPs (Figure 2, Table 1). Formation of two CPs has previously been observed in flagella and cilia that lack the RS: for example, flagella of *Chlamydomonas pf14* and *pf14pf6cpc1* mutants [20,23]; and nodal cilia of rabbit [43]. These observations suggest that CP assembly depends on the internal space of the axoneme. Our results, demonstrating a clear correlation between the CP assembly and the space size, lend strong support to the previous proposal (Table 1).

The distal end of the CP in *Chlamydomonas* flagella and *Tetrahymena* cilia is capped by a plate-like structure that is indirectly attached to the membrane through a spherical bead. In contrast, the proximal end of the CP, located near the transition zone, is associated with no detectable structure [12,13]. In some *Chlamydomonas* mutants with defects in the transition zone, as well as in isolated tracheal epithelial cells, the CP microtubules grow proximally into the centriole, suggesting that the proximal end is not the site of nucleation for CP growth [16,17]. Lechtreck et al. (2013) reported that, when gametes of a *Chlamydomonas* mutant lacking the CP are mated with wild type gametes, the mutant flagella in the fused cells start to produce the CP at the middle portion of the axoneme. This observation indicates that no organizing center is required for the CP assembly at either end of the flagellum. These authors also observed that the RS-lacking *pf14* axonemes contain two CPs (four microtubules) with correct polarity. Together with their findings, our observation of the extra CP in *bld12pf14* implies that the space size within the axoneme is an important factor that directs CP formation in the axoneme. Possibly, when the CP precursors are present in the axoneme, the CP may form spontaneously without interacting with any template or RSs. In such a case, only the available space and the amount of the precursors may limit the number of the CPs produced.

Nature of CP-RS interaction for the regulation of flagellar motility

A minor population of *bld12* axonemes having 10 or 11 outer doubles exhibited distortion in the circular arrangement of the outer doublets. This observation clearly indicates that the RS binds to the CP and the binding exerts force. Previous studies have provided substantial evidence that the CP and RS form a signal transduction pathway that modulates dynein activity through phosphorylation of a specific subunit of inner-arm dynein [44–46]. The RS is likely to chemically and mechanically control the dynein activity based on the interaction between the CP projections and the spokeheads [9,21]. However, whether or not the RS exerts mechanical force on the outer doublet has been unknown. Our present study is the first to show that RSs actually pull the outer doublet microtubules toward the axonemal center. Such a mechanical force may be at the center of the regulatory function of the CP/RS.

The CP-RS interaction should be transient and the binding strength is weak enough to allow such an interaction. This is because the *Chlamydomonas* CP rotates within the axoneme [40,41] and the RS slides over the CP for a certain distance as the axoneme propagates bending waves [47]. In this study, however, we showed that the interaction is still strong enough to distort the circular arrangement of the outer doublets. This finding prompts us to speculate that the mechanical force transmitted by the RS could change the relative position of the doublet microtubules and

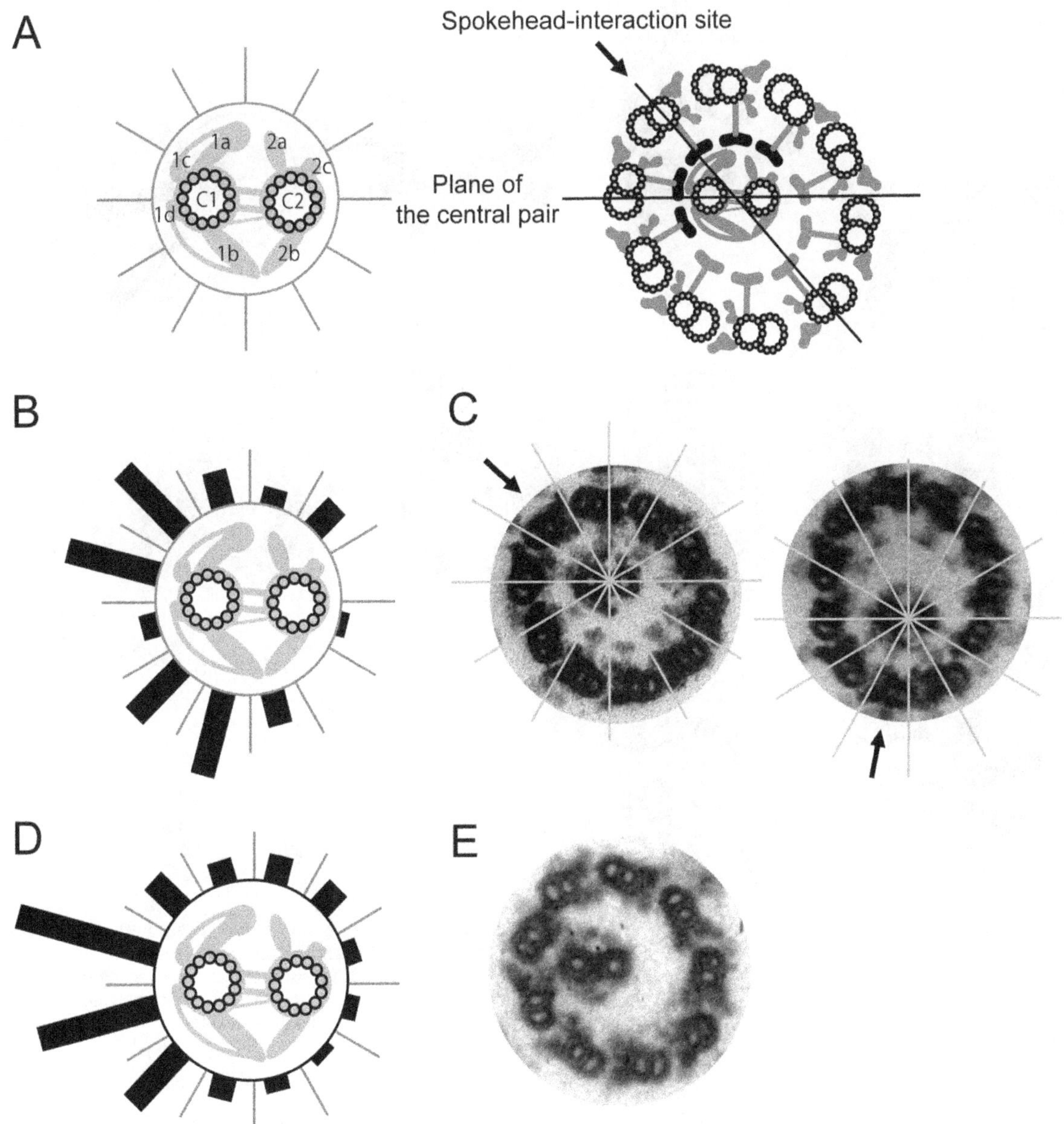

Figure 3. Frequency of CP-RS association in different sectors of the CPs. (A) The CP cross section was divided into 12 sectors by six lines including the one connecting the centers of the two central microtubules. The center of the consecutive spokeheads in contact with the CP was defined as the spokehead-interaction site. Major CP projections are designated [35]. (B) A polar histogram representing the distribution of the spokehead-interaction sites on the CP in cross section images of 10- or 11-doublet axonemes of *bld12* (n = 56). The length of each bar represents relative frequency to locate at the spokehead-interaction site. (C) Axoneme images that correspond to the two peaks in the histogram in (B). (D) Distribution of the CP region in contact with the wall of outer doublets in *pf14* axonemes (n = 80). The histogram suggests that the C1 microtubule is located on the outer surface on the helical CP. (E) A cross section image that represents the peak distribution in the histogram in (D). The differences between the distributions of *bld12* (B) and *pf14* (D) are significant (χ^2 test, $p<0.05$).

dyneins, and thereby transiently activate or inactivate the dyneins located in a particular region of the axoneme. A change in the dynein-microtubule positioning has also been postulated in the geometric clutch model of Lindemann [48]. The location of dynein molecules activated or inactivated by the CP/RS should propagate along the axoneme as the twisted CP rotates. In the cilia and flagella of multicellular organisms and some unicellular organisms, the CP neither assumes a helical shape nor rotates in the axoneme although sharing the structure and components with the *Chlamydomonas* CP [23,49,50]; in those organisms, the stationary CP determines the plane of axonemal beating possibly by activating dynein molecules on a particular side of the axoneme [51,52].

Our image analyses of mutant axonemes suggest that, while most area of the CP surface can bind to the RS, the two major projections on the C1 microtubule bind stronger than the other projections on the C2 microtubule. This asymmetric distribution of the RS binding affinities on the CP surface must be important for the signal generation by the rotating CP. Although the present study does not provide information as to whether the stronger binding promotes or inhibits dynein-driven microtubule sliding, previous studies suggest that the C1 microtubule or its projections

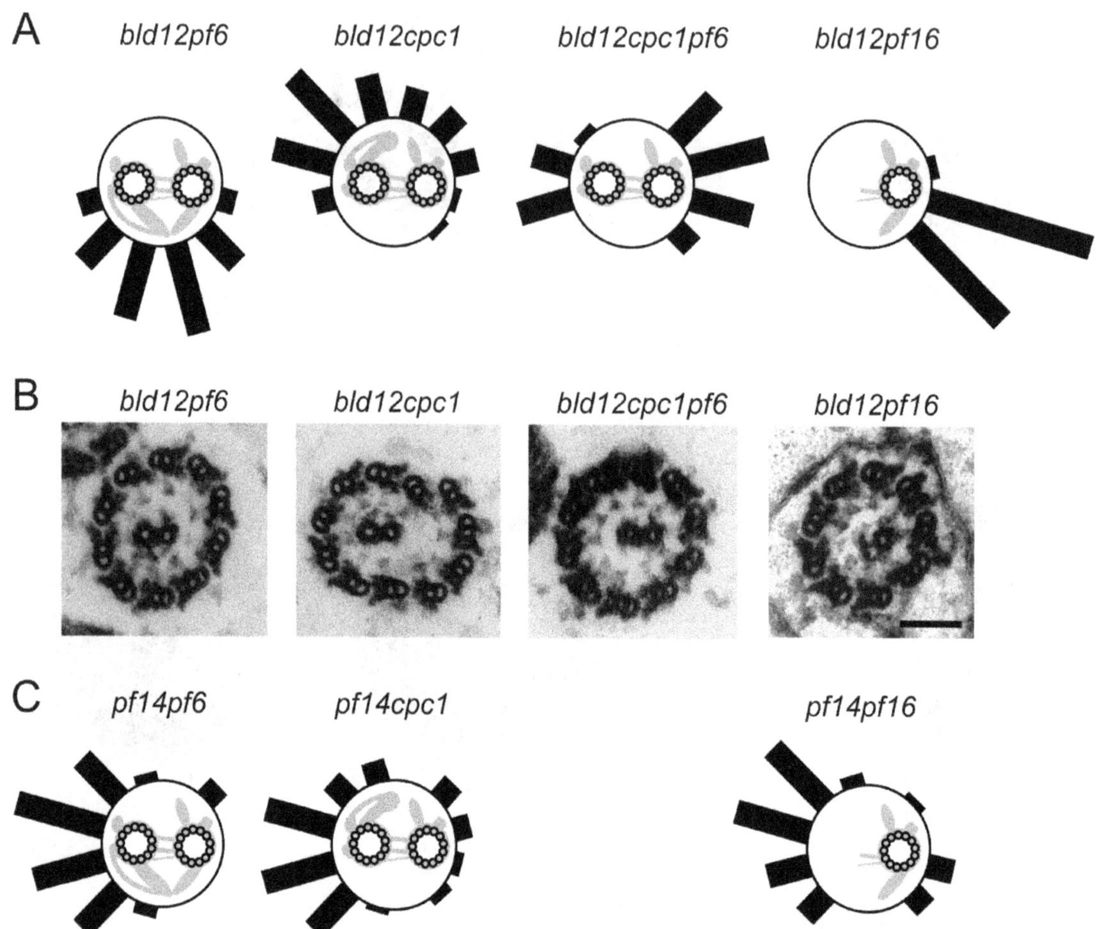

Figure 4. Removal of CP projections manifests weak CP-spokehead association. (A) Distributions of the spokehead-interaction sites on the CP in cross section images of 10- or 11-doublet axonemes of *bld12pf6*, *bld12cpc1*, *bld12cpc1pf6*, and *bld12pf16* (n = 59, 49, 40, and 41), which lack specific CP projections. (B) Cross section images that represent the peaks in the histograms in (A). (C) Distributions of contact sites on the CP surface with the outer doublet wall in cross section images of *pf14pf6*, *pf14cpc1*, and *pf14pf16* (n = 33, 69, and 46). The differences between the distributions of *bld12pf6* and *pf14pf6*; between *bld12cpc1* and *pf14cpc1*, and between *bld12pf16* and *pf14pf16*, are significant (χ^2 test, $p<0.05$ for each of the three pairs). These distribution patterns are similar to the pattern in *pf14*.

enhances microtubule sliding in axonemes [46,53–55]. Thus the doublet-pulling by stronger RS binding to the C1 surface may activate dyneins and promote microtubule sliding.

Our results are in apparent contradiction with a non-specific CP-RS interaction model recently proposed by Oda et al. (2014). These authors showed that the flagella of the *pf6* mutant lacking the C1a projection recovered motility if any one of three protein tags (hemagglutinin, biotin carboxyl carrier protein, and green fluorescent protein) was attached on top of the spokehead. Because the extent of motility recovery increased in the order of the size of the tag, they proposed that the added tag elongated the RS and compensated the loss of the C1a projection, possibly by enabling RSs to collide with the CP. The proposed physical CP-RS interaction must be non-specific since the three protein tags used are structurally unrelated. In contrast, our analysis of 10- or 11-doublet axoneme images showed that the C1a projection, together with the C1b projection, preferentially associates with the RSs among all the CP projections, favoring the view that the C1a projection pulls the outer doublet in a fairly specific manner. Both their results and our results must reflect some aspects of the CP-RS interaction, but their relationship is not understood. The molecular mechanism of the CP-RS interaction remains one of the most interesting problems in cilia/flagella motility studies.

An obvious question regarding the present study is whether or not the axonemes with aberrant numbers of outer doublets are motile. We may imagine that 8-doublet flagella are non-motile because they lack the CP, like the flagella of non-motile mutants such as *pf18* and *pf19* [56–58]. However, it is difficult to predict whether or not axonemes with 10 or 11 doublets can display some motility. Development of techniques that determine the number of axonemal microtubules under the microscope, or those that permit constant production of flagella with 10–11 doublets, may well provide answers. We can hope that the answer will provide a strong clue as to why motile cilia and flagella almost always contain nine outer doublets.

Acknowledgments

We thank Dr. David Mitchell for providing the *Chlamydomonas* mutant strain *cpc1*.

Author Contributions

Conceived and designed the experiments: YN RK MH. Performed the experiments: YN TA AN RK MH. Analyzed the data: YN TA AN RK MH. Contributed to the writing of the manuscript: YN RK MH.

References

1. Badano JL, Mitsuma N, Beales PL, Katsanis N (2006) The ciliopathies: an emerging class of human genetic disorders. Annual review of genomics and human genetics 7: 125–148.
2. Onoufriadis A, Shoemark A, Schmidts M, Patel M, Jimenez G, et al. (2014) Targeted NGS gene panel identifies mutations in RSPH1 causing primary ciliary dyskinesia and a common mechanism for ciliary central pair agenesis due to radial spoke defects. Human molecular genetics 23: 3362–3374.
3. Bui KH, Yagi T, Yamamoto R, Kamiya R, Ishikawa T (2012) Polarity and asymmetry in the arrangement of dynein and related structures in the *Chlamydomonas* axoneme. The Journal of cell biology 198: 913–925.
4. Bui KH, Sakakibara H, Movassagh T, Oiwa K, Ishikawa T (2008) Molecular architecture of inner dynein arms in situ in *Chlamydomonas reinhardtii* flagella. The Journal of cell biology 183: 923–932.
5. Pigino G, Bui KH, Maheshwari A, Lupetti P, Diener D, et al. (2011) Cryoelectron tomography of radial spokes in cilia and flagella. The Journal of cell biology 195: 673–687.
6. Nicastro D, Schwartz C, Pierson J, Gaudette R, Porter ME, et al. (2006) The molecular architecture of axonemes revealed by cryoelectron tomography. Science 313: 944–948.
7. Heuser T, Raytchev M, Krell J, Porter ME, Nicastro D (2009) The dynein regulatory complex is the nexin link and a major regulatory node in cilia and flagella. The Journal of cell biology 187: 921–933.
8. Carbajal-Gonzalez BI, Heuser T, Fu X, Lin J, Smith BW, et al. (2013) Conserved structural motifs in the central pair complex of eukaryotic flagella. Cytoskeleton (Hoboken, NJ) 70: 101–120.
9. Smith EF, Yang P (2004) The radial spokes and central apparatus: mechano-chemical transducers that regulate flagellar motility. Cell motility and the cytoskeleton 57: 8–17.
10. Kamiya R (2002) Functional diversity of axonemal dyneins as studied in *Chlamydomonas* mutants. International review of cytology 219: 115–155.
11. Mizuno N, Taschner M, Engel BD, Lorentzen E (2012) Structural studies of ciliary components. Journal of molecular biology 422: 163–180.
12. Dentler WL, Rosenbaum JL (1977) Flagellar elongation and shortening in *Chlamydomonas*. III. structures attached to the tips of flagellar microtubules and their relationship to the directionality of flagellar microtubule assembly. The Journal of cell biology 74: 747–759.
13. Dentler WL (1980) Structures linking the tips of ciliary and flagellar microtubules to the membrane. Journal of cell science 42: 207–220.
14. Allen RD (1969) The morphogenesis of basal bodies and accessory structures of the cortex of the ciliated protozoan *Tetrahymena pyriformis*. The Journal of cell biology 40: 716–733.
15. Dute R, Kung C (1978) Ultrastructure of the proximal region of somatic cilia in *Paramecium tetraurelia*. The Journal of cell biology 78: 451–464.
16. Dentler WL, LeCluyse EL (1982) The effects of structures attached to the tips of tracheal ciliary microtubules on the nucleation of microtubule assembly in vitro. Progress in clinical and biological research 80: 13–18.
17. Jarvik JW, Suhan JP (1991) The role of the flagellar transition region: Inferences from the analysis of a *Chlamydomonas* mutant with defective transition region structures. Journal of cell science 99: 731–740.
18. Silflow CD, Liu B, LaVoie M, Richardson EA, Palevitz BA (1999) Gamma-tubulin in *Chlamydomonas*: characterization of the gene and localization of the gene product in cells. Cell motility and the cytoskeleton 42: 285–297.
19. McKean PG, Baines A, Vaughan S, Gull K (2003) Gamma-tubulin functions in the nucleation of a discrete subset of microtubules in the eukaryotic flagellum. Current biology: CB 13: 598–602.
20. Lechtreck KF, Gould TJ, Witman GB (2013) Flagellar central pair assembly in *Chlamydomonas reinhardtii*. Cilia 2: 15.
21. Wirschell M, Nicastro D, Porter ME, Sale WS (2009) Chapter 9 - The Regulation of Axonemal Bending. In: Harris EH, Stern DB, Witman GB, editors. The Chlamydomonas Sourcebook (Second Edition). London: Academic Press. 253–282.
22. Yang P, Smith EF (2009) Chapter 7 - The Flagellar Radial Spokes. In: Harris EH, Stern DB, Witman GB, editors. The Chlamydomonas Sourcebook (Second Edition). London: Academic Press. 209–234.
23. Mitchell DR (2009) Chapter 8 - The Flagellar Central Pair Apparatus. In: Harris EH, Stern DB, Witman GB, editors. The Chlamydomonas Sourcebook (Second Edition). London: Academic Press. 235–252.
24. Witman GB, Plummer J, Sander G (1978) Chlamydomonas flagellar mutants lacking radial spokes and central tubules. Structure, composition, and function of specific axonemal components. The Journal of cell biology 76: 729–747.
25. Lechtreck KF, Delmotte P, Robinson ML, Sanderson MJ, Witman GB (2008) Mutations in Hydin impair ciliary motility in mice. The Journal of cell biology 180: 633–643.
26. Huang B, Ramanis Z, Luck DJ (1982) Suppressor mutations in *Chlamydomonas* reveal a regulatory mechanism for Flagellar function. Cell 28: 115–124.
27. Piperno G, Mead K, Shestak W (1992) The inner dynein arms I2 interact with a "dynein regulatory complex" in *Chlamydomonas* flagella. The Journal of cell biology 118: 1455–1463.
28. Omoto CK, Kung C (1979) The pair of central tubules rotates during ciliary beat in *Paramecium*. Nature 279: 532–534.
29. Oda T, Yanagisawa H, Yagi T, Kikkawa M (2014) Mechanosignaling between central apparatus and radial spokes controls axonemal dynein activity. The Journal of cell biology 204: 807–819.
30. Nakazawa Y, Hiraki M, Kamiya R, Hirono M (2007) SAS-6 is a cartwheel protein that establishes the 9-fold symmetry of the centriole. Current biology: CB 17: 2169–2174.
31. Gibbons IR, Grimstone AV (1960) On flagellar structure in certain flagellates. The Journal of biophysical and biochemical cytology 7: 697–716.
32. Cavalier-Smith T (1974) Basal body and flagellar development during the vegetative cell cycle and the sexual cycle of *Chlamydomonas reinhardii*. Journal of cell science 16: 529–556.
33. Kitagawa D, Vakonakis I, Olieric N, Hilbert M, Keller D, et al. (2011) Structural basis of the 9-fold symmetry of centrioles. Cell 144: 364–375.
34. van Breugel M, Hirono M, Andreeva A, Yanagisawa HA, Yamaguchi S, et al. (2011) Structures of SAS-6 suggest its organization in centrioles. Science (New York, NY) 331: 1196–1199.
35. Mitchell DR, Sale WS (1999) Characterization of a *Chlamydomonas* insertional mutant that disrupts flagellar central pair microtubule-associated structures. The Journal of cell biology 144: 293–304.
36. Dutcher SK (1995) Mating and tetrad analysis in *Chlamydomonas reinhardtii*. Methods in cell biology 47: 531–540.
37. Gorman DS, Levine RP (1965) Cytochrome f and plastocyanin: their sequence in the photosynthetic electron transport chain of *Chlamydomonas reinhardi*. Proceedings of the National Academy of Sciences of the United States of America 54: 1665–1669.
38. Witman GB (1986) [28] Isolation of Chlamydomonas flagella and flagellar axonemes. In: Richard BV, editor. Methods in enzymology: Academic Press. 280–290.
39. Piperno G, Huang B, Luck DJ (1977) Two-dimensional analysis of flagellar proteins from wild-type and paralyzed mutants of *Chlamydomonas reinhardtii*. Proceedings of the National Academy of Sciences of the United States of America 74: 1600–1604.
40. Kamiya R (1982) Extrusion and Rotation of the central-pair microtubules in detergent-treated *Chlamydomonas* flagella. Progress in clinical and biological research 80: 169–173.
41. Mitchell DR, Nakatsugawa M (2004) Bend propagation drives central pair rotation in *Chlamydomonas reinhardtii* flagella. The Journal of cell biology 166: 709–715.
42. Dutcher SK, Huang B, Luck DJ (1984) Genetic dissection of the central pair microtubules of the flagella of *Chlamydomonas reinhardtii*. The Journal of cell biology 98: 229–236.
43. Feistel K, Blum M (2006) Three types of cilia including a novel 9+4 axoneme on the notochordal plate of the rabbit embryo. Developmental dynamics: an official publication of the American Association of Anatomists 235: 3348–3358.
44. Habermacher G, Sale WS (1995) Regulation of dynein-driven microtubule sliding by an axonemal kinase and phosphatase in *Chlamydomonas* flagella. Cell motility and the cytoskeleton 32: 106–109.
45. Yang P, Sale WS (2000) Casein kinase I is anchored on axonemal doublet microtubules and regulates flagellar dynein phosphorylation and activity. The Journal of biological chemistry 275: 18905–18912.
46. Smith EF (2002) Regulation of flagellar dynein by the axonemal central apparatus. Cell motility and the cytoskeleton 52: 33–42.
47. Warner FD, Satir P (1974) The structural basis of ciliary bend formation. Radial spoke positional changes accompanying microtubule sliding. The Journal of cell biology 63: 35–63.
48. Lindemann CB, Kanous KS (1995) "Geometric clutch" hypothesis of axonemal function: key issues and testable predictions. Cell motility and the cytoskeleton 31: 1–8.
49. Tamm SL, Tamm S (1981) Ciliary reversal without rotation of axonemal structures in ctenophore comb plates. The Journal of cell biology 89: 495–509.
50. Gadelha C, Wickstead B, McKean PG, Gull K (2006) Basal body and flagellum mutants reveal a rotational constraint of the central pair microtubules in the axonemes of trypanosomes. Journal of cell science 119: 2405–2413.
51. Gibbons IR (1961) The relationship between the fine structure and direction of beat in gill cilia of a lamellibranch mollusc. The Journal of biophysical and biochemical cytology 11: 179–205.
52. Yoshimura M, Shingyoji C (1999) Effects of the central pair apparatus on microtubule sliding velocity in sea urchin sperm flagella. Cell structure and function 24: 43–54.

53. Brown JM, Dipetrillo CG, Smith EF, Witman GB (2012) A FAP46 mutant provides new insights into the function and assembly of the C1d complex of the ciliary central apparatus. Journal of cell science 125: 3904–3913.
54. Smith EF (2002) Regulation of flagellar dynein by calcium and a role for an axonemal calmodulin and calmodulin-dependent kinase. Molecular biology of the cell 13: 3303–3313.
55. Wargo MJ, Smith EF (2003) Asymmetry of the central apparatus defines the location of active microtubule sliding in *Chlamydomonas* flagella. Proceedings of the National Academy of Sciences of the United States of America 100: 137–142.
56. Randall J, Warr JR, Hopkins JM, McVittie A (1964) A SINGLE-GENE MUTATION OF *CHLAMYDOMONAS REINHARDII* AFFECTING MOTILITY: A GENETIC AND ELECTRON MICROSCOPE STUDY. Nature 203: 912–914.
57. Randall JT, Cavalier-Smith T, McVittie A, Warr JR, Hopkins JM (1967) Developmental and control processes in the basal bodies and flagella of *Chlamydomonas reinhardtii*. Dev Biol suppl 1: 43–83.
58. Warr JR, McVittie A, Randall JT, Hopkins JM (1966) Genetic control of flagellar structure in *Chlamydomonas reinhardtii*. Genet Res 7: 335–351.

22

Acetate Availability and Utilization Supports the Growth of Mutant Sub-Populations on Aging Bacterial Colonies

Jessica M. Bergman, Marie Wrande[¤], Diarmaid Hughes*

Department of Medical Biochemistry and Microbiology, Biomedical Center, Uppsala University, Uppsala, Sweden

Abstract

When bacterial colonies age most cells enter a stationary phase, but sub-populations of mutant bacteria can continue to grow and accumulate. These sub-populations include bacteria with mutations in *rpoB* (RNA polymerase β-subunit) or *rpoS* (RNA polymerase stress-response sigma factor). Here we have identified acetate as a nutrient present in the aging colonies that is utilized by these mutant subpopulations to support their continued growth. Proteome analysis of aging colonies showed that several proteins involved in acetate conversion and utilization were upregulated during aging. Acetate is known to be excreted during the exponential growth phase but can be imported later during the transition to stationary phase and converted to acetyl-CoA. Acetyl-CoA is used in multiple processes, including feeding into the TCA cycle, generating ATP via the glyoxylate shunt, as a source of acetyl groups for protein modification, and to support fatty acid biosynthesis. We showed that deletion of *acs* (encodes acetyl-CoA synthetase; converts acetate into acetyl-CoA) significantly reduced the accumulation of *rpoB* and *rpoS* mutant subpopulations on aging colonies. Measurement of radioactive acetate uptake showed that the rate of conversion decreased in aging wild-type colonies, was maintained at a constant level in the *rpoB* mutant, and significantly increased in the aging *rpoS* mutant. Finally, we showed that the growth of subpopulations on aging colonies was greatly enhanced if the aging colony itself was unable to utilize acetate, leaving more acetate available for mutant subpopulations to use. Accordingly, the data show that the accumulation of subpopulations of *rpoB* and *rpoS* mutants on aging colonies is supported by the availability in the aging colony of acetate, and by the ability of the subpopulation cells to convert the acetate to acetyl-CoA.

Editor: Martin G. Marinus, University of Massachusetts Medical School, United States of America

Funding: Diarmaid Hughes acknowledges support from The Swedish Science Research Council (Vetenskapsrådet), the Knut and Alice Wallenberg Foundation (RiboCore Project), SSF (Swedish Foundation for Strategic Research) and Vinnova (Swedish Innovation Agency). This work was supported by the Science for Life Mass Spectrometry Technology Platform in Uppsala, Sweden. The funders had no role in study design, data collection and analysis, decision to publish, or preparation of the manuscript. The authors thank Svenja N. Reinders, Eva Garcia Barreales, and Shanjun Gao, who each contributed to parts of this project as Masters students.

Competing Interests: The authors have declared that no competing interests exist.

* Email: diarmaid.hughes@imbim.uu.se

¤ Current address: Paul G. Allen School for Global Animal Health, Washington State University, Pullman, Washington, United States of America

Introduction

It was previously shown that mutants resistant to rifampicin, carrying mutations in *rpoB* (encoding the β-subunit of RNA polymerase) accumulate in aging colonies of *Salmonella* Typhimurium and *Escherichia coli* [1,2,3,4]. This accumulation occurs because many different rifampicin-resistant (Rif^R) mutants continue to grow after wild-type cells in the colony enter stationary phase [4]. The ability of mutants to postpone their entry into stationary phase is of general interest because natural environments are often nutrient-poor and bacteria can spend much of their time in a non-growth state, punctuated by periods of rapid growth when nutrient becomes available. Accordingly, mutants that can extend the length of their growth phase beyond the average may gain an advantage because they will increase as a proportion of the population prior to the next period of nutrient availability and rapid growth. Mutations in *rpoS*, encoding the general stress response sigma factor RpoS, have previously been shown to increase bacterial survival in long-term liquid and colony stationary phase [5,6,7,8].

In this manuscript we confirm that *rpoS* mutants also accumulate in stationary phase colonies. We also asked whether we could identify a specific nutrient that was important to support the continued growth of mutant subpopulations of cells in or on stationary phase bacterial colonies. We report here that the availability of acetate in aging colonies, and the ability of subpopulations to convert that acetate into acetyl-CoA, are critically important to support the growth of *rpoB* and *rpoS* mutants on wild-type stationary phase colonies.

Results and Discussion

Rif^R and RpoS mutants each accumulate as colonies age

We have previously shown that Rif^R mutants grow and accumulate in aging wild-type colonies of *S.* Typhimurium [4]. RpoS mutants have been shown by others to have a growth advantage (GASP phenotype), and to accumulate, in aging liquid cultures of *E. coli* [5,6]. We asked whether RpoS mutants would be selected in aging colonies of *Salmonella* and whether RpoS mutants would have a growth advantage if added onto aging colonies. Colonies of wild-type *S.* Typhimurium [9] were aged for 15 weeks on LB agar (LA) in sealed plastic bags. Sequence analysis revealed that a high proportion of the surviving bacteria (3/18 independent clones tested) carried mutations in *rpoS*. These RpoS mutants, including one with an in-frame stop codon, each had a growth advantage over the wild-type when spotted onto aging wild-type colonies (Table S1).

To confirm that loss of RpoS activity conferred a growth advantage when mutant bacteria were added onto an aging colony, we constructed a precise deletion of *rpoS* in *S.* Typhimurium 14028s by λ-Red recombineering [10,11]. We then measured and compared the relative ability of the wild-type and of two isogenic mutant strains (Δ*rpoS* and Rif^R *rpoB* P564L) to grow and accumulate when added as a subpopulation onto 24 h old wild-type colonies which were then allowed to age for a further 7 days. Each of the two mutants had a significant growth advantage relative to the wild-type on the aging colonies (Figure 1, Table S2). We concluded that the mutations, Δ*rpoS* and *rpoB* P564L, facilitate continued bacterial growth on aging colonies. These data are in agreement with a study of decades-old agar stabs of *S.* Typhimurium, where it was found that surviving cells in 10 of 27 examined vials had a mutated *rpoS* gene [12]. This suggests that loss or change of function of RpoS can confer an advantage in aging environments.

Carbon metabolism proteins are upregulated in aging colonies

We hypothesized that as bacterial colonies age the bacteria within them would upregulate proteins that were important to support the transition into stationary phase, and/or support continued growth. To gain insight into these phenotypes we made a shotgun proteome analysis of total protein prepared from 1 day-old and 7 day-old colonies of the wild-type, and of the isogenic *rpoB* P564L and Δ*rpoS* mutants [13]. Over 630 different proteins were identified in each of these samples (Table S3). The data were examined for proteins that were strongly upregulated during colony aging, and also present at a relatively high level in 7 day-old colonies. Many of the proteins fitting this profile are annotated as being involved in carbon utilization or carbon scavenging. Notably, these included aconitate hydratase (AcnA), identified as the single most abundant protein, and the most strongly upregulated protein, in 7 day-old colonies of the *rpoB* mutant strain. AcnA was also highly abundant and strongly upregulated in 7 day-old colonies of the wild-type and the RpoS mutant. In contrast, the level of AcnA in preparations from day 1 colonies of all three strains was very low. AcnA plays an important enzymatic role in the TCA cycle by being responsible for the isomerization of citrate and iso-citrate via cis-aconitate. It was also notable that four enzymes, Acs, AcnA, AceA, and AceB, that operate consecutively in the pathway of acetate utilization were upregulated in all three strains and present at relatively high levels in preparations from 7 day-old colonies (Table S3). In addition, GltA (citrate synthetase; required for catalyzing the interaction of oxaloacetate with acetyl-CoA to make citrate) was upregulated and present at a high level in 7 day-old colonies of all three genotypes studied. Transcriptome data from *E. coli* is in broad agreement with this proteome data and shows upregulation of mRNA for *acnA*, *aceA*, *aceB* and *gltA* in wild-type and an *rpoS* mutant aged for 7 days [8,14].

Taken together, the data from the shotgun proteome analysis suggested that in aging colonies, bacteria upregulated genes required for the utilization of acetate as a carbon source. The potential significance of this is that bacteria growing exponentially on good carbon sources produce and excrete acetate when the inflow of carbon is in excess of what can be processed with respect to central metabolic pathways [15,16,17,18]. During the transition to stationary phase, when the preferred carbon source has been consumed, bacteria then undergo a metabolic switch and import and consume the acetate that had previously been excreted during exponential growth [19]. Because the shotgun proteome analysis is not strictly quantitative an additional quantitative protein analysis was made to accurately determine the relative quantities of several key proteins involved in acetate utilization in colonies of wild-type and mutant strains.

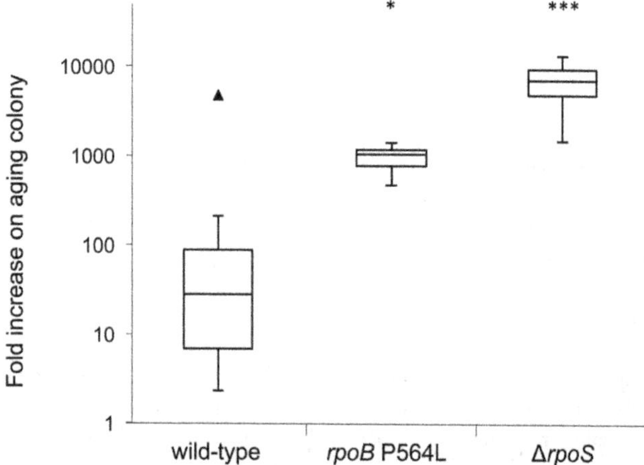

Figure 1. RpoB and RpoS mutants have a growth advantage on aging colonies. Fold increase in wild-type and mutant cells added to 24 h wild-type colonies and allowed to age for a further 7 days. The box plots show the first quartile, median, and third quartile values. Outlier indicated by a triangle. Statistical significance of differences in the distribution of values between strains, compared to the wild-type, is given in Table S2 and indicated in the figure by asterisks (* = 95% confidence interval, *** = 99.9% confidence interval).

Quantitative proteome analysis

Single Ion Monitoring mass spectrometry was used to quantify seven proteins associated with acetate utilization in mutant and wild-type bacterial colonies aged for 1, 3, 5 or 7 days, using a protocol described by Thermo Fisher Scientific [20]. The proteins quantified were: ActP, acetate/glycolate permease; AckA, acetate kinase; Pta, phosphate acetyltransferase; Acs, acetyl-CoA synthetase; AceA, isocitrate lyase; AceB, malate synthase A; and Pka (PatZ, YfiQ), peptidyl-lysine acetyltransferase (Figure 2). At least two peptides per protein were analyzed and each sample was measured in biological replicates. The concentration of each of these proteins was quantified and compared between the wild-type and the *rpoB* and *rpoS* mutant strains (Figure 3). Acs (Figure 3A) was upregulated in all colonies as a function of colony age, dramatically so in the *rpoB* mutant strain. Due to high variation, the apparent increase in Acs levels for the 7 day-old *rpoS* mutant was not significantly different from the 7 day-old wild-type. The glyoxylate shunt enzymes, AceA and AceB, were strongly upregulated as a function of colony age in each of the mutant strains (Figure 3B and 3C). Also for AceB levels, the aged *rpoS* samples showed a high variation, which made the apparent increase, compared to the wild-type insignificant (Figure 3C). AckA and Pta levels increased after day 1 in all colony types but the levels did not differ significantly between wild-type and mutant strains (Figure 3D and 3E). The levels of Pka and ActP were too low in all samples to be reliably quantified. This quantitative proteome data confirmed that key enzymes required for the capture of acetate (by converting it to acetyl-CoA), and acetate utilization via the glyoxylate shunt, are present at significantly higher levels in stationary phase colonies, and that the levels of Acs, AceA, AceB are higher in the *rpoB* and *rpoS* mutant strains than in the wild-type (Figure 3, panels A, B, C). This correlation suggested that the activity of one or more of these enzymes might be required to support the continued growth of *rpoB* and *rpoS* mutants on aging wild-type colonies. This hypothesis was tested genetically.

The ability to convert acetate to acetyl-CoA is important for subpopulation growth on aging wild-type colonies

We addressed whether the acetate that exponentially growing bacteria excrete might be important to support the growth of mutants in the stationary phase colony. Acetate can freely permeate the membrane [21,22,23] and the pathways of acetate metabolism have been thoroughly reviewed [24] and are outlined in Figure 2. To utilize acetate for protein acetylation or as a nutrient it must first be converted to acetyl-CoA. Conversion depends on either of two parallel pathways. The *acs* pathway (acetyl-CoA synthase) is a high affinity system used at low acetate concentrations whereas the constitutive *ackA-pta* pathway (acetate kinase, *ackA*, and phosphotransacetylase, *pta*) is used when acetate is available at relatively high concentrations [25]. Cells lacking both the *acs* and the *ackA-pta* pathways are unable to grow on acetate at any concentration [26].

Isogenic strains were constructed in which either *acs* alone or all three genes, *ackA, pta,* and *acs,* were deleted (labelled as Δ*apa* in Figure 4). These strains were otherwise wild-type, or carried either the *rpoB* P564L or Δ*rpoS* alleles. The ability of each of these strains to grow when added as a small population onto an aging wild-type colony was measured. Inactivation of the ability to convert acetate to acetyl-CoA reduced the growth advantage of each of the mutants but did not significantly affect the wild-type (Figure 4, Table S2). A significant reduction in mutant growth also occurred when only *acs* was knocked out. This shows that the ability to convert acetate to acetyl-CoA, dependent on an active *acs* gene, is crucial for these mutants subpopulations to grow and accumulate on aging wild-type colonies.

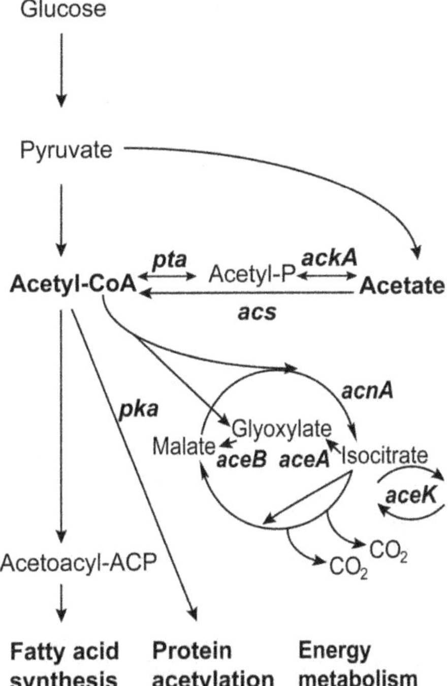

Figure 2. Outline of central metabolism indicating acetate production and utilization pathways. Outline of central metabolism showing the acetate synthesis and utilization pathways [24]. Gene names are shown in italics, substrates and products are referred to in the text.

The glyoxylate shunt is not required for *rpoB* and *rpoS* mutant growth on aging colonies

Growth on acetate as the sole carbon source requires the glyoxylate shunt, and thus the activities of the *aceB*, *aceA* and *aceK* gene products (Figure 2) to bypass the two CO_2 evolving steps of the TCA cycle, at the expense of energy [27]. This bypass allows the production of four-carbon compounds such as malate and oxaloacetate during growth on two-carbon substrates like acetate. We constructed isogenic strains lacking the genes of the glyoxylate shunt. Deletion of *aceBAK* did not significantly reduce the ability of *rpoB* or *rpoS* mutants to grow on aging colonies (Figure 4 and Table S2). This shows that the growth of subpopulations of *rpoB* and Δ*rpoS* mutants on aging wild-type colonies does not depend on the activity of the glyoxylate shunt.

Protein acetylation by Pka is not essential for mutant growth on aging colonies

Acetyl-CoA is also important as the source of donor acetyl groups for the acetylation of many proteins. Many of the enzymes involved in the central metabolic pathways of *E. coli* and *S.* Typhimurium are subject to lysine acetylation [28,29,30,31,32] and reversible acetylation has been shown to modulate enzyme activity [33,34]. In *Salmonella*, protein actetylation is dependent on the peptidyl-lysine acetyltransferase encoded by *pka* (synonyms: *yfiQ, pat*) [35,36,37]. We asked whether Pka activity was essential for the ability of the *rpoB* or *rpoS* mutants to grow on aging wild-type colonies. Isogenic wild-type and mutant strains with the *pka* gene deleted were constructed and assayed for their ability to grow on aging wild-type colonies. Deletion of *pka* had no significant affect on the ability of any of the mutant strains to grow on the aging wild-type colonies (Figure 4, Table S2).

Differential rate of acetate uptake and conversion in young and aged colonies

The genetic tests showed that Acs, the enzyme that converts acetate to acetyl-CoA, is critically important for the ability of subpopulations of *rpoB* and *rpoS* mutants to continue growing as wild-type colonies age. Because acetate itself can freely diffuse across the cell membrane, the enzymatic action of Acs functions to capture acetate inside the bacterial cell where it can then be used for a variety of biological purposes (energy generation, fatty acid synthesis, protein acetylation). We compared acetate capture by wild-type and mutant cells in aging colonies that had been exposed to C^{14}-acetate and measured the rate of the accumulation of C^{14} as a function of colony age. Colonies of wild-type, *rpoB* mutant and *rpoS* mutant were grown for either 24 h or 7 days. Colonies were suspended and washed in saline, C^{14}-acetate was added to the suspensions and samples were taken every 10 seconds for 1 minute. In cells from 1 day-old colonies the rate of C^{14} uptake was similar for all strains (Figure 5). In contrast, in 7 day-old colonies, the rate of uptake was reduced in the wild-type, it was unchanged

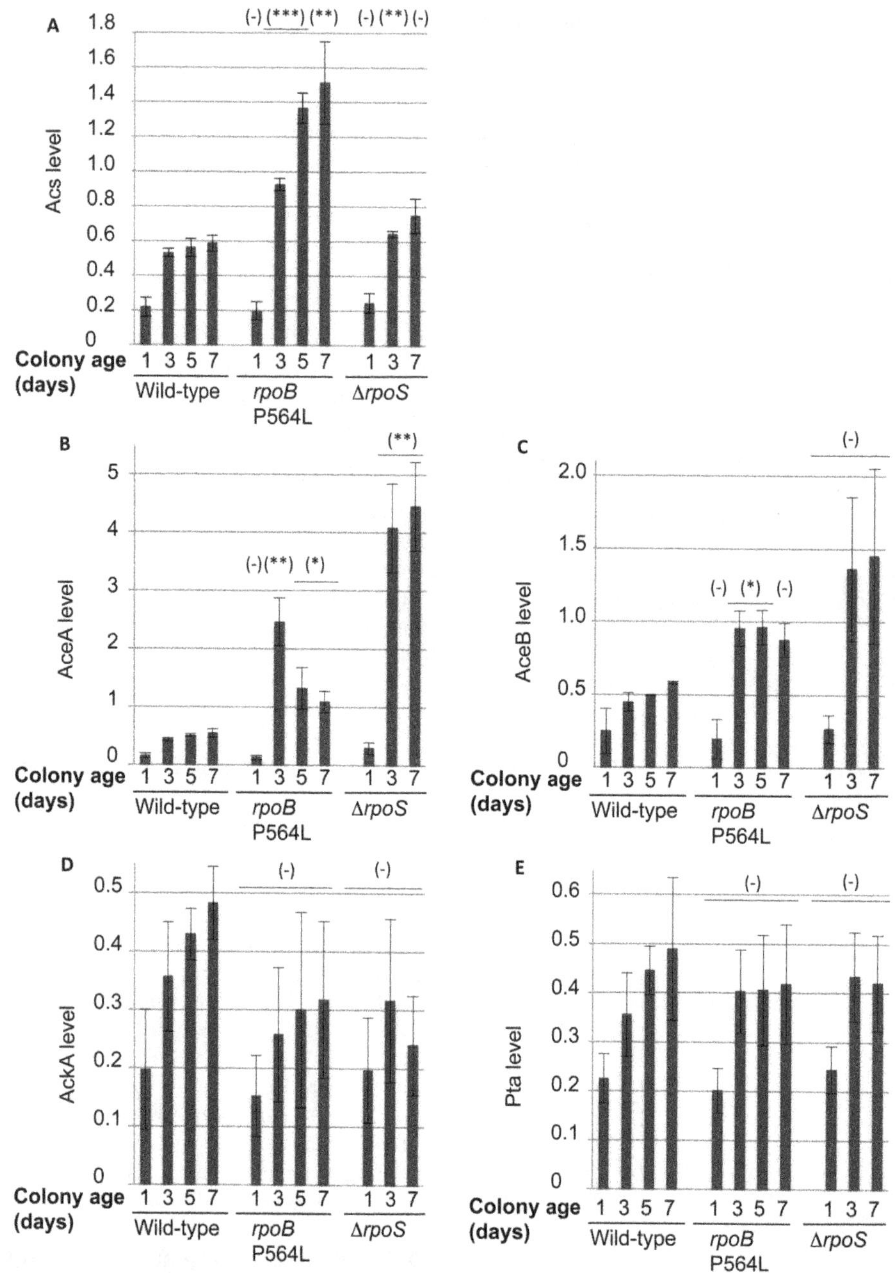

Figure 3. Protein quantification in aging colonies. Total protein was prepared from colonies of either wild-type, *rpoB* P564L mutant or Δ*rpoS* mutant, all grown for 1, 3, 5 and 7 days. Concentrations of the proteins of interest were quantified by Single Ion Monitoring mass spectrometry. Bars represent averages of two or three peptides per protein, measured in biological duplicates, each measured twice, with error bars representing standard deviation between the runs. Concentration values are in arbitrary units, normalized to total protein in the samples, to enable direct comparison between different days and different samples. The day 5 sample from the Δ*rpoS* mutant could not be analyzed. Statistical significance of differences, compared to wild-type samples of the same age, are indicated in the figure (- = not significant, * =95% confidence interval, ** =99% conficence interval, *** =99.9% confidence interval). **A.** Concentration of Acs, acetyl-CoA synthase. **B.** Concentration of AceA, isocitrate lyase. **C.** Concentration of AceB, malate synthase. **D.** Concentration of AckA, acetate kinase. **E.** Concentration of Pta, phosphotransacetylase.

in the *rpoB* mutant, and it increased almost 5-fold in the *rpoS* mutant (Figure 5).

The growth advantage of subpopulations is increased on aging colonies that are unable to convert acetate to acetyl-CoA

Bacteria in a wild-type colony excrete acetate during exponential growth, and then import and consume acetate during the transition into stationary phase. We hypothesized that if cells in the aging wild-type colony were unable to consume that acetate during its transition into stationary phase, that more acetate would then be available to support the growth of subpopulations of mutant cells. We tested this hypothesis by measuring the accumulation of isogenic wild-type and *rpoB* and *rpoS* mutants on aging colonies of a Δ*acs* mutant. The prediction was that the subpopulations of added cells would accumulate to greater

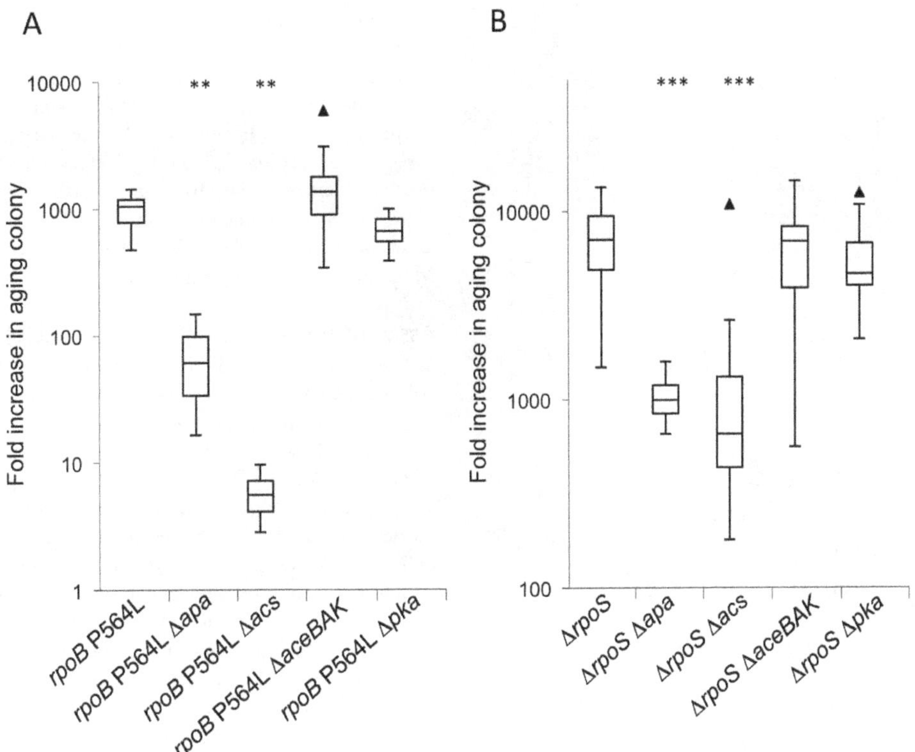

Figure 4. Influence of mutations affecting acetate metabolism on the growth advantage of *rpoB* P564L and Δ*rpoS* mutants. Fold increase in wild-type and mutant cells added to 24 h wild-type colonies and allowed to age for a further 7 days, as a function of acetate metabolism activity. **A**. Relative to the *rpoB* 564L mutant. **B**. Relative to the Δ*rpoS* mutant. The box plots show the first quartile, median, and third quartile values. Outliers are indicated as triangles. Δ*apa* indicates deletion of the three genes, *ackA-pta* and *acs*. Statistical significance of differences, compared to the *rpoB* or *rpoS* mutants, is given in Table S2 and is indicated in the figure by asterisks (** =99% confidence interval, *** =99.9% confidence interval).

numbers than when added onto aging wild-type colonies, because there should be a greater quantity of acetate available in aging colonies of a Δ*acs* mutant to support their growth. This prediction was borne out by the experiment (Figure 6, Table S4). All added subpopulations grew to a significantly greater number than when placed on aging wild-type colonies. Interestingly, the added wild-type subpopulation also showed enhanced accumulation on the Δ*acs* mutant. This data show that the relative availability of acetate is a critical factor, both enabling and limiting the growth of *rpoB* and *rpoS* mutant subpopulations on aging wild-type colonies.

Summary and conclusions

We have shown that mutations affecting RNA polymerase (Δ*rpoS* and *rpoB* P564L) facilitate continued growth of mutant cells on aging colonies of wild-type *S.* Typhimurium after most cells have entered stationary phase. We have also shown that the continued growth of these mutant subpopulations on aging colonies depends to a large extent on the availability of acetate and on the capacity of the mutant cells to import and convert that acetate into acetyl-CoA. We suggest from these data that the RNA polymerase status of individual bacteria in a population, in association with acetate availability, exerts a strong influence in determining whether *S.* Typhimurium enters stationary phase or continues to grow as the colony ages.

We have shown that the conversion of acetate to acetyl-CoA is important for the bacteria to be able to grow in the aging colony. However, since neither the glyoxylate shunt nor protein acetylation is required for growth, this suggests that the generated acetyl-CoA may be used for something else. Apart from the TCA cycle and protein acetylation, acetyl-CoA is also used in fatty acid synthesis [38]. It is tempting to speculate that the acetyl-CoA in the aging colonies may be utilized to synthesize fatty acids.

The phenomenon of continued growth of mutant sub-populations in stationary phase colonies (or liquid cultures) could have a general evolutionary significance. It illustrates the temporal and environmental dependency of the concept of relative fitness, and the interplay between spontaneous mutations, changing environmental selection pressures, and sibling rivalry, within clonal populations.

Materials and Methods

Bacterial strains and growth conditions

S. enterica serovar Typhimurium ATTC strain 14028 s for which the complete genome sequence is available [39] was used as wild-type and isogenic strains were constructed from it. Transductions were made using phage P22 HT [40]. Deletions were made by λ-red recombineering [10,11] replacing the relevant genes with an FRT-*tetRA*-FRT cassette as described previously [41]. The genotypes of the bacterial strains used are listed in Table 1. Oligonucleotides used for recombineering and/or DNA sequencing across deletion junctions are listed (Table S5). Drug resistance markers were removed from recombineered strains by site-specific recombination after transduction with a P22 phage lysate grown on a strain carrying the plasmid pCP20 (Amp^R) expressing Flp recombinase [42], leaving one FRT sequence at the site of deletion. Bacteria were grown at 37°C in Luria-Bertani

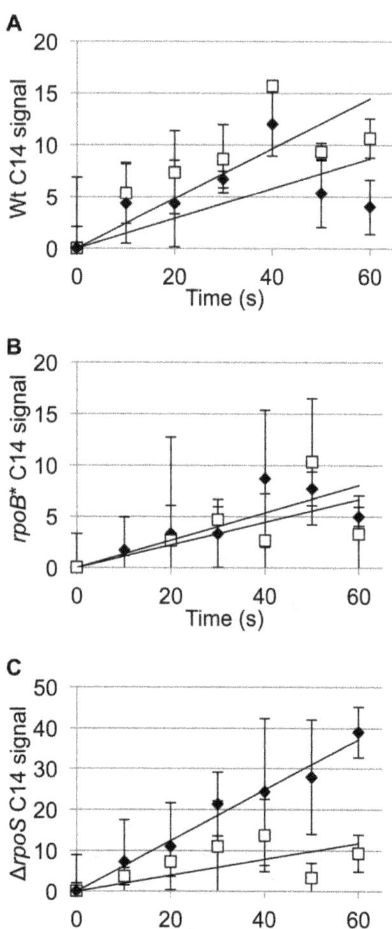

Figure 5. Acetate uptake rate in young and aged colonies. Colonies of 24 h (open squares) and 7 days (closed diamonds) were suspended in ^{14}C-acetate. Samples were taken every 10 seconds for one minute. **A.** Wild-type colonies. **B.** *rpoB* P564L colonies. **C.** Δ*rpoS* colonies. Each data point represents the average of three independent measurements, with standard deviations as error bars.

broth (LB) or on Luria Agar, LA (LB supplemented with 1.5% agar; Oxoid, Basingstoke, England; 0.2% glucose; 3 mM $CaCl_2$). Antibiotics were used at the following final concentrations; tetracycline 15 µg/mL, ampicillin 100 µg/mL, rifampicin 100 µg/mL. For colony-aging experiments agar plates were incubated at 37°C in sealed plastic bags to minimize dehydration.

PCR and DNA sequencing

The public genome sequences for *S. enterica* LT2 (NCBI accession number NC_003197) and *S. enterica* 14028 s (GenBank accession number CP001363) were used to design primers for PCR amplification and DNA sequencing. DNA samples for PCR were obtained by dispersing fresh bacterial colonies in 100 µl of sterile water, heating at 95°C for 5 min and then cooling on ice. PCR was performed using PuReTaq Ready-To-Go PCR beads (GE Healthcare, Uppsala, Sweden) according to the protocol of the manufacturer. Amplifications were carried out in 25 µl volumes containing 0.4 µM reverse and forward primers and 1 µl of DNA sample using a DNA engine PTC-200 thermocycler (SDS-diagnostics, Falkenberg, Sweden). PCR was initiated by denaturation at 95°C for 5 min, followed by 25 cycles of 95°C for 30 s, 56°C for 20 s, and 72°C for 1 min. Amplification products were visualized by agarose gel electrophoresis and ethidium bromide staining to assess the sizes of the gene fragments. Products were purified prior to sequencing using the QIAquick PCR purification kit (Qiagen, VWR International AB, Stockholm, Sweden) according to the manufacturer's instructions. PCR product concentrations were quantified using a Nanodrop NO-1000 spectrophotometer (Nanodrop, Wilmington, DE, USA). DNA sequencing of purified PCR products was performed at Macrogen Inc., Seoul, South Korea.

Assay of mutant growth on aging wild-type colonies

A fresh overnight culture was diluted in 0.9% NaCl and a 2 µl volume containing 100–1000 cfu was spotted onto an 82 mm diameter 0.2 µm pore size Protran BA83 nitrocellulose filter (Whatman, Dassel, Germany) on LA plates, to initiate growth of a wild-type colony. After 24 h incubation a 4 µl volume containing approximately 2×10^3 cfu of wild-type or an isogenic mutant, each marked with the same tetracycline resistance marker, *zhe*-8953::Tn*10*dTet, was spotted onto the 24 h colony. After this addition the colonies were incubated for an additional 7 days at 37°C in sealed plastic bags (the period of colony aging) after which the colony was suspended in 1 ml of 0.9% NaCl and appropriate dilutions were plated on LA and LA-Tet plates. The median fold increase in the number of mutant Tet^R cells over the 7-day aging period was calculated. Each colony aging competition experiment was biologically independent: separate independent cultures of wild-type bacteria were grown to initiate each independent colony; separate independent cultures of Tet^R-marked wild-type or mutant cells were grown for addition to each independent 24 h colony. Under these incubation conditions a bacterial cell undergoes up to 30 generations of growth and division within 24 h, reaching a population of approximately 10^9 cells in the colony [1,2,4]. Prior to 24 h the rate of growth slows significantly and the population initiates the process of entering the stationary phase. When incubation is continued beyond 24 h the population in the colony continues to grow but at a much slower rate, reaching approximately 10^{10} by 48–72 h. No significant further net increase in population size occurs if the incubation is continued longer [1,2,4].

Protein preparation and mass spectrometry measurements

Colonies of pure wild-type, pure *rpoB* P564L mutant or pure Δ*rpoS* mutant were initiated and aged. After the appropriate time, the colonies were cut out and suspended in 1 ml 0.9% NaCl. The cells were pelleted and lysed by the addition of 400 µl lysis buffer (20 mM HEPES pH 8.0, 9 M urea) containing a protease inhibitor cocktail added according to the manufacturers instructions (Complete, Mini, EDTA-free, Roche Diagnostics Scandinavia AB, Bromma, Sweden) and mixed by pipetting. The samples were sonicated with a 16 micron probe for 3×20 seconds, and cooled on ice 1 minute between each sonication. The lysates were centrifuged at 12000 g for 15 min at 4°C. Shotgun Proteome Analysis, and Quantitative Single Ion Monitoring were each carried out at the Science for Life Mass Spectrometry Technology Platform, Uppsala University, Sweden. An LTQ-Orbitrap Velos Pro ETD mass spectrometer (Thermo Fisher Scientific) was used for the mass spectrometry measurements and the PinPoint 1.3 software was used for the quantitation analysis [20].

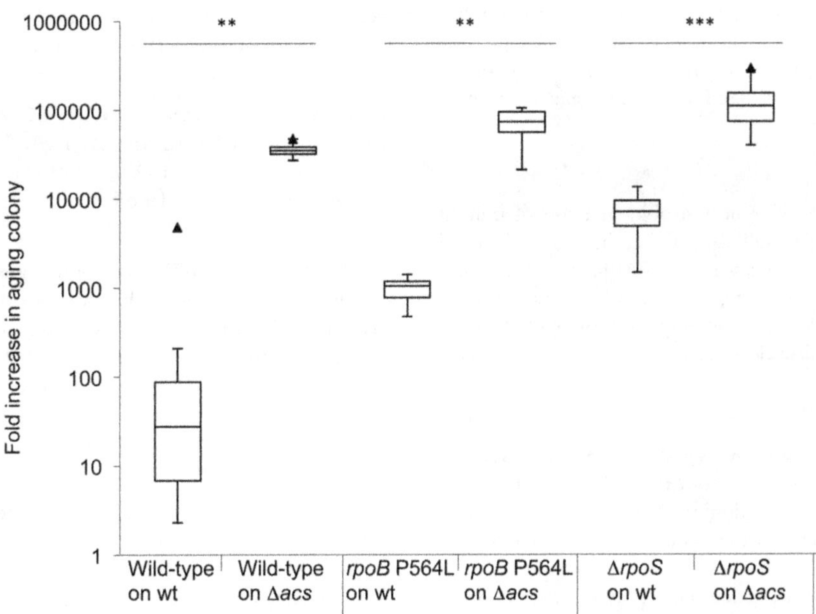

Figure 6. Influence of the *acs* status of the background colony on the growth advantage of Rif^R and RpoS mutants. Fold increase in wild-type and mutant cells added to 24 h wild-type or Δ*acs* colonies and allowed to age for a further 7 days. The box plots show the first quartile, median, and third quartile values. Outliers are indicated as triangles. Statistical significance of differences between strains is given in Table S2 and is indicated in the figure by asterisks (** = 99% confidence interval, *** = 99.9% confidence interval).

Uptake of ^{14}C-acetate

Colonies were initiated as described for the assay for the aged colonies, except that these colonies were pure wild-type, pure *rpoB* P564L or pure Δ*rpoS*. After the appropriate number of days of incubation at 37°C, the colonies were cut out and suspended in 10 ml 0.9% NaCl at room temperature. Immediately prior to initiating the assay a time zero sample was taken by transferring 750 µl cell suspension to 750 µl non-radioactive 2 mM sodium acetate at room temperature. To initiate the assay 10 µl (0.002 µCi) ^{14}C-acetate (Perkin Elmer, Upplands Väsby, Sweden, 50–62 mCi/mmol) was added to the cell suspension. Samples were then taken every 10 seconds for 1 minute by the removal of 750 µl cell suspension to 750 µl non-radioactive 2 mM sodium acetate chase solution at room temperature. The samples were chased for 8 minutes and then placed on ice. Cells were pelleted by centrifugation (1 min, 12000×g, 4°C) and washed three times in 500 µl 0.9% NaCl. The last pellet was re-suspended in 200 µl 0.9% NaCl and added to 4 ml scintillation fluid (Quicksafe Flow 2,

Table 1. Strain list.

Strains[a]	Genotype
TH6509	14028s Wild-type
TH6694	*zhe-8953*::Tn*10*dTet
TH7705	Δ*acs*
TH7141	*rpoB* P564L
TH7148	*zhe-8953*::Tn*10*dTet *rpoB* P564L
TH7718	*zhe-8953*::Tn*10*dTet Δ*aceBAK rpoB* P564L
TH8168	*zhe-8953*::Tn*10*dTet Δ*ackA-pta* Δ*acs rpoB* P564L
TH7722	*zhe-8953*::Tn*10*dTet Δ*acs rpoB* P564L
TH8486	*zhe-8953*::Tn*10*dTet Δ*pka rpoB* P564L
TH8082	Δ*rpoS*
TH8097	*zhe-8953*::Tn*10*dTet Δ*rpoS*
TH8163	*zhe-8953*::Tn*10*dTet Δ*aceBAK* Δ*rpoS*
TH8164	*zhe-8953*::Tn*10*dTet Δ*ackA-pta* Δ*acs* Δ*rpoS*
TH8516	*zhe-8953*::Tn*10*dTet Δ*acs* Δ*rpoS*
TH8487	*zhe-8953*::Tn*10*dTet Δ*pka* Δ*rpoS*

[a]All strains are isogenic to TH6509 *Salmonella enterica* serovar Typhimurium strain 14028 s.

Zinsser Analytic), vortexed, then placed in a scintillation counter. The uptake of ^{14}C-acetate in bacteria was analyzed by plotting radioactive incorporation as a function of time. The slope of the linear uptake curve was considered to be the specific acetate uptake ability of the strain.

Statistical analysis

The significance of the differences in fold increase of mutant subpopulations between different isogenic strains was calculated using the Mann-Whitney nonparametric test (http://vassarstats.net). The significance of differences in acetate uptake and in protein quantity was calculated by unpaired two-tailed t-tests (http://graphpad.com/quickcalcs/ttest1/).

Supporting Information

Table S1 High frequency of *rpoS* mutations among extremely aged colonies of *S. enterica* LT2. TH7792 *S. enterica* LT2 (*zfd-6825*::Tn*10*) colonies were aged for 15 weeks at 37°C, when surviving cells were examined for evidence of mutations affecting RNA polymerase. The *rpoA, rpoB, rpoC, and rpoS* genes from 18 independent clones were sequenced. No mutations were identified in *rpoA, rpoB*, or *rpoC*. However, in three of the 18 strains, mutations were identified in *rpoS*. The Tet^R phenotype had been lost in 5/18 strains but two of the RpoS mutants had retained Tet^R and were tested for growth advantage in a standard colony aging experiment. Each RpoS mutant had a growth advantage relative to the isogenic wild-type in the aging colony. Thus, under extreme aging conditions a significant fraction (3/18) of the surviving bacteria had acquired mutations in *rpoS*, and had also acquired a growth advantage in aging colonies.

Table S2 Inactivation of acetate utilization genes reduces the growth of *rpoS* and *rpoB* mutant subpopulations on aging wild-type[a] colonies. Colony competition experiments were made as described in Materials and Methods, using isogenic strains carrying the mutations listed.

Table S3 Proteomic analysis of wild-type and mutant strains (*rpoB* P564L, and Δ*rpoS*). The data show individual relative protein levels in Day 1 and Day 7 colonies, and the fold change over that time period.

Table S4 Inactivation of the *acs* gene in the background colony increases the growth of wild-type, *rpoB* and *rpoS* mutant subpopulations on aging colonies, compared to aging on wild-type background colonies. Colony competition experiments were made as described in Materials and Methods except that the background colony had the genotype Δ*acs*.

Table S5 Oligonucleotides used to make and confirm genetic deletions. List of oligonucleotides used in recombineering mutations into the chromosome, for PCR, and as primers for DNA sequencing.

Acknowledgments

This work was supported by the Science for Life Mass Spectrometry Technology Platform in Uppsala, Sweden. Data storage was obtained and supported by BILS (Bioinformatics Infrastructure for Life Sciences). We thank Svenja N. Reinders, Eva Garcia Barreales, and Shanjun Gao, who each contributed to parts of this project as Masters students.

Author Contributions

Conceived and designed the experiments: DH JMB MW. Performed the experiments: JMB MW. Analyzed the data: JMB MW DH. Wrote the paper: DH JMB MW.

References

1. Bjedov I, Tenaillon O, Gerard B, Souza V, Denamur E, et al. (2003) Stress-induced mutagenesis in bacteria. Science 300: 1404–1409.
2. Taddei F, Halliday JA, Matic I, Radman M (1997) Genetic analysis of mutagenesis in aging Escherichia coli colonies. Mol Gen Genet 256: 277–281.
3. Taddei F, Matic I, Radman M (1995) cAMP-dependent SOS induction and mutagenesis in resting bacterial populations. Proc Natl Acad Sci U S A 92: 11736–11740.
4. Wrande M, Roth JR, Hughes D (2008) Accumulation of mutants in "aging" bacterial colonies is due to growth under selection, not stress-induced mutagenesis. Proc Natl Acad Sci U S A 105: 11863–11868.
5. Zambrano MM, Siegele DA, Almiron M, Tormo A, Kolter R (1993) Microbial competition: Escherichia coli mutants that take over stationary phase cultures. Science 259: 1757–1760.
6. Finkel SE (2006) Long-term survival during stationary phase: evolution and the GASP phenotype. Nature reviews Microbiology 4: 113–120.
7. Notley-McRobb L, King T, Ferenci T (2002) rpoS mutations and loss of general stress resistance in Escherichia coli populations as a consequence of conflict between competing stress responses. J Bacteriol 184: 806–811.
8. Saint-Ruf C, Taddei F, Matic I (2004) Stress and survival of aging Escherichia coli rpoS colonies. Genetics 168: 541–546.
9. McClelland M, Sanderson KE, Spieth J, Clifton SW, Latreille P, et al. (2001) Complete genome sequence of Salmonella enterica serovar Typhimurium LT2. Nature 413: 852–856.
10. Yu D, Ellis HM, Lee EC, Jenkins NA, Copeland NG, et al. (2000) An efficient recombination system for chromosome engineering in Escherichia coli. Proc Natl Acad Sci U S A 97: 5978–5983.
11. Datsenko KA, Wanner BL (2000) One-step inactivation of chromosomal genes in Escherichia coli K-12 using PCR products. Proc Natl Acad Sci U S A 97: 6640–6645.
12. Sutton A, Buencamino R, Eisenstark A (2000) rpoS mutants in archival cultures of Salmonella enterica serovar typhimurium. J Bacteriol 182: 4375–4379.
13. Mann M, Kulak NA, Nagaraj N, Cox J (2013) The coming age of complete, accurate, and ubiquitous proteomes. Mol Cell 49: 583–590.
14. Saint-Ruf C, Garfa-Traore M, Collin V, Cordier C, Franceschi C, et al. (2014) Massive Diversification in Aging Colonies of Escherichia coli. J Bacteriol 196: 3059–3073.
15. el-Mansi EM, Holms WH (1989) Control of carbon flux to acetate excretion during growth of Escherichia coli in batch and continuous cultures. J Gen Microbiol 135: 2875–2883.
16. Farmer WR, Liao JC (1997) Reduction of aerobic acetate production by Escherichia coli. Applied and environmental microbiology 63: 3205–3210.
17. Majewski RA, Domach MM (1990) Simple constrained-optimization view of acetate overflow in E. coli. Biotechnology and bioengineering 35: 732–738.
18. Holms WH (1986) The central metabolic pathways of Escherichia coli: relationship between flux and control at a branch point, efficiency of conversion to biomass, and excretion of acetate. Current topics in cellular regulation 28: 69–105.
19. Pruss BM, Nelms JM, Park C, Wolfe AJ (1994) Mutations in NADH:ubiquinone oxidoreductase of Escherichia coli affect growth on mixed amino acids. J Bacteriol 176: 2143–2150.
20. Kiyonami R, Zeller M, Zabrouskov V (2011) Quantifying peptides in complex mixtures with high sensitivity and precision using a targeted approach with a hybrid linear ion trap-orbitrap mass spectrometer. In: #557. TSAN, editor: Thermo Fisher Scientific.
21. Repaske DR, Adler J (1981) Change in intracellular pH of Escherichia coli mediates the chemotactic response to certain attractants and repellents. J Bacteriol 145: 1196–1208.
22. Salmond CV, Kroll RG, Booth IR (1984) The effect of food preservatives on pH homeostasis in Escherichia coli. J Gen Microbiol 130: 2845–2850.
23. Kihara M, Macnab RM (1981) Cytoplasmic pH mediates pH taxis and weak-acid repellent taxis of bacteria. J Bacteriol 145: 1209–1221.
24. Wolfe AJ (2005) The acetate switch. Microbiology and molecular biology reviews: MMBR 69: 12–50.
25. Kumari S, Simel EJ, Wolfe AJ (2000) sigma(70) is the principal sigma factor responsible for transcription of acs, which encodes acetyl coenzyme A synthetase in Escherichia coli. J Bacteriol 182: 551–554.

26. Kumari S, Tishel R, Eisenbach M, Wolfe AJ (1995) Cloning, characterization, and functional expression of acs, the gene which encodes acetyl coenzyme A synthetase in Escherichia coli. J Bacteriol 177: 2878–2886.
27. Kornberg HL (1966) The role and control of the glyoxylate cycle in Escherichia coli. The Biochemical journal 99: 1–11.
28. Hu LI, Lima BP, Wolfe AJ (2010) Bacterial protein acetylation: the dawning of a new age. Mol Microbiol 77: 15–21.
29. Thao S, Escalante-Semerena JC (2011) Control of protein function by reversible Nvarepsilon-lysine acetylation in bacteria. Curr Opin Microbiol 14: 200–204.
30. Zhao S, Xu W, Jiang W, Yu W, Lin Y, et al. (2010) Regulation of cellular metabolism by protein lysine acetylation. Science 327: 1000–1004.
31. Yu BJ, Kim JA, Moon JH, Ryu SE, Pan JG (2008) The diversity of lysine-acetylated proteins in Escherichia coli. Journal of microbiology and biotechnology 18: 1529–1536.
32. Jones JD, O'Connor CD (2011) Protein acetylation in prokaryotes. Proteomics 11: 3012–3022.
33. Gardner JG, Grundy FJ, Henkin TM, Escalante-Semerena JC (2006) Control of acetyl-coenzyme A synthetase (AcsA) activity by acetylation/deacetylation without NAD(+) involvement in Bacillus subtilis. J Bacteriol 188: 5460–5468.
34. Starai VJ, Celic I, Cole RN, Boeke JD, Escalante-Semerena JC (2002) Sir2-dependent activation of acetyl-CoA synthetase by deacetylation of active lysine. Science 298: 2390–2392.
35. Starai VJ, Escalante-Semerena JC (2004) Identification of the protein acetyltransferase (Pat) enzyme that acetylates acetyl-CoA synthetase in Salmonella enterica. J Mol Biol 340: 1005–1012.
36. Wang Q, Zhang Y, Yang C, Xiong H, Lin Y, et al. (2010) Acetylation of metabolic enzymes coordinates carbon source utilization and metabolic flux. Science 327: 1004–1007.
37. Lima BP, Antelmann H, Gronau K, Chi BK, Becher D, et al. (2011) Involvement of protein acetylation in glucose-induced transcription of a stress-responsive promoter. Mol Microbiol 81: 1190–1204.
38. Cronan JE Jr, Waldrop GL (2002) Multi-subunit acetyl-CoA carboxylases. Progress in lipid research 41: 407–435.
39. Jarvik T, Smillie C, Groisman EA, Ochman H (2010) Short-term signatures of evolutionary change in the Salmonella enterica serovar typhimurium 14028 genome. J Bacteriol 192: 560–567.
40. Schmieger H, Backhaus H (1973) The origin of DNA in transducing particles in P22-mutants with increased transduction-frequencies (HT-mutants). Mol Gen Genet 120: 181–190.
41. Marcusson LL, Frimodt-Moller N, Hughes D (2009) Interplay in the selection of fluoroquinolone resistance and bacterial fitness. PLoS Pathog 5: e1000541.
42. Cherepanov PP, Wackernagel W (1995) Gene disruption in Escherichia coli: TcR and KmR cassettes with the option of Flp-catalyzed excision of the antibiotic-resistance determinant. Gene 158: 9–14.

The Impact of *spgM, rpfF, rmlA* Gene Distribution on Biofilm Formation in *Stenotrophomonas maltophilia*

Chao Zhuo*, Qian-yu Zhao, Shu-nian Xiao

State Key Laboratory of Respiratory Diseases, the first affiliated hospital of Guangzhou Medical College, Guangzhou, China

Abstract

Background: *Stenotrophomonas maltophilia* is emerging as one of the most frequently found bacteria in chronic pulmonary infection. Biofilm is increasingly recognized as a contributing factor to disease pathogenesis. In the present study, a total of 37 isolates of *S. maltophilia* obtained from chronic pulmonary infection patients were evaluated to the relationship between biofilm production and the relative genes expression.

Methods: The clonal relatedness of isolates was determined by pulse-field gel electrophoresis. Biofilm formation assays were performed by crystal violet assay, and confirmed by Electron microscopy analysis and CLSM analysis. PCR was employed to learn gene distribution and expression.

Results: Twenty-four pulsotypes were designated for 37 *S. maltophilia* isolates, and these 24 pulsotypes exhibited various levels of biofilm production, 8 strong biofilm-producing *S. maltophilia* strains with OD492 value above 0.6, 14 middle biofilm-producing strains with OD492 average value of 0.4 and 2 weak biofilm-producing strains with OD492 average value of 0.19. CLSM analysis showed that the isolates from the early stage of chronic infection enable to form more highly structured and multilayered biofim than those in the late stage. The prevalence of *spgM*, *rmlA*, and *rpfF* genes was 83.3%, 87.5%, and 50.0% in 24 *S. maltophilia* strains, respectively, and the presence of *rmlA*, *spgM* or *rpfF* had a close relationship with biofilm formation but did not significantly affect the mean amount of biofilm. Significant mutations of *spgM* and *rmlA* were found in both strong and weak biofilm-producing strains.

Conclusion: Mutations in *spgM* and *rmlA* may be relevant to biofilm formation in the clinical isolates of *S. maltophilia*.

Editor: Dipshikha Chakravortty, Indian Institute of Science, India

Funding: The work was supported by a grant from the Natural Science Foundation of China (no. 30971318) and a grant (no. 2009A030301011) from the Key Foundation of Guangdong province Bureau of Science and Technology. The funders had no role in study design, data collection and analysis, decision to publish, or preparation of the manuscript.

Competing Interests: The authors have declared that no competing interests exist.

* Email: chao_sheep@263.net

Introduction

Stenotrophomonas maltophilia is a Gram-negative opportunistic pathogen in hospitalized or compromised patients [1,2]. However, the role of this opportunistic pathogen as an innocent bystander or causative agent often remains unclear [3,4] and little is known about its virulence factors [5,6].

Biofilms, sessile structured bacterial communities exhibiting recalcitrance to antimicrobial compounds and persistence despite sustained host defenses, are increasingly recognized as a contributing factor to disease pathogenesis in the respiratory tract diseases associated with chronic bacterial infections [7]. *S. maltophilia* isolates are known to have the ability to form biofilms on both abiotic surfaces [8–10], and it is an intrinsic difference in biofilm formation among clinical isolates of *S. maltophilia*.

The molecular mechanisms underlying biofilm formation in *S. maltophilia* have not been extensively studied. Recently, mutants for *rmlA* gene and *rpfF* gene are reported to decrease biofilm formation [11,12]. Further, the *spgM* gene, encoding a bifunctional enzyme with both phosphoglucomutase (PGM) and phosphomannomutase activities, could be involved in biofilm-forming ability because of the homology with the *algC* gene that is responsible for the production of a PGM associated with LPS and alginate biosynthesis in *P. aeruginosa* [13].

However, it is still unclear that which gene mutation results in the change of biofilm formation among the three genes, and which key amino acid mutation determines the function changes of the protein.

In this study, we examined a set of 24 clinical isolates of *S. maltophilia* for biofilm formation traits in order to find significant differences, and evaluated the relationship between biofilm production and the detection of *rmlA*, *spgM*, and *rpfF* genes.

Materials and Methods

Ethics Statement

This study protocol was approved by the Ethics Committee of The First Affiliated Hospital of Guangzhou Medical University. All subjects signed written informed consent prior to the study. Patient information was anonymized and de-identified prior to analysis.

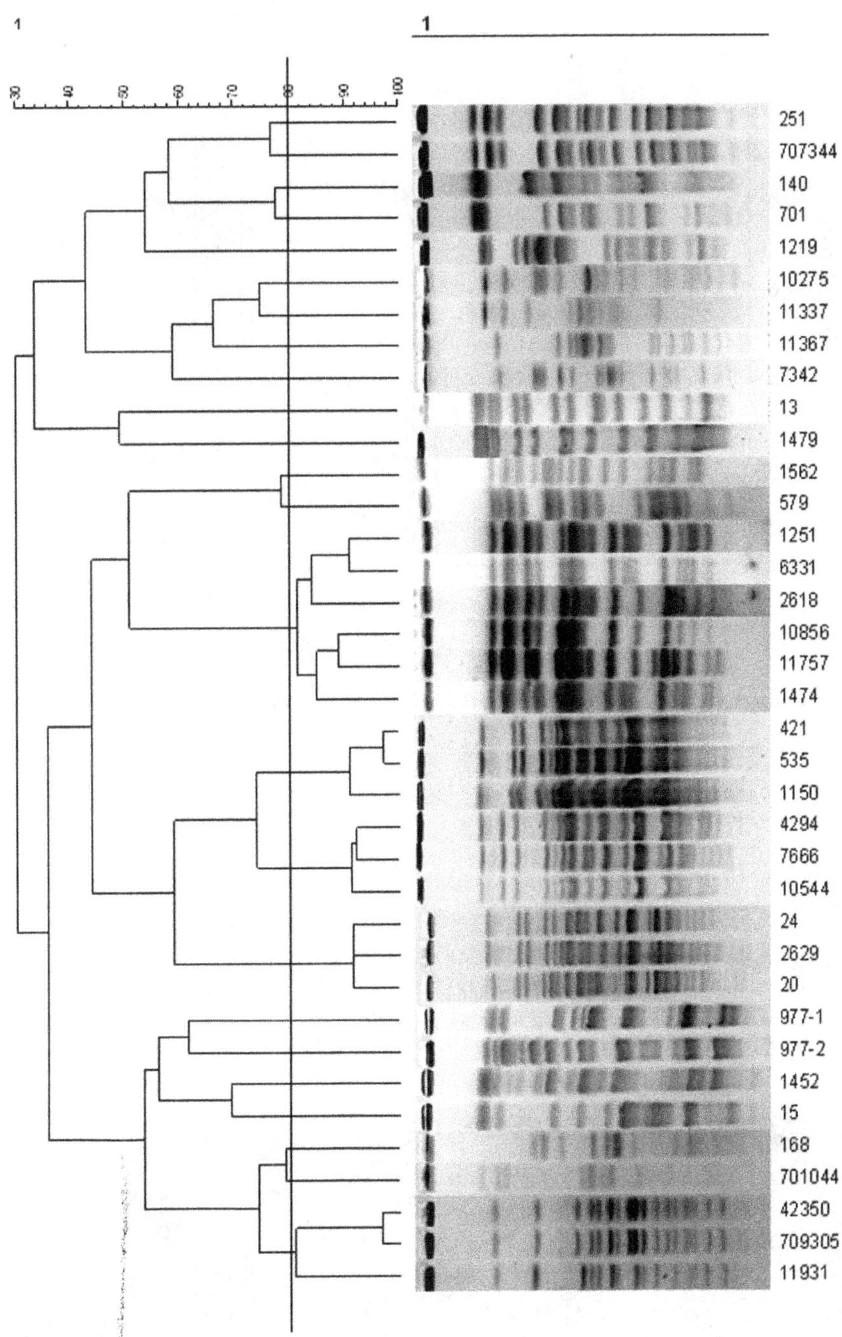

Figure 1. Dendrogram illustrating the percentage similarity of PFGE profiles in tested isolates of *S. maltophilia.*

Bacterial isolates and growth conditions

Overall, 37 *S. maltophilia* isolates and *S. maltophilia* ATCC13637 were investigated. All the strains collected from the sputa of patients attending the Guangzhou institute of respiratory diseases from 2010 to 2011. Among the 37 isolates, 4 sequential strains isolated from the same patient over a period of 2 years and 3 sequential strains isolated from another patient over a period of 1 year were investigated. The isolates were identified as *S. maltophilia* by biochemical tests using manual (API 20-NE System; BioMérieux, Marcy-L'Etoile, France) or automated (Vitek; BioMérieux) systems, then stored at −80°C until use when they were grown at 37°C in Mueller-Hinton agar (MHA; Oxoid) plates.

Genetic relatedness by PFGE and cluster analysis

Pulsed-field gel electrophoresis (PFGE) analysis of *Xba*I-digested genomic DNA was performed to determine the genetic relatedness of *S. maltophilia* isolates using a CHEF-Mapper XA System (Bio-Rad Laboratories, Hercules, CA, USA) as described by Seifert [14]. The interpreting criteria were described by Tenover [15] combining UPGMA (unweighted pair group method with hierarchic averages) method, Isolates were assigned the same pulsetype if the value of Dice coefficient of similarity was >80% [16].

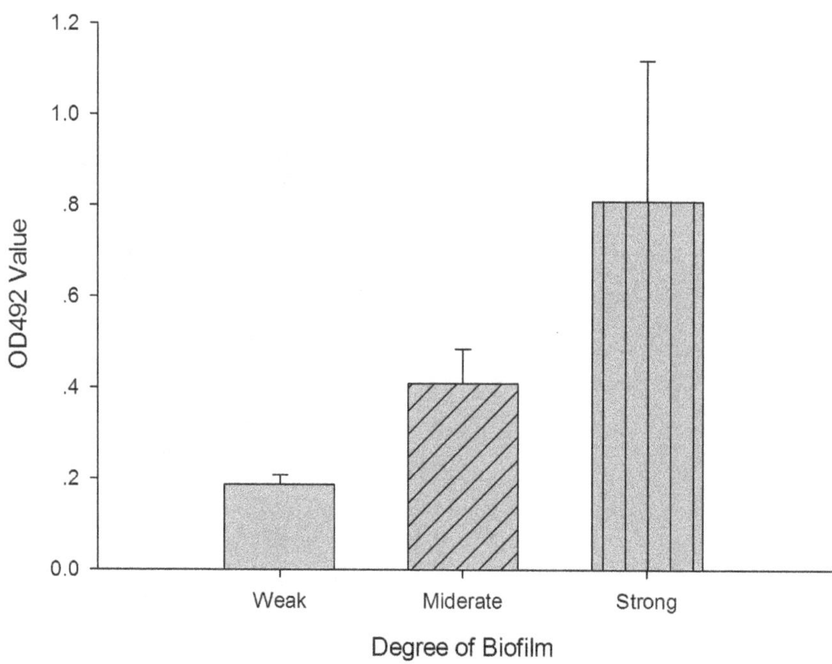

Figure 2. Histogram illustrating the ability of biofilm formation in tested isolates of *S. maltophilia.*

Biofilm formation assay

Overnight cultures in TSB were corrected with fresh TSB to an OD550 of 1.00 (corresponding to about 1×10^9 CFU/ml). Two-hundred microliters of 1:100 diluted inoculum were dispensed to each well of a sterile flatbottom polystyrene tissue culture 96-wells microtiter and incubated at 37°C for 24 h. Biofilm biomass was then measured by crystal violet assay. Briefly, biofilm samples were fixed for 1 h at 60°C, stained for 5 min at RT with 200 μl Hucker-modified crystal violet, then rinsed in standing water and allowed to dry. Biofilm samples were stained with 250 μl of 33% glacial acetic acid for 15 min, and the optical density at 492 nm (OD492) was read. Considering a low cut-off (ODc) represented by $3\times$SD above the mean OD of control wells, strains were classified into the following categories: no biofilm producer (OD≤ODc), weak biofilm producer (ODc<OD≤$2\times$ODc), moderate biofilm producer ($2\times$ODc<OD≤$4\times$ODc), and strong biofilm producer ($4\times$ODc<OD) [17].

The biofilm microstructure was detected by SEM and CLSM. SEM analyses were carried out with *S. maltophilia* strain S1219 and ATCC13637, selected because S1219 was the highest

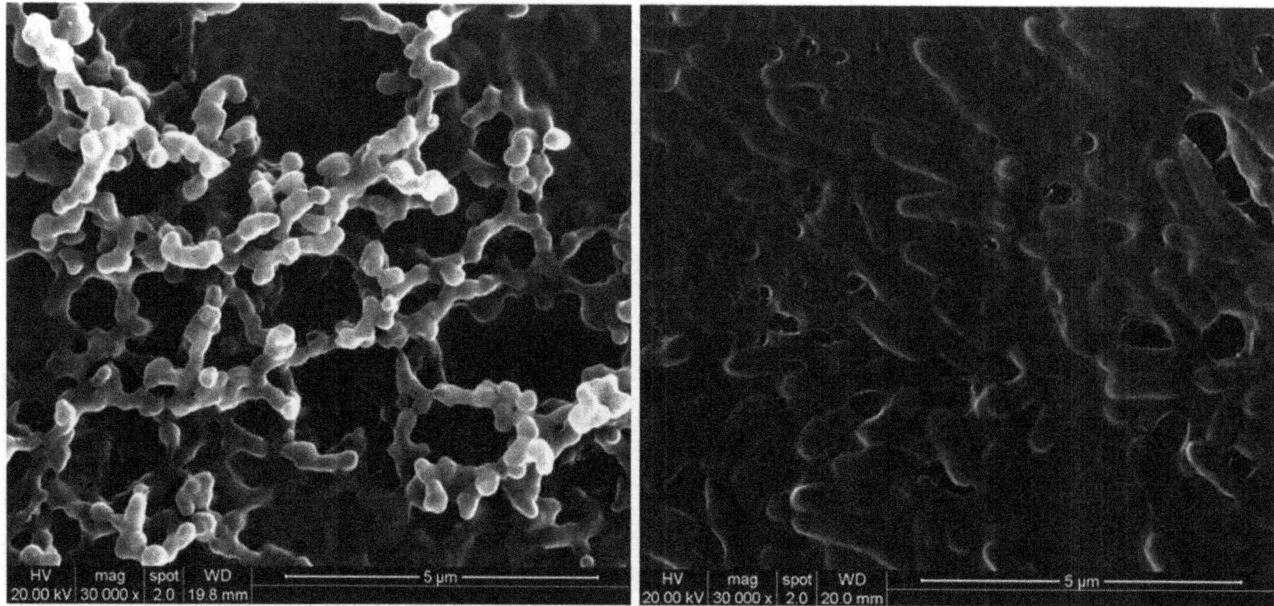

Figure 3. SEM of the strain Sm1219 of *S. maltophilia* biofilm mode in 8 h (left) and 24 h (right).

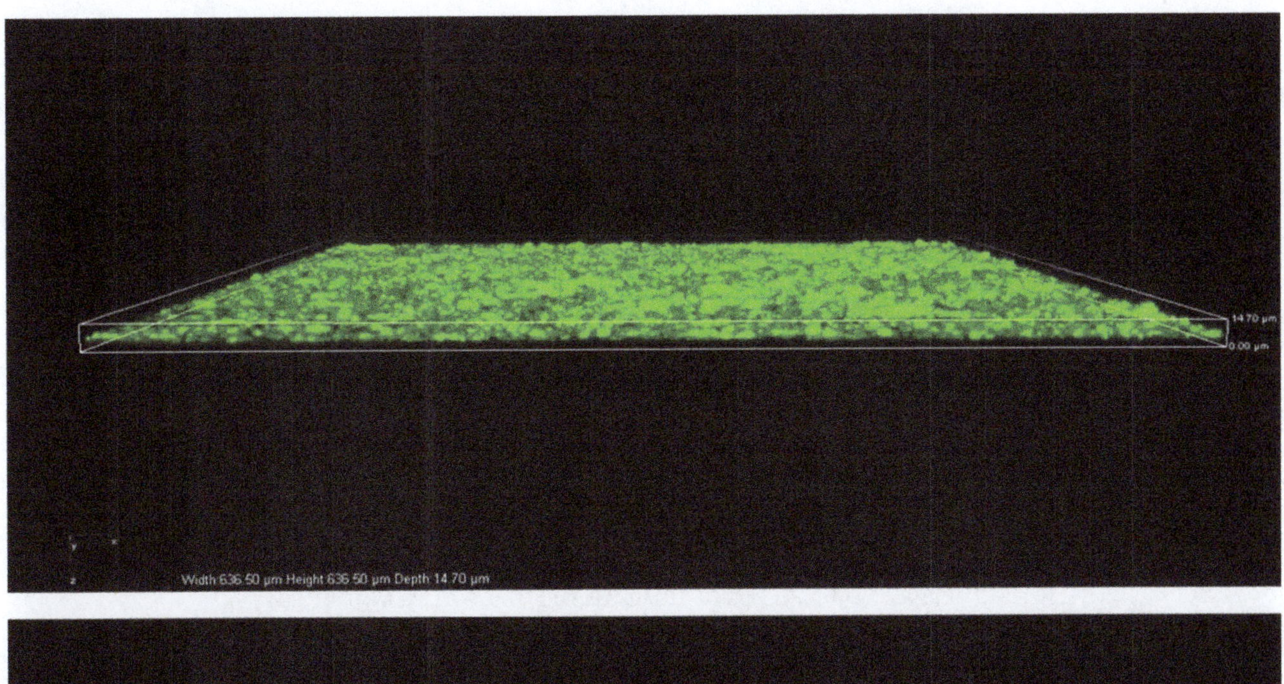

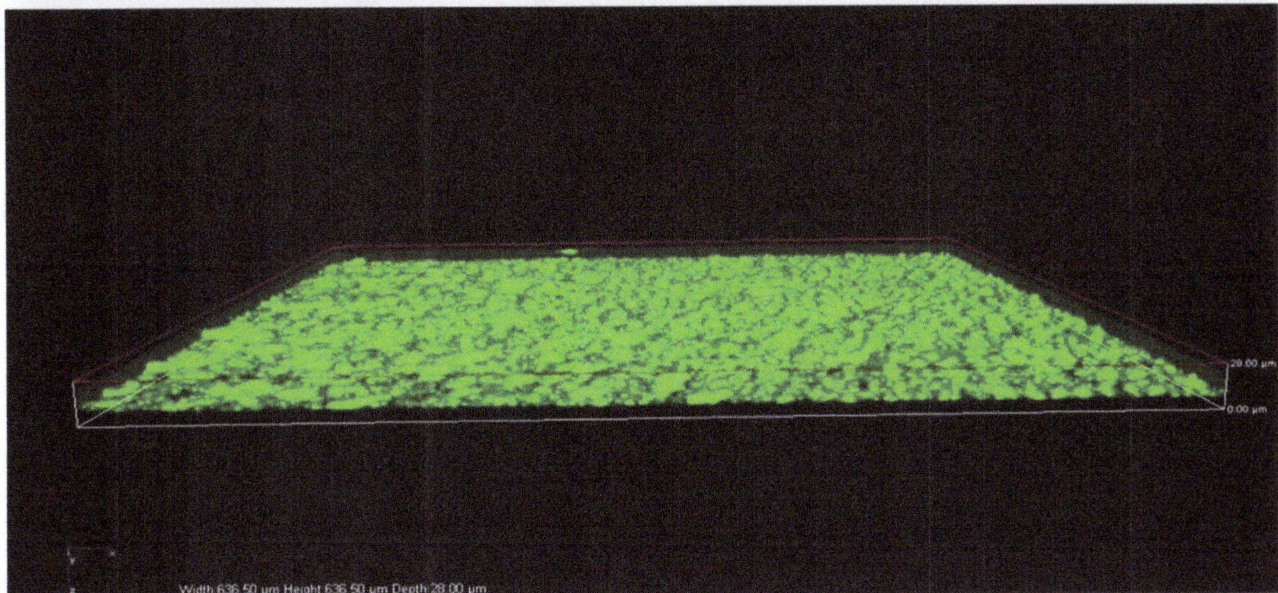

Figure 4. CLSM of the strain Sm1219 of *S. maltophilia* biofilm mode in 8 h (up) and 24 h (down).

producer of slime among 24 strains considered. Biofilms were allowed to grow on cover slip, and fixed with 2.5% glutaraldehyde for at least 40 min. After being washed with PBS, the samples were then dehydrated in a series of aqueous ethanol solutions (50 to 100%). The specimens were mounted on aluminum stubs with conductive carbon cement, allowed to dry for 3 h, and then coated with 15-nm Au film with an agar automatic sputter coater. After processing, samples were observed with a QUANTA 200 scanning electron microscope in the high-vacuum mode at 15 kV.

CLSM was performed as pervious report [18]. Briefly, Biofilm samples were fixed in 2.5% glutaraldehyde, and stained with FITC (fluorescein isothiocyanate). CLSM analysis was performed with an LSM 510 META laser scanning microscope attached to an Axioplan II microscope. The excitation wavelengths were 458 [Argon laser], and 543 nm [He-Ne laser], and emission wavelengths were 488 nm for FITC, respectively. Depth measurements were taken at regular intervals across the width of the device. To determine the structure of the biofilms, a series of horizontal (x-y) optical sections were taken throughout the full length of the biofilm. Confocal images of green (FITC) fluorescence were conceived using a track mode.

PCR-based genotyping for *rmlA*, *spgM*, and *rpfF*

Bacterial DNA was extracted and purified, then amplified and visualized on 2% agarose gel. PCR primers were respectively 5′ CGGAAAAGCAGAACATCG 3′ and 5′ GCAACTTGGTTT-CAATCACTT 3′ (1222 bp) for *rmlA*, 5′ ATACCGGGGTG-CGTTGAC 3′ and 5′ CATCTGCATGTGGATCTCGT 3′ (2750 bp) for *spgM* and, finally, 5′ CACGACAGTACAGGG-GACC 3′ and 5′ GGCAGGAATGCGTTGG 3′ (1140 bp) for *rpfF*. All PCRs products were sequenced and BLAST in PUBMED. Furthermore, the three gene expression was performed with Reverse transcriptase PCR according to pervious report [18].

```
S9110  1    MTQRKGIILAGGSGTRLYPITKGISKQLLPVYDKPMIYYPLSVLMLAGIRDVLIINTPHE  60
S2850  1    MTQRKGIILAGGSGTRLYPITKGISKQLLPVYDKPMIYYPLSVLMLAGIRDVLIINTPHE
S1219  1    MTQRKGIILAGGSGTRLYPITKGVSKQLLPVYDKPMIYYPLSVLMLAGIRDVLIINTPHE  60

S9110  61   QALFQALLGDGSQWGMNIQYAVQPSPDGLAQAYLIGRDFIDGKPSCLVLGDNIFHGHGLT  120
S2850  61   QALFQ LLGDGSQWGM+IQYAVQPSPDGLAQAYLIGRDF+ GKPSCLVLGDNIFHGHGL
S1219  61   QALFQQLLGDGSQWGMDIQYAVQPSPDGLAQAYLIGRDFVAGKPSCLVLGDNIFHGHGLR  120

S9110  121  DVLHSADQRQQGATVFGYWVNDPERYGVAEFDKNGKVVDLVEKPENPRSNYAVTGLYFYD  180
S2850  121  DVLHSADQRQQGATVFGYWVNDPERYGVAEFDKNGKVVDLVEKPENPRSNYAVTGLYFYD
S1219  121  EVLDSADQRSLGATVFGYWVNDPERYGVAEFDKNGKVIDLVEKPENPRSNYAVTGLYFYD  180

S9110  181  GNASDYAAELKPSPRGELEITDLNQRYLRDGQLQLEALGRGFAWLDTGTHQSLLEASNFI  240
S2850  181  GNASDYAAELKPSPRGELEITDLNQRYLRDGSLQLEALGRG+AWLDTGTHQSLLEASNFI
S1219  181  GNASDYAAELKPSPRGELEITDLNQRYLHEGSLHLEALGRGYAWLDTGTHQSLLEASNFI  240

S9110  241  ETIQTRQGLQVCCPEEIAFGKGWIDAAQLEALAAPLIKNGYGQYLHKLAVRGVV  294
S2850  241  ETIQTRQGLQVCCPEEIAFGKGWIDAAQLEALAAPLIKNGYGQYLHKLAVRGVV
S1219  241  ETIQTRQGLQVCCPEEIAFGKGWISAEQLEALAAPLIKNSYGQYLHKLALRGVV  294
```

Figure 5. Analysis of rmlA amino acid mutation in isolates with vary ability of biofilm forming.

Statistical analysis

All assays were carried out in triplicate and repeated twice, and the results are presented as means±SDs. The Pearson's correlation coefficient was calculated to determine the association between two variables. All P-values were based on two-tailed tests of significance, and a significance level of 0.05 or 0.01 was used. Statistical analysis of results was conducted with statistical software SPSS17.0.

Results

Clonal relatedness of isolates

According to the patterns of PFGE isolates, 24 pulsotypes were designated for 37 *S. maltophilia* isolates. Besides 19 individual pulsotypes, the remaining 18 isolates were classified into 5 pulsotypes, designated types A, B, C, D and E (Figure 1). Three out of 4 sequential strains isolated from one patient belong to type A, and the remaining 1 strain belong to one individual pulsotype (type T). Three sequential strains isolated from another patient belong to the same clone—type C. All the strains with individual pulsotypes and the 7 sequential strains were applied in further analysis.

Ability of the isolates of *S. maltophilia* in forming biofilm

All the 24 isolates of *S. maltophilia* were able to form biofilm, and the biofilm forming ability varied greatly among strains tested (OD492 range: 0.18–1.4). Eight isolates had higher ability of biofilm-producing (OD492, mean ± SD: 0.808±0.310), 14 had middle (OD492, mean ± SD: 0.408±0.215) and 2 had lower (OD492, mean ± SD: 0.186±0.036) (Figure 2).

Scanning electron micrographs (SEM) of the *S. maltophilia* biofilm mode of growth on polystyrene surface over 24 h of incubation was shown in figure 3. The biofilm covered just 70% of the surface, growing in specific clusters of cells (microcolonies), and a dense network of cells deeply embedded in an extracellular matrix. Spectrophotometric results were confirmed by Confocal Laser Scanning Microscopy (CLSM) analysis in figure 4. The biofilm formed by Sm1219 strain resulting to be the most complex, revealing a multilayered cell structure (58–63 μm, depth) embedded in an abundant extracellular polymeric substance (EPS) compared with *S. maltophilia* ATCC13637. Moreover, CLSM analysis revealed that the biofilm structure of the *S. maltophilia* isogenic sequential strains was significantly different in various isolated time. The 3 isogenic sequential strains belong to type C pulsotype displayed different ability in biofilm-formation were isolated from 11 months among his hospitalization. The first and second isolates revealed stronger ability in biofilm-formation with 0.56 and 0.60 of OD492 values, and the third isolate had weak ability in biofilm-formation with OD492 values of 0.19. Similarly, the 4 isolates of the other patient shown that the first and second isolates belong to type A pulsotype presented moderate ability in biofilm-formation with OD492 values of 0.40 to 0.42, while the fourth belong to type A presented weaker ability in biofilm-formation with OD492 values of 0.28. The third isolates (individual pulsotype) shown weak ability in biofilm-formation with D492 values of 0.20. All the two groups displayed that the strain isolated in early stage shown strong or middle biofilm-producing while the weak biofilm-producing in late stage. CLSM analysis also showed more highly structured, multilayered biofilm in early stage.

Different genotypes and gene mutation are associated to ability of biofilm formation

PCR-based typing of 24 *S. maltophilia* strains for *spgM*, *rmlA*, and *rpfF* genes showed an overall prevalence of 83.3%, 87.5%, and 50.0%, respectively. The presence of *rmlA*, *spgM* or *rpfF* had no close relationship with the ability of biofilm formation for each strain. However, considering the strain population as a whole, the presence of *spgM* and *rpfF* significantly improved biofilm formation. In particular, the strains with *spgM+/rpfF+/rmlA+*

Table 1. The relation of different genotypes and the ability of biofilm formation in *Stenotrophomonas maltophilia* isolates with individual pulsotype.

NO. strains	Pulsotypes	OD492 values	Degree of BF	spgM	rpfF	rmlA
57-9110	G	0.17	weak	−	−	+
11931	E	0.20	weak	+	−	−
1150#	C	0.19	weak	−	−	+
1562*	T	0.20	weak	−	−	+
11757*	A	0.28	moderate	+	+	+
20	B	0.44	moderate	+	+	+
997	I	0.36	moderate	+	−	−
1251*	A	0.40	moderate	+	+	+
1474*	A	0.42	moderate	+	+	+
1452	M	0.38	moderate	−	−	+
15-2850	L	0.32	moderate	+	−	+
168	K	0.40	moderate	+	−	+
10544	D	0.50	moderate	+	+	+
701044	Q	0.40	moderate	+	−	+
251	S	0.31	moderate	+	+	−
707344	R	0.42	moderate	−	−	+
701	F	0.53	moderate	+	+	+
140	J	0.56	moderate	+	−	+
1479	O	0.37	moderate	+	+	+
421#	C	0.56	moderate	+	−	+
535#	C	0.60	strong	+	−	+
13	N	0.62	strong	+	+	+
997	V	0.63	strong	+	−	+
11337	W	0.65	strong	+	+	+
10275	H	0.68	strong	−	+	+
11367	U	0.68	strong	+	−	+
7342	H	1.20	strong	+	+	+
1219	P	1.40	strong	+	+	+

BF: Biofilm;
+: Gene expression in the *S. maltophilia* isolate;
−: No gene expression in *S. maltophilia.*
*: Four sequential strains isolated from one patient.
#: Three sequential strains isolated from another patient.

genotype were easier to display strong or moderate biofilm-producer phenotype than the strains with *spgM−/rpfF−/rmlA+* genotype which was frequently detected in weak biofilm-producer phenotype (Table 1). Correlation analysis showed that the gene expression of *rmlA*, *spgM* and *rpfF* detected in the clinical strains were not associated to strong or moderate biofilm producers (For *rmlA*, Pearson $r = 0.32$, $P>0.05$; For *spgM*, Pearson $r = 0.36$, $P>0.05$; For *rpfF*, Pearson $r = 0.23$, $P>0.05$) (data not shown). In addition, significant amio acid mutation in *spgM* encoding gene and *rmlA* encoding gene were found between some strains of the strong and the weak biofilm-producer phenotype (Figure 5).

Discussion

A significant feature of *S. maltophilia* is its ability to form biofilms on surfaces including Teflon, glass, and plastics and on host tissues and biofilms have been estimated to be associated with 65% of hospital-acquired infections [26]. However, it is not yet known whether there are any variations in biofilm formation among clonally diverse clinical isolates of *S. maltophilia*, and there are any relationships between biofilm forming ability and mutation or expression of biofilm-relative genes such as *spgM*, *rmlA*, and *rpfF* gene, although it was reported that the *spgM* mutant formed more biofilm than that formed by the parental strain on polystyrene microtiter wells and the *rmlA* and *rmlC* mutants produced significantly more biofilm on glass than that produced by the wild type [12,25]. So, this study is designed to clarify these problems by analysis of morphology and genotype for biofilm-producing *S. maltophilia*.

Firstly, microtiter colorimetric assay for biofilm formation showed a wide range of biofilm formation ability, from biofilm-deficient phenotypes to those producing structurally complex biofilms. Morover, the images corresponded to three-dimensional reconstructions obtained from CLSM exhibited the native multicellular structures of biofilm in various forming ability, and CLSM analysis also showed that isolates from the early periods of chronic infection were able to form uniform flat biofilms or highly structured, multilayered and exopolysaccharide matrix-encased biofilms. On the contrary, isolates recovered from the late phase of chronic infection showed a significant reduction in adherence, lacking ability to form a mature biofilm. The variations in biofilm formation among the same or different patient may contribute to survival environment change of the bacteria. It was reported that he reduced efficiency in forming biofilm could be the consequences of *S. maltophilia* adaptation to a stressed environment such as CF lung [19–21], and oxidative stress and anaerobic conditions existed in COPD patient were also the risk factors to cause biofilm-producer reduced [22], and the similar support was found in *P.aeruginosa* [23].

In the present study we also focused our efforts on the relationship between biofilm formation and the presence of *rpfF*, *rmlA* and *spgM* genes. Overall, our results revealed the presence of *spgM* and *rpfF* significantly improved biofilm formation. The result was identified with Pompilio'report, which showed that *spgM* gene played a central role in biofilm formation of *S. maltophilia*, whose presence is significantly associated to a strong biofilm formation [9]. However, there were no regular mutation of *spgM* and *rmlA* gene between the strong and the weak biofilm formation in this study. Especially, 10 sites of amino acid mutation in *spgM* were found between the different phenotypes. *spgM* gene was shown to encode a bifunctional enzyme with both PGM and phosphomannomutase activities, an analysis of *S. maltophilia spgM* transposon insertion mutant strain JB12-23 showed that it formed more biofilm than that formed by its parental wild-type strain [13]. Unfortunately, it is uncertain that which site mutation in spgM might play a key role in the function change of SpgM by these reports. Similarly, the situation also appeared in *rmlA* gene. In this study, the *rmlA* genes shown more mutation sites than spgM gene among the clinical isolates with the three kinds of ability in biofilm formation, and only 15% amino acid sequences keep the homology between strong and weak ability groups. These results in this study may provide a clue for further study.

This study has several limitations. Firstly, it is difficult for us to observe dynamically the biofilm development of one strain such as adherence and twitching motility like previous report [23,24], since the devices commonly used for the confocal study of biofilms, such as flow-cells, capillary tubes or glass coverslides, may be relatively expensive, which caused it difficult to observe dynamically the biofilm development of one strain such as adherence and twitching motility. Furthermore, we failed to analyze the relationship between ability of biofilm formation and key mutation of *rmlA* and *spgM* genes due to many variations in the amino acid sequence of the *rmlA* and *spgM* genes respectively. We try to find other better methods to understand the impact of these genes distribution on biofilm formation in *S. maltophilia* in further study.

In conclusion, the results showed that clinical strains of *S. maltophilia* significantly differ in some phenotype and genotype in biofilm formation, and also added new insight and expand previous knowledge concerning the relative genes.

Acknowledgments

We thank qinqin wang (Sun Yat-sen University) for his experimental consideration and all the members of National surveillance of antimicrobial resistance program for supply of the isolates.

Author Contributions

Conceived and designed the experiments: CZ. Performed the experiments: QZ. Analyzed the data: QZ. Contributed reagents/materials/analysis tools: SX QZ. Wrote the paper: CZ.

References

1. Crossman LC, Gould VC, Dow JM, Vernikos GS, Okazaki A, et al. (2008) The complete genome, comparative and functional analysis of *Stenotrophomonas maltophilia* reveals an organism heavily shielded by drug resistance determinants. Genome Biol 9:R74.
2. Brooke JS. (2012) *Stenotrophomonas maltophilia*: an emerging global opportunistic pathogen. Clinical microbiology reviews 25: 2–41.
3. Goss CH, Ott K, Aitken ML, Rubenfeld GD. (2002) Detecting *Stenotrophomonas maltophilia* does not reduce survival of patients with cystic fibrosis. Am J Respir Crit Care Med 166:356–361.
4. Karpati F, Malmborg AS, Alfredsson H, Hjelte L, Strandvik B. (1994) Bacterial colonisation with *Xanthomonas maltophilia*–a retrospective study in a cystic fibrosis patient population. Infection 22:258–263.
5. de Oliveira-Garcia D, Dall'Agnol M, Rosales M, Azzuz AC, Martinez MB, et al. (2002) Characterization of flagella produced by clinical strains of *Stenotrophomonas maltophilia*. Emerg Infect Dis 8:918–923.
6. Figueirêdo PM, Furumura MT, Santos AM, Sousa AC, Kota DJ, et al. (2006) Cytotoxic activity of clinical *Stenotrophomonas maltophilia*. Lett Appl Microbiol 43:443–449.
7. Bjarnsholt T, Jensen PØ, Fiandaca MJ, Pedersen J, Hansen CR, et al. (2009) *Pseudomonas aeruginosa* biofilms in the respiratory tract of cystic fibrosis patients. Pediatr Pulmonol 44:547–558.
8. Di Bonaventura G, Spedicato I, D'Antonio D, Robuffo I, Piccolomini R.(2004) Biofilm formation by *Stenotrophomonas maltophilia*: modulation by quinolones, trimethoprim-sulfamethoxazole, and ceftazidime. Antimicrob Agents Chemother 48:151–160.
9. Di Bonaventura G, Pompilio A, Zappacosta R, Petrucci F, Fiscarelli E, et al.(2010) Excessive inflammatory response of DBA/2 mice to *Stenotrophomonas maltophilia* lung infection: implications in cystic fibrosis. Infect Immun 78:2466–2476.
10. Pompilio A, Piccolomini R, Picciani C, D'Antonio D, Savini V, et al. (2008) Factors associated with adherence to and biofilm formation on polystyrene by *Stenotrophomonas maltophilia*: the role of cell surface hydrophobicity and motility. FEMS Microbiol Lett 287:41–47.
11. Fouhy Y, Scanlon K, Schouest K, Spillane C, Crossman L, et al.(2007) Diffusible signal factor-dependent cell-cell signaling and virulence in the nosocomial pathogen *Stenotrophomonas maltophilia*. J Bacteriol 189:4964–4968.
12. Huang TP, Somers EB, Wong AC. (2006) Differential biofilm formation and motility associated with lipopolysaccharide/exopolysaccharide-coupled biosynthetic genes in *Stenotrophomonas maltophilia*. J Bacteriol 188:3116–3120.
13. McKay GA, Woods DE, MacDonald KL, Poole K. (2003) Role of phosphoglucomutase of *Stenotrophomonas maltophilia* in lipopolysaccharide biosynthesis, virulence, and antibiotic resistance. Infect Immun 71:3068–3075.
14. Seifert H, Gerner-Smidt P. (1995) Comparison of ribotyping and pulsed-field gel electrophoresis for molecular typing of *Acinetobacter* isolates. J Clin Microbiol 33: 1402–1407.
15. Tenover FC, Arbeit RD, Goering RV, Mickelsen PA, Murray BE, et al. (1995) Interpreting chromosomal DNA restriction patterns produced by pulsed-field gel electrophoresis: criteria for bacterial strain typing. J Clin Microbiol 33: 2233–2239.
16. Shi ZY, Liu PY, Lau YJ, Lin YH, Hu BS. (1996) Epidemiological typing of isolates from an outbreak of infection with multidrug-resistant *Enterobacter cloacae* by repetitive extragenic palindromic unit b1-primed PCR and pulsed-field gel electrophoresis. J Clin Microbio, 134: 2784–2790.
17. Stepanović S, Vuković D, Hola V, Di Bonaventura G, Djukić S, et al. (2007) Quantification of biofilm in microtiter plates: overview of testing conditions and practical recommendations for assessment of biofilm production by *staphylococci*. APMIS 115:891–899.
18. Pompilio A, Pomponio S, Crocetta V, Gherardi G, Verginelli F, et al. (2011) Phenotypic and genotypic characterization of *Stenotrophomonas maltophilia* isolates from patients with cystic fibrosis: Genome diversity, biofilm formation, and virulence. BMC Microbiology, 11:159
19. Drenkard E, Ausubel FM. (2002) *Pseudomonas* biofilm formation and antibiotic resistance are linked to phenotypic variation. Nature 416:740–743.
20. Mahenthiralingam E, Campbell ME, Speert DP. (1994) Nonmotility and phagocytic resistance of *Pseudomonas aeruginosa* isolates from chronically colonized patients with cystic fibrosis. Infect Immun 62:596–605.
21. Mahenthiralingam E, Campbell ME, Foster J, Lam JS, Speert DP. (1996) Random amplified polymorphic DNA typing of *Pseudomonas aeruginosa* isolates recovered from patients with cystic fibrosis. J Clin Microbiol 34:1129–1135.
22. Worlitzsch D, Tarran R, Ulrich M, Schwab U, Cekici A, et al. (2002) Effects of reduced mucus oxygen concentration in airway Pseudomonas infections of cystic fibrosis patients. J Clin Invest 109:317–325.
23. Lee B, Haagensen JA, Ciofu O, Andersen JB, Høiby N, et al. (2005). Heterogeneity of biofilms formed by nonmucoid *Pseudomonas aeruginosa* isolates from patients with cystic fibrosis. J Clin Microbiol 43:5247–5255.

24. Bridier A, Dubois-Brissonnet F, Boubetra A, Thomas V, Briandet R. (2010) The biofilm architecture of sixty opportunistic pathogens deciphered using a high throughput CLSM method. J Microbiol Methods 82:64–70.

25. Zgair AK, Chhibber S. (2011) Adhesion of *Stenotrophomonas maltophilia* to mouse tracheal mucus is mediated through flagella. J Med Microbiol 60:1032–1037

26. Potera C. (1999) Forging a link between biofilms and disease. Science 283:1837–1838.

24

A Sialoreceptor Binding Motif in the *Mycoplasma synoviae* Adhesin VlhA

Meghan May[1]*, Dylan W. Dunne[2], Daniel R. Brown[3]

1 Department of Biomedical Sciences, College of Osteopathic Medicine, University of New England, Biddeford, Maine, United States of America, **2** Department of Biological Sciences, Jess and Mildred Fisher College of Science and Mathematics, Towson University, Towson, Maryland, United States of America, **3** Department of Infectious Diseases and Pathology, College of Veterinary Medicine, University of Florida, Gainesville, Florida, United States of America

Abstract

Mycoplasma synoviae depends on its adhesin VlhA to mediate cytadherence to sialylated host cell receptors. Allelic variants of VlhA arise through recombination between an assemblage of promoterless *vlhA* pseudogenes and a single transcription promoter site, creating lineages of *M. synoviae* that each express a different *vlhA* allele. The predicted full-length VlhA sequences adjacent to the promoter of nine lineages of *M. synoviae* varying in avidity of cytadherence were aligned with that of the reference strain MS53 and with a 60-a.a. hemagglutinating VlhA C-terminal fragment from a Tunisian lineage of strain WVU1853^{T}. Seven different sequence variants of an imperfectly conserved, single-copy, 12-a.a. candidate cytadherence motif were evident amid the flanking variable residues of the 11 total sequences examined. The motif was predicted to adopt a short hairpin structure in a low-complexity region near the C-terminus of VlhA. Biotinylated synthetic oligopeptides representing four selected variants of the 12-a.a. motif, with the whole synthesized 60-a.a. fragment as a positive control, differed ($P<0.01$) in the extent they bound to chicken erythrocyte membranes. All bound to a greater extent ($P<0.01$) than scrambled or irrelevant VlhA domain negative control peptides did. Experimentally introduced branched-chain amino acid (BCAA) substitutions Val3Ile and Leu7Ile did not significantly alter binding, whereas fold-destabilizing substitutions Thr4Gly and Ala9Gly tended to reduce it ($P<0.05$). Binding was also reduced to background levels ($P<0.01$) when the peptides were exposed to desialylated membranes, or were pre-saturated with free sialic acid before exposure to untreated membranes. From this evidence we conclude that the motif P-X-(BCAA)-X-F-X-(BCAA)-X-A-K-X-G binds sialic acid and likely mediates VlhA-dependent *M. synoviae* attachment to host cells. This conserved mechanism retains the potential for fine-scale rheostasis in binding avidity, which could be a general characteristic of pathogens that depend on analogous systems of antigenically variable adhesins. The motif may be useful to identify previously unrecognized adhesins.

Editor: Mitchell F. Balish, Miami University, United States of America

Funding: This work was supported by the Robert M. Fisher Foundation (MM). The funder had no role in study design, data collection and analysis, decision to publish, or preparation of the manuscript.

Competing Interests: The authors have declared that no competing interests exist.

* Email: mmay3@une.edu

Introduction

The bacterial pathogen *Mycoplasma synoviae* is associated with a broad spectrum of clinical manifestations ranging from inapparent infection to systemic disease of poultry. Infection is most commonly associated with inflammatory lesions of the joints, respiratory and/or reproductive tract and results in reduced feed conversion and poor egg quality. Less commonly, *M. synoviae* can be found infecting additional tissues in galliform birds (*e.g.* spleen, liver, central nervous system, skeletal muscle, and eye) [1–4] and respiratory tissues or synovial membranes of distantly related avian species such as ducks, geese, pigeons, and sparrows [5].

Attachment to sialylated receptors on host cells is mediated by the *M. synoviae* variable lipoprotein hemagglutinin VlhA [6–7]. Previous analyses indicated that the *vlhA* gene family has been laterally transferred between *M. synoviae* and *Mycoplasma gallisepticum* possibly during coinfection of a shared avian host [8–9]. In *M. synoviae*, antigenic variants of this adhesin result from unidirectional recombination between a single expression site and a large reservoir of *vlhA* pseudogenes [10]. In contrast, altered expression in *M. gallisepticum* stems from the expansion and contraction of a poly-GAA repeat upstream of the promoters of each copy of *vlhA* [11]. The selective pressure of specific host immune responses to these antigens is thought to drive diversity in *vlhA* allele expression [10–13]. Despite the critical importance of cytadherence to the establishment and maintenance of infection, discrete VlhA types were demonstrated to have significantly different avidities for host cell binding, which can be quantified by agglutination of erythrocytes [14]. *M. synoviae*'s capacity for cytadherence maps surprisingly to a hypervariable C-terminal domain of VlhA called MSPA [15–16]. The precise means of attachment and how this capacity is retained despite such extensive sequence polymorphism and allele switching are not known. We sought to identify and characterize the specific motif that mediates adhesion of VlhA proteins to host cells.

PHM Residue	1	2	3	4	5	6	7	8	9	10	11	12
Strains/lineages												
F10-2-AS and K4907	P	K	V	T	F	D	V	A	Q	K	E	G
FMT	P	K	V	T	F	N	L	A	A	K	E	G
K5016	P	K	V	T	F	T	V	T	A	K	N	G
K5395	P	T	V	T	F	N	L	A	A	K	E	G
MS53	P	K	V	T	F	N	L	T	P	K	E	G
MS117, MS173, MS178	P	T	V	T	F	T	V	A	A	K	D	G
WVU1853^{T}/Florida and Tunisia	P	K	V	T	F	T	V	E	A	K	P	G
Preliminary consensus:	P	X	V	T	F	X	(B)	X	X	K	X	G
Site-directed Mutants												
FMT T4G	P	K	V	G	F	N	L	A	A	K	E	G
FMT A9G	P	K	V	T	F	N	L	A	G	K	E	G
FMT V3I	P	K	I	T	F	N	L	A	A	K	E	G
FMT L7I	P	K	V	T	F	N	I	A	A	K	E	G
FinalPHM:	P	X	(B)	T	F	X	(B)	X	A	K	X	G
FMT - scrambled:	L	A	F	G	A	V	K	K	T	P	E	N
Irrelevant peptide:	P	N	A	V	F	V	Q	Q	M	K	D	D

Figure 1. Aligned PHM and control peptide sequences. The putative hemagglutination motif (PHM) was deduced by aligning the adhesin protein VlhA allele present at the expression site of ten specimens of *M. synoviae* with a 60-a.a. hemagglutinating VlhA C-terminal fragment from the Tunisian lineage of strain WVU1853^{T}, then inspecting the alignment for contiguous residues inferred to be under stabilizing selection. Peptides representing five variants of the PHM, including strains having a>20-fold range in quantitative hemagglutination phenotypes [14], were synthesized. Directed mutations were introduced at selected residues relative to the PHM from strain FMT, which had only one difference (Thr6Asn) from the most common amino acid at each residue. The mutations Val3Ile and Leu7Ile were predicted to be inconsequential, while Thr4Gly and Ala9Gly were predicted to affect PHM structure and/or function. Negative control peptides used in erythrocyte membrane-binding assays are also shown. Functionally non-synonymous differences relative to the most common amino acid at each residue are shaded in black, synonymous differences are shaded in gray, and identical residues are not shaded. (B) = branched chain amino acid.

Materials and Methods

Identification and Structural Modeling of the Putative Hemagglutination Motif (PHM)

The predicted full-length VlhA sequences adjacent to the single transcription promoter of nine lineages of *M. synoviae* varying in avidity of cytadherence (F10-2AS, FMT, K4907, K5016, K5395, MS117, MS173, MS178, and a>30X-passaged Florida lineage of strain WVU1853^{T}) [14] were aligned with that of the reference strain MS53 [8] and with a 60-a.a. hemagglutinating VlhA C-terminal fragment from a ca. 12X-passaged Tunisian lineage of strain WVU1853^{T} [15] by using ClustalΩ [17]. The multiple alignment was manually inspected for conserved motifs, evident as contiguous residues inferred to be under stabilizing selection ($\omega<1$) by using Bayesian models of sequence evolution in the Selecton v2.4 software suite [18]. The secondary structures of full-length VlhA, MSPA and its C-terminal 60 residues, and of the putative hemagglutination motifs (PHMs) described were modeled using the Phyre2 suite of template-directed and *ab initio* protein structure prediction algorithms (http://www.sbg.bio.ic.ac.uk/phyre2) [19]. The effects of individual amino acid substitutions on peptide structural stability were predicted by applying the Site Directed Mutator algorithm (http://mordred.bioc.cam.ac.uk/~sdm/sdm.php) [20] to the.pdb files generated by Phyre2. Substitutions having stability scores ($\Delta\Delta G$) between -0.5 and 0.5 were predicted to be neutral, whereas those <-2 or>2 were predicted to be highly destabilizing. The potential to bind sialic acid (KEGG Compound C00270; PubChem.sdf 445063) or any other ligand in the KEGG Compound database was predicted by applying the eFindSite ligand binding site prediction algorithm (http://brylinski.cct.lsu.edu/) [21–22] also to the.pdb files generated by Phyre2.

Quantitative Binding of PHM Peptides

Twelve-a.a. peptides representing five variants of the PHM from strains FMT, K5016, K5395, MS53 and WVU1853^{T}, plus the whole 60-a.a. hemagglutinating fragment of the Tunisian lineage of strain WVU1853^{T}, were synthesized, biotinylated and lyophilized (Biomatik, Wilmington, DE). Purity of each lyophilized preparation was confirmed by HPLC to be 90–92% full-length peptide. Those strains were chosen because FMT, K5016, K5395 and the Florida lineage of WVU1853^{T} spanned a>20-fold range in quantitative hemagglutination phenotypes, and the entire *vlhA* locus sequence of the reference strain MS53 has been published. [8,14]. Peptides having single directed mutations introduced at the conserved residues 3 or 4, or non-conserved residues 7 or 9, were also synthesized using the strain FMT motif PKVTFNLAAKEG as a parent. FMT was chosen as the parent motif because it had only one difference (Thr6Asn) from the most commonly observed amino acid at each residue (Figure 1). The functionally synonymous substitutions Val3Ile and Leu7Ile (BLOSUM62 [23] scores>0) were predicted to be inconsequential, while non-synonymous Thr4Gly and Ala9Gly (BLOSUM62 scores ≤ 0) were predicted to affect PHM structure and/or function.

The capacity of the peptides to bind to native or desialylated chicken erythrocyte membranes was assessed quantitatively in an ELISA format. Microtiter plates were coated with 5% v/v suspensions of chicken erythrocytes (Lampire Biologicals,

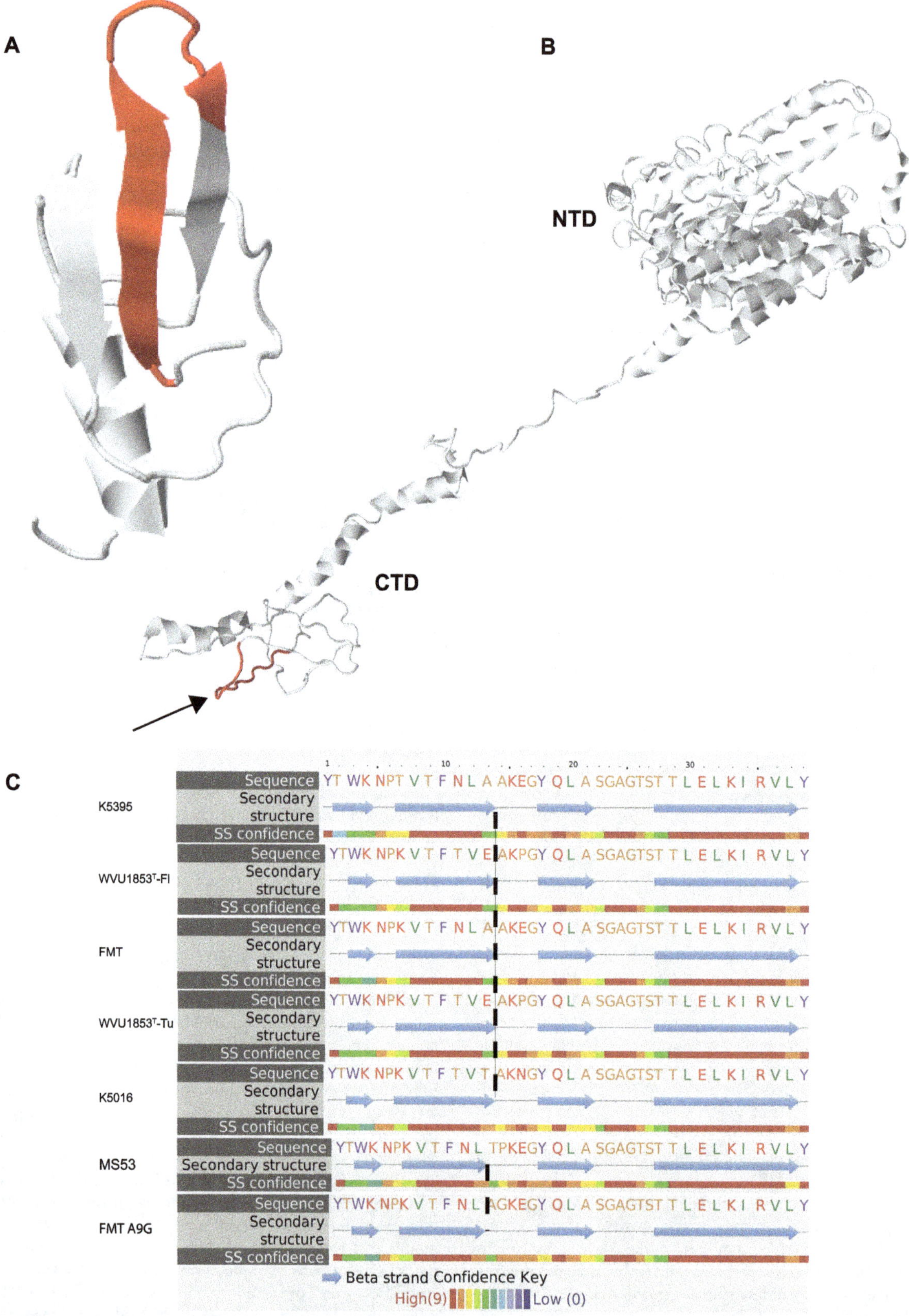

Figure 2. PHM structural predictions. (**A**) The putative hemagglutination motif (PHM; red) was predicted to adopt a hairpin structure of two anti-parallel β strands separated by a short disordered loop. (**B**) The motif (red, indicated by arrow) mapped to a low-complexity region near the carboxyterminal domain (CTD) of the *M. synoviae* adhesin protein VlhA cleavage product MSPA, shown here in the structure predicted for the Tunisian lineage of strain WVU1853^{T}. The N-terminal domain (NTD) of MSPA was predicted to have much greater 3-dimensional complexity. (**C**) The length of the disordered loop was predicted to be longer in PHM peptides that bound to avian erythrocyte membranes (representing Florida and

Tunisian lineages of strain WVU1853^T and strains FMT, K5016 and K5395) than in the reduced-binding peptide mutant FMT-Ala9Gly and the non-binding peptide representing strain MS53.

Pipersville, PA) diluted 1:3 in 0.5 M sodium bicarbonate lysis buffer, pH 10.0, to a total volume of 300 µL per well. Desialylated membranes were prepared by pre-treatment of the erythrocytes with 10 U/ml of sialidase purified from *Clostridium perfringens* (Sigma-Aldrich, St. Louis, MO) for 1 hr at 37°C. Following coating for 12 hr at 4°C, cellular debris including hemoglobin was removed by washing each well 3× with 300 µL of PBS, pH 7.4, and sealed plates were blocked 1 hr at 37°C with 300 µL per well of 5% v/v fetal bovine serum in PBS.

After washing the membrane-coated and blocked wells 3× with 300 µL of PBS, 50 µg of biotinylated peptide solubilized in 50 µL of water was added to each of duplicate wells and allowed to bind for 1 hr at 37°C. After washing each well 3× with 300 µL of PBS, bound peptides were detected using horseradish peroxidase-conjugated streptavidin (2 µg/mL, Sigma-Aldrich, St. Louis, MO) and the chromogenic substrate 3,3′,5,5′-tetramethylbenzidine (Thermo Fisher Scientific, Waltham, MA) with an acid stop followed by spectrophotometric analysis ($\lambda = 450$ nm). The hemagglutinating 60-mer of the Tunisian lineage of strain WVU1853^T served as the positive control peptide, and negative controls were a scrambled version of the PHM from strain FMT (LAFGAVKKTPEN) and an irrelevant peptide (PNAVFVQQMKDD) from a distant site in the expressed VlhA of the Florida lineage of strain WVU1853^T (GenBank AEA01932.1). The effect of pre-saturation with ligand was tested by first incubating the peptides in 250 mg/ml N-acetylneuraminic acid (Sigma-Aldrich, St. Louis, MO) in water without pH adjustment at a peptide: ligand molar ratio of $1{:}2\times10^4$ for 1 hr at 37°C.

Statistical Procedures

The effect of peptide sequence on extent of adherence to membranes ($n = 3$ independent replications of each treatment combination, with duplicate measurements of each peptide within replicate) was analyzed by ANOVA, with Tukey-Kramer Honestly Significant Difference (HSD) post-hoc comparisons used to group the means when the main effect was significant ($P<0.05$ or less). The effects of membrane pre-treatment with sialidase and peptide pre-saturation with sialic acid were analyzed by ANOVA, with HSD or Dunnett's post-hoc comparisons to the corresponding native specimens when the main effect was significant. Statistical analyses were performed using Origin 9 (OriginLab, Northampton, MA) software.

Motif Distribution in *M. synoviae* and *M. gallisepticum*

M. synoviae strain MS53 *vlhA* pseudogene sequences and *M. gallisepticum* strains R, F, WI01, NY01, NC06, CA06, VA94, NC95, NC08, and NC96 were obtained from GenBank (accession numbers NC_007294.1, NC_004829.2, NC_017503.1, NC_018410.1, NC_018409.1, NC_018411.1, NC_018412.1, NC_018406.1, NC_018407.1, NC_018413.1, and NC_018408.1, respectively). Occurrences of PHM-encoding sequences were totaled and normalized to the total length of *vlhA*-encoding sequence in each strain. Each member of the *vlhA* pseudogene reservoir of *M. synoviae* strain MS53 was used to construct a neighbor-joining tree (bootstrap n = 100) using ClustalW2 [24]. The designated outgroup was *vlhA* 4.02 from *M. gallisepticum*.

Results

Identification of the PHM

When the full-length expressed VlhA protein MSPA sequences of nine strains of *M. synoviae* that vary in avidity of cytadherence were aligned with MSPA of the reference strain MS53 [8] and a 60-a.a. hemagglutinating peptide derived from the C-terminus of MSPA expressed by the Tunisian lineage of strain WVU1853^T [15], an imperfectly conserved 12-a.a. motif was evident in all sequences (Figure 1). A total of seven different PHM sequence variants were evident among the 11 total sequences aligned.

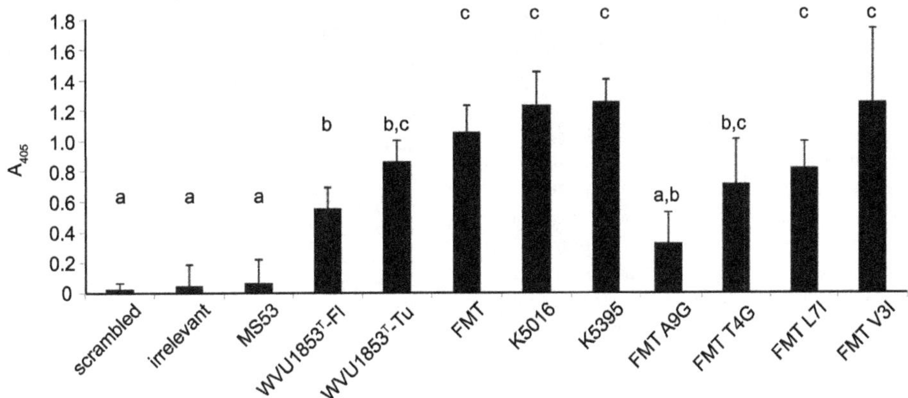

Figure 3. Erythrocyte membrane binding by PHM peptides. Bars depict mean ± standard error of the amount of synthetic peptide bound to avian erythrocyte membranes in an ELISA format (n = 3 independent replicates, with duplicate measurements of each peptide within replicate). The peptides represented variants of the putative hemagglutination motif (PHM) at the VlhA expression site of *M. synoviae* strains MS53, WVU1853^T (Florida and Tunisian lineages), FMT, K5016 and K5395, which spanned a>20-fold range in quantitative hemagglutination phenotypes [14]. The positive control was the Tunisian lineage of strain WVU1853^T, and negative controls were scrambled strain FMT peptide and an irrelevant peptide from a distant site in VlhA from the Florida lineage of strain WVU1853^T. Different letters above the bar indicate means that differ ($P<0.05$ or less) by Tukey-Kramer Honestly Significant Difference test. As predicted, the directed substitution Ala9Gly significantly reduced binding versus the parent peptide from strain FMT, and Thr4Gly tended to reduce binding, whereas Val3Ile and Leu7Ile did not significantly alter binding.

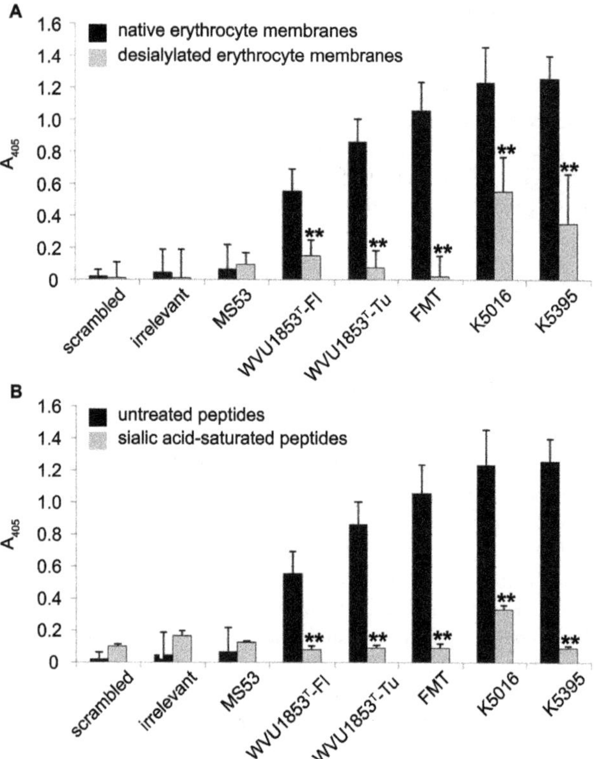

Figure 4. Effects of sialylation and desialylation on PHM peptide binding. Bars depict mean ± standard error of the amount of synthetic peptide bound to avian erythrocyte membranes in an ELISA format (n = 3 independent replicates, with duplicate measurements of each peptide within replicate). (**A**) Desialylation of erythrocyte membranes significantly reduced PHM peptide binding relative to native membranes (** = $P<0.01$) for all strains of *M. synoviae* except MS53, which bound to native or desialylated erythrocyte membranes at background levels. (**B**) Presaturation of PHM peptides with free sialic acid before exposure to native erythrocyte membranes significantly reduced binding relative to untreated peptides (** = $P<0.01$) for all strains except MS53, on which sialic acid had no effect.

Strains FMT, K5016, K5395 and MS53 all had unique PHM sequences; the sequences in strains F10-2-AS and K4907 were identical; the sequences in Florida and Tunisian lineages of WVU1853^T were identical; and the sequences in Argentine strains MS117, MS173 and MS178 were all identical. Six of twelve residues in the PHM were perfectly conserved across strains, two (residues 6 and 7) were conserved in polarity and hydrophobicity, respectively, and four were variable. Polar Asn_6 or Asp_6 were invariably paired with Leu_7, while Thr_6 was invariably paired with Val_7. The motif was predicted to adopt a short hairpin secondary structure of two anti-parallel beta strands, separated by a disordered loop of four or five residues, in a region of low structural complexity (regional structure prediction confidence < 70%) near the C-terminus of MSPA (Figure 2a, b). Fifty-three percent of residues in the full-length VlhA were modeled at>90% confidence [19], with the regions of greatest confidence being similar to the streptococcal adhesin emb (99.8% confidence) and the staphylococcal extracellular matrix-binding protein ebhA (99.4%). The degree of structural complexity in the C-terminus of MSPA was otherwise too low for the algorithms to predict binding of any specific ligand.

Synthetic biotinylated peptides representing the full-length 60-a.a. hemagglutinating fragment and four strain variants of its candidate 12-a.a. cytadherence motif (Figure 1) bound to chicken erythrocyte membranes in an ELISA format and could be detected by probing with horseradish peroxidase-conjugated streptavidin. Four of the peptides bound to membranes to a significantly greater extent ($P<0.05$) than scrambled or irrelevant control peptides did, but a peptide representing the corresponding motif from strain MS53 did not bind to membranes to any extent greater than background (Figure 3). Single neutral substitutions (predicted $\Delta\Delta G = -0.25$) experimentally introduced at conserved residue 3 (Val3Ile) or non-conserved residue 7 (Leu7Ile) did not alter binding to membranes with respect to the extent of binding by the parent motif of strain FMT, whereas the experimental destabilizing substitution Thr4Gly (predicted $\Delta\Delta G = -2.31$) tended to reduce binding (Figure 3). The motif of strain MS53 differs naturally from all others by Ala9Pro (BLOSUM62 = −1; predicted $\Delta\Delta G = -2.22$), and the even more destabilizing substitution Ala9Gly (predicted $\Delta\Delta G = -3.88$) nearly abolished binding when introduced into the parent motif of strain FMT ($P<0.05$; Figure 3). These effects correlated with a predicted change in length of the disordered loop in the hairpin secondary structure of the motif (Figure 2c).

Binding of the peptides to desialylated membranes was significantly reduced ($P<0.01$) relative to untreated membranes for all peptides except those representing strain MS53 and the scrambled and irrelevant controls (Figure 4a). When pre-incubated with free sialic acid, all peptides except the one representing strain MS53 and the scrambled and irrelevant controls had significantly diminished ($P<0.01$) capacity for membrane binding (Figure 4b). From this evidence we conclude that the composite amino acid motif P-X-(BCAA)-X-F-X-(BCAA)-X-A-K-X-G binds sialic acid and likely mediates VlhA-dependent *M. synoviae* attachment to sialylated receptors on the surface of avian erythrocytes.

PHM Distribution among *M. synoviae* Strain MS53 *vlhA* Pseudogenes and *Mycoplasma gallisepticum vlhA* Homologs

Candidate PHM sequences occurred in 45 of the 70 putative *vlhA* pseudogenes of *M. synoviae* strain MS53 [8], 39% of the time with no deviation from the consensus among the alleles expressed by the strains examined, 20% with a single deviation, and 17% with two deviations from consensus. Phylogenetic clustering of *vlhA* pseudogenes containing intact copies of the PHM did not correlate with their syntenic order in the strain MS53 genome (Figure S1). The PHM occurred at least 18-fold more frequently in strain MS53 (0.65 motifs/kb of *vlhA* sequence) than in the genomes of any of 10 strains of *M. gallisepticum* (0.014–0.037 motifs/kb of *vlhA* sequence), a species known to employ a different primary cytadherence mechanism [25] (Figure 5a). The rate of occurrence of imperfect PHMs was comparable between the two species (Figure 5b).

Discussion

One of the defining moments of many infections is the attachment of a disease-causing agent to its host. Understanding how the parasitic bacterial species *M. synoviae* colonizes a host cell's surface is paramount to understanding how to prevent infection. It is known that the protein family VlhA is responsible for attachment by *M. synoviae*, but the functional motifs of the adhesin and the molecular basis for rheostasis in binding avidity have not been characterized. Proteins in this family from multiple strains of *M. synoviae* have been identified as having a role in the attachment to host blood cells [7,15]. Khiari *et al.* [15] mapped

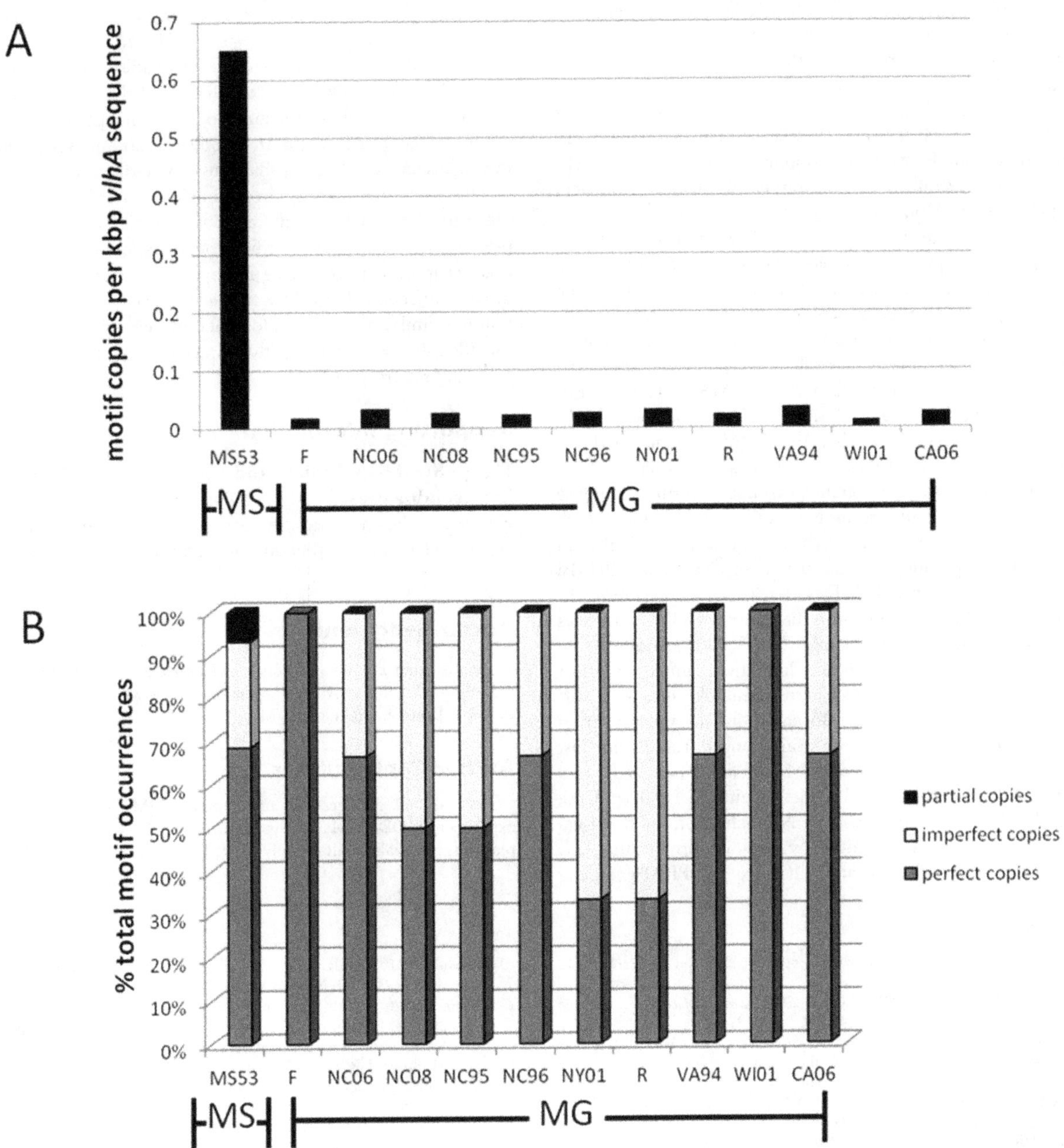

Figure 5. PHM distribution in *vlhA* genes and pseudogenes in *M. synoviae* and *M. gallisepticum*. (**A**) PHM-encoding sequence as a function of total kbp of *vlhA* sequence is elevated 18-fold in *M. synoviae* (MS) reference strain MS53, the only strain for which the entire *vlhA* locus sequence has been published, relative to 10 fully-sequenced strains of *M. gallisepticum* (MG), evidence that it is far more common in *M. synoviae*. (**B**) The relative proportions of perfect and imperfect PHM copies were comparable between strains of *M. synoviae* and *M. gallisepticum*.

the capacity for attachment to the carboxyterminus of VlhA, and we utilized that finding to identify a specific motif sufficient to mediate VlhA binding to sialylated host cells.

Sequence conservation across adherent strains enabled the identification of a 12-residue putative hemagglutination motif that could be characterized further. This motif was predicted to have remarkably little structural complexity, in contrast to the complex topology of sialic acid ligand-binding domains of other microbes [26–27]. While residues at PHM positions 3 and 7 were conserved, substitution with similar residues having BLOSUM62 scores>0 did not alter function. The conserved Thr residue at position 4 could be changed to the dissimilar residue Gly (BLOSUM62 $= -3$) without loss of function. It is thus likely that the binding mechanism will tolerate synonymous substitutions at positions 3 and 7, and nonsynonymous substitutions at position 4. Residue 9 was a conserved Ala in all adherent strains. Strain MS53, which has an unknown attachment phenotype but is an attenuated strain, had the nonsynonmymous substitution Ala9Pro (BLOSUM62 $= -1$). Changing the strain FMT peptide to Gly_9 (BLOSUM62 $= 0$) significantly diminished binding, and the strain MS53 peptide was non-adherent. Taken together, these results indicate that Ala_9 is critical to PHM domain function. Our results indicate that the composite amino acid motif P-X-(BCAA)-X-F-X-(BCAA)-X-A-K-X-G mediates MSPA binding to avian erythrocytes. The potential to accommodate all amino acids with BLOSUM62 scores>0 at PHM positions 3 and 7 (*i.e.*, Ala, Met, Thr and Met, Phe, respectively) rather than restricting the parameters to branched-chain amino acids (Ile, Leu, Val) merits further analysis.

Previous studies indicated that whole *M. synoviae* cells interact with sialylated host cell receptors in order to facilitate attachment. Extrapolation from PHM peptide-binding to whole cell attachment necessarily requires demonstration of peptide-sialic acid interactions. Desialylation of avian erythrocytes prior to antigen preparation resulted in significant losses of binding capacity for all PHM peptides except the scrambled and irrelevant controls and strain MS53, for which desialylation had no effect on binding. In a reciprocal experiment, pre-adsorption of peptides with free sialic acid prior to exposure to intact erythrocyte antigen similarly diminished binding capacity for all PHM peptides except the scrambled and irrelevant controls and strain MS53. These results indicate a specific interaction between sialic acid and the PHM and support the hypothesis that the PHM domain mediates attachment of whole *M. synoviae* cells to host sialoreceptors.

The occurrance of PHM domains was not uniform among the pseudogenes of *M. synoviae* strain MS53, the only strain for which the entire pseudogene reservoir has been sequenced [8]. A majority (69%) of pseudogenes had perfect or near-perfect PHMs, while 31% had no discernible PHMs. To provide some context for the distribution of PHM domains in the sample of VlhA sequences existing within *M. synoviae* strain MS53, we examined the frequency and distribution in an alternative sample of VlhA sequences that exist distributed across multiple strains of *M. gallisepticum*. In contrast to the 45 copies in *M. synoviae* strain MS53, sequenced *M. gallisepticum* strains ranged from having just a single copy of *vlhA* encoding a PHM domain (strains WI01 and F) up to a maximum of only 3 copies (strains R, NY01, NC06, CA06, VA94, NC95, and NC96). Normalization to the total amount of *vlhA* sequence within species confirmed that *M. synoviae* has a greatly elevated instance of PHM-encoding sequence relative to *M. gallisepticum*, and that the low frequency of PHM is consistent across strains of *M. gallisepticum*. The multiple independent cytadherence mechanisms of *M. gallisepticum* [28–32] may allow the decay of PHM domains within VlhA proteins, while selective pressure to retain the functional motif in the homologous proteins in *M. synoviae* is substantially greater due to the absence of other mechanisms of cytadherence.

This work describes a novel functional motif associated with adherence to sialic acid, and its distribution across *vlhA* pseudogenes. This very specific protein fragment pattern may be a target to design novel drug therapies or vaccines to alleviate or prevent infection due to *M. synoviae* as well as other pathogens that use similar mechanisms to attach to their hosts, and allows for the identification of currently unrecognized microbial adhesins targeting sialoreceptors.

Supporting Information

Figure S1 Distribution and relatedness of PHM-encoding pseudogenes. PHM-encoding pseudogenes (shaded) did not cluster together as a separate group from non-encoding pseudogenes. Relatedness of pseudogenes did not reflect gene synteny.

Acknowledgments

We thank Edan Tulman (University of Connecticut) for helpful discussions regarding *vlhA* loci in *M. gallisepticum*. This work was supported by the Robert M. Fisher Foundation (MM).

Author Contributions

Conceived and designed the experiments: MM DRB. Performed the experiments: DD MM. Analyzed the data: MM DRB. Contributed reagents/materials/analysis tools: MM. Wrote the paper: MM DRB.

References

1. Stipkovits L, Kempf I (1996) Mycoplasmoses in poultry. Rev Sci Tech. 15(4): 1495–525.
2. Senties-Cué G, Shivaprasad HL, Chin RP (2005) Systemic *Mycoplasma synoviae* infection in broiler chickens. Avian Pathol. 34(2): 137–42.
3. Chin RP, Meteyer CU, Yamamoto R, Shivaprasad HL, Klein PN (1991) Isolation of *Mycoplasma synoviae* from the brains of commercial meat turkeys with meningeal vasculitis. Avian Dis. 35(3): 631–7.
4. Lockaby SB, Hoerr FJ, Lauerman LH, Kleven SH (1998) Pathogenicity of *Mycoplasma synoviae* in broiler chickens. Vet Pathol. 35(3): 178–90.
5. Brown DR, May M, Bradbury JM, Balish MF, Calcutt MJ, et al. (2010) Genus I. *Mycoplasma*. In: Krieg NR, Ludwig W, Brown DR, Whitman WB, Hedlund BP, Paster BJ, Staley JT, et al., editors. Bergey's Manual of Systematic Bacteriology Volume 4. Springer, Inc.: New York, NY.
6. Manchee R, Taylor-Robinson D (1969) Utilization of neuraminic acid receptors by mycoplasmas. J Bacteriol. 98(3): 914–9.
7. Noormohammadi A, Markham P, Duffy M, Whithear K, Browning G (1998) Multigene families encoding the major hemagglutinins in phylogenetically distinct mycoplasmas. Infect Immun. 66(7): 3470–5.
8. Vasconcelos A, Ferreira H, Bizarro C, Bonatto S, Carvalho M, et al. (2005) Swine and poultry pathogens: the complete genome sequences of two strains of *Mycoplasma hyopneumoniae* and a strain of *Mycoplasma synoviae*. J Bacteriol. 187(16): 5568–77.
9. Szczepanek SM, Tulman ER, Gorton TS, Liao X, Lu Z, et al. (2010) Comparative genomic analyses of attenuated strains of *Mycoplasma gallisepticum*. Infect Immun. 78(4): 1760–71.
10. Noormohammadi A, Markham P, Kanci A, Whithear K, Browning G (2000) A novel mechanism for control of antigenic variation in the haemagglutinin gene family of mycoplasma synoviae. Mol Microbiol. 35(4): 911–23.
11. Glew MD, Baseggio N, Markham PF, Browning GF, Walker ID (1998) Expression of the pMGA genes of *Mycoplasma gallisepticum* is controlled by variation in the GAA trinucleotide repeat lengths within the 5' noncoding regions. Infect Immun. 66(12): 5833–41.
12. Citti C, Browning GF, Rosengarten R (2005) Phenotypic diversity and cell invasion in host subversion by pathogenic mycoplasmas. In: Blanchard A, Browning GF, editors. Mycoplasmas Molecular Biology Pathogenicity and Strategies for Control. Horizon Bioscience: Norfolk, UK.
13. Zimmerman C-U (2014). Current insights into phase and antigenic variation in mycoplasmas. In: Browning GF, Citti C, editors. Mollicutes Molecular Biology and Pathogenesis. Caister Academic Press: Norfolk, UK.
14. May M, Brown DR. (2011) Diversity of expressed vlhA adhesin sequences and intermediate hemagglutination phenotypes in *Mycoplasma synoviae*. J Bacteriol. 193(9): 2116–21.
15. Khiari AB, Guériri I, Mohammed RB, Mardassi BB (2010) Characterization of a variant vlhA gene of *Mycoplasma synoviae*, strain WVU 1853, with a highly divergent haemagglutinin region. BMC Microbiol. 10: 6. [doi: 1471-2180-10-6 [pii] 10.1186/1471-2180-10-6].
16. Noormohammadi A, Markham P, Whithear K, Walker I, Gurevich V, et al. (1997) *Mycoplasma synoviae* has two distinct phase-variable major membrane antigens, one of which is a putative hemagglutinin. Infect Immun. 65(7): 2542–7.
17. Sievers F, Wilm A, Dineen D, Gibson TJ, Karplus K, et al. (2011) Fast, scalable generation of high-quality protein multiple sequence alignments using Clustal Omega. Mol Syst Biol. 7: 539. [doi: 10.1038/msb.2011.75.]
18. May M, Brown DR (2009) Diversifying and stabilizing selection of sialidase and N-acetylneuraminate catabolism in *Mycoplasma synoviae*. J Bacteriol. 191(11): 3588–93.
19. Kelley LA, Sternberg MJ (2009) Protein structure prediction on the Web: a case study using the Phyre server. Nat Protoc. 4(3): 363–71.
20. Worth CL, Preissner R, Blundell TL (2011). SDM–a server for predicting effects of mutations on protein stability and malfunction. Nucleic Acids Res. 39(Web Server issue): W215–22. [doi: 10.1093/nar/gkr363].
21. Brylinski M, Feinstein WP (2013) eFindSite: improved prediction of ligand binding sites in protein models using meta-threading, machine learning and auxiliary ligands. J Comput Aided Mol Des. 27(6): 551–67.
22. Feinstein W, Brylinski M (2014) eFindSite: Enhanced fingerprint-based virtual screening against predicted ligand binding sites in protein models. Mol Inform. 33(2): 15 [doi: 10.1002/minf.201300143]
23. Henikoff S, Henikoff JG (1992) Amino acid substitution matrices from protein blocks. Proc Natl Acad Sci U S A. 89(22): 10915–9.

24. Larkin MA, Blackshields G, Brown NP, Chenna R, McGettigan PA, et al. (2007) Clustal W and Clustal X version 2.0. Bioinformatics. 23(21): 2947–8.
25. Papazisi L, Frasca S, Gladd M, Liao X, Yogev D, Geary SJ (2002) GapA and CrmA coexpression is essential for *Mycoplasma gallisepticum* cytadherence and virulence. Infect Immun. 70(12): 6839–45.
26. Tharakaraman K, Jayaraman A, Raman R, Viswanathan K, Stebbins NW, et al. (2013) Glycan receptor binding of the influenza A virus H7N9 hemagglutinin. Cell. 153(7): 1486–93.
27. Pang SS, Nguyen ST, Perry AJ, Day CJ, Panjikar S, et al. (2014) The three-dimensional structure of the extracellular adhesion domain of the sialic acid-binding adhesin SabA from *Helicobacter pylori*. J Biol Chem. 289(10): 6332–40.
28. Boguslavsky S, Menaker D, Lysnyansky I, Liu T, Levisohn S, et al. (2000) Molecular characterization of the *Mycoplasma gallisepticum* pvpA gene which encodes a putative variable cytadhesin protein. Infect Immun. 68(7): 3956–64.
29. Forsyth MH, Tourtellotte ME, Geary SJ (1992) Localization of an immunodominant 64 kDa lipoprotein (LP 64) in the membrane of *Mycoplasma gallisepticum* and its role in cytadherence. Mol Microbiol. 6(15): 2099–106.
30. Goh MS, Gorton TS, Forsyth MH, Troy KE, Geary SJ (1998) Molecular and biochemical analysis of a 105 kDa *Mycoplasma gallisepticum* cytadhesin (GapA). Microbiology. 144 (11): 2971–8.
31. Jenkins C, Geary SJ, Gladd M, Djordjevic SP (2007) The *Mycoplasma gallisepticum* OsmC-like protein MG1142 resides on the cell surface and binds heparin. Microbiology. 153(5): 1455–63.
32. May M, Papazisi L, Gorton TS, Geary SJ (2006) Identification of fibronectin-binding proteins in *Mycoplasma gallisepticum* strain R. Infect Immun. 74(3): 1777–85.

25

Characterization of Membrane Lipidome Changes in *Pseudomonas aeruginosa* during Biofilm Growth on Glass Wool

Hayette Benamara[1,2], Christophe Rihouey[1,2,3], Imen Abbes[1,2], Mohamed Amine Ben Mlouka[1,2,3], Julie Hardouin[1,2,3], Thierry Jouenne[1,2,3], Stéphane Alexandre[1,2]*

1 Polymères Biopolymères, Surfaces Laboratory - UMR 6270, CNRS - Université de Rouen, Mont-Saint-Aignan, France, **2** Normandie Université, Mont-Saint-Aignan, France, **3** PISSARO proteomic facility, IRIB - Université de Rouen, Mont-Saint-Aignan, France

Abstract

Bacteria cells within biofilms are physiologically distinct from their planktonic counterparts. In particular they are more resistant to detrimental environmental conditions. In this study, we monitored the evolution of the phospholipid composition of the inner and outer membranes of *P. aeruginosa* during the biofilm formation (i.e., from 1-, 2-, to 6-day-old biofilm). Lipidome analyses were performed by electrospray ionization mass spectrometry. In addition to the lipidomic analysis, the fatty acid composition was analysed by gas chromatography/mass spectrometry. We found that the lipidome alterations of the inner and the outer membranes varied with the biofilm age. These alterations in phospholipid compositions reflect a higher diversity in sessile organisms than in planktonic counterparts. The diversity is characterized by the presence of PE 30:1, PE 31:0 and PG 31:0 for the lower masses as well as PE 38:1, 38:2, 39:1, 39:2 and PG 38:0, 38:1, 38:2, 39:1, 39:2 for the higher masses. However, this lipidomic feature tends to disappear with the biofilm age, in particular the high mass phospholipids tend to disappear. The amount of branched chains phospholipids mainly located in the outer membrane decreased with the biofilm age, whereas the proportion of cyclopropylated phospholipids increased in both membranes. In bacteria present in oldest biofilms, i.e., 6-day-old, the phospholipid distribution moved closer to that of planktonic bacteria.

Editor: Eric Cascales, Centre National de la Recherche Scientifique, Aix-Marseille Université, France

Funding: Financial support came from the CNRS (Centre National de la Recherche Scientifique) and from the MESR (Ministère de l'Enseignement Supérieur et de la Recherche). H. Benamara and I. Abbes were supported by a MESR grant. The funders had no role in study design, data collection and analysis, decision to publish, or preparation of the manuscript.

Competing Interests: The authors have declared that no competing interests exist.

* Email: stephane.alexandre@univ-rouen.fr

Introduction

Pseudomonas aeruginosa is well known as an opportunistic pathogen that causes a variety of diseases in individuals predisposed to infections as the result of severe burns, wounds, urinary tract or corneal injury, or immunocompromised status [1] and is the leading cause of mortality and morbidity in cystic fibrosis (CF) patients [2]. This bacterium is characterised by its innate resistance to antibiotics due to a low outer membrane (OM) permeability and the presence of active efflux (Mex) systems [3,4].

P. aeruginosa can form biofilms which can be defined as biopolymer matrix-enclosed microbial populations adhering to each other and/or surfaces [5]. *P. aeruginosa* biofilms are involved in the pathogenesis of urinary, ventilator-associated pneumonia, peritoneal dialysis catheter infections, bacterial keratitis, otitis externa, burn wound infections and chronic bronchitis in CF patients [5]. Bacterial biofilm infections are particularly problematic because sessile bacteria are drastically more resistant to antimicrobials as compared with planktonic counterparts [6,7].

The protective mechanisms involved in biofilms resistance appear to be distinct from those that are responsible for conventional antibiotic resistance and it is becoming evident that biofilm resistance is multifactorial [8]. Poor antibiotic penetration, nutrient starvation, slow growth, adaptive stress responses and formation of persister cells are hypothesized to constitute a multi-layered defence [9]. The biofilm phenotype of *P. aeruginosa* appears regulated more at the translational and perhaps post-translational levels than at the transcriptional level, as highlighted by the discrepancies between transcriptomics [10,11] and proteomics [12,13].

Due to the importance of biofilms in industry, environment and for human health, the physiology of sessile micro-organisms has been widely investigated in the last decade. Initially, genomics investigations were performed by screening of biofilm defective mutants [14]. Then, transcriptomics [15] and proteomics [13] approaches were used to identify genes or proteins up- or down-regulated in sessile bacteria. More recently, metabolomics investigations were initiated to characterize the various metabolic states within biofilms [16].

Although, it is well known that inner membrane (IM) is a central component for microorganism survival, which are not insulated from extrinsic physical and chemical factors [17], few works were surprisingly devoted to lipidomics of biofilm organisms up to now. Abdul Lattif et *al* [18] compared the lipid profiles of biofilm and planktonic *Candida albicans* cells. They showed that significant differences existed in lipid composition according to the growth mode and the developmental phases. In a previous work, we pointed out a drastic decrease of the uneven chain phospholipids and an accumulation of long chain lipids in sessile *P. aeuginosa* cells as compared with planktonic counterparts, suggesting a higher lipid stability in the bilayer and a decrease in membrane fluidity [19].

The objective of the present study was to monitor the distribution of lipid classes during the biofilm growth (i.e. 1-, 2- to 6-day-old biofilms). Phospholipids quantification in *P. aeruginosa* IM enriched samples and OM enriched samples confirmed differences between biofilms and planktonic bacterial lipidomes but also demonstrated a biofilm age-dependence of the lipidomic alterations.

Materials and Methods

Bacterial strain and preculture

P. aeruginosa PAO1 strain was used. Bacteria were stocked in 30% (v/v) glycerol. Preculture was performed in a 50 mL flask containing 1 mL of bacteria stock suspensions and 10 mL of Muller-Hinton broth (MHB, Difco). The flask was incubated at 37°C on a rotary shaker (140 rpm) for 18 h.

Planktonic and biofilm cultures

For planktonic cultures, a preculture was used to inoculate (final concentration, 10^7 Colony Forming Units (CFU)/ml) 100 mL of Minimal Glucose Medium (MGM) of the following composition: 15 g/L Tris-HCl, 0.6 g/L Tris-base, 0.5 g/L NH_4Cl, 2 g/L yeast extract, 0.05 g/L $CaCl_2$, 0.05 g/L $MgSO_4$, 0.005 g/L $FeSO_4$, 0.005 g/L $MnSO_4$, 15 g/L glucose. Cultures were incubated on a rotary shaker (140 rpm) at 37°C for 24 h to reach the stationary phase of growth.

For biofilm cultures, 100 mL of MGM containing 2 g (total area 2800 cm^2) of sterile glass wool were inoculated at 10^7 CFU/mL from a pre-culture as previously described [12,19]. Biofilms were grown under slight agitation (20 rpm) at 37°C for 1, 2 or 6 days. The clear formation of a biofilm on glass wool has been checked by scanning electron microscopy. Support samples were taken after incubation for 1 day and rinsed twice with sterile phosphate buffer. Samples were fixed in a 2% glutaraldehyde, 0.1 M cacodylate buffer (pH 7.4) for 30 min and rinsed (3×10 min) in 0.2 M cacodylate buffer (pH 7.4). Samples were then dehydrated by passing them through the following ethanol series: 30%, 50%, 80%, each for 10 min; 100% ethanol, 2 10 min. Support samples were then dried at 37°C for 24 h. Once coated with gold - palladium (Sputtering Device), samples were examined using scanning electron microscopy (Cambridge S200).

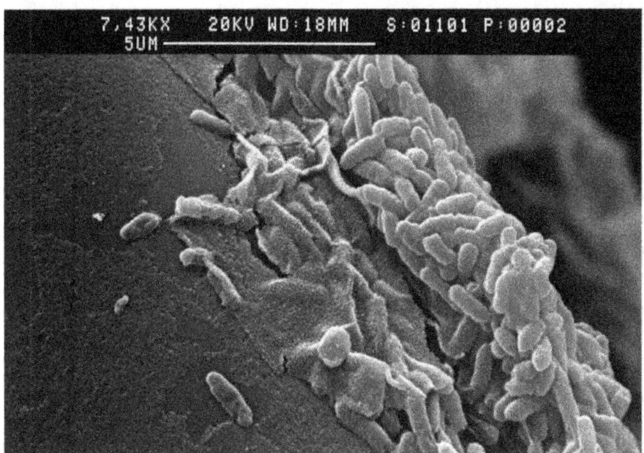

Figure 1. Formation of *P. aeruginosa* biofilms on glass wool after 24 h as observed by scanning electron microscopy. For complementary information on *P. aeruginosa* biofilms grown on glass wool, see [27].

Bacteria recovery

After 24 h of incubation, planktonic organisms were recovered by centrifugation (10 min at 2600×g at 4°C). Sessile bacteria were recovered as described previously [19]. Briefly, glass wool was aseptically removed after 1, 2 or 6 days of incubation, and washed twice in 0.1 M, pH 7 Phosphate Buffer Saline (PBS) to release weakly attached cells. It was then placed in sterile flasks containing 30 g of glass beads (diameter, 3 mm) and 50 mL of PBS. Bacteria were released from the substratum by vigorous shaking for 20 min. They were then harvested by centrifugation ((10 min at 2600×g) at 4°C) and resuspended in 5 ml of sterile Milli-Q water.

Membrane extractions

Bacterial IM enriched samples extraction was carried out following the spheroplast protocol first described by Mizuno and Kageyama [20]. After centrifugation (see above) bacterial pellets were washed in 10 mL of 20% (w/v) sucrose. After another centrifugation, pellets were weighted and resuspended in a digestion solution of the following composition: for 1.5 g bacteria wet weight, 18 mL of 20% sucrose, 9 mL of 2 M saccharose, 10 mL of 0.1 M Tris-HCl, 0.8 mL of 1% EDTA, 1.8 mL of 1% lysozyme, 1 µL of 1 mg/mL RNase and 5 µL of 20 mg/mL DNase. The solution was incubated at 37°C. Spheroplast formation was monitored by optical microscopy. When only ovoid forms were observed, the suspension was centrifuged at 30°C for 15 min at 5200×g to recover spheroplasts. The pellet was resuspended in 5 mL of 0.01 M PBS and was subjected to sonication (cycles of 30 s for 2 min). The suspension was then centrifuged at 30°C, for 20 min at 5200×g. The supernatant, containing IM, was diluted in 100 mM sodium carbonate, stirred at 4°C for 1 h to separate soluble and insoluble phase. Then it was ultracentrifuged (60,000×g for 1 h at 4°C) to harvest IM enriched samples. Pellets were washed twice with 40 mM Tris buffer (pH 7) and freezed at −20°C.

In order to harvest OM enriched samples, the digestion time was divided by a factor of 2. The suspension was then centrifuged at 30°C for 15 min at 5200×g and the supernatant containing mostly OM was collected and freezed at −20°C.

Outer membrane proteins detection in membrane extracts

In order to detect the cross-contamination in the different membrane extracts, OM proteins (OMP) fractions were identified using nanoLC-MS/MS. Proteins were digested according to the following procedure. Twenty-five µg of proteins were alkylated with 25 mM iodoacetamide for 45 min in the dark. The protein sample was mixed with SDS loading buffer (63 mM Tris-HCl, pH 6.8, 10 mM DTT, 2% SDS, 0.02% bromophenol blue, 10% glycerol), then loaded onto a SDS-PAGE stacking gel (7%). A short electrophoresis was performed (10 mA, 15 min). After migration, the gel were stained with Coomassie blue and destained with a solution containing 50% ethanol, 10% acetic acid and 40% deionized water. The revealed protein band was excised, washed with water, submitted to protein digestion with trypsin (0.5 µg per band). The digestion was achieved after 3 h at 37°C. Several steps

Table 1. Fatty acids profiles in planktonic and biofilm *P. aeruginosa* cells.

Fatty acid	retention time	mass (methyl ester fatty acid)	Planktonic (%)	1 day-old biofilm (%)	2 days-old biofilm (%)	6 days-old biofilm (%)
3-OH-C10:0	13:15	202	2.6	0.0	0.0	0.0
C12:0	14:11	214	2.7	0.3	0.0	0.2
2-OH-C12:0	15:53	230	1.2	0.2	0.0	0.0
3-OH-C12:0	16:16	230	2.0	2.8	0.4	3.6
12-CH3-C13:0	16:34	242	0.0	1.7	0.4	0.0
C14:0	17:02	242	1.3	2.3	0.2	0.9
13-CH3-C14:0	17:59	256	0.7	1.4	0.4	0.0
12-CH3-C14:0	18:03	256	0.0	21.0	6.3	0.0
C15:0	18:22	256	0.0	1.5	0.4	0.3
14-CH3-C15:0	19:12	270	0.0	2.3	1.2	0.2
C16:1	19:25	268	2.6	2.8	12.3	6.3
C16:0	19:42	270	64.9	28.4	39.5	45.0
15-CH3-C16:0	20:27	284	0.0	1.1	0.0	0.0
14-CH3-C16:0	20:34	284	0.0	5.1	0.4	0.0
C17:0cyc(9,10)	20:45	282	3.2	0.9	2.0	2.7
C17:0	20:53	284	0.0	1.9	1.9	0.1
16-CH3-C17:0	21:39	298	0.0	2.8	0.2	0.2
C18:1	21:51	296	13.0	15.1	30.0	34.2
C18:0	22:04	298	0.7	6.3	2.37	0.2
17-CH3-C18:0	22:53	312	0.0	2.0	0.16	0.1
C19:0cyc(11,12)	23:04	310	6.5	0.9	4.0	6.3
C20:0	24:18	326	0.0	2.8	0.1	0.0
Linear saturated fatty acids			**69.5**	**40.0**	**42.2**	**46.4**
Linear unsaturated fatty acids			**15.6**	**17.9**	**42.3**	**40.5**
Cyclic fatty acids			**9.7**	**1.7**	**5.9**	**9.0**
Branched chain fatty acids			**0.7**	**37.5**	**9.1**	**0.6**
Hydroxy fatty acids			**4.5**	**2.8**	**0.4**	**3.6**

of peptide extraction were performed in H_2O/ACN/TFA (49.5/49.5/1). The peptides were dried and stored at −20°C.

The nanoLC-MS/MS analysis of the eluates was repeated twice. All experiments were performed on a LTQ-Orbitrap Elite coupled to an Easy nLC II system (both from Thermo Scientific). Samples were injected onto an enrichment column (C_{18} PepMap100, Thermo Scientific). The separation was achieved with an analytical column needle (NTCC-360/100-5-153, NikkyoTechnos). The mobile phase consisted of H_2O/FA 0.1% (buffer A) and ACN/FA 0.1% (buffer B). Tryptic peptides were eluted at a flow rate of 300 nL/min, using a three-step linear gradient: from 2 to 40% B over 105 min, from 40 to 80% B in 4 min and at 80% B for 11 min.

The mass spectrometer was operated in positive ionization mode with a capillary voltage and a source temperature set at 1.6 kV and 275°C, respectively. The samples were analyzed using the collision induced dissociation. The first scan (MS spectra) was recorded in the Orbitrap analyzer (R = 60,000) with the mass range m/z 400–1800. Then, the 20 most intense ions were selected for MS^2 experiments. Singly charged species were excluded for MS(n) analysis. Dynamic exclusion of already fragmented precursor ions was applied for 30 s, with a repeat count of 1, a repeat duration of 30 s and an exclusion mass width of ±5 ppm. Fragmentation occurred in the linear ion trap analyzer with collision energy of 35. All measurements in the Orbitrap analyzer were performed with on-the-fly internal recalibration (lock mass) at m/z 445.12002 (polydimethylcyclosiloxane).

Raw data files were then processed using Proteome Discoverer 1.3 software (Thermo Scientific). Peak lists were searched using the MASCOT search engine (Matrix Science) against the *P. aeruginosa* PA01 database (http://www.pseudomonas.com) [21]. Database searches were performed with the following parameters: 2 missed cleavage sites allowed; variable modifications: carbamidomethylation of cystein, and oxidation of methionine. The parent ion and daughter ion tolerances were 10 ppm and 0.5 Da, respectively. Subcellular localizations were predicted by PSORTb V3.0.

IM and OM enriched membrane extracts from 3 independent bacterial cultures were analysed. In order to compare the results from the IM and OM enriched membrane extracts, peptide spectral match (PSM) values for the porin and structural outer membrane porin OprF precursor (PA14_41570) were used. The mean PSM ratio was calculated and the result is given with an error corresponding to a 95% confidence interval.

Table 2. Distribution of the predominant inner and outer membrane phosphatidylethanolamines and phosphatidylglycerols in planktonic and biofilms *Pseudomonas aeruginosa* (results are given with 95% confidence interval).

			PEs					PGs				
PL*	**R**	**R′**	**mass**	**plankt**	**BF 1 day**	**BF 2 days**	**BF 6 days**	**mass**	**plankt**	**BF 1 day**	**BF 2 days**	**BF 6 days**
32:1	C16:0	C16:1	688.49	4.8±0.6	6.6±0.5	6.6±0.3	6.0±0.5	719.49	4.4±2.0	4.7±0.8	4.8±0.9	4.9±1.1
	C14:0	C18:1		*5.4±0.6*	*5.6±0.5*	*7.3±0.5*	*7.3±0.5*		*n.d.*	*n.d.*	*n.d.*	*n.d.*
	CH3-C14:0 [traces]	C17:0cyc [traces]										
32:0	**C16:0**	**C16:0**	**690.51**	**8.6±0.6**	**8.2±0.5**	**11.0±0.3**	**8.8±0.5**	**721.50**	**5.6±2.0**	**7.2±0.8**	**4.4±0.9**	**4.4±1.1.**
	C14:0	**C18:0**		***7.0±0.5***	***5.4±0.5***	***13.6±0.5***	***5.6±0.5***		***n.d.***	***n.d.***	***n.d.***	***n.d.***
33:1	C16:0	C17:0cyc(9,10)	702.51	5.9±0.4	3.5±0.3	4.6±0.3	14.5±0.7	733.50	6.2±0.8	6.0±0.5	7.6±0.5	12.9±0.5
	CH3-C16:0	C16:1		*5.9±0.5*	*4.6±0.5*	*2.6±0.3*	*21.1±0.9*		*9.5±0.8*	*5.9±0.5*	*5.7±0.5*	*2.1±0.5*
33:0	**C16:0**	**CH3-C16:0**	**704.52**					**735.52**	**2.7±0.8**	**4.2±0.5**	**2.5±0.5**	**4.1±0.5**
	CH3-C14:0	**C18:0**		***2.6±0.8***	***2.5±0.7***	***2.1±0.5***	***2.6±0.6***		***8.2±0.5***	***2.2±0.7***	***3.0±0.5***	***3.1±0.6***
34:2	C16:1	C18:1	714.51	5.2±1.6	13.7±0.7	7.7±0.9	8.2±1.0	745.50	2.4±0.5	7.4±0.5	3.5±0.5	5.3±0.5
				6.2±0.8	*11.7±0.9*	*8.2±0.9*	*6.5±0.9*		*4.1±0.8*	*7.1±0.6*	*8.0±0.6*	*10.0±0.6*
34:1	**C16:0**	**C18:1**	**716.52**	**44.8±1.8**	**35.3±0.7**	**35.4±0.9**	**30.7±1.0**	**747.52**	**21.9±0.5**	**17.1±0.5**	**17.9±0.5**	**19.5±0.5**
	C16:1	**C18:0**		***45.8±1.0***	***31.4±0.9***	***45.2±0.9***	***26.9±0.8***		***25.9±0.8***	***23.4±0.7***	***33.4±0.7***	***28.2±0.8***
	CH3-C14:0 [traces]	**C19:0cyc [traces]**										
35:2	C16:1	C19:0cyc(11, 12)	728.52	4.2±0.8	4.0±0.5	4.6±0.5	4.6±0.5	759.59	n.d.	6.9±0.5	4.9±0.5	n.d.
	C17:0cyc(9,10)	C18:1		*4.0±1.0*	*5.9±0.5*	*2.6±0.5*	*4.3±0.5*		*3.6±0.5*	*8.8±0.5*	*4.3±0.5*	*6.7±0.5*
35:1	**C16:0**	**C19:0cyc(11, 12)**	**730.54**	**17.6±0.8**	**9.1±0.5**	**13.1±0.5**	**12.0±0.5**	**761.53**	**31.6±0.8**	**13.7±0.5**	**25.7±0.5**	**27.3±0.5**
	CH3-C16:0	**C18:1**		***15.0±1.0***	***12.8±0.8***	***5.2±0.5***	***11.8±0.9***		***23.5±0.8***	***20.4±0.8***	***13.9±0.5***	***22.9±0.9***
36:2	C18:1	C18:1	742.54	2.8±0.6	4.4±0.5	3.2±0.5	3.6±0.5	773.53	5.0±0.5	5.3±0.5	6.2±0.5	6.5±0.5
	C17:0cyc(9,10)	C19:0cyc(11, 12)		*2.7±0.6*	*5.1±0.5*	*2.3±0.5*	*3.4±0.5*		*5.0±0.5*	*10.4±0.5*	*5.2±0.5*	*7.6±0.5*
Sum				**93.9**	**84.8**	**86.2**	**88.5**		**74.8**	**67.2**	**71.3**	**78.6**
				91.9	***85.0***	***89.2***	***89.5***		***79.9***	***78.2***	***80.2***	***80.5***

Values in straight correspond to the internal membrane - *Values in italic correspond to the outer membrane.*
*Total number of C: Number of equivalent unsaturation – R and R′ represent the fatty acid moiety with no indication of their position.
Note that a cyclopropylation is equivalent to an unsaturation in terms of molecular mass (e.g. the molecular masses of .C17:0cyc and C17:1 are identical).

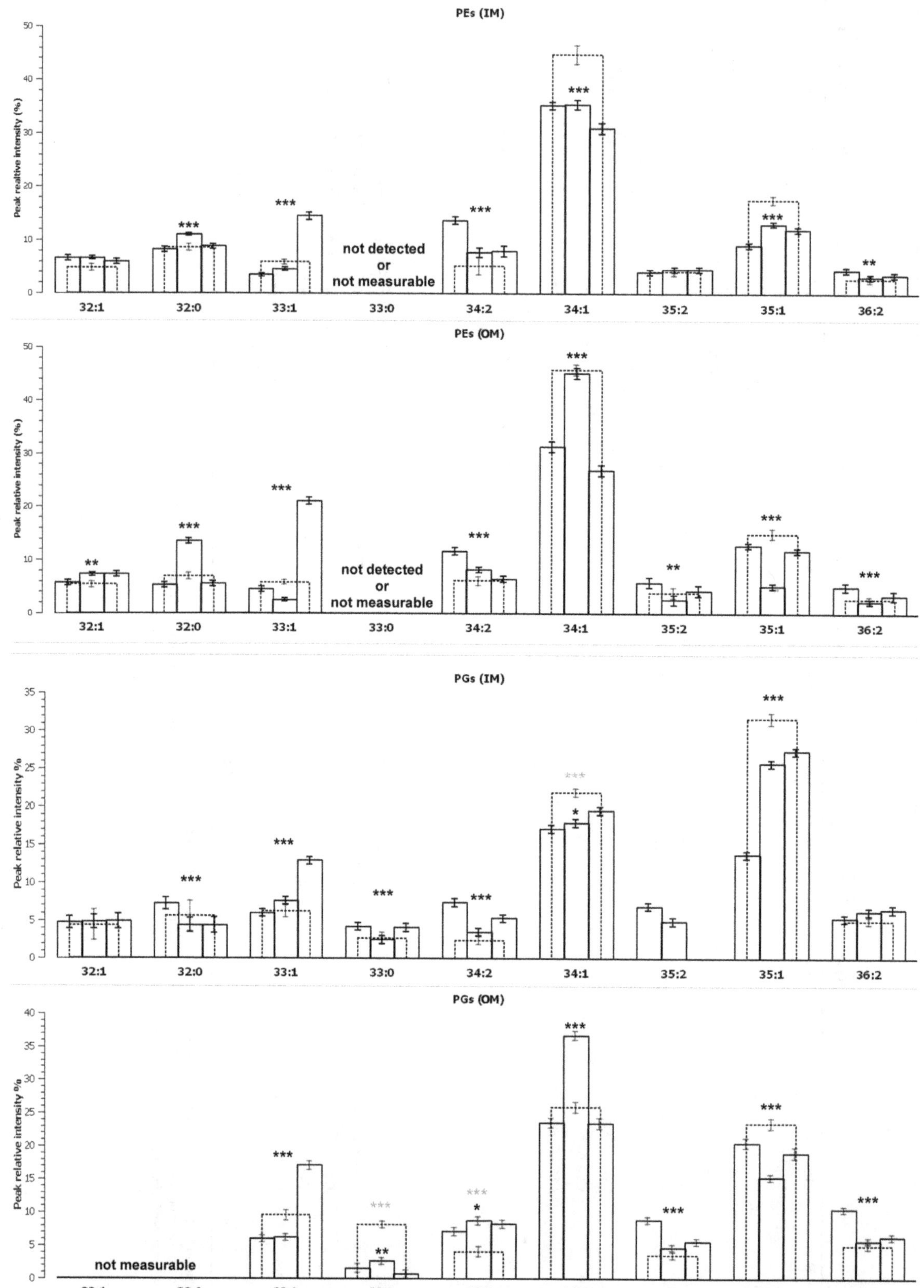

Figure 2. Evolution of the proportions of the main phosphatidyletnolamines and phosphatidylglycerols in *P. aeruginosa* inner and outer membranes. The proportions obtained for planktonic bacteria is represented in dashed lines. Error bars are given with a 95% confidence interval. Results of ANOVA test for the phospholipid proportions in sessile bacteria are shown (***, p-value<0,001; **, p-value<0,01; *, p-value<0,1 none p-value>0,1 (not statically relevant)). When different results of ANOVA test including the phospholipid proportions from planktonic bacteria are shown in grey.

Inner membrane proteins detection in membrane extracts

To evaluate the proportion of IM proteins in the membrane extracts, NADH oxidase assays were performed, NADH oxidase being described as an IM marker [22]. The overall protein concentration in the samples was measured using Bradford methods. Then, membrane samples containing 50 or 100 µg total protein was added in a mixture containing 50 mM Tris pH 7.5, 0.2 mM dithiothreitol, 0.12 mM NADH, in a volume of 3 ml. The decrease in absorbance of NADH at 340 nm was followed at room temperature in a Varian Cary 100Bio spectrophotometer for 30 min. Three different extracts obtained from 3 independent bacterial cultures were tested using 2 concentrations of total proteins per extract (16.6 and 33.3 µg/mL). The initial rates of NADH oxidation were measured and their mean values per µg of proteins were calculated. The mean ratio of the average rates between the OM enriched samples and the IM enriched samples was calculated (with an error corresponding to a 95% confidence interval).

Lipid extractions

Lipid extraction was carried out according to Bligh and Dyer protocol [23]. For 1 mL of membrane extract, 3.75 mL of chloroform: methanol (1:2 v/v) solution was added. The mixture was sonicated for 5 min and vortexed for 15 min until obtaining a milky-mixture. After adding 1.25 mL of chloroform, the mixture was vortexed for 1 min. A volume of 1.25 mL of 1 M NaCl aqueous solution was added and the mixture vortexed again for 15 min. Finally, the mixture was centrifuged (670×g for 10 min at 30°C) to separate organic and aqueous phase. The organic phase was recovered with a Pasteur pipette and 1.88 mL of chloroform was added to the aqueous phase. After stirring for 15 min, the mixture was centrifuged (670×g for 10 min at 30°C) to separate the two phases. The aqueous phase was removed and organic phases were mixed and evaporated under argon.

To degrade and eliminate proteolipids, 1 mL of methanol was added to lipid extracts. The mixture was vortexed for 5 min. Methanol was then evaporated. A chloroform: methanol (1:2 v/v) solution was added and the mixture was centrifuged (670×g for 10 min at 30°C), to sediment proteins. The organic phase was recovered and evaporated under argon. The lipid extracts were conserved at −20°C.

Mass spectrometry analyses of phospholipid extracts

Lipid extracts were analysed by Electrospray Ionization Mass Spectroscopy (ESI-MS) as previously described [19]. A Linear Quadruple Ion Trap (QTRAP, AB Sciex Instruments) spectrometer, equipped with a turbo spray ionization source heated to 300°C, was used. The potential applied during the acquisition was − 4500 V. The mass calibration and resolution were made according to the procedure of manufacturer specifications.

The concentrations of the lipid solutions were checked by the method described by Raheja et *al* [24] and by measuring the surface pressure of Langmuir film for each lipid extract [19]. Lipid extracts were diluted in a chloroform: methanol (80:20 v/v) solution. The volume of solvent was adjusted in order to obtain a 1 µM solution. The solution was subjected to sonication for 1 h to obtain stable and homogeneous solution. Samples were infused at 10 µL/min flow rate in negative ionization mode. Spectra were acquired at 1000 amu.sec^{-1} over the 50–1700 m/z range.

Phospholipid spectra were analysed via a home-made python script under the scientific data analysis software SciDaVis (http://scidavis.sourceforge.net/). Three IM enriched samples lipid extracts and two OM enriched samples lipid extracts were used for each culture condition. For each extracts, three spectra were analysed and all these results were combined. The mean value were calculated and the results are presented with error bars corresponding to a 95% confidence interval. Comparisons with the 3 biofilm growth conditions were analysed using one-way analysis of variance (ANOVA). In addition an ANOVA test including the planktonic mode of growth was performed. The ANOVA was performed using the script from Pr. H. Arsham (http://home.ubalt.edu/ntsbarsh/Business-stat/otherapplets/pvalues.htm).

Bacterial total fatty acid extraction

The fatty acid extraction was performed as described by Steger et *al* [25] and Hoffmann et *al* [26]. Planktonic and biofilm bacteria were harvested at 2600×g for 10 min at 4°C. The resulting pellets were resuspended into 2 mL Milli-Q water and then lyophilised for 2 days at −62°C. Five to 10 mg of lyophilised bacteria were used in each extraction. For the saponification reaction, 1 mL of sodium hydroxide (3.75 M) in 1:1 methanol-water was first used at 80°C for 30 min to release fatty acids. The second step, i.e., the transesterification (fatty acids methylation) was performed with 2 mL of 10% hydrochloric acid in methanol at 80°C for 10 min. This second step allows the decrease of fatty acid polarity and to obtain stable molecules. In the last step, fatty acids were separated using 1.5 mL of 1:1 hexane-methyl tert-butyl ether mixture. Organic phases containing fatty acids were then washed with 3 mL of 0.3 M sodium hydroxide, and submitted to an additional extraction with 1 mL of hexane. Fatty acid methyl esters (FAMEs) were thus obtained.

Fatty acids analysis by Gas Chromatography-Mass Spectrometry (GC-MS)

FAMEs were analysed by GC-MS. A capillary column (30 m×0.25 µm×250 µm id) was used to separate fatty acids. Injection temperature was 250°C. Oven temperature ranged from 60 to 300°C at 8°C/min rate. One hundred microlitres of cyclohexane were added to 100 µL of samples. $MgSO_4$ was then added to obtain dry samples. A volume of 0.5 µL of the mixture was injected with splitless injector flow. Resulting chromatogram picks were then identified by mass spectrometry.

Results and Discussion

Biofilm formation

After one day, *P. aeruginosa* formed a biofilm as shown on **figure 1**. Individual bacteria were still observed, but a biofilm is clearly already formed. We previously showed that biofilms cultured in this manner produce an extensive canopy after two days [27].

Fatty acids profiles in planktonic and biofilm *P. aeruginosa* cells

P. aeruginosa lipids were saponified in order to get the fatty acids composition in planktonic and biofilm bacteria at various times of biofilm formation (**Table 1**). The list of fatty acids accords well with that previously published for sessile *P. aeruginosa* cells [28]. The notable exceptions are 16-CH_3-C17:0, 17-CH_3-C18:0 and C20:0, which were here observed whereas previously absent [28]. However, we found those fatty acid in only 1-day-old biofilm organisms. The study by Chao et al [28] was conducted using 2-day-old biofilms, and this may explain this discrepancy. This suggests that the diversity in fatty acid composition is higher in young biofilm bacteria than in older (e.g., 6-day-old) ones.

In most samples, the two main fatty acid constituents were the saturated fatty acid C16:0 and the unsaturated fatty acid C18:1, in accordance with observations by Chao et *al* [28]. The ratio between C16:0 and C18:1, we observed, was around 5 in planktonic cells but decreased to 1.9 in 1-day-old and to 1.3 in 6-day-old biofilm cells. This is in contradiction with previous data [28] showing a mean ratio of 0.5 whatever the sample. Another discrepancy came from branched chains and hydroxy fatty acids. Thus, for 2-days-old biofilms, branched chains fatty acids represent here 9.1% of the total fatty acids while they were previously undetected [28]. It should also be noticed that for 1-day-old biofilm cells, the anteiso methylated-C14:0 itself represented 21% of the total fatty acids. In contrary in 2-days-old biofilms, hydroxy fatty acids proportion was here 0.4% while it was 19.9% in the previous investigation [28]. However in both studies, the proportion of cyclopropylated fatty acids increased during the biofilm formation, reaching in 6-day-old biofilm cells, the amount observed in planktonic counterparts.

These results illustrate the experimental difficulties in fatty acid analysis. Another difficulty is in the interpretation of the results. Indeed, while some fatty acids came from phospholipids hydrolysis, others may have lipopolysaccharide, and/or lipoproteins, as origin. We then monitored the evolution of the total phospholipids in the inner and outer membrane of the sessile bacteria. The main advantage of such approach is that lipid extracts are used for the ESI-MS analysis [29]. This procedure is simple and do not need further chemical modifications (e.g., hydrolysis) or extraction. Moreover, the amounts of lipid required is low. The risk of contamination or error analysis is consequently limited.

Evolution of phospholipid distribution during biofilm formation

The phospholipid composition of IM and OM enriched samples were monitored over a 1 to 6 days incubation period and compared with that of planktonic bacteria. Since it is difficult to fully separate OM from IM we evaluated the cross-contamination in the samples. Membrane samples were analysed with nanoLC-MS/MS. It clearly showed a higher proportion of OMPs in the OM enriched samples than in the IM enriched ones as expected. Among these OMPs, we used the major porin and structural outer membrane porin OprF precursor (PA14_41570) to evaluate the cross-contamination within the two samples as already described [30]. The ratio of PA14_41570 amount between the OM and IM enriched samples was 1.7±0.3.

In the case of the IM protein samples, due to the sample complexity and the low expression level of each protein, we evaluated the cross-contamination by the measure of the NADH oxidase activity [22,30]. Actually we observed NADH oxidase activity in both samples. However the enzymatic activity in the OM enriched samples was twice lower than in IM enriched samples. The oxidase activity ratio between the OM enriched samples and the IM enriched samples was 0.5±0.1.

These results demonstrate the efficacy of the membrane fragments enrichment. It shows that the proportion of IM is 2 times higher in the IM enriched sample compared to OM enriched samples and that the proportion of OM is 1.7 times higher in the OM enriched sample compared to IM enriched samples.

Phosphatidylglycerol (PG) and phosphatidylethalonamine (PE) were the most detected phospholipids. Phosphatidic acid, cardiolipin and ornithine lipids (OL) were also detected. However the peak intensity for these phospholipids were very low and to observe any significant changes. Then, We consequently focused on the changes in PGs and PEs amounts (**table 2**). In the text and table 2, phospholipids are identified as C:U with C being the total number of carbon atoms in the aliphatic chains and U being the equivalent number of unsaturation. The equivalent number of unsaturation accounts for either the presence of unsaturation or the presence of a cyclopropyl group.

In all samples, predominant PEs and PGs have a total number of from 32 to 36 carbon atoms in their aliphatic chains. However, it should be noticed that PEs and PGs with lower and higher numbers of carbon atoms were also detected.

In IM and OM enriched samples, the proportions of PEs and PGs with a total number of carbon atoms in their aliphatic chains below 32 were double in sessile bacteria (5% for PEs and 9% for PGs) compared to planktonic counterparts (2.4% for PEs and 5% for PGs). These ratios remained constant during biofilm growth. PE 30:0 and PG 30:0 were found in similar proportions in all samples whereas PE 30:1 and its cyclopropylated form PE 31:1, as well as PE 31:0 were only observed in sessile organisms. PG 30:1 and PG 31:0 are present in both planktonic and sessile cells. However their proportions are double for sessile bacteria. PE 31:0 and PG 31:0 are constituted of one 12-CH3-C14:0 and one C16:0 chain. While these phospholipids were found in MS spectra from biofilm lipid extracts, their proportion did not evolve with biofilm age. The presence of these phospholipids do not correlate with the high level of 2-CH3-C14:0 found in the fatty acid analysis from 1-day-old and 2-day-old biofilm lipid extracts. It might be explained by the presence of OLs in lipid extracts. Indeed, in 1-day-old biofilm IM enriched samples, we observed a peak corresponding to OL 31:0 with an ester-linked 12-CH3-C14:0 fatty acid chain. This phospholipid was absent in all other samples. However, we detected the presence of OL 32:0 and OL 34:1. In biofilm lipid extracts, OL 32:0 peaks were 10 times lower in IM enriched samples and 20 times lower in OM enriched samples compared to PE 34:1 peaks. In 1-day-old biofilm IM enriched samples, the intensity of the of OL 31:0 peak was 3 times higher than that of OL 32:0. The presence of these OLs in the samples is not surprising since they were already found to represent between 2 to 15% of the total *P. aeruginosa* lipids, the most abundant structures containing ester-linked C16:0, C18:0 and C18:1 fatty acid chains [31]. Some ester-linked C15:0 fatty acid chain which was likely 12-CH3-C14:0 were also detected. Consequently, we can conclude that the presence of OL 31:0 in the IM enriched samples at the early stage of biofilm growth may explain the high proportion of 12-CH3-C14:0 in the fatty acid distribution. However we cannot exclude the existence of other lipids, unidentified at that time and containing such fatty acid chain.

In the case of phospholipids with a total number of carbon atoms in their aliphatic chains above 37, the proportions of PEs and PGs were respectively 6 and 3 time higher in 1-day-old biofilms lipid extracts compared to planktonic bacteria lipid extracts in both membrane samples. With the exception of PE and PG 39:2 detected at very low levels (i.e., <1% and 3% respectively) in planktonic samples, PE and PG 38:1, 38:2, 39:1 and 39:2 as well as PG 38:0 were detected in the 1-day-old biofilm lipid extracts. However the proportions of these phospholipids decreased with the biofilm age. These results are coherent with fatty acid analysis in which C20:0 was clearly detected (3%) in 1-day-old and slightly detected (0.1%) in 2-day-old biofilms lipid extracts. These data show that the overall phospholipid diversity is higher in biofilms than in planktonic bacterias. The diversity decreased with the biofilm age especially for high mass phospholipids.

Table 2 gives the proportions of predominant phospholipids. Uneven numbered phospholipids with one equivalent number of unsaturation (PEs and PGs 33:1 and 35:1) evolved differently in IM and OM enriched samples (**figure 2 and table 2**). In the IM enriched samples, PE 33:1, PG 33:1 and PG 35:1 increased during the biofilm formation to reach a maximum in 6-day-old sessile cells. From 1- to 6-day-old biofilm lipid extracts, the peak intensity of PE 33:1 was increased by a factor of 4 and the peak intensity of PGs 33:1 and 35:1 by a factor of 2. This may be easily explained by the increase in cyclopropylated chains observed in the fatty acid analysis. However, it should be noticed that the peak intensity of PE 32:1, PG 32:1 and PG 34:1 do not change with the biofilm age. This observation shows that those phospholipids are synthesised continuously during the biofilm growth. In the OM enriched samples, the evolution of PE 33:1, PG 33:1 and PG 35:1 is more complex with minimum proportions in 2-day-old biofilms lipid extracts. This may be explained by the competition between the production kinetics of phospholipids with a branched chain and that of phospholipids with a cyclopropylated group. This hypothesis is reinforced by the fatty acid analysis which showed that 14-CH3-C16:0 and 15-CH3-C16:0 (which may be parts of PE and PG 33:1) were abundant in 1-day-old biofilm cells and rapidly disappeared with the biofilm age. A similar remark may be done for 17-CH3-C18:0 which may be part of PE and PG 35:1. Conversely, C17:0cyc(9,10) and C19:0cyc(11,12) which may be parts of PE (or PG) 33:1 and PE (or PG) 35:1 respectively, increased during the biofilm formation. This may suggest that in 1-day-old biofilms, branched chains phospholipids proportions are relatively high in OM compared to IM. The proportions of these phospholipids decreased with the biofilm age. In addition, like in IM, cyclopropylated chains phospholipids increased continuously during the biofilm formation. The combination of these alterations explains the evolution of PEs and PGs 33:1 and 35:1 in OM enriched samples.

In 2-day-old biofilm OM enriched samples, the proportions of PE 34:1 and PG 34:1 exhibited an optimum value, while the proportions of PE 35:1 and PG 35:1 exhibited a minimum value. We also observed that in most cases even numbered phospholipids bearing 2 unsaturations continuously decreased during the biofilm growth. All these results show an obvious diminution of unsaturated phospholipids during the biofilm growth. This observation disagrees with the fatty acid analysis but accords well with Chao *et al* results [28].

Conclusion

Compared with that of planktonic organisms, the lipid distribution in biofilm bacterial membranes exhibited alterations in terms of acyl chains length, nature of the phosphate substituent and nature of the polar head group of glycerophospholipids (PG and PE), confirming previous observations [19]. An outstanding information brought by the present study is that these lipidome alterations are biofilm age dependent, changes being higher in the early phase of the biofilm formation. It has been shown that the diversity of glycerophospholipids within the membrane is the result of numerous combinations of acyl chains and head groups and is crucial to maintain the bilayer structure, the cell permeability and the membrane fluidity [32]. In conclusion, the alterations of the IM and OM lipid compositions which we observed here reflect a bacterial adaptation to the environment conditions prevailing in biofilm.

In mature biofilms (6-day-old), the phospholipid distribution seemsclose to that of planktonic bacteria with some notable exception, however, e.g. PE and PG 33:1 (C17:0cyc(9,10) and C16:0 fatty acid chains) which are more abundant in 6-day-old biofilms. This observation might suggest a progressive "bacteria conditioning" to a return to the planktonic mode.

Acknowledgments

We are grateful to C. Bourhis-Loutelier for the GC-MS spectra analysis.

Author Contributions

Conceived and designed the experiments: SA HB JH TJ. Performed the experiments: HB CR IA MABM JH SA. Analyzed the data: SA HB JH. Wrote the paper: HB SA TJ.

References

1. Petrova OE, Sauer K (2009) A novel signaling network essential for regulating *Pseudomonas aeruginosa* biofilm development, PLoS Pathog. 5(11): e1000668.
2. Tummler B, Koopmann U, Grothues D, Weissbrodt H, Steinkamp G, et al. (1991) Nosocomial acquisition of Pseudomonas aeruginosa by cystic fibrosis patients. J Clin Microbiol. 29: 1265–1267.
3. Lambert PA (2002) Mechanisms of antibiotic resistance in *Pseudomonas aeruginosa*. J R Soc Med 95: 22–26.
4. Strateva T, Yordanov D (2009) *Pseudomonas aeruginosa* – a phenomenon of bacterial resistance. J Med Microbiol. 58: 1133–1148.
5. Costerton JW, Stewart PS, Greenberg EP (1999) Bacterial biofilms: a common cause of persistent infections. Science. 284: 1318–1322.
6. Mah TF, O'Toole GA (2001) Mechanisms of biofilm resistance to antimicrobial agents. Trends Microbiol. 9: 34–39.
7. Schierholz JM, Beuth J, Pulverer G (1999) Adherent bacteria and activity of antibiotics, J. Antimicrob. Chemother. 43: 158–160.
8. Drenkard E (2003) Antimicrobial resistance of *Pseudomonas aeruginosa* biofilms. Microbes Infect. 5: 1213–1219.
9. Stewart PS (2002) Mechanisms of antibiotic resistance in bacterial biofilms. Int J Med Microbiol. 292: 107–113.
10. Whiteley M, Bangera MG, Bumgarnes RE, Parsek MR, Teitzeil GM, et al. (2001) Gene expression in *Pseudomonas aeruginosa* biofilms. Nature 413: 860–864.
11. Schembri MA, Kjærgaard K, Klemm P (2001) Global gene expression in *Escherichia coli* biofilms. Mol. Microbiol. 48: 253–267.
12. Vilain S, Cosette P, Hubert M, Lange C, Junter GA, et al. (2004) Comparative proteomic analysis of planktonic and immobilized *Pseudomonas aeruginosa* cells: a multivariate statistical approach. Anal Biochem 329: 120–130.
13. Junter GA, Jouenne T (2004) Immobilized viable microbial cells: from the process to the proteome em leader or the cart before the horse. Biotechnol Adv. 22: 633–658.
14. Folstom JP, Richards L, Pitts B, Roe F, Ehrlich GD, et al. (2010) Physiology of *Pseudomonas aeruginosa* in biofilms as revealed by transcriptome analysis. BMC Microbiology, 10:294. doi: 10.1186/1471-2180-10-294.
15. Zhang B, Powers R (2012) Analysis of bacterial biofilm using NMR-based metabolomics. Future Med. Chem. 4: 1273–1306.
16. Denich TJ, Beaudette LA, Lee H, Trevors JT (2003) Effect of selected environmental and physico-chemical factors on bacterial cytoplasmic membranes. J. Microbiol. Methods. 52: 149–182.
17. O'Toole GA, Pratt LA, Watnick PI, Newman DK, Weaver VB, et al. (1999) Genetic approaches to study of biofilms. Methods Enzymol. 310: 91–109.
18. Abdul Lattif A, Mukherjee PK, Chandra J, Roth MR, Welti R, et al. (2011) Lipidomics of *Candida albicans* biofilms reveals phase-dependent production of phospholipid molecular classes and role for lipid rafts in biofilm formation. Microbiology. 157: 3232–3242.
19. Benamara H, Rihouey C, Jouenne T, Alexandre S (2011) Impact of the biofilm mode of growth on the inner membrane phospholipid composition and lipid domains in *Pseudomonas aeruginosa*. Biochim Biophys Acta. 1808: 98–105.
20. Mizuno T, Kageyama M (1978) Separation and characterization of the outer membrane of *Pseudomonas aeruginosa*, J Biochem. 84: 179–191.
21. Winsor GL, Lam DK, Fleming L, Lo R, Whiteside MD, et al. (2011). Pseudomonas Genome Database: improved comparative analysis and population genomics capability for Pseudomonas genomes. Nucleic Acids Res. 39(Database issue): D596–600.
22. Osborn MJ, Gander JE, Parisi E, Carson J (1972) Mechanism of assembly of the outer membrane of *Salmonella typhimurium*: Isolation and characterization of cytoplasmic and outer membrane, J. Biol. Chem. 247: 3962–3972.
23. Bligh EG, Dyer WJ (1959) A rapid method for total lipid extraction and purification. Can J Biochem Physiol. 37: 911–917.
24. Raheja RK, Kaur C, Singh A, Bhatia IS (1973) New colorimetric method for the quantitative estimation of phospholipids without acid digestion. J Lipid Res. 14: 695–697.

25. Steger K, Jarvis A, Smars S, Sundh I (2003) Comparison of signature lipid methods to determine microbial community structure in compost. J Microbiol Methods. 55: 371–382.
26. Hoffmann M, Wagner M, Abbadi A, Fulda M, Feussner I (2008) Metabolic engineering of_3-very long chain polyunsaturated fatty acid production by an exclusively acyl-CoA-dependent pathway. J. Biol. Chem. 283: 22352–22362.
27. Vilain S, Cosette P, Zimmerlin I, Dupont JP, Junter GA, et al. (2004) Biofilm proteome: homogeneity or versatility? J Proteome Res. 3(1): 132–6.
28. Chao J, Wolfaardt GM, Arts MT (2011) Characterization of *Pseudomonas aeruginosa* fatty acid profiles in biofilms and planktonic cultures. Can. J. Microbiol. 56: 1028–1039.
29. Milne S, Ivanova P, Forrester J, Alex Brown H (2006) Lipidomics: an analysis of cellular lipids by ESI-MS. Methods 39(2): 92–103.
30. Vasseur P, Soscia C, Voulhoux R, Filloux A (2007) PelC is a *Pseudomonas aeruginosa* outer membrane lipoprotein of the OMA family of proteins involved in exopolysaccharide transport. Biochimie 89: 903–915.
31. Kawai Y, Yano I, Kaneda K, Yabuuch E (1988) Ornithine-containing lipids of some *Pseudomonas* species. Eur. J. Biochem. 175: 633–641.
32. Dowhan W (1997) Molecular basis for membrane phospholipid diversity: why are there so many lipids? Annu Rev Biochem. 66: 199–232.

Biohybrid Polymer-Antimicrobial Peptide Medium against *Enterococcus faecalis*

Lea H. Eckhard[1], Asaf Sol[2], Ester Abtew[3], Yechiel Shai[4], Abraham J. Domb[3], Gilad Bachrach[2]❡, Nurit Beyth[1]*❡

1 Department of Prosthodontics, the Hebrew University – Faculty of Dental Medicine, Jerusalem, Israel, **2** Institute of Dental Science, the Hebrew University – Faculty of Dental Medicine, Jerusalem, Israel, **3** Institute for Drug Research, School of Pharmacology, Faculty of Medicine, the Hebrew University, Jerusalem, Israel, **4** Department of Biological Chemistry, the Weizmann Institute of Science, Rehovot, Israel

Abstract

Antimicrobial peptides (AMPs) are conserved evolutionary components of the innate immune system that are being tested as alternatives to antibiotics. Slow release of AMPs using biodegradable polymers can be advantageous in maintaining high peptide levels for topical treatment, especially in the oral environment in which dosage retention is challenged by drug dilution with saliva flow and by drug inactivation by salivary enzymatic activity. *Enterococcus faecalis* is a multidrug resistant nosocomial pathogen and a persistent pathogen in root canal infections. In this study, four ultra-short lipopeptides (C16-KGGK, C16-KLLK, C16-KAAK and C16-KKK) and an amphipathic α-helical antimicrobial peptide (Amp-1D) were tested against *E. faecalis*. The antibacterial effect was determined against planktonic bacteria and bacteria grown in biofilm. Of the five tested AMPs, C16-KGGK was the most effective. Next C16-KGGK was formulated with one of two polymers poly (lactic acid co castor oil) (DLLA) or ricinoleic acid-based poly (ester-anhydride) P(SA-RA). Peptide-synthetic polymer conjugates, also referred to as biohybrid mediums were tested for antibacterial activity against *E. faecalis* grown in suspension and in biofilms. The new formulations exhibited strong and improved anti- *E. faecalis* activity.

Editor: Tilmann Harder, University of New South Wales, Australia

Funding: This research was supported by The Legacy Heritage Clinical Research Initiative of the Israel Science Foundation (grant No.1764/11) (NB). The funders had no role in study design, data collection and analysis, decision to publish, or preparation of the manuscript.

Competing Interests: The authors have declared that no competing interests exist.

* Email: nuritb@ekmd.huji.ac.il

❡ These authors contributed equally to this work.

Introduction

The widespread use of antibiotics leads to the emergence of more resistant and virulent strains of microorganisms. Consequently, the development of new antimicrobial agents becomes paramount for novel treatment options [1]. Moreover, bacterial resistance to antibiotics becomes even more complicated when dealing with bacterial biofilms. Interestingly, potent antimicrobial components against a wide range of pathogens can be found in the innate immune system of various organisms including humans. An example is host-defense cationic antimicrobial peptides (AMPs) which are conserved evolutionary components that possess the capacity to kill invading microbes [2]. It is generally accepted that AMP-mediated killing typically occurs through microbial membrane disruption resulting in irreparable damage.

AMPs exhibit broad-spectrum activity against a wide range of microorganisms including Gram-positive and Gram-negative bacteria, protozoa, yeast, fungi and viruses [4]. Furthermore, whereas conventional antibiotics are becoming less effective, bacteria do not appear to develop resistance to AMPs.

AMPs function and mode of action is a direct derivative of their structure and electric charge. They differ in amino acid length (12–50 aa), sequence and dimensional structure, they are composed of about 50% hydrophobic amino acids and their electric charge is positive. This unique amphipathic structure allows AMPs to bind to the negatively charged outer surfaces of microorganisms and to disrupt and permeate their cell membranes [3]. Consequently, their advantage over conventional antibiotics is a non-specific but antimicrobial selective mode of action. Several resistance strategies to avoid AMPs function were reported. These included, degradation with extracellular proteases [4], altering the net surface charge [5], transporting AMPs into the cytoplasm and degrading them [6] and exporting AMPs by efflux pumps [7].

AMPs are one of the reasons why humans stay healthy [2]. In humans, one of the known potent AMPs groups is the defensins. Specifically, human beta-defensin 3 (hBD3). hBD3 is considered the most potent β-defensin peptide described so far [8]. An additional group of potent antimicrobial peptides, produced in bacteria and fungi are the lipopeptides. They are composed of specific lipophilic moieties attached to anionic peptides. Unfortunately, native lipopeptides are non-cell selective and thus can be toxic to mammalian cells. Interestingly, all of the structural advantages of the native AMPs can be recruited to synthesize improved antimicrobial agents, *e.g.* ultra-short lipopeptides and amphipathic α-helical antimicrobial peptide (Amp-1D). Recently, it was reported that ultra-short lipopeptides composed of only four amino acids conjugated to an aliphatic acids chain (16C, palmitate) can achieve potent antimicrobial activity without compromising biocompatibility.

Enterococcus faecalis is a microorganism residing in the gastrointestinal tract. None the less, *E. faecalis* can cause life-threatening infections such as: endocarditis, bacteremia, urinary tract infection and meningitis. These complications are mostly associated with the acquisition of resistance to antibiotics. In dentistry, *E. faecalis* is considered a persistent root canal pathogen [9].

Integration of the structural and functional properties of peptides and proteins with the versatility of synthetic polymers has gained significant interest in material design and application [10]. Peptides and proteins have unique structures that convey their ability to function in specific biological activities. Hybrid molecules of peptides conjugated to polymers can be used for various applications with the advantages of being resistant to enzymatic cleavage and less cytotoxic to human cells [11]. Peptide-synthetic polymer conjugates, also referred to as biohybrid medium, consist of biologically relevant peptides and synthetic polymers, aiming to combine the advantages of the two components, namely biological function (biological component) and process-ability (synthetic component). A slow release mechanism can enable high concentration maintenance of therapeutic agents for prolonged periods of time. Examples of biodegradable polymers that were previously described as efficient controlled delivery mediums include the fatty acid-based polymer poly (lactic acid co castor oil) and the ricinoleic acid-based poly (ester-anhydride) [11–15]. Herein, biohybrid medium consisted of two different assembled components, including a polymer matrix, which was responsible for the sustained release function and an antimicrobial agent, *i.e.* AMPs, which was responsible for the potency of the formulation. In the present study novel formulations of biodegradable polymers integrating AMPs were evaluated against *E. faecalis* in planktonic bacteria and biofilm growth.

Materials and Methods

Test materials

Antimicrobial peptides. Human recombinant β-defensin 3 (hBD3) (GIINTLQKYY CRVRGGRCAV LSCLPKEEQI GKCSTRGRKC CRRKK) was obtained from PeproTech (Lot #0108210, Rocky Hill, NJ, USA). Five different synthetic AMP candidates were tested. amphipathic α-helical antimicrobial peptide (Amp-1D) and four ultra short lipopeptides which were synthesized, purified and confirmed as described before [16,17].

Biodegradable polymer synthesis. Poly (lactic acid co castor oil) (DLLA) and ricinoleic acid-based poly (ester-anhydride) P(SA-RA) were synthesized as previously describe [11–13,15,18]. Briefly, a poly (ester-anhydride) copolymer of sebacic acid (SA) and ricinoleic acid (RA) in a 3:7 w/w ratio [P(SA-RA) 3:7] was synthesized by transesterification followed by anhydride melt condensation. SA was used as supplied by Sigma-Aldrich (St. Louis, MO, USA) without any additional purification. RA (>98%) was isolated from castor oil by fractional precipitation based on salt-solubility. Poly (DL lactic acid co castor oil) 4:6 and 3:7 designated P(DLLA:CO) 4:6 and P(DLLA:CO) 3:7 was prepared using racemic mixture (DL) lactic acid. The synthesized polymers were characterized by Infrared (IR) spectroscopy and nuclear magnetic resonance (NMR) spectroscopy. Gel permeation chromatography (GPC) was used to estimate the molecular weight.

Formulation of AMP-based biohybrid media. The peptide powder was mixed with a pasty polymer to form a uniform homogeneous paste. A novel formulation of peptide and biodegradable polymer was prepared at a ratio of 100 µg peptide integrated in 100 mg polymer. The two ingredients were mixed manually with a spatula.

Preparation of bacterial suspension

E. faecalis (ATCC #v583), was cultured overnight in 5 ml brain-heart infusion (BHI) (Difco, Detroit, MI, USA) broth supplemented with 2 mg/ml vancomycin (Sigma-Aldrich), at 37°C under aerobic conditions. The top 4 ml were transferred to a fresh test tube and the optical density (OD) was determined according to the specific experiment.

Antibacterial activity

Minimal inhibitory concentration. The antibacterial activity of the peptides was examined using the microdilution assay [19]. Briefly, the bacterial suspension (at OD 0.3) was diluted at a ratio of 1:1000. Aliquots of 150 µl of bacterial suspension were added to 50 µl of peptide dilutions in phosphate buffered saline (PBS) (Sigma-Aldrich) (in triplicate for each concentration) in a 96-well plate (Nunc 96 microtiter plates, Roskilde, Denmark). The optical density (595 nm) in each well was recorded every 20 min using a microplate reader (VERSAmax tunable microplate reader, molecular devices, Sunnyvale, CA, USA) at 37°C for 18–24 hrs. The minimal inhibitory concentration (MIC) was determined as the concentration which inhibited visible growth after 18–24 hrs.

Antibacterial activity of controlled released peptide. A total 10 mg of formulation was placed on the side walls of each of 6 wells in a 96 microtiter plate and then 270 µl of medium (BHI supplemented with vancomycin) were added Every 24 hrs the medium was collected and transferred to a new set of 6 wells in the same 96-well-plate and fresh medium was added to the 6 original wells containing the tested formulation. After one week, a 10 µl volume of *E. faecalis* suspension was added to each of the 6 wells and bacterial outgrowth was recorded. The plate was incubated at 37°C in a VERSAmax microplate reader and turbidity (OD_{650} nm) changes were recorded, every 20 min for 18–24 hrs.

Antibiofilm activity

Antibiofilm activity was tested on *E. faecalis* biofilms grown for 72 hrs. Biofilm was formed in microtiter plates (24 well plates for the ATP bioluminescence assay and 96 well plates for the crystal violet biomass assay and confocal laser spectroscopy). Saliva was collected from one donor and DL-Dithiolthreitol (DTT) (Thermo Scientific, Abu-Gosh, Israel) was added to 2.5 mM. The suspension was kept at 4°C for 10 min and then centrifuged for 15 min at 6,500×g. The supernatant was transferred to a fresh sterile tube and diluted to 25% with sterile double distilled water (DDW). The diluted saliva was disinfected using a 0.2 µm vacuum-driven filter (0.22 µm, 250 µl, Jet biofil, Belgium). Wells in the microtiter plate were coated with clarified saliva by adding the saliva to the wells for 1 hr at 37°C (150 µl of saliva in the 24 well plate and 50 µl in the 96 well plate). Unbound saliva was removed and the wells were washed gently with PBS. The polymer peptide formulations were placed on the side walls of the wells and 10 µl of bacterial suspension (prepared as described above) were placed in the center of each well not touching the coated sidewall. The saliva coating was used to cover the entire well surface, followed by formulation placement on the sidewall of the wells. The bacterial inoculum was placed in the center of each well, not touching the formulation. After 1 hr incubation at 37°C BHI broth was added (1 ml in the 24 well plate and 100 µl in the 96 well plate). BHI broth was added every 24 hrs during 72 hrs. After 3 days the medium was discarded and the wells were washed gently with PBS. Bacterial metabolism in the attached biofilm was assessed using ATP bioluminescence. Biofilm mass was measured using crystal violet as described below.

ATP bioluminescence. Bacterial killing was evaluated by measuring intracellular ATP levels, an energy parameter commonly used as an indicator of cell injury and viability [19]. The 72 hr biofilm formed on the bottom of the wells was scraped using a pipette tip and collected into a set of 15 ml tubes. The cells were then centrifuged (6,500×g, 5 min), resuspended in 1 ml Lysis Buffer (2 mM DTT, 2 mM trans 1,2 Diaminocyclohexane NNNN Tetraacetic acid, 0.5 mM EDTA, 1% Triton, 25 mM Tris, 25 mM K_2HPO_4, 10% glycerol) and transferred to a 2 ml microcentrifuge tube containing glass beads (Lysing Matrix tubes, 0.1 mm silica spheres; MP Biomedicals, Eschwege, Germany). The cells were disrupted with the aid of a FastPrep cell disrupter (MP Biomedicals, Irvine, CA, USA). The tube was centrifuged for 10 min (4°C, 13,400×g). ATP levels were determined using an ATP bioluminescence assay kit (CLS 2, Roch Diagnostics, Mannheim, Germany). In a 96 microtiter plate designed for luminescence assay (Thermo Scientific, NUNC, 96-well optical Btm Plt white, Rochester, NY, USA) a 100 µl volume of the samples was added to 6 wells for each tested group. Then 100 µl luciferase (from the kit) were added to the same wells. The plate was inserted in a GENios reader (TECAN, Salzburg, Austria) and luminescence was measured using the Magelan program (TECAN, V6.6, 2009). ATP calibration was performed using ATP and luciferase from the kit.

Crystal violet. The total biofilm yield was assessed using crystal violet staining as follows. Biofilm fixation was performed using 200 µl methanol (MERCK, Darmstadt, Germany) that were added to each well for 20 min. The biofilm was then stained using 200 µl 1% crystal violet (Merck) for 20 min. Then the wells were washed gently 3x with PBS, and 200 µl of 30% acetic acid (GADOT, Netanya, Israel) were added to the wells. The acetic acid was transferred to wells of a new 96-well microtiter plate that was placed in a microplate reader and absorbance (OD_{595} nm) was measured.

Confocal microscopy. Confocal laser scanning microscopy (CLSM) was used to explore the vitality of bacteria in the different depth layers of the biofilm. Bacteria were stained using a live/dead kit (Live/Dead BacLight viability kit, Molecular Probes, OR, USA) as described before [20]. Briefly, wells were washed, incubated for 15 min in a solution containing propidium iodide and SYTO 9 and washed again. To read the results directly, the wells were coated with emulsion oil to prevent dehydration. Fluorescence emission was detected using a Zeiss LSM 410 confocal laser scanning microscope (Carl Zeiss Microscopy, Jena, Germany). Red fluorescence was measured at 630 nm and green fluorescence at 520 nm; objective lenses: x60/oil, 1.4 numerical aperture. Horizontal plane (x-y axes) optical sections were made at 700 µm intervals from the surface outwards and images were displayed individually. The biofilm was quantified by measuring the area occupied by the bacteria with the aid of Image Pro 4.5 software (Media Cybernetics, Rockville, MD, USA).

Statistical analysis

The presented data are the mean and standard deviation of triplicates of a representative experiment repeated three times. The growth mean, and multiple comparisons of growth inhibition by AMP (compared with the growth of untreated bacteria) were calculated from each growth curve using Student's t-test. The level of significance was $p<0.01$.

Results

Antibacterial activity

Minimal inhibitory concentration. The MICs results for each of the tested AMPs are summarized in Table 1. Growth of *E. faecalis* was not inhibited by hBD3 at concentrations of up to 20 µg/ml. Amp-1D did not affect bacterial growth at concentrations of up to 25 µg/ml. C16-KGGK, C16-KKK, C16-KAAK and C16-KLLK completely inhibited bacterial growth at concentrations ranging between 5 and 25 µg/ml. The most potent AMP was the lipopeptide C16-KGGK that caused complete growth inhibition at 5 µg/ml (Fig. 1). As a result, all further formulations were tested using the C16-KGGK lipopeptide. Surprisingly, at concentrations below 5 µg/ml the growth of *E. faecalis* was not inhibited by C16-KGGK but was actually accelerated.

Sustained release and anti-*E. faecalis* activity. Release of the C16-KGGK lipopeptide from two biodegradable polymers was monitored over one week in two modes. In the first, the antibacterial action of C16-KGGK released into the medium that came in contact with the formulation every 24 hrs was measured (see Fig. 2A, B). In the second, the bacteria were added to the wells with C16-KGGK that was released from the polymer and accumulated for one week (see Fig. 2C). The anti-*E. faecalis* activity of C16-KGGK released from each formulation was reflected by: 1. The final optical density of the treated bacteria, which was lower than that of the untreated ones. 2. The slope of the curve (generation time, see Fig. 2C), which was more moderate in the treated bacteria.

The anti-*E. faecalis* activity in the medium exposed to the formulation for an entire week (see Fig. 2D, E) generated a longer generation time, especially with P(SA:RA). Bacteria treated with the DLLA formulation exhibited a 27% reduction in growth and those treated with P(SA:RA) a 60% reduction compared with the non treated bacteria.

Anti-biofilm effect

Crystal violet dye. Crystal violet was used to stain and measure biofilm mass so that inhibition of biofilm formation in the presence of C16-KGGK formulated with P(SA-RA) (Fig. 3A) or DLLA (Fig. 3B) or in the presence of the soluble tested AMPs (Fig. 3C) could be determined. A significant anti-biofilm effect was obtained with C16-KGGK using both formulations but not with the soluble C16-KGGK (Fig. 3A–B). The vehicle formulation itself does not possess anti-biofilm activity. From the other peptides tested in suspension, only C16-KKK showed anti-biofilm activity (Fig. 3C).

ATP bioluminescence assay. The level of ATP indicates the active metabolism of a cell. ATP levels in *E. faecalis* biofilms treated with soluble C16-KGGK were relatively low compared with that in the untreated control (*E. faecalis* alone) (see Fig. 4). The P(SA:RA) formulation (without C16-KGGK) also reduced bacterial viability. As opposed to P(SA:RA), DLLA had the reverse effect and the luminescence values were much higher than that of the positive control. These findings led to the question whether the luminescence values are derived from bacterial number, the metabolic status or both. In addition, a similar experiment was performed in which the biofilm was first grown for 48 hrs and then the tested materials were added to verify if C16-KGGK can affect an already constructed biofilm. The results were similar to those above (where the materials were added immediately after inoculating the bacteria). This may indicate that the formulation and the peptide (each) have an anti-metabolic effect even after the biofilm is formed.

Table 1. Anti *E. faecalis* MICs of the AMPs investigated.

AMP	Amino acid sequence	MIC [μg/ml]
hBD3	GIINTLQKYY CRVRGGRCAV LSCLPKEEQI GKCSTRGRKC CRRKK	>20
Amp-1D	LKLLKKLLKKLLKLL-NH_2	>25
C16-KGGK	$CH_3(CH_2)_{14}$CO–KGGK-NH_2	4–5
C16-KKK	$CH_3(CH_2)_{14}$CO–KKK-NH_2	6–12.5
C16-KAAK	$CH_3(CH_2)_{14}$CO–KAAK-NH_2	12.5–25
C16-KLLK	$CH_3(CH_2)_{14}$CO–KLLK-NH_2	6–12.5

Underlined amino acids are D-enantiomers.

Bacterial vitality. To test the vitality of the bacteria within the biofilm by a different, independent method, live/dead staining followed by confocal microscopic analysis was performed. The differences between the four tested groups are clearly evident for both P(SA:RA) and DLLA incorporated C16-KGGK (Fig. 5). Soluble C16-KGGK induced death in the biofilm bacteria (Fig. 5A–B, *E. faecalis* + KGGK). However, C16-KGGK in both formulations was more effective than the soluble peptide alone. The P(SA:RA) polymer had a strong inhibitory activity against biofilm formation as seen by the reduction in bacterial load.

Discussion

Antimicrobial peptides are one of nature's solutions to bacterial invasion. Their nonspecific mode of action, which is based on physical membrane disruption, is effective against various bacteria

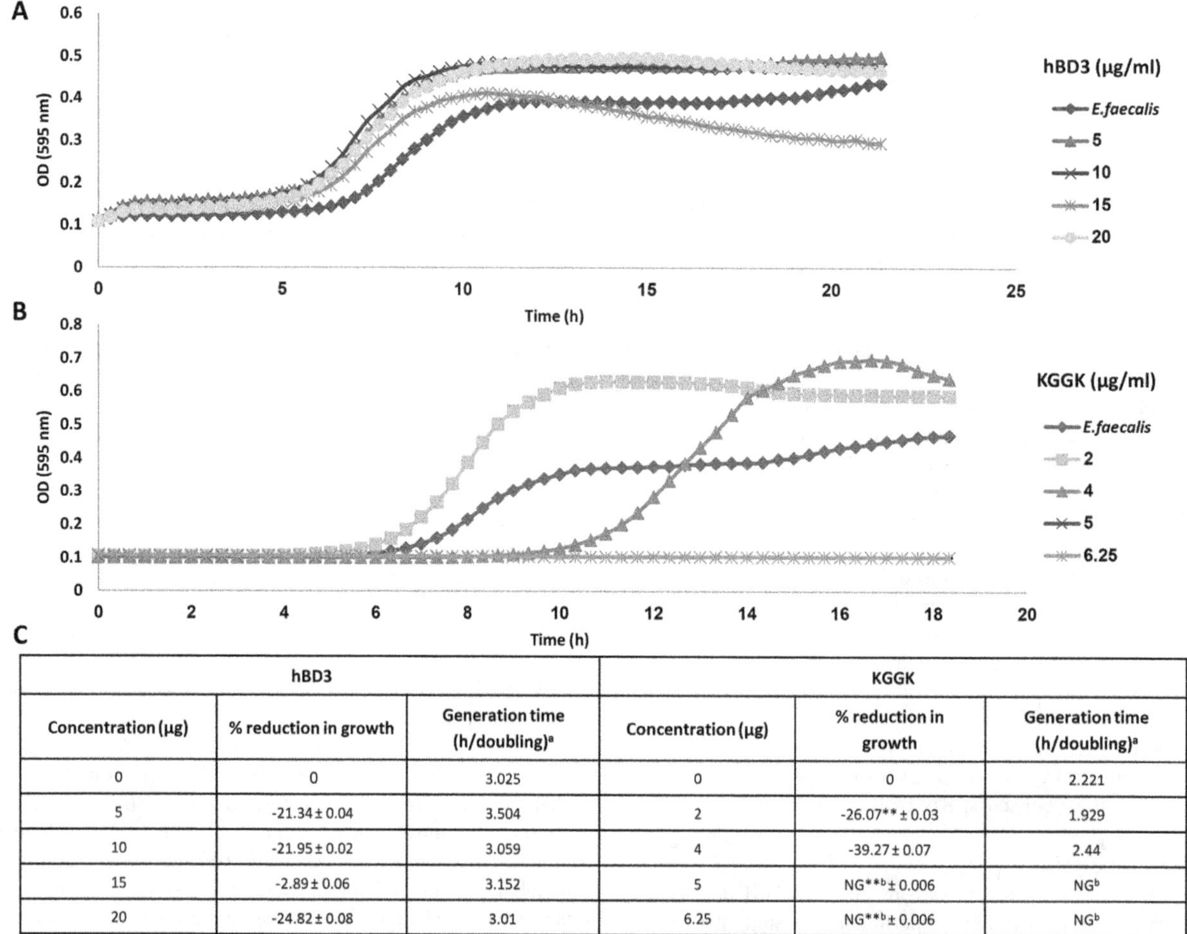

hBD3			KGGK		
Concentration (μg)	% reduction in growth	Generation time (h/doubling)[a]	Concentration (μg)	% reduction in growth	Generation time (h/doubling)[a]
0	0	3.025	0	0	2.221
5	-21.34 ± 0.04	3.504	2	-26.07** ± 0.03	1.929
10	-21.95 ± 0.02	3.059	4	-39.27 ± 0.07	2.44
15	-2.89 ± 0.06	3.152	5	NG**[b] ± 0.006	NG[b]
20	-24.82 ± 0.08	3.01	6.25	NG**[b] ± 0.006	NG[b]

[a] calculated at logarithmic phase at each tested condition ** $P<0.01$ compared with control without AMP, determined using Student's t-test [b] No growth

Figure 1. *E. faecalis* growth is inhibited by C16-KGGK but not by hBD3. Growth of *E. faecalis* was measured (see Materials and Methods) in the presence of increasing concentrations of hBD3 (A) or of the lipopeptide KGGK (B). Percent growth inhibition was calculated compared with that of untreated bacteria during the logarithmic phase of the non treated bacteria. Generation time was calculated from each curve using the section representing the exponential growth phase (C).

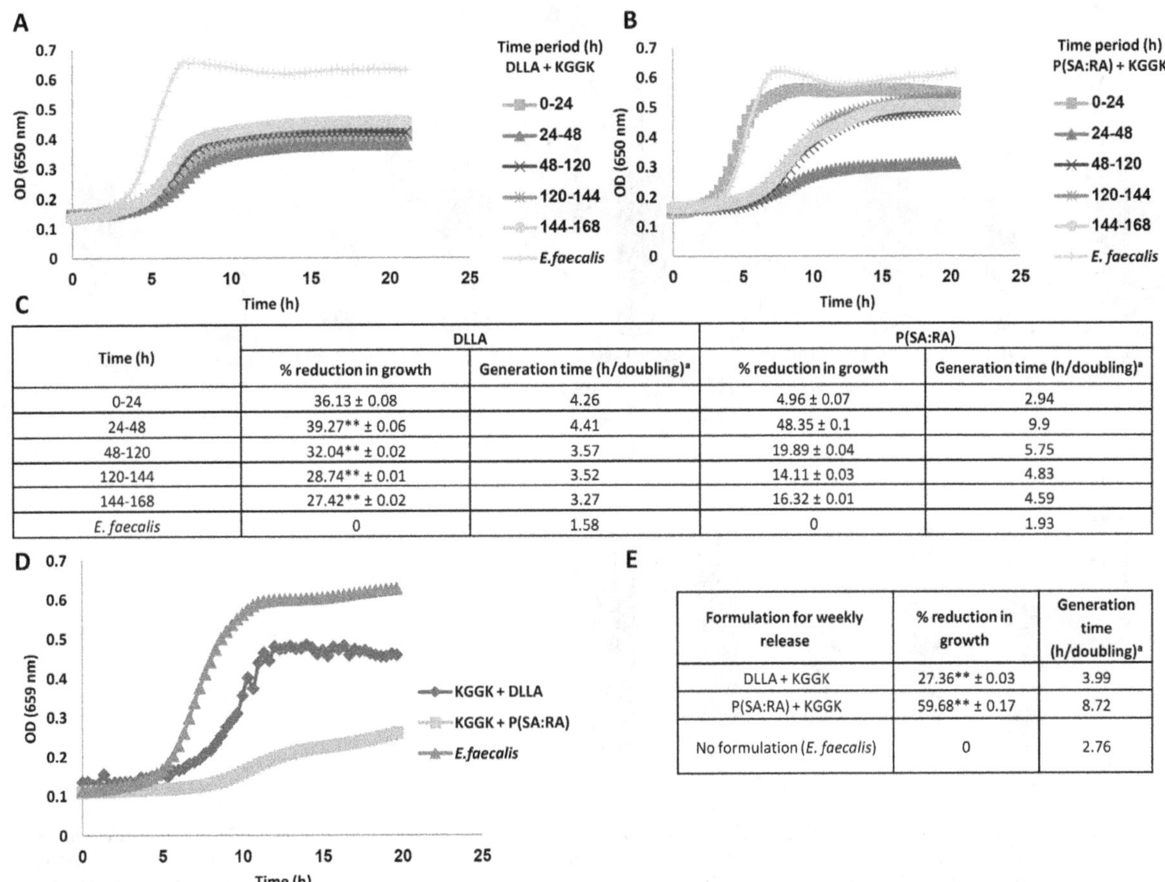

C

Time (h)	DLLA		P(SA:RA)	
	% reduction in growth	Generation time (h/doubling)[a]	% reduction in growth	Generation time (h/doubling)[a]
0-24	36.13 ± 0.08	4.26	4.96 ± 0.07	2.94
24-48	39.27** ± 0.06	4.41	48.35 ± 0.1	9.9
48-120	32.04** ± 0.02	3.57	19.89 ± 0.04	5.75
120-144	28.74** ± 0.01	3.52	14.11 ± 0.03	4.83
144-168	27.42** ± 0.02	3.27	16.32 ± 0.01	4.59
E. faecalis	0	1.58	0	1.93

E

Formulation for weekly release	% reduction in growth	Generation time (h/doubling)[a]
DLLA + KGGK	27.36** ± 0.03	3.99
P(SA:RA) + KGGK	59.68** ± 0.17	8.72
No formulation (*E. faecalis*)	0	2.76

[a] calculated at logarithmic phase at each tested condition ** P<0.01 compared with control without AMP, determined using Student's t-test

Figure 2. Growth inhibition of *E. faecalis* by KGGK released from P(SA:RA) or from DLLA. The side walls of 6 wells from line A of a 96 microwell plate were coated with the tested formulation (100 µg peptide+100 mg polymer, ratio 1:1000). Fresh medium was added to the first line of wells and was transferred every 24 hrs to a new line below for a week. Then the bacteria were added to the tested wells and the plate was incubated at 37°C in a VERSAmax microplate reader and OD_{650} in each well was followed automatically for 20 hrs. (A, C) KGGK+DLLA. (B, C) KGGK+ P(SA:RA) (D, E) weekly release of both formulations. Percent growth inhibition calculated compared with that of the non- treated bacteria during the logarithmic phase of the non treated bacteria. Generation time was calculated from each curve using the section representing the exponential growth phase (C, E).

and is less likely to induce bacterial resistance than antibiotics. Recently, synthetic AMPs mimicking these strategic antibacterial agents have been gaining interest. Combining sustained release and an antimicrobial compound holds many advantages and has proved itself in the past. In this study a potent antimicrobial agent was identified against *E. faecalis*, and then incorporated in two candidate biodegradable polymers. The most efficient of the six investigated antimicrobial peptides was the lipopeptide C16-KGGK.

The antibacterial effect in this study was tested against *E. faecalis*. We chose this bacterium as an example of a pathogen that causes severe nosocomial infections and as an example of a strongly forming biofilm bacterium. *E. faecalis* can grow and survive in a wide range of environments (wide range of temperatures and pH) affording it the ability to surmount many obstacles [9]. Interestingly, root canal treated teeth are about nine times more likely to harbor *E. faecalis* than are primary infections. *E. faecalis* has been found in root canal-treated teeth in 30% to 90% of the cases. This frustrating rate of post treatment disease is mainly attributed to the limitations of the present technology that offers no tool to combat intra-canal infection following the cleaning and shaping stage of the endodontic treatment [21].

The tested antibacterial peptides were first assayed in suspension against planktonic *E. faecalis*. Although hBD3 was previously reported as being a highly potent antibacterial AMP against *E. faecalis* [8,22–24], in the present study it showed an antibacterial effect against *E. faecalis* only when used at high concentrations. This may be due to the differences in *E. faecalis* strains and the hBD3 chemical synthesis. As hBD3 is a costly peptide, high concentrations are predestined to be irrelevant as a conventional therapeutic agent and thus were not tested further. Screening of the AMPs' MICs demonstrated that the C16-KGGK lipopeptide was the most potent against *E. faecalis* and it was further investigated and formulated into biodegradable polymers. Interestingly, in some experiments at low concentrations bacterial growth was not inhibited but rather accelerated. As this phenomenon may compromise the antimicrobial effect, further investigation of the peptides' mode of release is required. The exact mechanism of this opposite outcome is unknown, but the main assumption is that somehow the bacteria overcome lower concentrations of the lipopeptide and show accelerated growth compared with the untreated bacteria. This phenomenon needs to be considered when dealing with the amount of peptides that are released from the polymer. The new biohybrid medium incorpo-

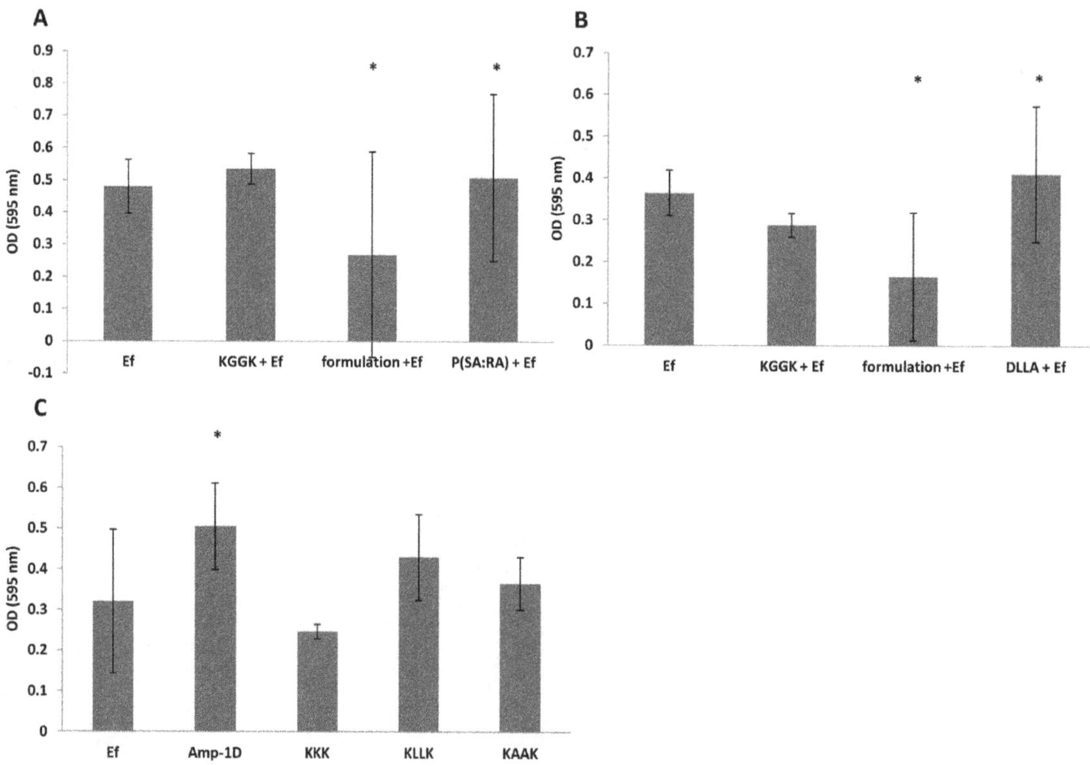

Figure 3. Effect of the antimicrobial peptides on the development of *E. faecalis* biofilms. *E. faecalis* biofilms were grown in 96 microtiter plate wells for 72 hrs in the presence of KGGK formulated with P(SA:RA) (panel A), or formulated with DLLA (B,) or with soluble peptides (C). *Ef* represents the non-treated bacteria, *KGGK+EF* - bacteria treated only with peptide, *formulation+EF* - bacteria treated with polymer and peptide and *Ef + polymer* - bacteria treated only with polymer control). The biofilm was stained with 1% crystal violet measured at OD 595 nm (see Materials and Methods). The optical density of the polymers alone without the bacteria was subtracted from the results of the biofilm that came in contact with the formulation and the polymer.

rating C16-KGGK results in an anti- *E. faecalis* effect when tested against planktonic bacteria. Indeed calculation of bacterial number (using a calibration curve) revealed that the final bacterial load was lower by one order of magnitude in the treated wells. Additionally, the slope of the curve representing the bacterial growth rate (generation time) was more moderate in the treated bacteria, showing that the peptide is released into the medium. The generation time of the bacteria treated with each of the

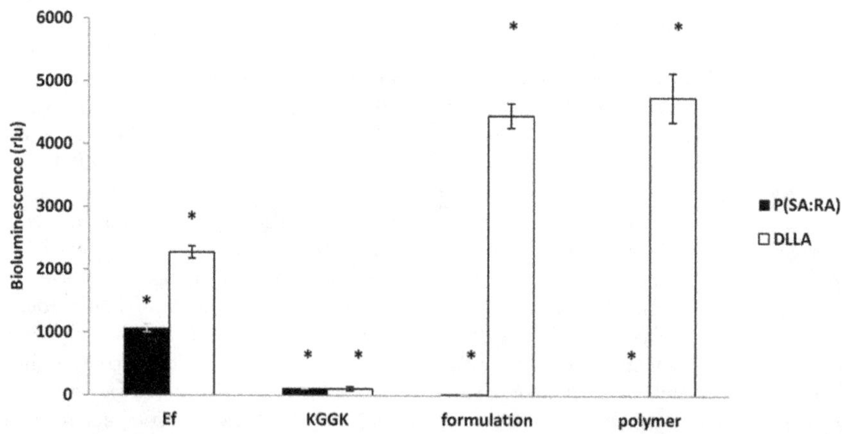

Figure 4. Effect of KGGK incorporated in biodegradable polymer on ATP in *E. faecalis* biofilm. Biofilm was exposed to the formulation for 72 hrs and ATP was measured as described in Materials and Methods. *Ef* represents the untreated bacteria as control, *KGGK* - bacteria treated only with peptide; *formulation* - bacteria treated with sustained release peptide; *polymer* - bacteria treated only with polymer as control.

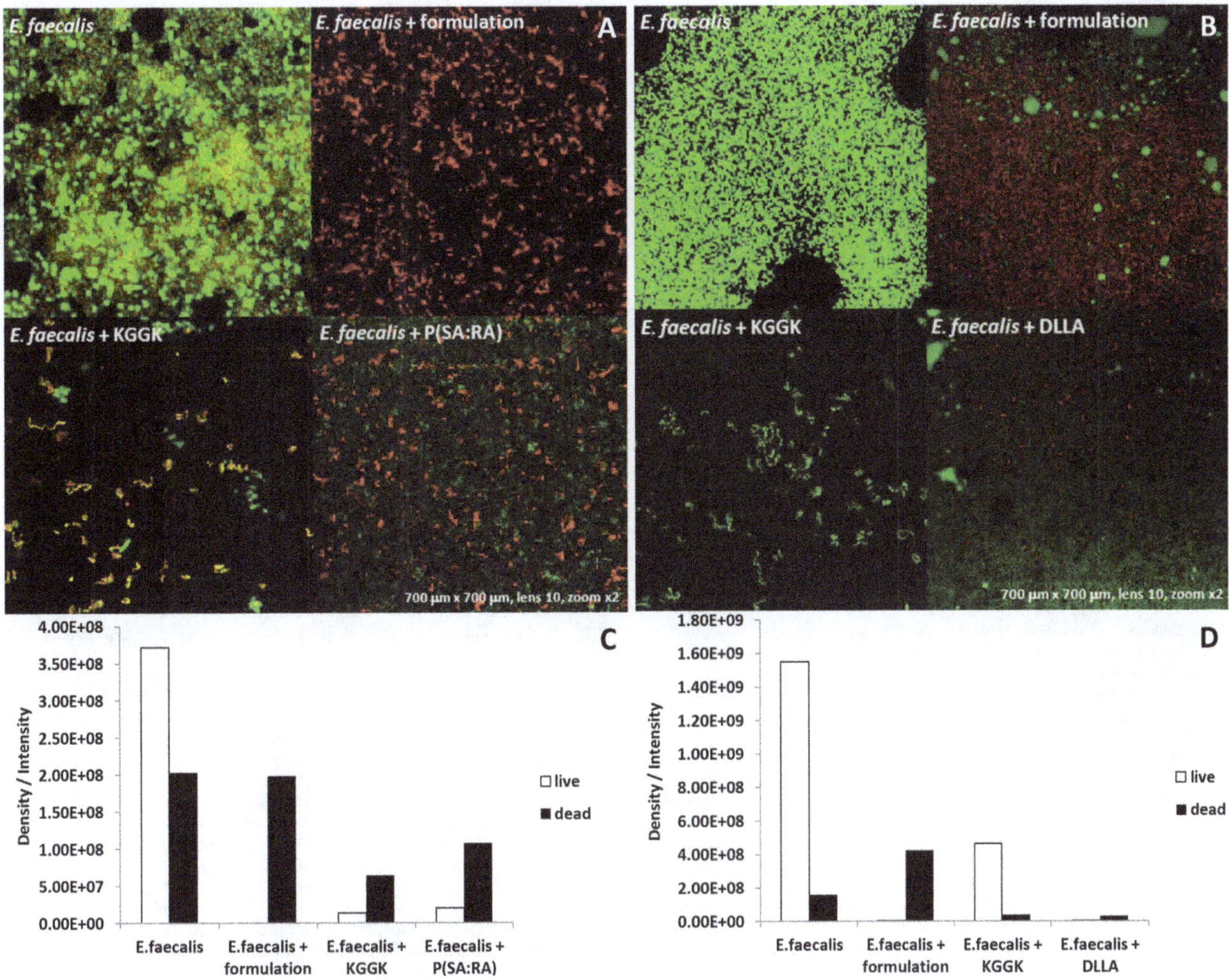

Figure 5. Live/dead assay. *E. faecalis* came in contact with the examined materials for 72 hrs to form biofilm. The medium was discarded and the wells were washed gently with PBS. The live bacteria were stained with green dye, the dead bacteria were stained with a red dye. A 5 ml volume of each dye from the dead/live dying kit was added to 450 µl PBS using an Eppendorf and 30 µl of the solution were added in each well. Images were taken using an Olympus confocal microscope [A, B]. The black column represents the dead bacteria, the white column represents the live bacteria. The biofilm was quantified by measuring the area occupied by the bacteria with the aid of Image Pro 4.5 software (Media Cybernetics) [C, D].

formulations and especially with P(SA:RA) was longer compared with that of the non-treated bacteria.

Bacteria grow naturally as biofilm, especially *E. faecalis* within the root canal. Moreover, *E. faecalis* is known to form biofilms that greatly increase its resistance to phagocytosis, antibodies and antimicrobials [21]. Therefore, in the second part of the study the anti-biofilm effect was tested. Three approaches were used to test the activity of the soluble AMPs and the new controlled release C16-KGGK lipopeptide formulations against *E. faecalis* biofilms. In the first, crystal violet was used to stain and measure biofilm mass. In the second, an ATP bioluminescence assay was performed and used as a viability indicator. In the third, the vitality of bacteria grown in a biofilm was tested using a dead/live stain. All three experiments revealed inhibition of biofilm formation when *E.* faecalis was exposed to the novel formulation. The three aspects examined were the amount of biofilm, its metabolic state and bacterial viability. Interestingly, the formulations were effective against a biofilm in the process of formation and against an established biofilm (mature biofilm). This is an important finding considering the fact that mature biofilm is much harder to treat because of its virulence factors. Moreover, we specifically tested the formulations' potency against ATCC v583 strain due to its high known resistance to several antibiotics (among them vancomycin), compared to other strains such as ATCC 29212 [25]. It can be suggested that a formulation that was shown to be active against ATCC v583 is likely to be potent against other *E. faecalis* strains.

The polymer candidates which contain fatty acids have several advantages over other biodegradable polymers such as: flexibility, low melting point, improved handling and provide better degradation and release profiles [26]. As previously reported, biodegradable polyanhydrides and polyesters are useful materials for controlled drug delivery. They have a hydrophobic backbone with hydrolytically labile anhydride and/or ester that may be hydrolyzed to dicarboxylic acids and hydroxy acid monomers when placed in an aqueous medium. Fatty acids are suitable candidates for the preparation of biodegradable polymers, as they are natural body components and hydrophobic, and thus may retain an encapsulated drug for longer time periods when used as drug carriers [11]. Moreover, it was shown that these polymers are biocompatible [15]. As described before, this polymer-peptide interaction may hold many advantages over the peptide by itself, such as improved solubility, reduced immunogenicity, increased stability against degradation and prolonged biological activity [27]. Two different polymers were tested as delivery media and led to different results in their activity and mode of action. In the sustained release experiments, DLLA showed similar bacterial kinetic growth curves whereas P(SA:RA) did not, indicating that the two have separate modes of release. Furthermore, in the ATP bioluminescence assay, DLLA presented higher levels of luminescence and accordingly higher levels of ATP, suggesting that this polymer elevates the metabolic state of the biofilm, compared with P(SA:RA) which had the opposite effect. In the live/dead assay, the main difference between the two polymers is that in P(SA:RA) a larger amount of dead bacteria appeared, reinforcing our previous findings that P(SA:RA) itself may be an antibacterial agent. Thus, P(SA:RA) is apparently a more suitable delivery medium for this purpose.

Within the sustained release field wider experiments should be performed in order to learn the exact amount of peptide released from the polymer and the kinetics of the release. Furthermore, the period tested in this study was one week so that additional experiments should be performed using various periods of time.

Conclusions

Synthetic AMPs were shown to have an effective antimicrobial activity against *E. faecalis*. A peptide that allows selective killing of *E. faecalis* would be a good candidate for endodontic treatment. Here, we show that a synthetic lipopetide can be highly effective against *E. faecalis*. Moreover, this lipopeptide when formulated in a biohybrid polymer medium has an increased antibiofilm effect. Thus, the novel effective formulation presented here can be advantageous in root canal treatment for the prevention of endodontic failure due to *E. faecalis*.

Author Contributions

Conceived and designed the experiments: LHE NB GB AS. Performed the experiments: LHE. Analyzed the data: LHE NB GB AS. Contributed reagents/materials/analysis tools: AJD EA YS NB GB AS. Contributed to the writing of the manuscript: LHE NB GB AS.

References

1. Blondelle SE, Perez-Paya E, Houghten RA (1996) Synthetic combinatorial libraries: novel discovery strategy for identification of antimicrobial agents. Antimicrob Agents Chemother 40: 1067–1071.
2. Boman HG (2003) Antibacterial peptides: basic facts and emerging concepts. J Intern Med 254: 197–215.
3. Reddy KV, Yedery RD, Aranha C (2004) Antimicrobial peptides: premises and promises. Int J Antimicrob Agents 24: 536–547.
4. Devine DA, Marsh PD, Percival RS, Rangarajan M, Curtis MA (1999) Modulation of antibacterial peptide activity by products of *Porphyromonas gingivalis* and *Prevotella* spp. Microbiology-Uk 145: 965–971.
5. Peschel A, Otto M, Jack RW, Kalbacher H, Jung G, et al. (1999) Inactivation of the dlt operon in *Staphylococcus aureus* confers sensitivity to defensins, protegrins, and other antimicrobial peptides. J Biol Chem 274: 8405–8410.
6. Shelton CL, Raffel FK, Beatty WL, Johnson SM, Mason KM (2011) Sap transporter mediated import and subsequent degradation of antimicrobial peptides in *Haemophilus*. PLoS Pathog 7: e1002360.
7. Nikaido H (1996) Multidrug efflux pumps of gram-negative bacteria. J Bacteriol 178: 5853–5859.
8. Joly S, Maze C, McCray PB Jr, Guthmiller JM (2004) Human beta-defensins 2 and 3 demonstrate strain-selective activity against oral microorganisms. J Clin Microbiol 42: 1024–1029.
9. Murray BE (1990) The life and times of the *Enterococcus*. Clin Microbiol Rev 3: 46–65.
10. Al-Tahami K, Singh J (2007) Smart polymer based delivery systems for peptides and proteins. Recent Pat Drug Deliv Formul 1: 65–71.
11. Shikanov A, Vaisman B, Krasko MY, Nyska A, Domb AJ (2004) Poly(sebacic acid-co-ricinoleic acid) biodegradable carrier for paclitaxel: In vitro release and in vivo toxicity. Journal of Biomedical Materials Research Part A 69A: 47–54.
12. Shikanov A, Domb AJ (2006) Poly(sebacic acid-co-ricinoleic acid) biodegradable injectable in situ gelling polymer. Biomacromolecules 7: 288–296.
13. Shikanov A, Ezra A, Domb AJ (2005) Poly(sebacic acid-co-ricinoleic acid) biodegradable carrier for paclitaxel-effect of additives. Journal of Controlled Release 105: 52–67.
14. Slager J, Tyler B, Shikanov A, Domb AJ, Shogen K, et al. (2009) Local Controlled Delivery of Anti-Neoplastic RNAse to the Brain. Pharmaceutical Research 26: 1838–1846.
15. Vaisman B, Motiei M, Nyska A, Domb AJ (2010) Biocompatibility and safety evaluation of a ricinoleic acid-based poly(ester-anhydride) copolymer after implantation in rats. Journal of Biomedical Materials Research Part A 92A: 419–431.
16. Makovitzki A, Avrahami D, Shai Y (2006) Ultrashort antibacterial and antifungal lipopeptides. Proc Natl Acad Sci U S A 103: 15997–16002.
17. Papo N, Oren Z, Pag U, Sahl HG, Shai Y (2002) The consequence of sequence alteration of an amphipathic alpha-helical antimicrobial peptide and its diastereomers. J Biol Chem 277: 33913–33921.
18. Krasko MY, Shikanov A, Ezra A, Domb AJ (2003) Poly(ester anhydride)s prepared by the insertion of ricinoleic acid into poly(sebacic acid). Journal of Polymer Science Part a-Polymer Chemistry 41: 1059–1069.
19. Soren L, Nilsson M, Nilsson LE (1995) Quantitation of Antibiotic Effects on Bacteria by Bioluminescence, Viable Counting and Quantal Analysis. Journal of Antimicrobial Chemotherapy 35: 669–674.
20. Beyth N, Yudovin-Farber I, Perez-Davidi M, Domb AJ, Weiss EI (2010) Polyethyleneimine nanoparticles incorporated into resin composite cause cell death and trigger biofilm stress in vivo. Proc Natl Acad Sci U S A 107: 22038–22043.
21. Stuart CH, Schwartz SA, Beeson TJ, Owatz CB (2006) *Enterococcus faecalis*: its role in root canal treatment failure and current concepts in retreatment. J Endod 32: 93–98.
22. Lee JK, Chang SW, Perinpanayagam H, Lim SM, Park YJ, et al. (2013) Antibacterial Efficacy of a Human beta-Defensin-3 Peptide on Multispecies Biofilms. Journal of Endodontics 39: 1625–1629.
23. Lee JK, Park YJ, Kum KY, Han SH, Chang SW, et al. (2013) Antimicrobial efficacy of a human -defensin-3 peptide using an *Enterococcus faecalis* dentine infection model. International Endodontic Journal 46: 406–412.
24. Lee SH, Baek DH (2012) Antibacterial and Neutralizing Effect of Human beta-Defensins on *Enterococcus faecalis* and *Enterococcus faecalis* Lipoteichoic Add. Journal of Endodontics 38: 351–356.
25. Swenson JM, Clark NC, Sahm DF, Ferraro ML, Doern G, et al. (1995) Molecular Characterization and Multilaboratory Evaluation of *Enterococcus-Faecalis* Atcc-51299 for Quality-Control of Screening-Tests for Vancomycin and High-Level Aminoglycoside Resistance in *Enterococci*. J Clin Microbiol 33: 3019–3021.
26. Jain JP, Sokolsky M, Kumar N, Domb AJ (2008) Fatty acid based biodegradable polymer. Polymer Reviews 48: 156–191.
27. Krishna OD, Kiick KL (2010) Protein- and peptide-modified synthetic polymeric biomaterials. Biopolymers 94: 32–48.

27

Antinematode Activity of Violacein and the Role of the Insulin/IGF-1 Pathway in Controlling Violacein Sensitivity in *Caenorhabditis elegans*

Francesco Ballestriero[1], Malak Daim[1], Anahit Penesyan[2], Jadranka Nappi[1], David Schleheck[3], Paolo Bazzicalupo[4], Elia Di Schiavi[4¤], Suhelen Egan[1]*

1 School of Biotechnology and Biomolecular Sciences and Centre for Marine Bio-Innovation, University of New South Wales, Sydney, New South Wales, Australia, **2** Department of Chemistry and Biomolecular Sciences, Macquarie University, Sydney, New South Wales, Australia, **3** Biology Department, University of Konstanz, Konstanz, Germany, **4** Institute of Genetics and Biophysics "Adriano Buzzati Traverso", National Research Council, Naples, Italy

Abstract

The purple pigment violacein is well known for its numerous biological activities including antibacterial, antiviral, antiprotozoan, and antitumor effects. In the current study we identify violacein as the antinematode agent produced by the marine bacterium *Microbulbifer* sp. D250, thereby extending the target range of this small molecule. Heterologous expression of the violacein biosynthetic pathway in *E. coli* and experiments using pure violacein demonstrated that this secondary metabolite facilitates bacterial accumulation in the nematode intestine, which is accompanied by tissue damage and apoptosis. Nematodes such as *Caenorhabditis elegans* utilise a well-defined innate immune system to defend against pathogens. Using *C. elegans* as a model we demonstrate the DAF-2/DAF-16 insulin/IGF-1 signalling (IIS) component of the innate immune pathway modulates sensitivity to violacein-mediated killing. Further analysis shows that resistance to violacein can occur due to a loss of DAF-2 function and/or an increased function of DAF-16 controlled genes involved in antimicrobial production (*spp-1*) and detoxification (*sod-3*). These data suggest that violacein is a novel candidate antinematode agent and that the IIS pathway is also involved in the defence against metabolites from non-pathogenic bacteria.

Editor: Raffi V. Aroian, UMASS Medical School, United States of America

Funding: These authors have no support or funding to report.

Competing Interests: The authors have declared that no competing interests exist.

* Email: s.egan@unsw.edu.au

¤ Current address: Institute of Bioscience and BioResources, National Research Council, Naples, Italy

Introduction

Parasitic nematodes are an important group of human, animal and plant pathogens representing a major threat not only to public health, but also to livestock and agricultural industries around the globe [1]. Yet heavy reliance on the few available chemotherapeutic agents has resulted in the development of nematode resistance and little progress has been made in the search for new treatments [2,3]. Thus there is an urgent requirement for the discovery of new antinematode compounds that can be developed into chemotherapeutic drugs.

The nematode *C. elegans* is a powerful model organism used broadly across the fields of cellular biology, developmental biology and neurobiology, and more recently also as a model organism for the study of host-microbial interactions with a focus on pathogenesis and drug discovery [4,5]. In a recent functional screen of genomic libraries of marine bacteria, a number of fosmid clones expressing high toxicity towards *C. elegans* were identified [6]. One of the highly toxic clones (designated 20G8) with a sequence-insert originating from the marine bacterium *Microbulbifer* sp. D250, expressed a violet pigment. Genetic analysis of the insert revealed that the clone 20G8 contained genes encoding for the synthesis of the indole-antibiotic violacein (*vioA-E*) [6], suggesting that this metabolite is responsible for its toxic phenotype.

Violacein is produced by several bacterial species thriving in a range of habitats such as terrestrial, marine, fresh water and glacier environments. Some of the known violacein producing organisms include *Chromobacterium violaceum* [7], *Collimonas* sp. [8], *Duganella* sp. [9], *Janthinobacterium lividum* [10,11] and *Pseudoalteromonas* spp. [12,13]. Violacein exhibits several biological activities with ecological relevance. Firstly, violacein has been suggested to be involved in oxidative stress resistance in *C. violaceum* [14]. Secondly, violacein producing bacteria (namely *J. lividum*) have been implicated in the natural defence of amphibians to fungal disease [15]. Finally, Matz and colleagues [12] have demonstrated that in the marine bacterium *Pseudoalteromonas tunicata*, violacein production can act as an antipredator defence mechanism against protozoan grazers. In addition to its ecological significance, violacein has also gained increasing importance for its potential medical and industrial applications. The biological activities of this compound include antioxidant, leishmanicidal, trypanocidal, antifungal, antiviral, antibacterial and antiprotozoal effects, as well as antitumoral and apoptosis-inducing activities in mammalian cancer cells (reviewed in [16]).

Although violacein has a broad range of activities, the direct molecular and cellular targets remain unknown. In addition there is limited understanding of its activity in multicellular eukaryotes and whether metazoans have the capacity to mount a defence to neutralize its activity or not. Whilst lacking adaptive immune strategies, nematodes, such as *C. elegans*, are suitable models to investigate metazoan defence as they posses a sophisticated innate immune system that protects against toxic microorganisms [17]. Key features of the nematodes defence include the conserved immune regulatory pathways, *i.e.*, the p38 mitogen activated protein kinase (MAPK) and the insulin/IGF-1 signalling (IIS) pathways [18]. In particular the IIS pathway with the gene regulators DAF-2 and DAF-16, is increasingly recognised for its important role in stress response, aging and immune homeostasis across nematodes, insects and mammals [18,19]. Furthermore, the IIS pathway is known to play a key role in the innate immune response against different pathogen-induced stresses, including colonization [17,20,21] and bacterial virulence factors [22].

In *C. elegans* the binding of insulin to DAF-2 triggers a phosphorylation cascade that results in activation of PDK-1 (3-phosphoinositide-dependent kinase 1) and eventual retention of the DAF-16 transcriptional activator in the cytoplasm [23]. De-activation or loss of DAF-2 function allows DAF-16 to move to the nucleus where it enhances the expression of genes including among others *sod-3* (superoxide dismutase), *spp-1* (SaPosin-like Protein) and *lys-7*, which are involved in detoxification, antimicrobial peptide expression and antimicrobial lysozyme production, respectively [24,25,26]. Moreover recent data indicates that the canonical IIS signalling diverges at PDK-1 into a second arm of the pathway mediated by the protein WWP-1 (WW domain protein 1) [22]. In the present study, we hypothesised that *C. elegans* makes use of this immune response pathway not only in the situation of infection by pathogenic organisms, but also to neutralize the effect of toxic bacterial secondary metabolites, such as violacein that originate from non-pathogens. Furthermore studying the mechanisms in which *C. elegans* mediates resistance to bacterial metabolites may shed further light into their molecular/cellular targets. To address this hypothesis, we first confirm that violacein is responsible for the toxic activity against *C. elegans* in clone 20G8 and its parental strain *Microbulbifer* sp. D250. We further show that the expression of enzymes that synthesize violacein in *E. coli* facilitates bacterial accumulation in the host intestine and induces apoptosis in the nematode. Finally we demonstrate that the IIS immune pathway modulates *C. elegans* sensitivity to violacein toxicity, most likely via the control of genes involved in detoxification and antimicrobial production.

Materials and Methods

Strains and culture conditions

All bacterial strains and vectors used in this study are listed in Table 1. Bacteria were grown in Luria broth (LB10), nematode growth medium (NGM) [27] or marine broth (Difco Laboratories, Maryland) [28] as indicated, and stored in 30% (v/v) glycerol at −80°C. Solid medium was prepared by the addition of 19 g of agar (Oxoid, Australia) per litre of culture fluid. All strains were grown at 25°C. Where required (see Table 1), chloramphenicol (12.5 µg/ml), kanamycin (100 µg/ml), and L-arabinose (0.02%, w/v) were added to the media. *C. elegans* strains (listed in Table 2) were maintained at 20°C on NGM agar plates spread with *E. coli* OP50 as a food source [29,30]. *C. elegans* strains were stored in glycerol (70:30 vol/vol) at −80°C [30].

Fosmid analysis and transposon mutant library screening

A transposon mutant library of the antinematode fosmid clone 20G8 was generated using an *in vitro* transposon mutagenesis kit (EZ-Tn5 insertion kit; Epicentre) following the manufacturers' instructions. The DNA fosmid sequence for clone 20G8 is available from the National Center for Biotechnology Information (NCBI) public database (GenBank) via accession number JX523957. The subsequent library of 96 *E. coli* transposon mutants was replicated on LB10 Omnitray plates (Nunc, Denmark), and screened for loss of toxic activity towards *C. elegans* as previously described [31]. Clones that were partially or totally grazed by the nematodes were chosen for further characterization in the nematode killing assay (below). The disrupted genes were identified by outward sequencing from the transposon using the KAN-2 forward and reverse primers (Epicentre) (KAN-2 Forward Primer 5' ACCTACAACAAAGCTCTCATCAACC 3', KAN-2 Reverse Primer 5' GCAATGTAACATCAGAGATTTTGAG 3') and sequences were subjected to BLAST analysis [32].

Purification and identification of violacein as the antinematode agent produced by *Microbulbifer* sp. D250

Violacein was purified from an overnight culture of *Microbulbifer* sp. D250, cells were collected by centrifugation and the cell pellets repeatedly extracted with 100% methanol. The resulting (pooled) crude extract was applied to a C18 solid phase extraction column (pre-packed C18 columns, 10 g, Alltech); after several washing steps with methanol:water (20 to 60% methanol), the violacein was eluted with 100% methanol. The violacein fraction from the solid phase extraction was further purified by preparative high performance liquid chromatography (HPLC), when using a Prodigy ODS3 column (Phenomenex, 150×4.6 mm, 5 µm particle size) and a methanol gradient from 0–100% methanol.

Liquid chromatography electrospray ionisation ion-trap mass spectrometry (LC-MS/MS) was employed to confirm that the purified purple pigment (see above) is violacein. Briefly, the LC-MS/MS was performed on a LCQ Deca SP Iontrap-MS/MS system (Thermo Finnigan). Up to 20 µl of the extract were loaded on a Nucleosil C18 column (125×3 mm, 5 µm particle size, Macherey-Nagel, Germany). The mobile phases used were (A) water acidified with 0.1% formic acid and (B) acetonitrile, at a flow rate of 0.2 ml/min. The gradient program was started at 20% B, and after 3 minutes, increased to 100% B over 13 min, and maintained at 100% B. For the MS conditions, the electrospray voltage was -5kV with a current of 12 µA; the sheath gas flow rate was 34l/min; the capillary was maintained at 275°C. For the MS/MS fragmentation, the mass width for isolation of precursor ions was 1.0Da, and the relative collision energy set at 40%. The MS chromatograms were recorded in the positive ion mode. The purified purple pigment eluted at 13.9 min, with an absorption scan corresponding to violacein as observed by HPLC-diode array detection (maxima at 260, 378 and 570 nm), and this peak in the MS exhibited a protonated molecule ($[M+H]^+$) corresponding to violacein (MW 343Da; mass of the observed $[M+H]^+$ ion, 344 Da); the MS/MS fragmentation pattern of this $[M+H]^+$ ion also corresponded to violacein, with ions observed at (% basepeak) 344 (64), 326 (45), 316 (100), 301 (49), 299 (24), 273 (5), 251 (20), 211 (4), 183 (2), 158 (1) and 132 (1) Da (e.g, due to loss of water, elimination of CO and nitrogen species, and cleavages at the rings).

Table 1. Bacterial strains and vectors used in this study.

Strain/Vector	Relevant characteristic or genotype	Source or reference
E. coli EPI300-T1[R]	F-*mcrA* Δ(*mrrhsd*RMS*mcr*BC) φ80d*lac*ZΔM15Δ*lac*X 74 *rec*A1 *end*A1 *ara*D139 Δ(*ara, leu*) 7697*gal*U *gal*K λ- *rps*L *nup*G *trf*A *ton*A *dhfr*	Epicentre
E. coli 20G8	Fosmid 20G8 cloned in *EPI300-T1*[R]; Cm[r]	This study
E. coli 20G8*vioA*[-]	Fosmid 20G8 mutated in *vioA* gene and cloned in *EPI300-T1*[R]; Cm[r], Kan[r]	This study
E. coli 20G8*vioB*[-]	Fosmid 20G8 mutated in *vioB* gene and cloned in *EPI300-T1*[R]; Cm[r], Kan[r]	This study
E. coli 20G8*vioC*[-]	Fosmid 20G8 mutated in *vioC* gene and cloned in *EPI300-T1*[R]; Cm[r], Kan[r]	This study
E. coli 20G8*vioD*[-]	Fosmid 20G8 mutated in *vioD* gene and cloned in *EPI300-T1*[R]; Cm[r], Kan[r]	This study
E. coli OP50	Uracil auxotroph	[29]
Microbulbifer sp. D250	Wild type strain	[62]
Microbulbifer sp. D250 dv2	D250 strain mutated in *vioB* gene; Kan[r]	This study
pLof/Tn10 KM	Mini-Tn10 (Kan[r]); Amp [r]	[34]
pCC1FOS[a]	Fosmid backbone for genomic library; Cm[r]	Epicentre

[a]Copy number inducible by arabinose.

Transposon mutagenesis of *Microbulbifer* sp. D250

To generate mutants of strain D250 that are unable to produce violacein, a random transposon mutagenesis was performed using the Tn-10-Km^R mini-transposon systems as described previously [33]. Briefly, a spontaneous streptomycin (Sm) resistant mutant of *Microbulbifer* sp. D250 was generated (D250-Sm^R) and used as the recipient. *E. coli* containing the Tn10 based delivery plasmid pLOF/Km, encoding a kanamycin (Km) resistance gene marker [34], was used as a transposon donor. Donor cells were conjugated with Sm resistant recipient cells of isolate D250 (D250-Sm^R) on filter discs in 1:1 ratio and incubated for 12 hours at 30°C. The conjugation mix was resuspended in marine broth and serial dilutions of this mixture were spread on, marine agar medium supplemented with 200 μg/ml Sm and 100 μg/ml Km. Transposon mutants were allowed to grow at room temperature for 48 to 72 hours and visually screened for the loss of purple pigmentation; the selected transposon mutants and wild type cells were extracted with methanol (as described above) and loss of pigmentation in the mutants confirmed by measuring the absence of absorbance at 575 nm. The DNA flanking the transposon insertion in the relevant mutants was sequenced using a "pan-handle" method as described previously [33].

Table 2. *C. elegans* strains used in this study.

Strain name	Genotype/allele designation	Relevant characteristics	Source or reference
N2 Bristol	*C. elegans* wild isolate	Wild type isolate	CGC[a]
CU1546	*smIs34*	ced-1p::ced-1::GFP + rol-6(su1006)	CGC[a]
CB1370	*daf-2(e1370)* III	Mutated in the insulin-like receptor DAF-2. Temperature sensitive dauer constitutive	CGC[a]
IU10	*daf-16(mgDf47)* I; *rrf-3(pk1426)* II	Mutated in the FOXO-family transcription factor DAF-16	CGC[a]
TJ356	*zIs356* IV	Integrated DAF-16::GFP roller strain. Daf-c, Rol, fluorescent DAF-16::GFP. Overexpression of DAF-16	CGC[a]
JT9609	*pdk-1 (Sa680)* x	Mutation in the gene encoding for 3-phosphoinositide-dependent protein kinase	CGC[a]
RB1178	*wwp-1(ok1102) I.*	Mutation in the gene encoding for the WW domain protein 1	CGC[ab]
TM127	*daf-2(e1370)* III; *sod-3(sj134)* X	Double mutant in the insulin-like receptor DAF-2 and in the superoxide dismutase SOD-3	CGC[a]
MQ876	*daf-2(e1370)* III; *lys-7(ok1384)* V	Double mutant in the insulin-like receptor DAF-2 and in the putative antimicrobial lysozyme LYS-7	CGC[a]
MQ513	*daf-2(e1370)* III; *spp-1(ok2703)* III	Double mutant in the insulin-like receptor DAF-2 and in the antimicrobial peptide caenopore SPP-1	CGC[a]
GA186	*sod-3(tm760)* X	Mutated in the iron/manganese superoxide dismutase SOD-3	CGC[a]
RB1286	*lys-7(ok1384)* V	Mutated in the putative antimicrobial lysozyme LYS-7	CGC[a]
RB2045	*spp-1(ok2703)* III	Mutated in the antimicrobial peptide caenopore SPP-1	CGC[a]

[a]*Caenorhabditis* Genetics Center, the University of Minnesota.
[b]*C. elegans* Gene Knockout Project http://www.celeganskoconsortium.omrf.org.

Nematode killing assay

E. coli clones were pre-grown overnight at 37°C in LB10, and 10 µl of the cultures were spread onto 3.5 cm diameter LB10 agar plates supplemented with selective antibiotic and L-arabinose as required, followed by incubation at 25°C for four days. L4-stage nematodes were added to the bacterial lawns (30 to 40 per plate), incubated at 20°C, and scored for live and dead nematodes every 24 hours for 20 days. In order to avoid multiple generations of nematodes on the same plate, which may have lead to errors when scoring, the nematodes where transferred to a fresh plate each day. A random non-toxic *E. coli* clone was used as a negative control under the same conditions. A nematode was considered dead when it failed to respond to touch. Since the *C. elegans daf-2* and *pdk-1* mutants are temperature sensitive (at 20°C 15% of the population enter in the resistant dauer stage), all the *daf-2* and *pdk-1* mutant assays were carried out at 15°C. Control *C. elegans* strains were also tested at 15°C in order to avoid temperature bias in nematode's life span. Each assay was carried out in independent triplicate plates.

Since marine agar does not support *C. elegans* growth due to its high osmolarity, the standard nematode killing assay had to be modified in order to assess the toxicity of the marine isolates (*i.e.*, strain D250 and violacein deficient mutant dV2). The marine bacteria were pre-grown overnight at 20°C in liquid marine broth, and 10 µL of the cultures were spread onto 2 cm paper filters (0.22 µm, Millipore, Maryland) and the filters placed on 3.5 cm marine agar plates supplemented with selective antibiotics as required, followed by incubation at 20°C for four days. On day four, the paper filters were removed from marine agar plates and placed onto 3.5 cm LB10 agar plates, to which the L4-stage nematodes were added (30 to 40 per plate). p values were calculated on the pooled data of all of the experiments done in each set by using the log- rank (Mantel–Cox) method [35,36] with the Prism software version 6.0c (GraphPad Software, La Jolla, CA, USA). A $p<0.05$ was considered significant.

Violacein dose response assay

In order to estimate the dose response of *C. elegans* towards violacein, a nematode killing assay was carried out using 96 well agar plates as previously described [31]. *E. coli* clone 20G8 mutated in *vioA* gene (hereafter referred to as 20G8*vioA*$^-$ mutant, see Table 1) was pre-grown overnight in LB10 at 37°C and inoculated (2 µL) to each LB10 agar well supplemented with selective antibiotic and L-arabinose as required. Plates were thereafter incubated at 25°C for four days to grow the bacterial lawns, and then each 5 µL of the purified violacein preparation (in methanol, see above) was added at various concentrations to the side of the bacterial lawn and the methanol was allowed to evaporate before the violacein was gently mixed with the lawn by the addition of 10 µl of sterile water. *E. coli* mutant clone 20G8*vioA*$^-$ alone, and pure methanol added to a 20G8*vioA*$^-$ lawn, were used as negative controls. Methanol was allowed to evaporate before L4-stage nematodes were added (5–15 per well). The lethal concentration at which 50% of the nematodes were killed (LC_{50}) was calculated by inference with the Prism software version 6.0c (GraphPad Software, La Jolla, CA, USA). The assays were carried out in independent triplicate plates. In order to determine if there was a significant difference in the survival of nematodes, a Students t-test was performed on the number of surviving nematodes at day seven. A $p<0.05$ was considered significant.

Heat killing of bacterial strains

Bacterial lawns pre-grown on agar plates were heat killed at 65°C for one hour, and the plates cooled to 20°C before adding nematodes. In order to confirm that heat killed bacteria were dead, samples of the bacterial lawns were streaked on LB10 agar plates before the nematode killing assay, and 20 days after the start of the incubation with nematodes. These plates were each incubated at 37°C for 24 to 48 hours and subsequently checked for an absence of bacterial colonies.

Microscopy

Overnight liquid cultures of each bacterial strain were spread on 3.5 cm LB10 agar plates and grown for four days. Nematode strains were exposed to bacteria for 24 hours and up to four days, depending on the assay, and placed on a microscope slide that had been immersed in 0.1 M sodium azide solution. Nematodes were examined under an Olympus DP70 digital camera system (Japan) with differential interference contrast (DIC) microscope optics. In order to detect the GFP signal indicative of an apoptosis induction, the CED-1::GFP reporter nematodes (strain CU546, Table 2) were exposed to a bacterial lawn of the 20G8 clone and 20G8*vioA*$^-$ and 20G8*vioC*$^-$ mutants for 24 hours. Nematodes were visualized under epifluorescence microscopy, and the percentage of nematodes with a GFP signal calculated for each treatment.

Results

Violacein is a toxic metabolite that mediates antagonistic interactions between the bacterium *Microbulbifer* sp. D250 and *C. elegans*

Screening of a random transposon mutant library of the fosmid clone 20G8 for the loss of toxic activity towards *C. elegans* resulted in four mutant clones (Tables 1 and 3 and Figure 1). The nematode killing assay demonstrated that the killing phenotype of the four mutant clones was significantly reduced ($p<0.0001$) when compared to the wild type clone 20G8 (Figure 1). The activity of mutant 20G8*VioA*$^-$ was similar to the negative control ($p = 0.803$) and was therefore considered non-toxic for *C. elegans*. Sequencing of the four non-active mutants revealed that in all cases transposons had inserted in open reading frames with high sequence identity to the *vioABCDE* gene cluster of *C. violaceum*, which has previously been shown to be involved in the synthesis of violacein [37] (Table 3). Specifically, two mutants that had lost the violet pigmentation, had insertions in the *vioA* and *vioB* genes, and the two clones that expressed a grey pigmentation typical of violacein precursors, were mutated in the *vioC* and *vioD* genes (Table 3 and Figure 1) [37].

To further support the involvement of violacein in nematode toxicity, *Microbulbifer* sp. D250, the parent organism for the fosmid clone 20G8, was shown to rapidly kill *C. elegans* ($p<$ 0.0001) when compared to negative control OP50 (Figure 2A). In contrast, the nematode's life span was significantly improved when exposed to a violacein deficient mutant dV2 ($p<0.0001$, Figure 2A). Sequencing of the transposon insertion site in strain dV2 supported the loss of violacein production, by demonstrating that the homologue to the violacein biosynthesis gene *vioB* (GenBank AFT64168) had been disrupted in the dV2 mutant. Notably, survival of the nematodes fed with the dV2 mutant strain was significantly reduced ($p<0.0001$) compared to the negative control using *E. coli* OP50 (see Figure 2A), which might indicate that other antinematode activities are also present in this bacterium. However, the involvement of violacein in toxicity towards *C. elegans* was further confirmed by the direct chemical

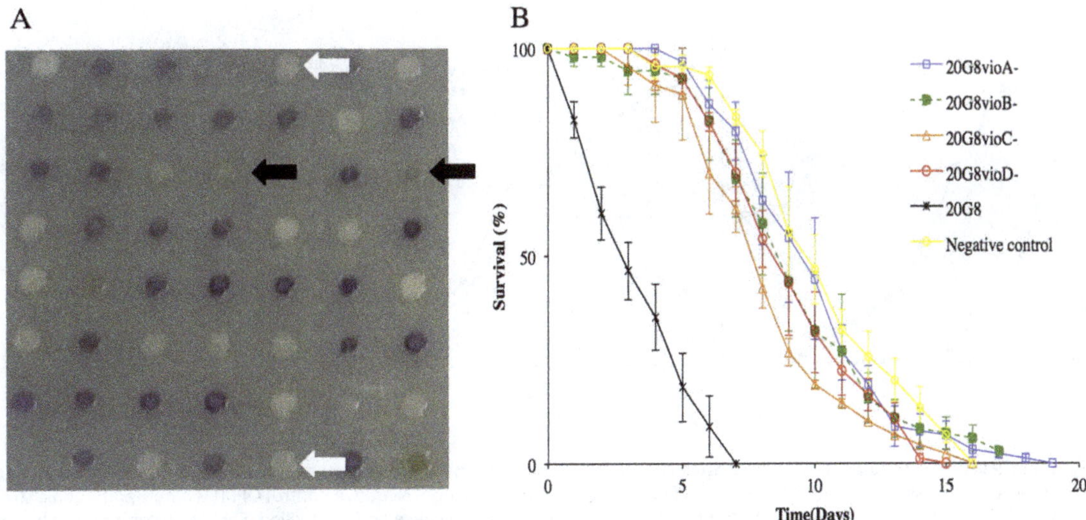

Figure 1. Characterization of the transposon mutant library of violacein producing clone 20G8. (A) Several mutants exhibit changes in violet pigmentation typical of the presence of violacein: 20G8*vioA*$^-$ and 20G8*vioB*$^-$ mutants have lost pigmentation (white arrows). 20G8*vioC*$^-$ and 20G8*vioD*$^-$ mutants express green-grey pigmentation (black arrows). (B) Nematode killing assay (N2 animals vs bacterial strains 20G8 clone and 20G8*vioABCD*$^-$ mutants). Survival kinetics of nematodes fed with either the 20G8 clone or the 20G8*vioABCD*$^-$ mutants deficient in violacein production. The killing phenotype of the four mutant clones is significantly reduced ($p<0.0001$) when compared to the wild type clone 20G8. A randomly chosen clone from the library with no activity is used as a negative control. Each data point represents means ± the standard error of three replicate plates. *p* values were calculated on the pooled data of all of the plates in each experiment by using the log-rank (Mantel–Cox) method.

identification of the purified violet pigment produced by *Microbulbifer* sp. D250 wild type strain. The pigment was extracted from cultures, purified and analysed by liquid chromatography-mass spectrometry (LC-MS/MS). The purified purple pigment eluted as one peak in the liquid chromatography, and this peak represented violacein, as was identified by the matching mass of the protonated molecule ($[M+H]^+$) and its characteristic MS/MS fragmentation pattern (see Material and Methods).

Finally, we used pure violacein directly in the toxicity assays (see Material and Methods). The 50% survival (LC_{50}) of nematodes exposed to violacein falls between the range of 7.5 µM and 75 µM of pure violacein (LC_{50} = 31.13 µM calculated from Figure 2B), when added to a viable bacterial lawn of the violacein non-producing mutant 20G8*vioA*$^-$ (Figure 2B). Interestingly, the survival of nematodes improved when violacein preparations were added to lawns of heat killed 20G8*vioA*$^-$ bacteria compared to violacein added to lawns of viable 20G8*vioA*$^-$ cells (*i.e.* $p = 0.001$

Table 3. Summary of the effects of violacein-producing clone 20G8 and its violacein deficient mutants on *C. elegans*.

Compound predicted to be expressed by the mutant[a]	**L-tryptophan**	**prodeoxyviolacein**	**proviolacein**	**violacein**
Proposed gene function involved in the biosynthesis[a,b]	*vioAB*: tryptophan 2-monooxygenase	*vioD*: hydroxylase	*vioC*: monooxygenases	N/A
Mutant/clone name and NCBI accession number	20G8*vioA*$^-$(AFT64169) 20G8*vioB*$^-$(AFT64168)	20G8*vioD*$^-$ (AFT64166)	20G8*vioC*$^-$ (AFT64167)	20G8 (JX523957)
BLASTp analysis[c]	VioA tryptophan 2-monooxygenase; VioB polyketide synthase	VioD hydroxylase	VioC monooxygenase	N/A
Nematode survival (*p* values)[d]	$p<0.0001$	$p<0.0001$	$p<0.0001$	N/A
Colonization[e]	0%	N/A	0%	76%±1.6
Apoptosis[f]	0%	N/A	0%	70.3%±5.7
Clone pigmentation	white	grey	grey	violet

[a]According to [37,45].
[b]Genes were ordered based on enzymatic activity in violacein biosynthesis and not on the locus position within the cluster [37].
[c]Only hits with 100% query coverage, 100% identity and E value of 0 were considered.
[d]*p* values of the survival of nematodes exposed to violacein mutants compared to the wild type clone 20G8 are reported.
[e]Percentage of alive nematodes colonized after four days.
[f]Percentage of CU1546 *C. elegans* strain displaying GFP signal indicative of apoptosis induction. All studies were carried out in *C. elegans* strain N2 except for apoptosis studies where CU1546 strain was employed. Each data represents means ± the standard error of three replicates. N/A = not applicable.

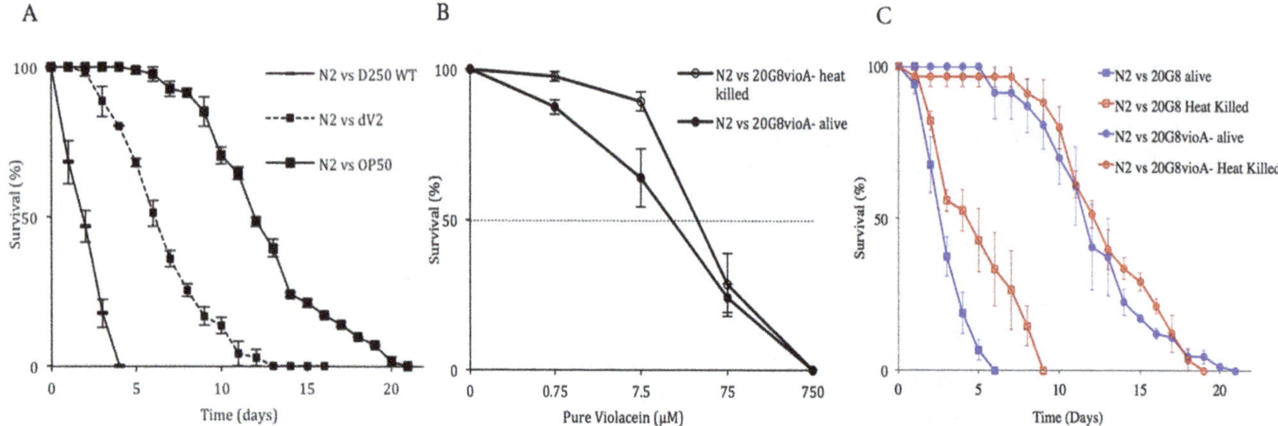

Figure 2. Nematode killing assay and dose response assay (N2 animal vs the alive and heat killed 20G8 clone and D250 strains). (A) Killing kinetics of *Microbulbifer* sp. D250 and dV2 mutant deficient in violacein production. Negative control OP50 is a non-pathogenic strain of *E. coli*. (B) Dose response of *C. elegans* to pure violacein added to 20G8*vioA*$^-$ mutant bacteria alive and heat killed. Each point on the graph represents the average survival of worms after seven days exposure to violacein (C) Kinetics of nematode killing when nematodes are fed either the live or heat killed 20G8 clone or 20G8*vioA*$^-$ mutant. Each data point represents means ± the standard error of three replicate plates. *p* values were calculated on the pooled data of all of the plates in each experiment by using the log-rank (Mantel–Cox) method.

and $p = 0.043$ for nematodes exposed to 0.75 μM and 7.5 μM of pure violacein, respectively).

Bacterial accumulation in the nematode intestine is involved in the killing activity of the violacein-producing *E. coli* clone

Given that the presence of live bacteria contributed to the decrease in the survival of nematodes in the presence of violacein (see above), we looked for evidence of bacterial accumulation in the nematodes intestine. The gut of 76% (n = 47) of the nematodes exposed to violacein-producing *E. coli* clone 20G8 had an accumulation of bacterial cells expressing a violet pigmentation in their intestine (Figure 3A). In contrast, nematodes exposed for four days to the violacein deficient mutants 20G8*vioA*$^-$ (n = 42) and 20G8*vioC*$^-$ (n = 57) showed no accumulation of bacteria in the intestinal lumen (Figure 3B and C). Microscopic analysis further showed tissue damage with enlargement of the intestinal lumen and enlargement of extracellular regions in nematodes fed with the 20G8 clone (arrowheads in Figure 3A). These data suggest that accumulation of bacteria in the intestinal lumen is one factor in the killing activity of clone 20G8. In order to further assess this, a nematode killing assay was performed using live and heat killed 20G8 cells. Heat inactivation of bacterial cells showed that nematode survival significantly increases ($p<0.0001$) in the presence of heat killed 20G8 cells compared to viable 20G8 bacteria (Figure 2C). In contrast, there was no significant effect ($p>0.05$) of heat killing on the survival of nematodes exposed to the violacein deficent mutant 20G8*vioA*$^-$ (Figure 2C). In order to confirm that bacterial cells were successfully inactivated by heat, they were streaked on LB10 agar plates and monitored for growth. Bacterial growth was not detected confirming that the heat treatment was sufficient to kill the bacteria. Together these data suggest that the toxic effect of violacein is significantly increased by the presence of live violacein-producing bacteria that accumulate in the intestine.

Violacein-producing bacteria induce apoptosis in *C. elegans*

To further elucidate the process of killing, transgenic nematodes with a GFP marker for apoptosis (CU1546 strain, Table 2) were exposed to a bacterial lawn of the violacein producing clone 20G8 and the violacein deficient mutant clones 20G8*vioA*$^-$ and 20G8*vioC*$^-$ for 24 hours, and thereafter visualised using differential interference contrast (DIC) and epifluorescence microscopy. The GFP signal was detected in somatic cells only in animals incubated with the 20G8 clone (70.3% of animals, n = 20), but not with 20G8*vioA*$^-$ and 20G8*vioC*$^-$ mutants (0%, n = 13 and n = 15 respectively) (Figure 4) showing that only the violacein-producing bacteria are capable of inducing apoptosis.

Mutations in the IIS pathway influence the sensitivity of *C. elegans* to violacein-producing bacteria and pure violacein

The IIS pathway with gene regulators DAF-2 and DAF-16 plays an important role in *C. elegans* pathogen defence [18]. We therefore questioned if *C. elegans* uses this pathway to also protect itself against violacein-mediated killing. To address this question, *C. elegans* loss of function mutants in the genes *daf-2* (CB1370), *pdk-1* (JT9609), *daf-16* (IU10) and *wwp-1* (RB1178), and a DAF-16 over-expressing strain (TJ356) (Table 2) were fed with the violacein-producing clone 20G8 in the nematode killing assay. The life span of *C. elegans* strains carrying loss of function mutations in *daf-2* and *pdk-1* and *wwp-1* were significantly increased compared to both wild type animals and *daf-16* mutant ($p<0.0001$) (Figure 5A, 5B and 5C). In contrast the *C. elegans* IU10 strain with loss of function mutation in the FOXO-family transcription factor DAF-16 displayed significantly reduced ($p<0.0005$) survival compared to wild type N2 animals (Figure 5A). Whereas viability of transgenic DAF-16::GFP nematodes that overexpress DAF-16 was significantly improved compared to both wild type and *daf-16* mutant animals ($p<0.0001$) (Figure 5A).

Similar results were observed for the *C. elegans* strains fed with the violacein deficient mutant (20G8*vioA*$^-$) supplemented with pure violacein in a dose response assay (Figure 5D). The toxic effect of violacein was dose dependent in wild type N2 nematodes and in *daf-2* and *daf-16* mutant animals (significant difference when nematodes of the same strain were exposed to 7.5, 75 or 750 μM of pure violacein, $p<0.05$). In these assays the lethal dose of violacein required to kill 50% of the population was higher for *daf-2* mutants (75 μM$<LC_{50}<$750 μM) than for the wild type

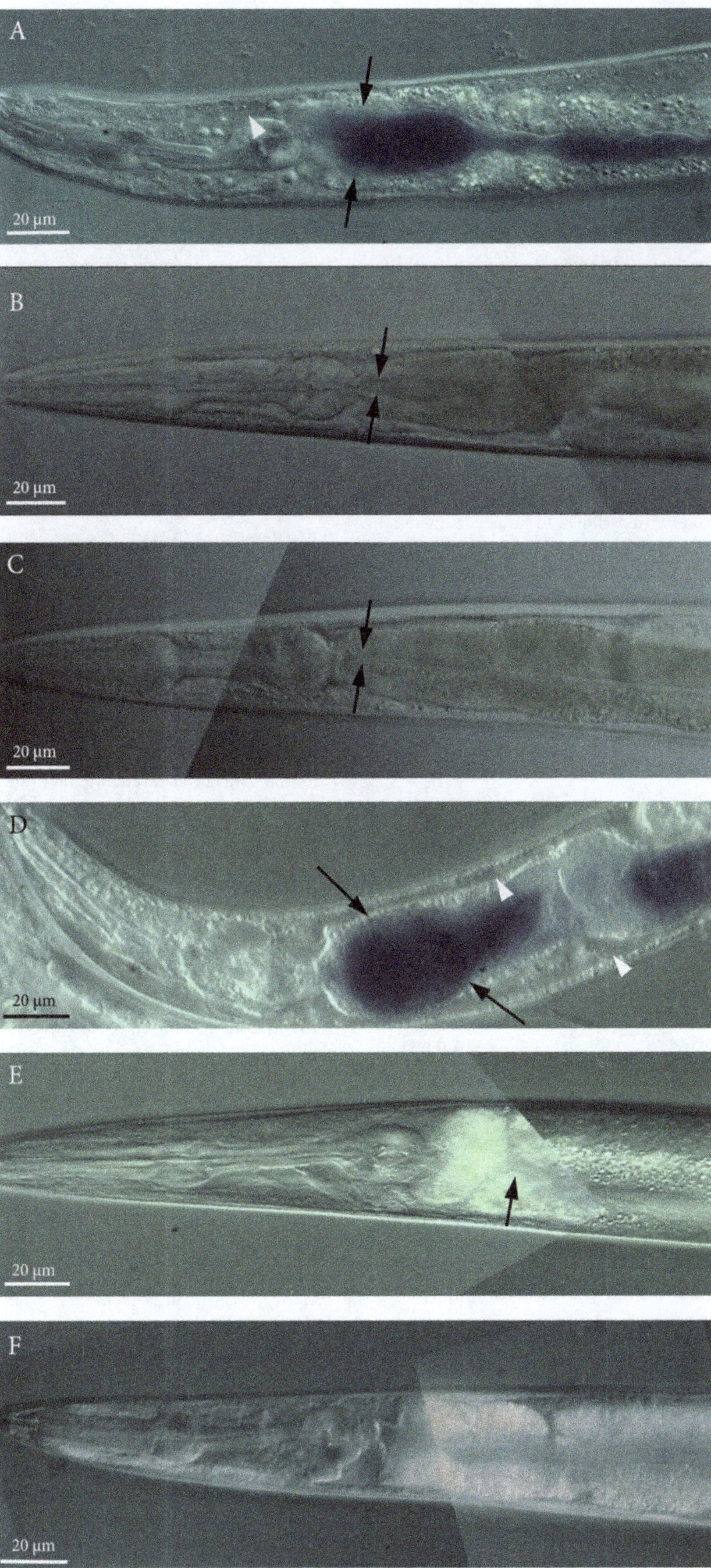
A
20 µm
B
20 µm
C
20 µm
D
20 µm
E
20 µm
F
20 µm

Figure 3. Visualization of bacterial accumulation in the nematode intestine by the *E. coli* clones (wild type animals, the *daf-2* and *daf-16* null mutant animals and the DAF-16 overexpressing nematodes vs the *E. coli* clone 20G8 and 20G8*vioA*⁻ and 20G8*vioC*⁻ mutant clones). All images present *C. elegans* anterior to the left and show the pharynx and first part of the intestinal lumen by differential interference contrast (DIC) microscopy. Accumulation by 20G8 cells in the nematodes intestine (arrows in panels A and D) and extensive enlargement of extracellular regions (white arrowheads panels A and D) in N2 wild type (panel A) and *daf-16* mutant (panel D) animals. No change in phenotype was observed in N2 nematodes fed with violacein deficient mutant clones 20G8*vioA*⁻ (negative control-panel B) and 20G8*vioC*⁻ (panel C). Similarly, no bacterial cells or enlargement of extracellular regions were detected in *daf-2* and DAF-16 over-expressing mutant nematodes exposed to the violacein producing clone 20G8 (arrows in panel E and F respectively). Each panel was assembled from multiple photomicrographs taken with the same magnification and same acquisition settings.

(7.5 μM<LC_{50}<75 μM) or the *daf-16* mutant animals (7.5 μM< LC_{50}<75 μM) (Figures 2B and 5D), further indicating that sensitivity or resistance to violacein is at least partially mediated by the IIS pathway.

C. elegans daf-2 mutant and DAF-16 overexpressing strains are resistant to intestinal accumulation by the violacein-producing clone 20G8

Given that bacterial intestinal accumulation is in part responsible for the toxic phenotype of violacein producing cells and that the IIS pathway has previously been shown to influence bacterial accumulation in the nematode intestine [21], we aimed to determine if elements of the IIS pathway mediate resistance to violacein by preventing bacterial accumulation.

Firstly, we compared the sensitivity of various *C. elegans* strains to violacein when fed viable or heat killed bacteria. The presence of viable bacterial cells significantly increased the sensitivity of *daf-16* mutant animals to violacein compared to when the mutant was exposed to heat killed bacteria (significant difference when pure violacein was added to viable or heat killed bacteria $p<0.05$, Figure 5D). In contrast, heat killed bacteria had no or little impact on the survival of *daf-2* mutant ($p>0.05$, Figure 5D).

Secondly, we assessed the ability of *E. coli* clone 20G8 to accumulate in the intestine of the various *C. elegans* strains. Ninety one percent of *daf-16* mutant nematodes (n = 35) showed evidence of bacterial accumulation when exposed to 20G8 cells, in contrast no or little bacterial accumulation was present in the *daf-2*-loss of function and DAF-16 over-expressing mutants under the same treatment (0% n = 45 and 1.1% n = 31, respectively). Enlargement of extracellular regions was present in wild type and *daf-16* mutant nematodes (90% n = 20 and 92% n = 25, respectively) fed with clone 20G8 (arrowheads in Figure 3 panel A and panel D respectively) while *daf-2*-loss of function and DAF-16 over-expressing mutant animals showed normal anatomical structures and no tissue injury was detected (Figure 3 panel E 0% n = 19 and Figure 3 panel F 0% n = 16, respectively). Together these data indicate that the resistance to violacein by *daf-2* null mutants and DAF-16 over-expressing *C. elegans* strains is, in part, due to their improved ability to prevent accumulation of bacteria in the presence of violacein.

Loss of function in IIS controlled genes results in increased sensitivity to violacein

In *C. elegans* DAF2/DAF16 controls the expression of various effector genes including those relevant for detoxification and antimicrobial activity such as the superoxidase dismutase gene *sod-3* and antimicrobial genes *spp-1* and *lys-7* [23,38]. Thus given that the precise molecular target/s for violacein in *C. elegans* are unknown we sought to determine which, if any, of these relevant downstream genes are required for the increased resistance to violacein observed in *daf-2* null and DAF-16 over-expressing strains. Specifically we chose to test violacein sensitivity in *C. elegans* mutants defective in *sod-3, spp-1* and *lys-7* (Table 2) because of the previous reported involvement of these genes in immunity to bacterial accumulation [39,40,41]. We found that *daf-2;spp-1* and *daf-2;sod-3* double mutants displayed significantly reduced survival compared to the single mutant *daf-2* ($p<0.0001$, Figure 6A) when exposed to the 20G8 clone in a nematode killing assay. No reduction in viability was detected in the *daf-2;lys-7* double mutant when compared to the single mutant *daf-2* ($p = 0.937$, Figure 6A). Interestingly a single mutation in gene *spp-1* significantly reduced the nematode's life span when compared to wild type animals ($p<0.0001$), while the viability of the nematode was not affected by mutations in the *lys-7* and *sod-3* genes ($p>0.05$, Figure 6B). These data indicate that resistance to violacein in *daf-2* mutants is at least in part driven by SPP-1 and SOD-3, with the antimicrobial LYS-7 having little or no involvement.

Discussion

Violacein has antinematode activity

In this study, we identified violacein as the metabolite responsible for the antinematode activity of *Microbulbifer* sp. D250. Violacein is arguably best known for its antibacterial properties and its activity as a potentially novel therapeutic against a range of tumors [16]. However, to the best of our knowledge, and with the exception of studies related to cancer therapy, this is the first report of violacein toxicity towards a multicellular eukaryote, thus adding to the list of biological functions for this natural metabolite.

Genetic analysis identified an operon of five conserved biosynthesis genes *vioA-E* that have been identified across all violacein-producing strains studied to date [7,12,42,43]. Interestingly, mutations in *vioC* or *vioD* result in a grey colony pigmentation, reminiscent of the accumulation of the violacein precursor pro-violacein [7,43,44]. Pro-violacein differs from violacein by the absence of one oxo-group in the C15 position (indolyl instead of indolone, see Table 3 and reference [7]). This minor chemical difference, however, has a substantial impact on the toxicity, as the grey-pigmented *vioC* mutant did not kill nematodes in our study. Similar observation have also been recently made for *E. coli* K12 strains producing either violacein or pro-violacein, which were and were not resistant, respectively, to protozoan predation [45].

C. elegans is intrinsically more resistant to the effects of violacein (LC_{50}>30 μM) than other bacterial grazers such as flagellates and amoebae, which were found to be effective at concentrations of 10 μM [46] and 1 μM [12] respectively. Nevertheless, violacein appears to be more potent towards *C. elegans* than toxins derived from known bacterial pathogens. For example, small phenazine molecules, including phenazine-1-carboxylic acid recently identified from *Pseudomonas aeruginosa*, are toxic to *C. elegans* only at concentrations greater than 70 μM. The effective concentration of violacein against *C. elegans* observed here is also similar to that of recent studies investigating novel antihelminthic therapies. For example, in a screen of existing drug leads Taylor *et al* [47]

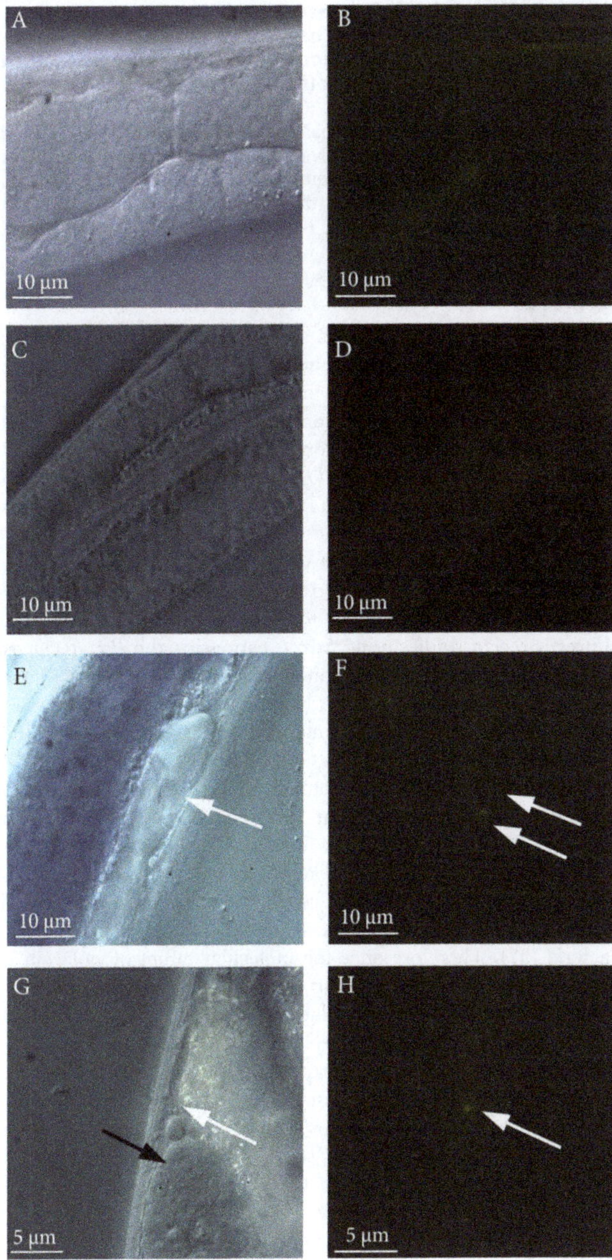

Figure 4. Apoptosis analysis in CED-1::GFP transgenic nematode strain CU1546. Visualization of GFP apoptosis marker inside the nematodes exposed to the violacein deficient mutant clones (20G8*vioA*$^-$ panels A and B; 20G8*vioC*$^-$ panels C and D) or violacein producing clone (20G8, panels E and F and G and H). Images show a section of the intestine and the gonads of *C. elegans* by DIC (panels A, C, E, G) and by epifluorescence microscopy (panels B, D, F, H). A ring-shaped GFP signal was detected in somatic cells of CU1546 nematodes exposed to 20G8 clone (two white arrows in panel F and white arrow in panel H) but not to 20G8*vioA*$^-$ and 20G8*vioC*$^-$ mutants (panel B and D). The ring-shaped GFP signal is associated with the expression of CED-1 receptor on the cell wall of engulfing cells only when phagocytosis of apoptotic cells is taking place. The ring-shaped GFP signal is apoptosis-specific and is visible in somatic cells of the nematodes (white arrows in E and G), the germ line (gonad) of the nematode is shown in panel G (black arrow). A non-specific fluorescence signal is visible in panels B and D.

identified 18 candidate compounds having detectable phenotypes against *C. elegans* with EC_{50} ranging between 0.7 µM and > 192 µM, including those already approved as cancer therapeutics such as Dasatinib (LC_{50} 22.3 µM) and Flavopiridol (LC_{50} 48.3 µM). Despite its many biological activities, the toxicity of violacein on (non-tumoral) mammalian cells is quite low. Recently, it has been reported that intraperitoneal doses of violacein of up to 1 mg kg^{-1} are not toxic to mouse blood, kidneys, or liver, thus enabling a potential *in vivo* use of violacein and its derivatives as a therapeutic compound with few side effects [48]. Given the relatively low effective concentration of violacein towards *C. elegans* determined here, we propose that the investigation of violacein as an antiparasitic compound is a reasonable prospect and that further studies regarding the activity of violacein against model parasitic nematodes, could reinforce the therapeutic potential of this drug.

Bacterial accumulation in the intestine of *C. elegans* is a key factor involved in the toxicity of violacein

We observed that in the presence of violacein the otherwise non-pathogenic *E. coli* has the ability to accumulate in the intestine and eventually kill *C. elegans*. The exact mechanism by which violacein treatment leads to bacterial accumulation and reduced nematode viability is yet to be determined, however recent reports have demonstrated a link between nematode longevity and intestinal colonization [21]. Specifically, Portal-Celhay *et al* [21] showed that the capacity to control bacterial accumulation in the gut was dependent on the immunological status and age of the individual animal. Heavy bacterial accumulation has also been shown to reduce the lifespan of the nematodes depending on the bacterial strain used [21,49]. Thus it is possible that exposure to violacein compromises the nematode's defence resulting in a reduced capacity to control bacteria in the gut and, thus, increasing the mortality rate. This is supported by similar observations recently made in various *Bacillus* species, in which treatment with the Bacillus pore-forming crystal protein (Cry PFP) seemingly sensitizes *C. elegans* to bacterial infection [50]. An alternative explanation is that the presence of violacein allows bacteria to penetrate the intestinal tissue resulting in a lethal infection. Whilst we did not observe bacteria within the tissue of nematodes exposed to violacein, previous reports have suggested internal infection as a possible cause of death in older nematodes [49,51] and so this alternative possibility should not be dismissed.

Violacein induces apoptosis in *C. elegans*

In addition to increased susceptibility to bacterial accumulation, violacein is capable of inducing apoptosis in *C. elegans* (Figure 4). Previous studies have revealed that apoptosis is also involved in violacein-mediated cell death in mammalian cell lines [16] and amoebae [12]. Thus although the exact molecular target of violacein in the eukaryotic cell is yet to be elucidated, the induction of an apoptosis-like cell death mechanism in multiple, distantly related eukaryotic systems (mammalian cells, amoeba and nematodes) suggest that an ancient, common eukaryotic cell process may be an additional target of violacein-driven toxicity.

The insulin/IGF-1 signalling pathway contributes to the native defence against violacein

The innate immune response, including the IIS pathway with regulators DAF-2 and DAF-16, is a key component of the nematodes first line of defence particularly against pathogens [18,19,52]. Here we add to that knowledge by demonstrating that the IIS pathway also contributes to the nematodes native defence

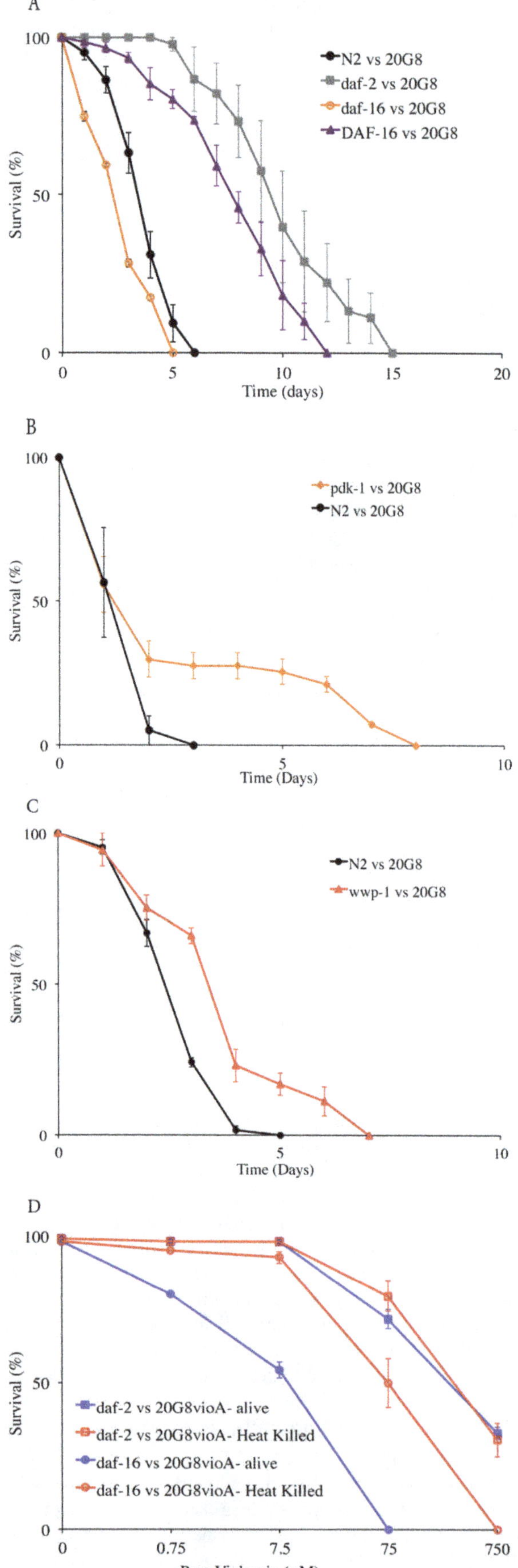

Figure 5. Nematode killing assay and dose response assay (wild type animals and *daf-2, daf-16, pdk-1, wwp-1* and DAF-16 mutant nematodes vs 20G8 clone). The survival of nematodes fed with 20G8 clone was measured for (A) *daf-2, daf-16* and DAF-16 mutant animals. (B) *pdk-1* mutant animals (C) *wwp-1* mutant animals (D) Dose response of *C. elegans daf-2* and *daf-16* strains to pure violacein added to 20G8*vioA*$^-$ mutant bacteria alive and heat killed. Each data point represents means ± the standard error of three replicate plates *p* values were calculated on the pooled data of all of the plates in each experiment by using a Students *t*-test on the number of surviving nematodes at day seven.

against secondary metabolites derived from non-pathogenic bacteria. The *C. elegans daf-2* and *pdk-1* deficient animals were significantly more resistant to the toxicity of violacein and the associated bacterial accumulation as compared to the wild type strain. These finding are consistent with the "long-lived" phenotype of the *C. elegans daf-2* mutant [53], which is also known to mediate the immune defence to bacterial infections [52,54].

Since the molecular target of violacein-mediated toxicity in *C. elegans* remains to be elucidated the mechanisms involved in the nematode immune response towards violacein is unknown. However recent studies using *E. coli* expressing the *Pseudomonas aeruginosa* translational inhibitor exotoxin A (ToxA), have demonstrated that *C. elegans* induces an immune response towards ToxA, which the nematode detects indirectly via the toxin-mediated damage [55]. Others have also demonstrated activation of immunity and detoxification genes in response to damage to a variety of cellular functions [56], many of which could result from exposure to bacterial toxins. Thus such an effector-triggered immunity is likely to be widespread in animals and may function to enable bacteriovorus organisms such as *C. elegans* to discriminate between commensal and pathogenic bacteria [57]. Therefore once the molecular target of violacein is established it will be of interest to determine if *C. elegans* responds directly to the presence of violacein or rather to the associated inhibition of, or damage to, specific cellular functions.

Identifying genes under DAF-2/DAF-16 control that are involved in the increased resistance to violacein may provide further insight into the molecular/cellular target of this compound. Indeed assessment of violacein sensitivity of selected *C. elegans daf-2* double mutant strains in the current study indicates that while the antimicrobial lysozyme (LYS-7) is not involved in violacein resistance, both the superoxide dismutase SOD-3 and the antimicrobial peptide SPP-1 seem to play a role in host defence against violacein. The potential role of SOD-3 is consistent with previous findings that violacein can cause oxidative stress in human cancer cell lines [58]. However other studies have indicated that violacein acts as an antioxidant for the producing bacteria [59]. This apparent contradiction may result from the same molecule producing different responses depending on the target cells. *Spp-1* is expressed in the nematode intestinal cells and has previously been shown to be involved in the immune response against pathogens *Salmonella typhimurium* [41] and *P. aeruginosa* [60]. Results from this study support the idea that SPP-1 could also control the violacein induced intestinal accumulation of *E. coli* in *C. elegans* and further suggest a wider target spectrum for this peptide.

Despite the large body of knowledge surrounding the IIS pathway of the innate immune response in *C. elegans* and its role in stress resistance and pathogenesis, there is a paucity of information regarding its role in response to bacterial toxins. In one of the few studies to date, Chen *et al.* [22] demonstrated that

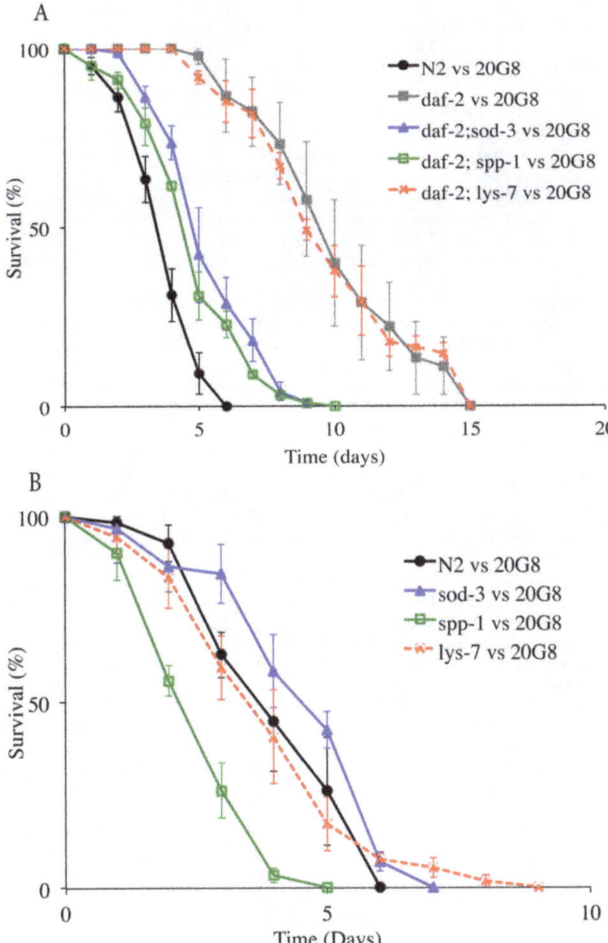

Figure 6. Nematode killing assay (wild type animals and *daf-2, daf-2;spp-1, daf-2;lys-7, daf-2;sod-3, spp-1, lys-7, sod-3* mutant nematodes vs the 20G8 clone). (A) The survival of the nematode was tested using *C. elegans* double mutants *daf-2;sod-3, daf-2;spp-1, daf-2;lys-7* and (B) the single mutant animals *sod-3, spp-1,* and *lys-7.* Each data point represents means ± the standard error of three replicate plates. *p* values were calculated on the pooled data of all of the plates done in each experiment by using the log-rank (Mantel–Cox) method and the values are provide in the text.

the IIS pathway is involved in the cellular defence of *C. elegans* against *Bacillus thuringensis* crystal pore-forming toxins (Cry PFT). It appears likely that *C. elegans* uses this conserved pathway as a general bacterial toxin defence mechanism, as, similar to our findings for violacein, reduction in DAF-2 signalling confers resistance to Cry PFT, which is at least in part, dependent on DAF-16 function. However the newly described second branch of the IIS pathway mediated by WWP-1 appears to play a different role in defence against violacein, as in contrast to the hypersensitivity towards Cry PFT [22], *wwp-1* mutant animals were more resistant to violacein than wild type nematodes (Figure 5C). These contrasting observations may speak to differences in how the two toxins affect target cells and will be the subject of future studies investigating the cellular response of *C. elegans* to bacterial toxins.

Conclusion

In their natural habitat nematodes, including *C. elegans*, are major predators for bacteria that, in turn, have evolved a number of predatory defence mechanism including the production of inhibitory metabolites [61]. This work presents, for the first time, violacein as an antinematode compound and assesses toxicity and the cellular response to violacein exposure in a *C. elegans* animal model. We found that this small molecule facilitates intestinal accumulation of the *E. coli* host cells and stimulates apoptotic activity. Mutations in a number of genes in the IIS pathway (*daf-2 and pdk-1*) and over-expression of another (*daf-16*) can confer significant resistance to violacein toxicity, providing evidence that *C. elegans* uses the DAF-2/DAF-16 innate immune signalling pathway to defend itself against this and potentially other noxious bacterial metabolites. Finally we demonstrate that defence against violacein requires, at least in part, the function of the antimicrobial peptide SPP-1 and possibly superoxide dismutase SOD-3. It is expected that the added knowledge will assist in the design of future mutational and/or gene expression studies aimed at determining the mechanisms of violacein toxicity in the nematode model including its direct molecular target/s. Such studies will not only prove important for further understanding the molecular basis for microbial antagonistic interactions but also for the development of microbial secondary metabolites such as violacein as novel chemotherapeutics.

Acknowledgments

We thank the IGB Open Laboratory for facilities and resources, the *C. elegans* gene knockout project and the *Caenorhabditis* Genetics Center (CGC) at the University of Minnesota for provision of *C. elegans* samples. We further thank Silvana Castro for technical assistance and A/Prof. Torsten Thomas for revision of the draft manuscript.

Author Contributions

Conceived and designed the experiments: FB PB EDS SE. Performed the experiments: FB MD AP JN DS. Analyzed the data: FB DS PB EDS SE. Contributed reagents/materials/analysis tools: PB EDS SE. Wrote the paper: FB AP DS PB EDS SE.

References

1. Holden-Dye L, Walker RJ (2007) Anthelmintic drugs. WormBook 2: 1–13.
2. Kaplan RM (2004) Drug resistance in nematodes of veterinary importance: a status report. Trends Parasitol 20: 477–481.
3. Prichard RK, Geary TG (2008) Drug discovery: Fresh hope to can the worms. Nature 452: 157–158.
4. Jones AK, Buckingham SD, Sattelle DB (2005) Chemistry-to-gene screens in *Caenorhabditis elegans*. Nat Rev Drug Discov 4: 321–330.
5. Sifri CD, Begun J, Ausubel FM (2005) The worm has turned-microbial virulence modeled in *Caenorhabditis elegans*. Trends Microbiol 13: 119–127.
6. Penesyan A, Ballestriero F, Daim M, Kjelleberg S, Thomas T, et al. (2012) Assessing the effectiveness of functional genetic screens for the identification of bioactive metabolites. Mar Drugs 11: 40–49. doi: 10.3390/md11010040.
7. Hoshino T (2011) Violacein and related tryptophan metabolites produced by *Chromobacterium violaceum*: biosynthetic mechanism and pathway for construction of violacein core. Appl Microbiol Biotechnol 91: 1463–1475.
8. Hakvag S, Fjarvik E, Klinkenberg G, Borgos S, Josefsen K, et al. (2009) Violacein producing *Collimonas* sp. from the sea surface microlayer of costal waters in trondelag, Norway. Mar Drugs 7: 576–588.
9. Aranda S, Montes-Borrego M, Landa BB (2011) Purple-pigmented violacein-producing *Duganella* spp. inhabit the rhizosphere of wild and cultivated olives in southern Spain. Microb Ecol 62: 446–459.
10. Pantanella F, Berlutti F, Passariello C, Sarli S, Morea C, et al. (2007) Violacein and biofilm production in *Janthinobacterium lividum*. J Appl Microbiol 102: 992–999.

11. Lu Y, Wang L, Xue Y, Zhang C, Xing X-H, et al. (2009) Production of violet pigment by a newly isolated psychrotrophic bacterium from a glacier in Xinjiang, China. Biochem Eng J 43: 135–141.
12. Matz C, Webb JS, Schupp PJ, Phang SY, Penesyan A, et al. (2008) Marine biofilm bacteria evade eukaryotic predation by targeted chemical defense. PLoS One 3: e2744.
13. Yang LH, Xiong H, Lee OO, Qi SH, Qian PY (2007) Effect of agitation on violacein production in *Pseudoalteromonas luteoviolacea* isolated from a marine sponge. Lett Appl Microbiol 44: 625–630.
14. Konzen M, De Marco D, Cordova CA, Vieira TO, Antonio RV, et al. (2006) Antioxidant properties of violacein: possible relation on its biological function. Bioorg Med Chem 14: 8307–8313.
15. Becker MH, Brucker RM, Schwantes CR, Harris RN, Minbiole KP (2009) The bacterially produced metabolite violacein is associated with survival of amphibians infected with a lethal fungus. Appl Environ Microbiol 75: 6635–6638.
16. Durán M, Ponezi A, Faljoni-Alario A, Teixeira MS, Justo G, et al. (2012) Potential applications of violacein: a microbial pigment. Med Chem Res 21: 1524–1532.
17. Marsh EK, May RC (2012) *Caenorhabditis elegans*, a model organism for investigating immunity. Appl Environ Microbiol 78: 2075–2081.
18. Tan MW, Shapira M (2011) Genetic and molecular analysis of nematode-microbe interactions. Cell Microbiol 13: 497–507.
19. Yanase S, Yasuda K, Ishii N (2002) Adaptive responses to oxidative damage in three mutants of *Caenorhabditis elegans* (age-1, mev-1 and daf-16) that affect life span. Mech Ageing Dev 123: 1579–1587.
20. Zhang X, Zhang Y (2009) Neural-immune communication in *Caenorhabditis elegans*. Cell Host Microbe 5: 425–429.
21. Portal-Celhay C, Bradley ER, Blaser MJ (2012) Control of intestinal bacterial proliferation in regulation of lifespan in *Caenorhabditis elegans*. BMC Microbiol 12: 49.
22. Chen CS, Bellier A, Kao CY, Yang YL, Chen HD, et al. (2010) WWP-1 is a novel modulator of the DAF-2 insulin-like signaling network involved in pore-forming toxin cellular defenses in *Caenorhabditis elegans*. PLoS One 5: e9494. doi: 9410.1371/journal.pone.0009494.
23. Mukhopadhyay A, Oh SW, Tissenbaum HA (2006) Worming pathways to and from DAF-16/FOXO. Exp Gerontol 41: 928–934.
24. Back P, Matthijssens F, Vlaeminck C, Braeckman BP, Vanfleteren JR (2010) Effects of sod gene overexpression and deletion mutation on the expression profiles of reporter genes of major detoxification pathways in *Caenorhabditis elegans*. Exp Gerontol 45: 603–610.
25. Durai S, Karutha Pandian S, Balamurugan K (2011) Changes in *Caenorhabditis elegans* exposed to *Vibrio parahaemolyticus*. J Microbiol Biotechnol 21: 1026–1035.
26. Roeder T, Stanisak M, Gelhaus C, Bruchhaus I, Grotzinger J, et al. (2010) Caenopores are antimicrobial peptides in the nematode *Caenorhabditis elegans* instrumental in nutrition and immunity. Dev Comp Immunol 34: 203–209.
27. Sulston J, Hodgkin J (1988) The nematode *Caenorhabditis elegans*. New York: Cold Spring Harbor Lab Press.
28. Marden P, Tunlid A, Malmcrona-Friberg K, Odham G, Kjelleberg S (1985) Physiological and morphological changes during short term starvation of marine bacteria isolates. Arch Microbiol 142: 326–332.
29. Brenner S (1974) The genetics of *Caenorhabditis elegans*. Genetics 77: 71–94.
30. Stiernagle T (2006) Maintenance of *C. elegans*. WormBook 1:11101895/wormbook11011. Available: http://www.wormbook.org/chapters/www_strainmaintain/strainmaintain.html. Accessed 2014 Sep 18.
31. Ballestriero F, Thomas T, Burke C, Egan S, Kjelleberg S (2010) Identification of compounds with bioactivity against the nematode *Caenorhabditis elegans* by a screen based on the functional genomics of the marine bacterium *Pseudoalteromonas tunicata* D2. Appl Environ Microbiol 76: 5710–5717.
32. Altschul SF, Gish W, Miller W, Myers EW, Lipman DJ (1990) Basic local alignment search tool. J Mol Biol 215: 403–410.
33. Egan S, James S, Holmström C, Kjelleberg S (2002) Correlation between pigmentation and antifouling compounds produced by *Pseudoalteromonas tunicata*. Environ Microbiol 4: 433–442.
34. Herrero M, Delorenzo V, Timmis K (1990) Transposon vectors containing non-antibiotic resistance selection markers for cloning and stable chromosomal insertion of foreign genes in gram-negative bacteria. J Bacteriol 172: 6557–6567.
35. Harrington D (2005) Linear Rank Tests in Survival Analysis. Encyclopedia of Biostatistics. Hoboken, NJ: John Wiley & Sons, Ltd.
36. Mantel N (1966) Evaluation of survival data and two new rank order statistics arising in its consideration. Cancer Chemo Rep Part 1 50: 163–170.
37. Balibar CJ, Walsh CT (2006) In vitro biosynthesis of violacein from L-tryptophan by the enzymes VioA-E from *Chromobacterium violaceum*. Biochemistry 45: 15444–15457.
38. Murphy CT, McCarroll SA, Bargmann CI, Fraser A, Kamath RS, et al. (2003) Genes that act downstream of DAF-16 to influence the lifespan of *Caenorhabditis elegans*. Nature 424: 277–283.
39. Mallo GV, Kurz CL, Couillault C, Pujol N, Granjeaud S, et al. (2002) Inducible antibacterial defense system in *C. elegans*. Curr Biol 12: 1209–1214.
40. Chavez V, Mohri-Shiomi A, Maadani A, Vega LA, Garsin DA (2007) Oxidative stress enzymes are required for DAF-16-mediated immunity due to generation of reactive oxygen species by *Caenorhabditis elegans*. Genetics 176: 1567–1577.
41. Alegado RA, Tan MW (2008) Resistance to antimicrobial peptides contributes to persistence of *Salmonella typhimurium* in the *C. elegans* intestine. Cell Microbiol 10: 1259–1273.
42. August PR, Grossman TH, Minor C, Draper MP, MacNeil IA, et al. (2000) Sequence analysis and functional characterization of the violacein biosynthetic pathway from *Chromobacterium violaceum*. J Mol Microbiol Biotechnol 2: 513–519.
43. Zhang X, Enomoto K (2011) Characterization of a gene cluster and its putative promoter region for violacein biosynthesis in *Pseudoalteromonas* sp. 520P1. Appl Microbiol Biotechnol 90: 1963–1971.
44. Jiang PX, Wang HS, Xiao S, Fang MY, Zhang RP, et al. (2012) Pathway redesign for deoxyviolacein biosynthesis in *Citrobacter freundii* and characterization of this pigment. Appl Microbiol Biotechnol 94: 1521–1532.
45. Ahmetagic A, Philip DS, Sarovich DS, Kluver DW, Pemberton JM (2011) Plasmid encoded antibiotics inhibit protozoan predation of *Escherichia coli* K12. Plasmid 66: 152–158.
46. Matz C, Deines P, Boenigk J, Arndt H, Eberl L, et al. (2004) Impact of violacein-producing bacteria on survival and feeding of bacterivorous nanoflagellates. Appl Environ Microbiol 70: 1593–1599.
47. Taylor CM, Martin J, Rao RU, Powell K, Abubucker S, et al. (2013) Using existing drugs as leads for broad spectrum anthelmintics targeting protein kinases. PLoS Pathog 9: e1003149.
48. Bromberg N, Dreyfuss JL, Regatieri CV, Palladino MV, Durán N, et al. (2010) Growth inhibition and pro-apoptotic activity of violacein in Ehrlich ascites tumor. Chem-Biol Interact 186: 43–52.
49. Garigan D, Hsu AL, Fraser AG, Kamath RS, Ahringer J, et al. (2002) Genetic analysis of tissue aging in *Caenorhabditis elegans*: a role for heat-shock factor and bacterial proliferation. Genetics 161: 1101–1112.
50. Kho MF, Bellier A, Balasubramani V, Hu Y, Hsu W, et al. (2011) The pore-forming protein Cry5B elicits the pathogenicity of *Bacillus* sp. against *Caenorhabditis elegans*. PLoS One 6: e29122.
51. Gems D, Riddle DL (2000) Genetic, behavioral and environmental determinants of male longevity in *Caenorhabditis elegans*. Genetics 154: 1597–1610.
52. Garsin DA, Villanueva JM, Begun J, Kim DH, Sifri CD, et al. (2003) Long-lived *C. elegans* daf-2 mutants are resistant to bacterial pathogens. Science 300: 1921.
53. Kenyon C, Chang J, Gensch E, Rudner A, Tabtiang R (1993) A *C. elegans* mutant that lives twice as long as wild type. Nature 366: 461–464.
54. Kawli T, He F, Tan MW (2010) It takes nerves to fight infections: insights on neuro-immune interactions from *C. elegans*. Dis Model Mech 3: 721–731.
55. McEwan DL, Kirienko NV, Ausubel FM (2012) Host translational inhibition by *Pseudomonas aeruginosa* Exotoxin A Triggers an immune response in *Caenorhabditis elegans*. Cell Host Microbe 11: 364–374.
56. Melo JA, Ruvkun G (2012) Inactivation of conserved *C. elegans* genes engages pathogen- and xenobiotic-associated defenses. Cell 149: 452–466.
57. Kleino A, Silverman N (2012) UnZIPping mechanisms of effector-triggered immunity in animals. Cell Host & Microbe 11: 320–322.
58. de Carvalho DD, Costa FT, Duran N, Haun M (2006) Cytotoxic activity of violacein in human colon cancer cells. Toxicol In Vitro 20: 1514–1521.
59. Konzen M, De Marco D, Cordova CA, Vieira TO, Antonio RV, et al. (2006) Antioxidant properties of violacein: possible relation on its biological function. Bioorg Med Chem 14: 8307–8313.
60. Evans EA, Kawli T, Tan MW (2008) *Pseudomonas aeruginosa* suppresses host immunity by activating the DAF-2 insulin-like signaling pathway in *Caenorhabditis elegans*. PLoS Pathog 4: e1000175. doi: 1000110.1001371/journal.-ppat.1000175.
61. Jousset A (2012) Ecological and evolutive implications of bacterial defences against predators. Environ Microbiol 14: 1830–1843.
62. Penesyan A, Marshall-Jones Z, Holmström C, Kjelleberg S, Egan S (2009) Antimicrobial activity observed among cultured marine epiphytic bacteria reflects their potential as a source of new drugs. FEMS Microbiol Ecol 69: 113–124.

PERMISSIONS

All chapters in this book were first published in PLOS ONE, by The Public Library of Science; hereby published with permission under the Creative Commons Attribution License or equivalent. Every chapter published in this book has been scrutinized by our experts. Their significance has been extensively debated. The topics covered herein carry significant findings which will fuel the growth of the discipline. They may even be implemented as practical applications or may be referred to as a beginning point for another development.

The contributors of this book come from diverse backgrounds, making this book a truly international effort. This book will bring forth new frontiers with its revolutionizing research information and detailed analysis of the nascent developments around the world.

We would like to thank all the contributing authors for lending their expertise to make the book truly unique. They have played a crucial role in the development of this book. Without their invaluable contributions this book wouldn't have been possible. They have made vital efforts to compile up to date information on the varied aspects of this subject to make this book a valuable addition to the collection of many professionals and students.

This book was conceptualized with the vision of imparting up-to-date information and advanced data in this field. To ensure the same, a matchless editorial board was set up. Every individual on the board went through rigorous rounds of assessment to prove their worth. After which they invested a large part of their time researching and compiling the most relevant data for our readers.

The editorial board has been involved in producing this book since its inception. They have spent rigorous hours researching and exploring the diverse topics which have resulted in the successful publishing of this book. They have passed on their knowledge of decades through this book. To expedite this challenging task, the publisher supported the team at every step. A small team of assistant editors was also appointed to further simplify the editing procedure and attain best results for the readers.

Apart from the editorial board, the designing team has also invested a significant amount of their time in understanding the subject and creating the most relevant covers. They scrutinized every image to scout for the most suitable representation of the subject and create an appropriate cover for the book.

The publishing team has been an ardent support to the editorial, designing and production team. Their endless efforts to recruit the best for this project, has resulted in the accomplishment of this book. They are a veteran in the field of academics and their pool of knowledge is as vast as their experience in printing. Their expertise and guidance has proved useful at every step. Their uncompromising quality standards have made this book an exceptional effort. Their encouragement from time to time has been an inspiration for everyone.

The publisher and the editorial board hope that this book will prove to be a valuable piece of knowledge for researchers, students, practitioners and scholars across the globe.

LIST OF CONTRIBUTORS

Stephanie Speck, Alexandre Courtiol and Karin Müller
Leibniz Institute for Zoo and Wildlife Research, Berlin, Germany

Christof Junkes and Margitta Dathe
Leibniz Institute of Molecular Pharmacology, Berlin, Germany

Martin Schulze
Institute for Reproduction of Farm Animals Schoenow e. V., Bernau, Germany

Hui-Qi Zhou, Lu-Wen Ning, Hui-Xiong Zhang and Feng-Biao Guo
Center of Bioinformatics and Key Laboratory for NeuroInformation of the Ministry of Education, University of Electronic Science and Technology of China, Chengdu, China

Kurt G. Vandervoort
Physics and Astronomy Department, California State Polytechnic University, Pomona, California, United States of America

Graciela Brelles-Mariño
Biological Sciences Department, California State Polytechnic University, Pomona, California, United States of America

Ariel Alperstein, Brittany Ulrich, Denise M. Garofalo, Ruth Dreisbach, Hannah Raff and Kelly Sheppard
Chemistry Department, Skidmore College, Saratoga Springs, New York, United States of America

Ryan M. Chanyi and Susan F. Koval
Department of Microbiology and Immunology, University of Western Ontario, London, Ontario, Canada

François Guérin and Vincent Cattoir
CHU de Caen, Service de Microbiologie, F-14033 Caen, France
Université de Caen Basse-Normandie, EA4655 (équipe "Antibio-résistance"), F-14032 Caen, France

Marc Galimand and Patrice Courvalin
Institut Pasteur, Unité des Agents Antibactériens, Paris, France

Fabrice Tuambilangana
CHU de Caen, Service de Microbiologie, F-14033 Caen, France

Michelle M. Nerandzic
Research Service, Louis Stokes Cleveland Veterans Affairs Medical Center, Cleveland, Ohio,
United States of America

Christopher W. Fisher
STERIS Corporation, Healthcare Group, Mentor, Ohio, United States of America

Curtis J. Donskey
Research Service, Louis Stokes Cleveland Veterans Affairs Medical Center, Cleveland, Ohio
United States of America
Geriatric Research, Education and Clinical Center, Cleveland Veterans Affairs Medical
Center, Cleveland, Ohio, United States of America

Mariana Claudia Allievi, María Mercedes Palomino, Mariano Prado Acosta, Leonardo Lanati, Sandra Mónica Ruzal and Carmen Sánchez-Rivas
Departamento de Química Biológica, Facultad de Ciencias Exactas y Naturales, Universidad de Buenos Aires, IQUIBICEN-CONICET, Buenos Aires, Argentina

Hironobu Koseki, Akihiko Yonekura, Takayuki Shida, Itaru Yoda, Hidehiko Horiuchi, Makoto Osaki and Masato Tomita
Department of Orthopedic Surgery, Graduate School of Biomedical Sciences, Nagasaki University, Nagasaki, Japan

Yoshitomo Morinaga and Katsunori Yanagihara
Department of Laboratory Medicine, Graduate School of Biomedical Sciences, Nagasaki University, Nagasaki, Japan

Hideyuki Sakoda
Division of Medical Devices, National Institute of Health Sciences, Tokyo, Japan

Marco Archetti
School of Biological Sciences, University of East Anglia, Norwich, United Kingdom

Claudia Barberis, Marisa Almuzara, Angela Famiglietti and Carlos Vay
Instituto de Fisiopatología y Bioquímica Clínica, Hospital de Clínicas José de San Martín
Facultad de Farmacia y Bioquímica, Universidad de Buenos Aires, Ciudad Autónoma de
Buenos Aires, Argentina

Olivier Join-Lambert
Laboratoire de Microbiologie, Hopital Necker-Enfants malades, Paris, France
Université Paris Descartes, Paris, France

María Soledad Ramírez
Instituto de Microbiología y Parasitología Médica, Universidad de Buenos Aires-Consejo Nacional de Investigaciones Científicas y Tecnológicas, Ciudad Autónoma de Buenos Aires
Argentina

Diana M. Mitchell
Department of Biochemistry, Cellular and Molecular Biology, University of Tennessee, Knoxville, Tennessee, United States of America
Department of Biological Sciences, University of Idaho, Moscow, Idaho, United States of America

Lindsey R. Uehlein-Klebanow
Department of Biochemistry, Cellular and Molecular Biology, University of Tennessee, Knoxville, Tennessee, United States of America
Department of Molecular Genetics, Lerner Research Institute, Cleveland, Ohio, United States of America

Joshua N. Bembenek
Department of Biochemistry, Cellular and Molecular Biology, University of Tennessee, Knoxville, Tennessee, United States of America

Katharina Mir and Steffen Schober
Institute of Communications Engineering, Ulm University, Ulm, Germany

Clara Torres-Barceló, Flor I. Arias-Sánchez, Marie Vasse, Johan Ramsayer and Oliver Kaltz
Institut des Sciences de l'Evolution, CNRS-Universite´ Montpellier 2, Montpellier, France

Michael E. Hochberg
Institut des Sciences de l'Evolution, CNRS-Universite´ Montpellier 2, Montpellier, France
Santa Fe Institute, Santa Fe, New Mexico, United States of America
Wissenshaftskolleg zu Berlin, Berlin, Germany

Diana Gutiérrez, Patricia Ruas-Madiedo, Beatriz Martínez, Ana Rodríguez and Pilar García
Instituto de Productos Lácteos de Asturias (IPLA-CSIC), Villaviciosa, Asturias, Spain

Seyhan Yazar and George E. C. Gooden
Centre for Ophthalmology and Visual Science, University of Western Australia, Lions Eye Institute, Perth, Western Australia, Australia

David A. Mackey
Centre for Ophthalmology and Visual Science, University of Western Australia, Lions Eye Institute, Perth, Western Australia, Australia
School of Medicine, Menzies Research Institute Tasmania, University of Tasmania, Hobart, Tasmania, Australia

Alex W. Hewitt
Centre for Ophthalmology and Visual Science, University of Western Australia, Lions Eye Institute, Perth, Western Australia, Australia
School of Medicine, Menzies Research Institute Tasmania, University of Tasmania, Hobart, Tasmania, Australia
Centre for Eye Research Australia, University of Melbourne, Department of Ophthalmology, Royal Victorian Eye and Ear Hospital, Melbourne, Victoria, Australia

Helen E. Barnes, Guohong Liu, Christopher Q. Weston, Paula King, Long K. Pham, Shannon Waltz, Kimberly T. Helzer, Laura Day, Dan Sphar, Robert T. Yamamoto and R. Allyn Forsyth
FLIR Systems, Inc., La Jolla, California, United States of America

Ningfei Lei, Jun Li and Shijun Ni
Chengdu University of Technology, Chengdu, China

Jinsong Chen
Sichuan Normal University, Chengdu, China

Yu-Kuo Tsai and Ling Ma
Institute of Infectious Diseases and Vaccinology, National Health Research Institutes, Miaoli, Taiwan

Ci-Hong Liou and Jung-Chung Lin
Division of Infectious Diseases and Tropical Medicine, Department of Internal Medicine, Tri-Service General Hospital, National Defense Medical Center, Taipei, Taiwan

Chang-Phone Fung
Section of Infectious Diseases, Department of Medicine, Taipei Veterans General Hospital and National Yang-Ming University, Taipei, Taiwan

Feng-Yee Chang
Division of Infectious Diseases and Tropical Medicine, Department of Internal Medicine, Tri-Service General Hospital, National Defense Medical Center, Taipei, Taiwan
Taiwan Centres for Disease Control, Taipei, Taiwan

L. Kristopher Siu
Institute of Infectious Diseases and Vaccinology, National Health Research Institutes
Miaoli, Taiwan

Division of Infectious Diseases and Tropical Medicine, Department of Internal Medicine, Tri-Service General Hospital, National Defense Medical Center, Taipei, Taiwan
Graduate Institute of Basic Medical Science, China Medical University, Taichung, Taiwan

Zhiping Wang
Environmental Biotechnology Laboratory, The University of Hong Kong, Hong Kong
School of Environmental Science and Engineering, Shanghai Jiao Tong University, Shanghai, China

Feng Guo and Tong Zhang
Environmental Biotechnology Laboratory, The University of Hong Kong, Hong Kong

Lili Liu
Environmental Biotechnology Laboratory, The University of Hong Kong, Hong Kong
State Environmental Protection Key Laboratory of Environmental Risk Assessment and Control on Chemical Process, East China University of Science and Technology, Shanghai, China

Masafumi Hirono, Yuki Nakazawa, Tetsuro Ariyoshi and Akira Noga
Department of Biological Sciences, Graduate School of Science, University of Tokyo, Bunkyo-ku, Tokyo, Japan

Ritsu Kamiya
Department of Biological Sciences, Graduate School of Science, University of Tokyo, Bunkyo-ku, Tokyo, Japan
Department of Life Science, Faculty of Science, Gakushuin University, Toshima-ku, Tokyo, Japan

Jessica M. Bergman, Marie Wrande and Diarmaid Hughes
Department of Medical Biochemistry and Microbiology, Biomedical Center, Uppsala University, Uppsala, Sweden

Chao Zhuo, Qian-yu Zhao and Shu-nian Xiao
State Key Laboratory of Respiratory Diseases, the first affiliated hospital of Guangzhou Medical College, Guangzhou, China

Meghan May
Department of Biomedical Sciences, College of Osteopathic Medicine, University of New England, Biddeford, Maine, United States of America

Dylan W. Dunne
Department of Biological Sciences, Jess and Mildred Fisher College of Science and Mathematics, Towson University, Towson, Maryland, United States of America

Daniel R. Brown
Department of Infectious Diseases and Pathology, College of Veterinary Medicine, University of Florida, Gainesville, Florida, United States of America

Imen Abbes, Hayette Benamara and Stéphane Alexandre
Polyméres Biopolyméres, Surfaces Laboratory - UMR 6270, CNRS - Université de Rouen, Mont-Saint-Aignan, France
Normandie Université, Mont-Saint-Aignan, France

Christophe Rihouey, Mohamed Amine Ben Mlouka, Julie Hardouin and Thierry Jouenne
Polyméres Biopolyméres, Surfaces Laboratory - UMR 6270, CNRS - Universite´ de Rouen, Mont-Saint-Aignan, France
Normandie Université, Mont-Saint-Aignan, France
PISSARO proteomic facility, IRIB - Université de Rouen, Mont-Saint-Aignan, France

Nurit Beyth and Lea H. Eckhard
Department of Prosthodontics, the Hebrew University – Faculty of Dental Medicine, Jerusalem, Israel

Asaf Sol and Gilad Bachrach
Institute of Dental Science, the Hebrew University – Faculty of Dental Medicine, Jerusalem, Israel

Ester Abtew and Abraham J. Domb
Institute for Drug Research, School of Pharmacology, Faculty of Medicine, the Hebrew University, Jerusalem, Israel

Yechiel Shai
Department of Biological Chemistry, the Weizmann Institute of Science, Rehovot, Israel

Suhelen Egan, Francesco Ballestriero, Malak Daim and Jadranka Nappi
School of Biotechnology and Biomolecular Sciences and Centre for Marine Bio-Innovation, University of New South Wales, Sydney, New South Wales, Australia

Anahit Penesyan
Department of Chemistry and Biomolecular Sciences, Macquarie University, Sydney, New South Wales, Australia

David Schleheck
Biology Department, University of Konstanz, Konstanz, Germany

Paolo Bazzicalupo and Elia Di Schiavi
Institute of Genetics and Biophysics "Adriano Buzzati Traverso", National Research Council, Naples, Italy

Index

www.ingramcontent.com/pod-product-compliance
Lightning Source LLC
Chambersburg PA
CBHW080525200326
41458CB00012B/4337
9781682865934